John Wiley & Sons, Inc.
Publishers Since 1807

READ IMPORTANT LICENSE INFORMATION

INSTRUCTOR'S SOLUTIONS MANUAL

CALCULUS
SEVERAL VARIABLES
10th Edition

THE WILEY BICENTENNIAL—KNOWLEDGE FOR GENERATIONS

$\mathcal{E}$ach generation has its unique needs and aspirations. When Charles Wiley first opened his small printing shop in lower Manhattan in 1807, it was a generation of boundless potential searching for an identity. And we were there, helping to define a new American literary tradition. Over half a century later, in the midst of the Second Industrial Revolution, it was a generation focused on building the future. Once again, we were there, supplying the critical scientific, technical, and engineering knowledge that helped frame the world. Throughout the 20th Century, and into the new millennium, nations began to reach out beyond their own borders and a new international community was born. Wiley was there, expanding its operations around the world to enable a global exchange of ideas, opinions, and know-how.

For 200 years, Wiley has been an integral part of each generation's journey, enabling the flow of information and understanding necessary to meet their needs and fulfill their aspirations. Today, bold new technologies are changing the way we live and learn. Wiley will be there, providing you the must-have knowledge you need to imagine new worlds, new possibilities, and new opportunities.

Generations come and go, but you can always count on Wiley to provide you the knowledge you need, when and where you need it!

WILLIAM J. PESCE
PRESIDENT AND CHIEF EXECUTIVE OFFICER

PETER BOOTH WILEY
CHAIRMAN OF THE BOARD

INSTRUCTOR'S SOLUTIONS MANUAL

Garret Etgen
University of Houston

to accompany

CALCULUS
SEVERAL VARIABLES

10th Edition

Saturnino Salas

Einar Hille

Garret Etgen
University of Houston

1807
WILEY
2007

John Wiley & Sons, Inc.

Bicentennial Logo Design: Richard Pacifico

To order books or for customer service, please call 1-800-CALL-WILEY (225-5945).

ISBN-13   978-0-470-11931-0

10 9 8 7 6 5 4 3 2 1

Printed and bound by Lightning Source.

CONTENTS

CHAPTER 13

SECTION 13.1

1.

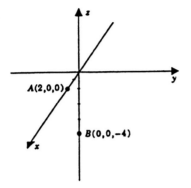

length $\overline{AB}$: $2\sqrt{5}$

midpoint: $(1, 0, -2)$

2.

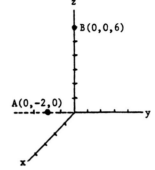

length $\overline{AB}$: $2\sqrt{10}$

midpoint: $(0, -1, 3)$

3.

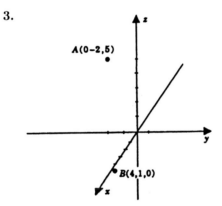

length $\overline{AB}$: $5\sqrt{2}$

midpoint: $\left(2, -\frac{1}{2}, \frac{5}{2}\right)$

4.

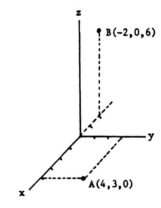

length $\overline{AB}$: 9

midpoint: $\left(1, \frac{3}{2}, 3\right)$

5. $z = -2$

6. $y = 1$

7. $y = 1$

8. $z = -2$

9. $x = 3$

10. $x = 3$

11. $x^2 + (y - 2)^2 + (z + 1)^2 = 9$

12. $(x - 1)^2 + y^2 + (z + 2)^2 = 16$

13. $(x - 2)^2 + (y - 4)^2 + (z + 4)^2 = 36$

14. $x^2 + y^2 + z^2 = 9$

15. $(x - 3)^2 + (y - 2)^2 + (z - 2)^2 = 13$

16. $(x - 2)^2 + (y - 3)^2 + (z + 4)^2 = 16$

17.
$$x^2 + y^2 + z^2 + 4x - 8y - 2z + 5 = 0$$
$$x^2 + 4x + 4 + y^2 - 8y + 16 + z^2 - 2z + 1 = -5 + 4 + 16 + 1$$
$$(x + 2)^2 + (y - 4)^2 + (z - 1)^2 = 16$$

center: $(-2, 4, 1)$,   radius: 4

18. Rewrite as $x^2 - 4x + 4 + y^2 + z^2 - 2z + 1 = -1 + 4 + 1 = 4$
$\implies (x - 2)^2 + y^2 + (z - 1)^2 = 4$ center $(2, 0, 1)$;   radius 2

19. $(2, 3, -5)$ **20.** $(2, -3, 5)$ **21.** $(-2, 3, 5)$

22. $(2, -3, -5)$ **23.** $(-2, 3, -5)$ **24.** $(-2, -3, 5)$

25. $(-2, -3, -5)$ **26.** $(0, 3, 5)$ **27.** $(2, -5, 5)$

28. $(2, 3, 3)$ **29.** $(-2, 1, -3)$ **30.** $(6, -3, -3)$

31. $d(PR) = \sqrt{14}$, $d(QR) = \sqrt{45}$, $d(PQ) = \sqrt{59}$; $[d(PR)]^2 + [d(QR)]^2 = [d(PQ)]^2$

32. Let the vertices be (x_i, y_i, z_i), $i = 1, 2, 3$.   Then

$$\left(\frac{x_1 + x_2}{2}, \frac{y_1 + y_2}{2}, \frac{z_1 + z_2}{2}\right) = (5, -1, 3); \quad \left(\frac{x_2 + x_3}{2}, \frac{y_2 + y_3}{2}, \frac{z_2 + z_3}{2}\right) = (4, 2, 1);$$

$$\left(\frac{x_1 + x_3}{2}, \frac{y_1 + y_3}{2}, \frac{z_1 + z_3}{2}\right) = (2, 1, 0)$$

Solving simultaneously gives vertices $(3, -2, 2), (7, 0, 4), (1, 4, -2)$.

33. The sphere of radius 2 centered at the origin, together with its interior.

34. The exterior of the sphere of radius 3 centered at the origin.

35. A rectangular box in the first octant with sides on the coordinate planes and dimensions $1 \times 2 \times 3$, together with its interior.

36. A cube of side length 4, together with its interior; the origin in at the center of the cube.

37. A circular cylinder with base the circle $x^2 + y^2 = 4$ and height 4, together with its interior.

38. $x^2 + y^2 + z^2 = 4$ and $x^2 + y^2 + z^2 = 9$ are concentric spheres; Ω is the region between the two spheres.

39. Let $B = (x, y, z)$. Then

$$\frac{x+2}{2} = 1 \Longrightarrow x = 0, \quad \frac{y+3}{2} = 2 \Longrightarrow y = 1, \quad \frac{z+4}{2} = 3 \Longrightarrow z = 2.$$

Therefore $B = (0, 1, 2)$.

41. Let $P_1 = (x, y, z)$ be the trisection point closest to A. Then

$$\overrightarrow{AP_1} = \tfrac{1}{3}\overrightarrow{AB} \Longrightarrow (x - a_1, y - a_2, z - a_3) = \tfrac{1}{3}(b_1 - a_1, b_2 - a_2, b_3 - a_3).$$

Solving for x, y, z gives $(x, y, z) = \left(\dfrac{2a_1 + b_1}{3}, \dfrac{2a_2 + b_2}{3}, \dfrac{2a_3 + b_3}{3} \right)$.

Similarly, if $P_2 = (x, y, z)$ is the trisection point closest to B, then

$$(x, y, z) = \left(\dfrac{a_1 + 2b_1}{3}, \dfrac{a_2 + 2b_2}{3}, \dfrac{a_3 + 2b_3}{3} \right).$$

42. The points on the line segment $\overline{AB}$ are given by $x = 1 + t$, $y = -2 + 3t$, $z = \sqrt{2} - \sqrt{2}\,t$, $0 \le t \le 1$. The line segment $\overline{AP}$ has length 3 if

$$\sqrt{t^2 + (3t)^2 + (-\sqrt{2}\,t)^2} = \sqrt{12t^2} = 2t\sqrt{3} = 3 \Longrightarrow t = \tfrac{1}{2}\sqrt{3}.$$

Thus, the point P on the line segment $\overline{AB}$ that is 3 units from A has coordinates:

$$1 + \frac{1}{2}\sqrt{3}, \quad -2 + \frac{3}{2}\sqrt{3}, \quad \sqrt{2} - \frac{\sqrt{2}}{2}\sqrt{3}.$$

43. Substituting the coordinates of the points into the equation $Ax + By + Cz + D = 0$, we get the equations

$$Ax_0 + D = 0, \; By_0 + D = 0, \; Cz_0 + D = 0 \quad \text{which implies} \quad Ax_0 = By_0 = Cz_0.$$

Therefore, we have

$$Ax + \frac{Ax_0}{y_0}y + \frac{Ax_0}{z_0}z + D = 0 \quad \text{or} \quad \frac{x}{x_0} + \frac{y}{y_0} + \frac{z}{z_0} + \frac{D}{Ax_0} = 0.$$

Substituting the point $(x_0, 0, 0)$ into this equation gives

$$\frac{x}{x_0} + \frac{y}{y_0} + \frac{z}{z_0} = 1.$$

44. Substituting the coordinates of the points into the equation $Ax + By + Cz + D = 0$, we get the equations

$$Ax_0 + By_0 + D = 0, \; Ax_0 + Cz_0 + D = 0, \; By_0 + Cz_0 + D = 0$$

which implies $A = \frac{Cz_0}{x_0}$, $B = \frac{Cz_0}{y_0}$. Therefore, we have

$$\frac{Cz_0}{x_0}x + \frac{Cz_0}{y_0}y + Cz + D = 0 \quad \text{or} \quad \frac{x}{x_0} + \frac{y}{y_0} + \frac{z}{z_0} + \frac{D}{Cz_0} = 0.$$

Substituting the point $(x_0, y_0, 0)$ into this equation gives

$$\frac{x}{x_0} + \frac{y}{y_0} + \frac{z}{z_0} = 2.$$

45. (i) $a_3 \neq 0$ The line through the origin and (a_1, a_2, a_3) is given by $x = a_1 t$, $y = a_2 t$, $z = a_3 t$, t any real number. The line intersects the plane $z = z_0$ at the point Q where $t = z_0/a_3$. The coordinates of Q are: $\frac{a_1}{a_3} z_0$, $\frac{a_2}{a_3} z_0$, z_0.

(ii) $a_3 = 0$ If $z_0 \neq 0$, the line does not intersect the plane. If $z_0 = 0$, the line lies in the plane.

46. Using the arguments in Exercise 45 we have (*i*) $a_1 \neq 0$: $Q = \left(x_0, \frac{a_2}{a_1} x_0, \frac{a_3}{a_1} x_0 \right)$.   (*ii*) $a_1 = 0$: If $x_0 \neq 0$, the line does not intersect the plane. If $x_0 = 0$, the line lies in the plane.

47. The ray that emanates from the origin and passes through the point (a_1, a_2, a_3) is given by $x = a_1 t$, $y = a_2 t$, $z = a_3 t$, $t \geq 0$. The ray intersects the sphere $x^2 + y^2 + z^2 = 1$ at the point Q where

$$a_1^2 t^2 + a_2^2 t^2 + a_3^2 t^2 = 1 \implies t = \frac{1}{\sqrt{a_1^2 + a_2^2 + a_3^2}}.$$

The coordinates of Q are: $\dfrac{a_1}{\sqrt{a_1^2 + a_2^2 + a_3^2}}$, $\dfrac{a_2}{\sqrt{a_1^2 + a_2^2 + a_3^2}}$, $\dfrac{a_3}{\sqrt{a_1^2 + a_2^2 + a_3^2}}$.

48. It follows from Exercise 47 that the points where the line $x = a_1 t$, $y = a_2 t$, $z = a_3 t$, intersects the sphere $x^2 + y^2 + z^2 = 1$ are: $\dfrac{\pm a_1}{\sqrt{a_1^2 + a_2^2 + a_3^2}}$, $\dfrac{\pm a_2}{\sqrt{a_1^2 + a_2^2 + a_3^2}}$, $\dfrac{\pm a_3}{\sqrt{a_1^2 + a_2^2 + a_3^2}}$.

SECTION 13.2

1. $\overrightarrow{PQ} = (3, 4, -2)$; $\| \overrightarrow{PQ} \| = \sqrt{29}$ **2.** $\overrightarrow{PQ} = (-2, 6, 0)$; $\| \overrightarrow{PQ} \| = 2\sqrt{10}$

3. $\overrightarrow{PQ} = (0, -2, -1)$; $\| \overrightarrow{PQ} \| = \sqrt{5}$ **4.** $\overrightarrow{PQ} = (4, 3, -8)$; $\| \overrightarrow{PQ} \| = \sqrt{89}$

5. $2\mathbf{a} - \mathbf{b} = (2 \cdot 1 - 3, 2 \cdot [-2] - 0, 2 \cdot 3 + 1) = (-1, -4, 7)$

6. $2\mathbf{b} + 3\mathbf{c} = (6, 0, -2) + (-12, 6, 3) = (-6, 6, 1)$

7. $-2\mathbf{a} + \mathbf{b} - \mathbf{c} = [-(2\mathbf{a} - \mathbf{b})] - \mathbf{c} = (1 + 4, 4 - 2, -7 - 1) = (5, 2, -8)$

8. $\mathbf{a} + 3\mathbf{b} - 2\mathbf{c} = (1, -2, 3) + 3(3, 0, -1) - 2(-4, 2, 1) = (18, -6, -2)$.

9. $3\mathbf{i} - 4\mathbf{j} + 6\mathbf{k}$ **10.** $3\mathbf{i} + 5\mathbf{j} + \mathbf{k}$

11. $-3\mathbf{i} - \mathbf{j} + 8\mathbf{k}$ **12.** $14\mathbf{i} + 4\mathbf{j} - 12\mathbf{k}$

13. 5 **14.** $\sqrt{2}$ **15.** 3

16. $\sqrt{41}$ **17.** $\sqrt{6}$ **18.** $\sqrt{2}$

19. (a) $\mathbf{a}$, $\mathbf{c}$, and $\mathbf{d}$ since $\mathbf{a} = \frac{1}{3}\mathbf{c} = -\frac{1}{2}\mathbf{d}$
(b) $\mathbf{a}$ and $\mathbf{c}$ since $\mathbf{a} = \frac{1}{3}\mathbf{c}$
(c) $\mathbf{a}$ and $\mathbf{c}$ both have direction opposite to $\mathbf{d}$

20. Let R be the point $R(x, y, z)$. Then $\vec{QR} = (x - 3, y + 1, z - 1)$ and $\vec{OP} = (1, 4, -2)$.

$$\vec{QR} = \vec{OP} \Longrightarrow x - 3 = 1, \ y + 1 = 4, \ z - 1 = -2 \Longrightarrow x = 4, \ y = 3, \ z = -1.$$

21. $\vec{RQ} = (3 - x, -1 - y, 1 - z)$ and $\vec{OP} = (1, 4, -2)$.

$$\vec{RQ} = \vec{OP} \Longrightarrow 3 - x = 1, \ -1 - y = 4, \ 1 - z = -2 \Longrightarrow x = 2, \ y = -5, \ z = 3.$$

22. $\vec{RQ} = (3 - x, -1 - y, 1 - z) = 3 \, \vec{OP} = (3, 12, -6) \Longrightarrow 3 - x = 3, \ -1 - y = 12, \ 1 - z = -6$

$$\Longrightarrow x = 0, \ y = -13, \ z = 7.$$

23. $\vec{RQ} = (3 - x, -1 - y, 1 - z) = -2 \, \vec{OP} = (-2, -8, 4) \Longrightarrow 3 - x = -2, \ -1 - y = -8, \ 1 - z = 4$

$$\Longrightarrow x = 5, \ y = 7, \ z = -3.$$

24. $\|\mathbf{a}\| - \|\mathbf{b}\| \le \|\mathbf{a} - \mathbf{b}\|$ since

$$\|\mathbf{a}\| = \|(\mathbf{a} - \mathbf{b}) + \mathbf{b}\| \le \|\mathbf{a} - \mathbf{b}\| + \|\mathbf{b}\|.$$

Similarly $\|\mathbf{b}\| - \|\mathbf{a}\| \le \|\mathbf{b} - \mathbf{a}\| = \|\mathbf{a} - \mathbf{b}\|.$

25. $\|\mathbf{a}\| = 5;$ $\dfrac{\mathbf{a}}{\|\mathbf{a}\|} = \left(\dfrac{3}{5}, -\dfrac{4}{5}, 0 \right)$ **26.** $\left(\dfrac{-2}{\sqrt{13}}, \dfrac{3}{\sqrt{13}} \right)$

27. $\|\mathbf{a}\| = 3;$ $\dfrac{\mathbf{a}}{\|\mathbf{a}\|} = \dfrac{1}{3}\mathbf{i} - \dfrac{2}{3}\mathbf{j} + \dfrac{2}{3}\mathbf{k}$ **28.** $\left(\dfrac{2}{3}, \dfrac{1}{3}, \dfrac{2}{3} \right)$

29. $\|\mathbf{a}\| = \sqrt{14};$ $-\dfrac{\mathbf{a}}{\|\mathbf{a}\|} = \dfrac{1}{\sqrt{14}}\mathbf{i} - \dfrac{3}{\sqrt{14}}\mathbf{j} - \dfrac{2}{\sqrt{14}}\mathbf{k}$ **30.** $-\dfrac{2}{\sqrt{5}}\mathbf{i} + \dfrac{1}{\sqrt{5}}\mathbf{k}$

31. (i) $\mathbf{a} - \mathbf{b}$ (ii) $-(\mathbf{a} + \mathbf{b})$ (iii) $\mathbf{a} - \mathbf{b}$ (iv) $\mathbf{b} - \mathbf{a}$

32. (a) $6\mathbf{i} + 3\mathbf{j} + 12\mathbf{k}$

(b) $A(1, 1, 1) + B(-1, 3, 2) + C(-3, 0, 1) = (4, -1, 1).$

Solve simultaneously to get $A = \frac{26}{7}, \quad B = -\frac{11}{7}, \quad C = \frac{3}{7}$

33. (a) $\mathbf{a} - 3\mathbf{b} + 2\mathbf{c} + 4\mathbf{d} = (2\mathbf{i} - \mathbf{k}) - 3(\mathbf{i} + 3\mathbf{j} + 5\mathbf{k}) + 2(-\mathbf{i} + \mathbf{j} + \mathbf{k}) + 4(\mathbf{i} + \mathbf{j} + 6\mathbf{k})$

$$= \mathbf{i} - 3\mathbf{j} + 10\mathbf{k}$$

(b) The vector equation

$$(1, 1, 6) = A(2, 0, -1) + B(1, 3, 5) + C(-1, 1, 1)$$

implies

$$1 = 2A + B - C,$$
$$1 = \phantom{2A + {}} 3B + C,$$
$$6 = -A + 5B + C.$$

Simultaneous solution gives $A = -2, \quad B = \frac{3}{2}, \quad C = -\frac{7}{2}.$

34. $\alpha = -12$

35. $\|3\mathbf{i} + \mathbf{j}\| = \|\alpha\mathbf{j} - \mathbf{k}\| \implies 10 = \alpha^2 + 1$ so $\alpha = \pm 3$

36. $\frac{1}{3}(\mathbf{i} - 2\mathbf{j} + 2\mathbf{k})$

37.

$$\|\alpha\mathbf{i} + (\alpha - 1)\mathbf{j} + (\alpha + 1)\mathbf{k}\| = 2 \implies \alpha^2 + (\alpha - 1)^2 + (\alpha + 1)^2 = 4$$
$$\implies 3\alpha^2 = 2 \quad \text{so} \quad \alpha = \pm\frac{1}{3}\sqrt{6}$$

38. $\frac{2}{\sqrt{6}}(\mathbf{i} + 2\mathbf{j} - \mathbf{k}) = \frac{\sqrt{6}}{3}(\mathbf{i} + 2\mathbf{j} - \mathbf{k})$

39. $\pm\frac{2}{13}\sqrt{13}\,(3\mathbf{j} + 2\mathbf{k})$ since $\|\alpha(3\mathbf{j} + 2\mathbf{k})\| = 2 \implies \alpha = \pm\frac{2}{13}\sqrt{13}$

40. (i) $\mathbf{c} = \mathbf{a} + \frac{1}{2}(\mathbf{b} - \mathbf{a}) = \frac{1}{2}(\mathbf{a} + \mathbf{b})$ (ii) $\mathbf{a} + \mathbf{c} = \frac{1}{2}(\mathbf{a} + \mathbf{b}) \implies \mathbf{c} = \frac{1}{2}(\mathbf{b} - \mathbf{a})$

41. (a) Since $\|\mathbf{a} - \mathbf{b}\|$ and $\|\mathbf{a} + \mathbf{b}\|$ are the lengths of the diagonals of the parallelogram, the parallelogram must be a rectangle.

(b) Simplify

$$\sqrt{(a_1 - b_1)^2 + (a_2 - b_2)^2 + (a_3 - b_3)^2} = \sqrt{(a_1 + b_1)^2 + (a_2 + b_2)^2 + (a_3 + b_3)^2}.$$

the result is $a_1 b_1 + a_2 b_2 + a_3 b_3 = 0$.

42. (a) If $\alpha > 0$, then

$$\|\mathbf{a} + \alpha\mathbf{a}\| = \|(1 + \alpha)\mathbf{a}\| = (1 + \alpha)\|\mathbf{a}\| = \|\mathbf{a}\| + \alpha\|\mathbf{a}\| = \|\mathbf{a}\| + \|\alpha\mathbf{a}\|.$$

(b) The equation does not necessarily hold if $\alpha < 0$. For example, if $\alpha = -1$,

$$0 = \|\mathbf{a} + (-1)\mathbf{a}\| \neq \|\mathbf{a}\| + \|(-1)\mathbf{a}\| = 2\|\mathbf{a}\|.$$

43. Let $P = (x_1, y_1, z_1)$, $Q = (x_2, y_2, z_2)$, and $M = (x_m, y_m, z_m)$. Then

$$(x_m, y_m, z_m) = (x_1, y_1, z_1) + \frac{1}{2}(x_2 - x_1, y_2 - y_1, z_2 - z_1) \implies \mathbf{m} = \mathbf{p} + \frac{1}{2}(\mathbf{q} - \mathbf{p}).$$

44. $\mathbf{r} - \mathbf{p} = 2(\mathbf{q} - \mathbf{r}) \implies 3\mathbf{r} = \mathbf{p} + 2\mathbf{q} \implies \mathbf{r} = \frac{1}{3}\mathbf{p} + \frac{2}{3}\mathbf{q}$

SECTION 13.3

1. $\mathbf{a} \cdot \mathbf{b} = (2)(-2) + (-3)(0) + (1)(3) = -1$

2. $\mathbf{a} \cdot \mathbf{b} = (4)(-2) + (2)(2) + (-1)(1) = -5$

3. $\mathbf{a} \cdot \mathbf{b} = (2)(1) + (-4)(1/2) + (0)(0) = 0$

4. $\mathbf{a} \cdot \mathbf{b} = (-2)(3) + (0)(0) + (5)(1) = -1$

5. $\mathbf{a} \cdot \mathbf{b} = (2)(1) + (1)(1) - (2)(2) = -1$

6. $\mathbf{a} \cdot \mathbf{b} = (2)(1) + (3)(4) + (1)(0) = 14$

7. $\mathbf{a} \cdot \mathbf{b}$

8. $\mathbf{a} \cdot (\mathbf{a} - \mathbf{b}) + \mathbf{b} \cdot (\mathbf{b} + \mathbf{a}) = \mathbf{a} \cdot \mathbf{a} - \mathbf{a} \cdot \mathbf{b} + \mathbf{b} \cdot \mathbf{b} + \mathbf{b} \cdot \mathbf{a} = \|\mathbf{a}\|^2 + \|\mathbf{b}\|^2$

9. $(\mathbf{a} - \mathbf{b}) \cdot \mathbf{c} + \mathbf{b} \cdot (\mathbf{c} + \mathbf{a}) = \mathbf{a} \cdot \mathbf{c} - \mathbf{b} \cdot \mathbf{c} + \mathbf{b} \cdot \mathbf{c} + \mathbf{b} \cdot \mathbf{a} = \mathbf{a} \cdot (\mathbf{b} + \mathbf{c})$

10. $\mathbf{a} \cdot (\mathbf{a} + 2\mathbf{c}) + (2\mathbf{b} - \mathbf{a}) \cdot (\mathbf{a} + 2\mathbf{c}) - 2\mathbf{b} \cdot (\mathbf{a} + 2\mathbf{c}) = (\mathbf{a} + 2\mathbf{b} - \mathbf{a} - 2\mathbf{b}) \cdot (\mathbf{a} + 2\mathbf{c}) = \mathbf{0}$

11. (a) $\mathbf{a} \cdot \mathbf{b} = (2)(3) + (1)(-1) + (0)(2) = 5$

$\mathbf{a} \cdot \mathbf{c} = (2)(4) + (1)(0) + (0)(3) = 8$

$\mathbf{b} \cdot \mathbf{c} = (3)(4) + (-1)(0) + (2)(3) = 18$

(b) $\|\mathbf{a}\| = \sqrt{5}, \quad \|\mathbf{b}\| = \sqrt{14}, \quad \|\mathbf{c}\| = 5.$ Then,

$$\cos \sphericalangle\,(\mathbf{a}, \mathbf{b}) = \frac{\mathbf{a} \cdot \mathbf{b}}{\|\mathbf{a}\|\,\|\mathbf{b}\|} = \frac{5}{(\sqrt{5})(\sqrt{14})} = \frac{1}{14}\sqrt{70},$$

$$\cos \sphericalangle\,(\mathbf{a}, \mathbf{c}) = \frac{8}{(\sqrt{5})(5)} = \frac{8}{25}\sqrt{5},$$

$$\cos \sphericalangle\,(\mathbf{b}, \mathbf{c}) = \frac{18}{(\sqrt{14})(5)} = \frac{9}{35}\sqrt{14}.$$

(c) $\mathbf{u_b} = \dfrac{1}{\sqrt{14}}(3\mathbf{i} - \mathbf{j} + 2\mathbf{k}), \quad \mathrm{comp_b}\,\mathbf{a} = \mathbf{a} \cdot \mathbf{u_b} = \dfrac{1}{\sqrt{14}}(6 - 1) = \dfrac{5}{14}\sqrt{14},$

$\mathbf{u_c} = \dfrac{1}{5}(4\mathbf{i} + 3\mathbf{k}), \quad \mathrm{comp_c}\,\mathbf{a} = \mathbf{a} \cdot \mathbf{u_c} = \dfrac{8}{5}$

(d) $\mathrm{proj_b}\,\mathbf{a} = (\mathrm{comp_b}\,\mathbf{a})\,\mathbf{u_b} = \dfrac{5}{14}(3\mathbf{i} - \mathbf{j} + 2\mathbf{k}), \quad \mathrm{proj_c}\,\mathbf{a} = (\mathrm{comp_c}\,\mathbf{a})\,\mathbf{u_c} = \dfrac{8}{25}(4\mathbf{i} + 3\mathbf{k})$

12. (a) $\mathbf{a} \cdot \mathbf{b} = 5, \quad \mathbf{a} \cdot \mathbf{c} = -3, \quad \mathbf{b} \cdot \mathbf{c} = 4$

(b) $\cos \sphericalangle\,(\mathbf{a}, \mathbf{b}) = \frac{1}{6}\sqrt{10}, \quad \cos \sphericalangle\,(\mathbf{a}, \mathbf{c}) = -\frac{3}{10}, \quad \cos \sphericalangle\,(\mathbf{b}, \mathbf{c}) = \frac{2}{15}\sqrt{10}$

(c) $\mathrm{comp_b}\mathbf{a} = \frac{5}{3}, \quad \mathrm{comp_c}\mathbf{a} = -\frac{3}{10}\sqrt{10}$

(d) $\mathrm{proj_b}\mathbf{a} = \frac{5}{9}(2\mathbf{i} - \mathbf{j} + 2\mathbf{k}), \quad \mathrm{proj_c}\mathbf{a} = -\frac{3}{10}(3\mathbf{i} - \mathbf{k})$

13. $\mathbf{u} = \cos\dfrac{\pi}{3}\mathbf{i} + \cos\dfrac{\pi}{4}\mathbf{j} + \cos\dfrac{2\pi}{3}\mathbf{k} = \dfrac{1}{2}\mathbf{i} + \dfrac{1}{2}\sqrt{2}\mathbf{j} - \dfrac{1}{2}\mathbf{k}$

14. $\mathbf{v} = 2(\cos\dfrac{\pi}{4}\mathbf{i} + \cos\dfrac{\pi}{4}\mathbf{j} + \cos\dfrac{\pi}{2}\mathbf{k}) = \sqrt{2}\mathbf{i} + \sqrt{2}\mathbf{j}.$

15. $\cos\theta = \dfrac{(3\mathbf{i} - \mathbf{j} - 2\mathbf{k}) \cdot (\mathbf{i} + 2\mathbf{j} - 3\mathbf{k})}{\|3\mathbf{i} - \mathbf{j} - 2\mathbf{k}\|\,\|\mathbf{i} + 2\mathbf{j} - 3\mathbf{k}\|} = \dfrac{7}{\sqrt{14}\sqrt{14}} = \dfrac{1}{2}, \quad \theta = \dfrac{\pi}{3}$

16. $\cos\theta = \dfrac{(2\mathbf{i} - 3\mathbf{j} + \mathbf{k}) \cdot (-3\mathbf{i} + \mathbf{j} + 9\mathbf{k})}{\|2\mathbf{i} - 3\mathbf{j} + \mathbf{k}\| \cdot \|-3\mathbf{i} + \mathbf{j} + 9\mathbf{k}\|} = 0 \implies \theta = \dfrac{\pi}{2}$

17. Since $\|\mathbf{i} - \mathbf{j} + \sqrt{2}\,\mathbf{k}\| = 2$,   we have $\cos\alpha = \frac{1}{2}$, $\cos\beta = -\frac{1}{2}$, $\cos\gamma = \frac{1}{2}\sqrt{2}$.
The direction angles are $\frac{1}{3}\pi$, $\frac{2}{3}\pi$, $\frac{1}{4}\pi$.

18. $\alpha = \arccos\dfrac{1}{2} = \dfrac{\pi}{3}$, $\beta = \arccos 0 = \dfrac{\pi}{2}$, $\gamma = \arccos\dfrac{-\sqrt{3}}{2} = \dfrac{5}{6}\pi$

19. $\theta = \arccos\dfrac{\mathbf{a}\cdot\mathbf{b}}{\|\mathbf{a}\|\|\mathbf{b}\|} = \arccos\left(\dfrac{-9}{\sqrt{231}}\right) \cong 2.2$ radians   or $126.3°$

20. $\theta = \arccos\dfrac{\mathbf{a}\cdot\mathbf{b}}{\|\mathbf{a}\|\|\mathbf{b}\|} = \arccos\dfrac{12}{\sqrt{13}\sqrt{52}} \cong 1.09$ radians or $62.5°$.

21. $\theta = \arccos\dfrac{\mathbf{a}\cdot\mathbf{b}}{\|\mathbf{a}\|\|\mathbf{b}\|} = \arccos\left(\dfrac{-13}{5\sqrt{10}}\right) \cong 2.54$ radians   or $145.3°$

22. $\theta = \arccos\dfrac{\mathbf{a}\cdot\mathbf{b}}{\|\mathbf{a}\|\|\mathbf{b}\|} = \arccos\dfrac{-4}{\sqrt{11}\sqrt{2}} \cong 2.59$ radians or $148.5°$

23. angles: $38.51°,\ 95.52°,\ 45.97°$;   perimeter : $\cong 15.924$

24. $\|\mathbf{a}\| = \sqrt{41}$;

$\cos\alpha = \dfrac{2}{\sqrt{41}}$, $\cos\beta = \dfrac{6}{\sqrt{41}}$, $\cos\gamma = \dfrac{-1}{\sqrt{41}}$

$\alpha \cong 71.8°$ $\beta \cong 20.4°$, $\gamma \cong 99.0°$

25. $\|\mathbf{a}\| = \sqrt{1^2 + 2^2 + 2^2} = 3$;

$\cos\alpha = \dfrac{1}{3}$, $\cos\beta = \dfrac{2}{3}$, $\cos\gamma = \dfrac{2}{3}$

$\alpha \cong 70.5°$ $\beta \cong 48.2°$, $\gamma \cong 48.2°$

26. $\|\mathbf{a}\| = \sqrt{50}$;

$\cos\alpha = \dfrac{3}{\sqrt{50}}$, $\cos\beta = \dfrac{5}{\sqrt{50}}$, $\cos\gamma = \dfrac{-4}{\sqrt{50}}$

$\alpha \cong 64.9°$ $\beta \cong 45°$, $\gamma \cong 124.4°$

27. $\|\mathbf{a}\| = \sqrt{3^2 + (12)^2 + 4^2} = 13$;

$\cos\alpha = \dfrac{3}{13}$, $\cos\beta = \dfrac{12}{13}$ $\cos\gamma = \dfrac{4}{13}$

$\alpha \cong 76.7°$ $\beta \cong 22.6°$, $\gamma \cong 72.1°$

28. $\cos\alpha \cong -0.8835,\ \alpha \cong 152.067°;\ \cos\beta \cong -0.3313,\ \beta \cong 109.347°;\ \cos\gamma \cong 0.3313, \gamma \cong 70.653°$

29. $2\mathbf{i} + 5\mathbf{j} + 2x\,\mathbf{k} \perp 6\mathbf{i} + 4\mathbf{j} - x\,\mathbf{k} \Longrightarrow 12 + 20 - 2x^2 = 0 \Longrightarrow x^2 = 16 \Longrightarrow x = \pm 4$

30. $(x\,\mathbf{i} + 11\,\mathbf{j} - 3\,\mathbf{k})\cdot(2x\,\mathbf{i} - x\,\mathbf{j} - 5\,\mathbf{k}) = 0 \implies 2x^2 - 11x + 15 = 0$

$\implies x = 3,\quad x = \dfrac{5}{2}$

31. $\cos\dfrac{\pi}{3} = \dfrac{\mathbf{c}\cdot\mathbf{d}}{\|\mathbf{c}\|\,\|\mathbf{d}\|}$, $\dfrac{1}{2} = \dfrac{2x+1}{x^2+2}$, $x^2 = 4x$; $x = 0,\quad x = 4$

32. $(\mathbf{i} + x\mathbf{j} + \mathbf{k}) \cdot (2\mathbf{i} - \mathbf{j} + y\mathbf{k}) = 0 \implies 2 - x + y = 0$

$1 + x^2 + 1 = 4 + 1 + y^2 \implies x^2 - y^2 = 3 \implies x = \dfrac{7}{4}, \quad y = -\dfrac{1}{4}$

33. (a) The direction angles of a vector always satisfy

$$\cos^2 \alpha + \cos^2 \beta + \cos^2 \gamma = 1$$

and, as you can check,

$$\cos^2 \tfrac{1}{4}\pi + \cos^2 \tfrac{1}{6}\pi + \cos^2 \tfrac{2}{3}\pi \neq 1.$$

(b) The relation

$$\cos^2 \alpha + \cos^2 \tfrac{1}{4}\pi + \cos^2 \tfrac{1}{4}\pi = 1$$

gives

$$\cos^2 \alpha + \tfrac{1}{2} + \tfrac{1}{2} = 1, \quad \cos \alpha = 0, \quad a_1 = \|\mathbf{a}\| \cos \alpha = 0.$$

34. $\gamma = \tfrac{1}{3}\pi \;\; \text{or} \;\; \gamma = \tfrac{2}{3}\pi$

35. Let $\theta_1, \theta_2, \theta_3$ be the direction angles of $-\mathbf{a}$. Then

$$\theta_1 = \arccos\left[\dfrac{(-\mathbf{a} \cdot \mathbf{i})}{\|-\mathbf{a}\|}\right] = \arccos\left[-\dfrac{(\mathbf{a} \cdot \mathbf{i})}{\|\mathbf{a}\|}\right] = \arccos\left(-\cos \alpha\right) = \pi - \arccos\left(\cos \alpha\right) = \pi - \alpha.$$

Similarly $\theta_2 = \pi - \beta$ and $\theta_3 = \pi - \gamma$.

36. If $\mathbf{v} = a\mathbf{i} + a\mathbf{j} + a\mathbf{k}$, then $\alpha = \beta = \gamma = \cos^{-1} \dfrac{a}{a\sqrt{3}} = \cos^{-1}\left(\dfrac{1}{\sqrt{3}}\right) \cong 54.7°.$

37. Set $\mathbf{u} = a\mathbf{i} + b\mathbf{j} + c\mathbf{k}$. The relations

$$(a\mathbf{i} + b\mathbf{j} + c\mathbf{k}) \cdot (\mathbf{i} + 2\mathbf{j} + \mathbf{k}) = 0 \quad \text{and} \quad (a\mathbf{i} + b\mathbf{j} + c\mathbf{k}) \cdot (3\mathbf{i} - 4\mathbf{j} + 2\mathbf{k}) = 0$$

give

$$a + 2b + c = 0 \qquad 3a - 4b + 2c = 0$$

so that $b = \tfrac{1}{8}a$ and $c = -\tfrac{5}{4}a.$

Then, since $\mathbf{u}$ is a unit vector,

$$a^2 + b^2 + c^2 = 1, \quad a^2 + \left(\dfrac{a}{8}\right)^2 + \left(\dfrac{-5a}{4}\right)^2 = 1, \quad \dfrac{165}{64}a^2 = 1.$$

Thus, $a = \pm\dfrac{8}{165}\sqrt{165}$ and $\mathbf{u} = \pm\dfrac{\sqrt{165}}{165}(8\mathbf{i} + \mathbf{j} - 10\mathbf{k}).$

38. $\pm\mathbf{k}, \;\; \pm\dfrac{\sqrt{13}}{13}(3\mathbf{i} - 2\mathbf{j})$

39.

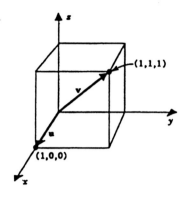

We take $\mathbf{u} = \mathbf{i}$ as an edge and $\mathbf{v} = \mathbf{i} + \mathbf{j} + \mathbf{k}$ as a diagonal of a cube. Then,

$$\cos\theta = \frac{\mathbf{u} \cdot \mathbf{v}}{\|\mathbf{u}\|\,\|\mathbf{v}\|} = \frac{1}{3}\sqrt{3},$$

$$\theta = \cos^{-1}\left(\tfrac{1}{3}\sqrt{3}\right) \cong 0.96 \text{ radians.}$$

40. Take $\mathbf{a} = \mathbf{i} + \mathbf{j}$, $\mathbf{b} = \mathbf{i} + \mathbf{j} + \mathbf{k}$.

$$\theta = \cos^{-1}\frac{\mathbf{a} \cdot \mathbf{b}}{\|\mathbf{a}\|\,\|\mathbf{b}\|} = \cos^{-1}\frac{2}{\sqrt{2}\cdot\sqrt{3}} = \cos^{-1}\left(\frac{\sqrt{6}}{3}\right) \cong 0.62 \text{ radians.}$$

41. (a) $\mathbf{proj}_\mathbf{b}\,\alpha\mathbf{a} = (\alpha\mathbf{a}\cdot\mathbf{u_b})\mathbf{u_b} = \alpha(\mathbf{a}\cdot\mathbf{u_b})\mathbf{u_b} = \alpha\,\mathbf{proj}_\mathbf{b}\,\mathbf{a}$

(b) $\qquad\qquad \mathbf{proj}_\mathbf{b}\,(\mathbf{a} + \mathbf{c}) = [(\mathbf{a}+\mathbf{c})\cdot\mathbf{u_b}]\,\mathbf{u_b}$

$$= (\mathbf{a}\cdot\mathbf{u_b} + \mathbf{c}\cdot\mathbf{u_b})\mathbf{u_b}$$

$$= (\mathbf{a}\cdot\mathbf{u_b})\mathbf{u_b} + (\mathbf{c}\cdot\mathbf{u_b})\mathbf{u_b} = \mathbf{proj}_\mathbf{b}\,\mathbf{a} + \mathbf{proj}_\mathbf{b}\,\mathbf{c}$$

42. (a) If $\beta > 0$, then $\mathbf{u}_{\beta\mathbf{b}} = \mathbf{u_b}$ and

$$\mathbf{proj}_{\beta\mathbf{b}}\mathbf{a} = (\mathbf{a}\cdot\mathbf{u}_{\beta\mathbf{b}})\mathbf{u}_{\beta\mathbf{b}} = (\mathbf{a}\cdot\mathbf{u_b})\mathbf{u_b} = \mathbf{proj}_\mathbf{b}\mathbf{a}.$$

If $\beta < 0$, then $\mathbf{u}_{\beta\mathbf{b}} = -\mathbf{u_b}$ and

$$\mathbf{proj}_{\beta\mathbf{b}}\mathbf{a} = (\mathbf{a}\cdot\mathbf{u}_{\beta\mathbf{b}})\mathbf{u}_{\beta\mathbf{b}} = [\mathbf{a}\cdot(-\mathbf{u_b})](-\mathbf{u_b}) = (\mathbf{a}\cdot\mathbf{u_b})\mathbf{u_b} = \mathbf{proj}_\mathbf{b}\mathbf{a}.$$

(b) If $\beta > 0$,

$$\text{comp}_{\beta\mathbf{b}}\mathbf{a} = (\mathbf{a}\cdot\mathbf{u}_{\beta\mathbf{b}}) = (\mathbf{a}\cdot\mathbf{u_b}) = \text{comp}_\mathbf{b}\mathbf{a}.$$

If $\beta < 0$,

$$\text{comp}_{\beta\mathbf{b}}\mathbf{a} = (\mathbf{a}\cdot\mathbf{u}_{\beta\mathbf{b}}) = [\mathbf{a}\cdot(-\mathbf{u_b})] = -(\mathbf{a}\cdot\mathbf{u_b}) = -\text{comp}_\mathbf{b}\mathbf{a}.$$

43. (a) $\mathbf{a}\cdot\mathbf{b} = \mathbf{a}\cdot\mathbf{c} \Longrightarrow \mathbf{a}(\mathbf{b}-\mathbf{c}) = 0 \Longrightarrow \mathbf{a}\perp(\mathbf{b}-\mathbf{c})$.

For $\mathbf{a} \neq \mathbf{0}$ the following statements are equivalent:

$$\mathbf{a}\cdot\mathbf{b} = \mathbf{a}\cdot\mathbf{c}, \quad \mathbf{b}\cdot\mathbf{a} = \mathbf{c}\cdot\mathbf{a},$$

$$\mathbf{b}\cdot\frac{\mathbf{a}}{\|\mathbf{a}\|} = \mathbf{c}\cdot\frac{\mathbf{a}}{\|\mathbf{a}\|}, \quad \mathbf{b}\cdot\mathbf{u_a} = \mathbf{c}\cdot\mathbf{u_a}$$

$$(\mathbf{b}\cdot\mathbf{u_a})\mathbf{u_a} = (\mathbf{c}\cdot\mathbf{u_a})\mathbf{u_a},$$

$$\mathbf{proj}_\mathbf{a}\,\mathbf{b} = \mathbf{proj}_\mathbf{a}\,\mathbf{c}.$$

Thus, $\mathbf{a}\cdot\mathbf{b} = \mathbf{a}\cdot\mathbf{c}$ implies only that the projection of $\mathbf{b}$ on $\mathbf{a}$ equals the projection of $\mathbf{c}$ on $\mathbf{a}$.

(b) $\mathbf{b} = (\mathbf{b} \cdot \mathbf{i})\mathbf{i} + (\mathbf{b} \cdot \mathbf{j})\mathbf{j} + (\mathbf{b} \cdot \mathbf{k})\mathbf{k} = (\mathbf{c} \cdot \mathbf{i})\mathbf{i} + (\mathbf{c} \cdot \mathbf{j})\mathbf{j} + (\mathbf{c} \cdot \mathbf{k})\mathbf{k} = \mathbf{c}$

$\quad\quad\quad$ (13.3.14) $\quad\quad\quad\quad\quad\quad\quad\quad\quad\quad\quad$ (13.3.14)

44. (a) $\|\mathbf{a} + \mathbf{b}\|^2 = \|\mathbf{a}\|^2 + \|\mathbf{b}\|^2 + 2\mathbf{a} \cdot \mathbf{b} = \|\mathbf{a}\|^2 + \|\mathbf{b}\|^2 \implies \mathbf{a} \perp \mathbf{b}.$

$\quad\quad$ (b) $\|\mathbf{a} - \mathbf{b}\|^2 = \|\mathbf{a}\|^2 + \|\mathbf{b}\|^2 - 2\mathbf{a} \cdot \mathbf{b} = \|\mathbf{a}\|^2 + \|\mathbf{b}\|^2 \implies \mathbf{a} \perp \mathbf{b}.$

45. (a) $\|\mathbf{a} + \mathbf{b}\|^2 - \|\mathbf{a} - \mathbf{b}\|^2 = (\mathbf{a} + \mathbf{b}) \cdot (\mathbf{a} + \mathbf{b}) - (\mathbf{a} - \mathbf{b}) \cdot (\mathbf{a} - \mathbf{b})$

$\quad\quad\quad\quad\quad = [(\mathbf{a} \cdot \mathbf{a}) + 2(\mathbf{a} \cdot \mathbf{b}) + (\mathbf{b} \cdot \mathbf{b})] - [(\mathbf{a} \cdot \mathbf{a}) - 2(\mathbf{a} \cdot \mathbf{b}) + (\mathbf{b} \cdot \mathbf{b})] = 4(\mathbf{a} \cdot \mathbf{b})$

$\quad\quad$ (b) The following statements are equivalent:

$$\mathbf{a} \perp \mathbf{b}, \quad \mathbf{a} \cdot \mathbf{b} = 0, \quad \|\mathbf{a} + \mathbf{b}\|^2 - \|\mathbf{a} - \mathbf{b}\|^2 = 0, \quad \|\mathbf{a} + \mathbf{b}\| = \|\mathbf{a} - \mathbf{b}\|.$$

$\quad\quad$ (c) By (b), the relation $\|\mathbf{a} + \mathbf{b}\| = \|\mathbf{a} - \mathbf{b}\|$ gives $\mathbf{a} \perp \mathbf{b}$. The relation $\mathbf{a} + \mathbf{b} \perp \mathbf{a} - \mathbf{b}$ gives

$$0 = (\mathbf{a} + \mathbf{b}) \cdot (\mathbf{a} - \mathbf{b}) = \|\mathbf{a}\|^2 - \|\mathbf{b}\|^2 \quad \text{and thus} \quad \|\mathbf{a}\| = \|\mathbf{b}\|.$$

The parallelogram is a square since it has two adjacent sides of equal length and these meet at right angles.

46. $|\mathbf{a} \cdot \mathbf{b}| = \|\mathbf{a}\|\|\mathbf{b}\||\cos\theta| = \|\mathbf{a}\|\|\mathbf{b}\| \quad$ iff $\quad \theta = 0 \quad$ or $\quad \theta = \pi$

47. $\|\mathbf{a} + \mathbf{b}\|^2 = (\mathbf{a} + \mathbf{b}) \cdot (\mathbf{a} + \mathbf{b}) = \mathbf{a} \cdot \mathbf{a} + 2\mathbf{a} \cdot \mathbf{b} + \mathbf{b} \cdot \mathbf{b} = \|\mathbf{a}\|^2 + 2\mathbf{a} \cdot \mathbf{b} + \|\mathbf{b}\|^2$

$\quad\quad$ $\|\mathbf{a} - \mathbf{b}\|^2 = (\mathbf{a} - \mathbf{b}) \cdot (\mathbf{a} - \mathbf{b}) = \mathbf{a} \cdot \mathbf{a} - 2\mathbf{a} \cdot \mathbf{b} + \mathbf{b} \cdot \mathbf{b} = \|\mathbf{a}\|^2 - 2\mathbf{a} \cdot \mathbf{b} + \|\mathbf{b}\|^2$

Add the two equations and the result follows.

48. $\|\mathbf{a}\|^2 = \|\mathbf{b}\|^2 \implies \|\mathbf{a}\|^2 - \|\mathbf{b}\|^2 = 0 \implies (\mathbf{a} + \mathbf{b}) \cdot (\mathbf{a} - \mathbf{b}) = 0 \implies (\mathbf{a} + \mathbf{b}) \perp (\mathbf{a} - \mathbf{b})$

49. Let $\mathbf{c} = \|\mathbf{b}\|\mathbf{a} + \|\mathbf{a}\|\mathbf{b}$. Then

$$\frac{\mathbf{a} \cdot \mathbf{c}}{\|\mathbf{a}\|\,\|\mathbf{c}\|} = \|\mathbf{a}\|\,\|\mathbf{b}\| + \mathbf{a} \cdot \mathbf{b} = \frac{\mathbf{b} \cdot \mathbf{c}}{\|\mathbf{b}\|\,\|\mathbf{c}\|}$$

50. $\cos t = \dfrac{\mathbf{a} \cdot \beta\mathbf{b}}{\|\mathbf{a}\|\,\|\beta\mathbf{b}\|} = \left(\dfrac{\beta}{|\beta|}\right)\left(\dfrac{\mathbf{a} \cdot \mathbf{b}}{\|\mathbf{a}\|\,\|\mathbf{b}\|}\right) = -\cos\theta \implies t = \pi - \theta$

51. Existence of decomposition:

$$\mathbf{a} = (\mathbf{a} \cdot \mathbf{u_b})\mathbf{u_b} + [\mathbf{a} - (\mathbf{a} \cdot \mathbf{u_b})\mathbf{u_b}].$$

Uniqueness of decomposition: suppose that

$$\mathbf{a} = \mathbf{a}_\| + \mathbf{a}_\perp = \mathbf{A}_\| + \mathbf{A}_\perp.$$

Then the vector $\mathbf{a}_\| - \mathbf{A}_\| = \mathbf{A}_\perp - \mathbf{a}_\perp$ is both parallel to $\mathbf{b}$ and perpendicular to $\mathbf{b}$. Therefore it is zero. Consequently $\mathbf{A}_\| = \mathbf{a}_\|$ and $\mathbf{A}_\perp = \mathbf{a}_\perp$.

52. (a) $\|\mathbf{u}_r\|^2 = \cos^2\theta + \sin^2\theta = 1, \quad \|\mathbf{u}_\theta\|^2 = \sin^2\theta + \cos^2\theta = 1.$

$\quad\quad\quad$ $\mathbf{u}_r \cdot \mathbf{u}_\theta = -\cos\theta\sin\theta + \sin\theta\cos\theta = 0, \quad$ so $\quad \mathbf{u}_r \perp \mathbf{u}_\theta.$

(b) $P = (r\cos\theta, r\sin\theta) = r\mathbf{u}_r,$ so $\overrightarrow{OP}$ has same direction as $\mathbf{u}_r$.

To see that $\mathbf{u}_\theta$ is $90°$ counterclockwise from $\mathbf{u}_r$, check the sign of the coefficient in all four quadrants.

53. Place center of sphere at the origin.

$$\overrightarrow{P_1Q} \cdot \overrightarrow{P_2Q} = (-\mathbf{a} + \mathbf{b}) \cdot (\mathbf{a} + \mathbf{b})$$

$$= -\|\mathbf{a}\|^2 + \|\mathbf{b}\|^2$$

$$= 0.$$

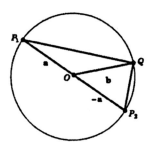

54. Take λ arbitrary, $\mathbf{b} \ne \mathbf{0}$.

$$0 \le \|\mathbf{a} - \lambda\mathbf{b}\|^2 = (\mathbf{a} - \lambda\mathbf{b}) \cdot (\mathbf{a} - \lambda\mathbf{b}) = \mathbf{a} \cdot \mathbf{a} - \lambda(\mathbf{b} \cdot \mathbf{a}) - \lambda(\mathbf{a} \cdot \mathbf{b}) + \lambda^2(\mathbf{b} \cdot \mathbf{b})$$

$$= \|\mathbf{a}\|^2 - 2\lambda(\mathbf{a} \cdot \mathbf{b}) + \lambda^2\|\mathbf{b}\|^2$$

Setting $\lambda = (\mathbf{a} \cdot \mathbf{b})/\|\mathbf{b}\|^2$ we have

$$0 \le \|\mathbf{a}\|^2 - 2\frac{|(\mathbf{a} \cdot \mathbf{b})|^2}{\|\mathbf{b}\|^2} + \frac{|(\mathbf{a} \cdot \mathbf{b})|^2}{\|\mathbf{b}\|^2}$$

$$0 \le \|\mathbf{a}\|^2\|\mathbf{b}\|^2 - |(\mathbf{a} \cdot \mathbf{b})|^2.$$

Thus $|(\mathbf{a} \cdot \mathbf{b})|^2 \le \|\mathbf{a}\|^2\|\mathbf{b}\|^2$ and $|(\mathbf{a} \cdot \mathbf{b})| \le \|\mathbf{a}\| \cdot \|\mathbf{b}\|$.

PROJECT 13.3

1. (a) $W = \mathbf{F} \cdot \mathbf{r}$ (b) 0   (c) $\|\mathbf{F}\| \mathbf{i} \cdot (b - a)\mathbf{i} = \|\mathbf{F}\|(b - a)$

2. (a) $W = (15\cos 0° \, \mathbf{i} + 15\sin 0° \, \mathbf{j}) \cdot (50\,\mathbf{i}) = 15(50) = 750$ joules

   (b) $W = (15\cos 30° \, \mathbf{i} + 15\sin 30° \, \mathbf{j}) \cdot (50\,\mathbf{i}) = \left(\frac{1}{2}15\sqrt{3}\right)50 \cong 649.5$ joules

   (c) $W = (15\cos 45° \, \mathbf{i} + 15\sin 45° \, \mathbf{j}) \cdot (50\,\mathbf{i}) = \left(\frac{1}{2}15\sqrt{2}\right)50 \cong 530.3$ joules

3. (a) $W_1 = \mathbf{F}_1 \cdot \mathbf{r} = \|\mathbf{F}_1\|\|\mathbf{r}\|\cos\theta;$ $W_2 = \mathbf{F}_2 \cdot \mathbf{r} = \|\mathbf{F}_2\|\|\mathbf{r}\|\cos(-\theta) = \|\mathbf{F}_2\|\|\mathbf{r}\|\cos\theta$

   Therefore $W_2 = \dfrac{\|\mathbf{F}_2\|}{\|\mathbf{F}_1\|}W_1.$

   (b) $W_1 = \|\mathbf{F}_1\|\|\mathbf{r}\|\cos\pi/3 = \frac{1}{2}\|\mathbf{F}_1\|\|\mathbf{r}\|;$ $W_2 = \|\mathbf{F}_2\|\|\mathbf{r}\|\cos\pi/6 = \frac{1}{2}\sqrt{3}\|\mathbf{F}_2\|\|\mathbf{r}\|$

   Therefore $W_2 = \sqrt{3}\dfrac{\|\mathbf{F}_2\|}{\|\mathbf{F}_1\|}W_1.$

4. Since the object returns to its starting point, the total displacement is zero, so the work done is zero.

SECTION 13.4

1. $(\mathbf{i} + \mathbf{j}) \times (\mathbf{i} - \mathbf{j}) = [\mathbf{i} \times (\mathbf{i} - \mathbf{j})] + [\mathbf{j} \times (\mathbf{i} - \mathbf{j})] = (\mathbf{0} - \mathbf{k}) + (-\mathbf{k} - \mathbf{0}) = -2\,\mathbf{k}$

2. $\mathbf{0}$

3. $(\mathbf{i} - \mathbf{j}) \times (\mathbf{j} - \mathbf{k}) = [\mathbf{i} \times (\mathbf{j} - \mathbf{k})] - [\mathbf{j} \times (\mathbf{j} - \mathbf{k})] = (\mathbf{j} + \mathbf{k}) - (\mathbf{0} - \mathbf{i}) = \mathbf{i} + \mathbf{j} + \mathbf{k}$

4. $\mathbf{j} \times (2\,\mathbf{i} - \mathbf{k}) = \mathbf{j} \times 2\,\mathbf{i} - \mathbf{j} \times \mathbf{k} = -2\,\mathbf{k} - \mathbf{i} = -\mathbf{i} - 2\,\mathbf{k}$

5. $(2\,\mathbf{j} - \mathbf{k}) \times (\mathbf{i} - 3\,\mathbf{j}) = [2\,\mathbf{j} \times (\mathbf{i} - 3\,\mathbf{j})] - [\mathbf{k} \times (\mathbf{i} - 3\,\mathbf{j})] = (-2\,\mathbf{k}) - (\mathbf{j} + 3\,\mathbf{i}) = -3\,\mathbf{i} - \mathbf{j} - 2\,\mathbf{k}$

   or

   $$(2\,\mathbf{j} - \mathbf{k}) \times (\mathbf{i} - 3\,\mathbf{j}) = \begin{vmatrix} \mathbf{i} & \mathbf{j} & \mathbf{k} \\ 0 & 2 & -1 \\ 1 & -3 & 0 \end{vmatrix} = \mathbf{i} \begin{vmatrix} 2 & -1 \\ -3 & 0 \end{vmatrix} - \mathbf{j} \begin{vmatrix} 0 & -1 \\ 1 & -3 \end{vmatrix} + \mathbf{k} \begin{vmatrix} 0 & 2 \\ 1 & -3 \end{vmatrix} = -3\,\mathbf{i} - \mathbf{j} - 2\,\mathbf{k}$$

6. $\mathbf{i} \cdot (\mathbf{j} \times \mathbf{k}) = \mathbf{i} \cdot \mathbf{i} = 1$ 7. $\mathbf{j} \cdot (\mathbf{i} \times \mathbf{k}) = \mathbf{j} \cdot (-\mathbf{j}) = -1$ 8. $(\mathbf{j} \times \mathbf{i}) \cdot (\mathbf{i} \times \mathbf{k}) = (-\mathbf{k}) \cdot (-\mathbf{j}) = 0$

9. $(\mathbf{i} \times \mathbf{j}) \times \mathbf{k} = \mathbf{k} \times \mathbf{k} = \mathbf{0}$ 10. $\mathbf{k} \cdot (\mathbf{j} \times \mathbf{i}) = \mathbf{k} \cdot (-\mathbf{k}) = -1$ 11. $\mathbf{j} \cdot (\mathbf{k} \times \mathbf{i}) = \mathbf{j} \cdot (\mathbf{j}) = 1$

12. $\mathbf{j} \times (\mathbf{k} \times \mathbf{i}) = \mathbf{j} \times \mathbf{j} = \mathbf{0}$

13. $(\mathbf{i} + 3\,\mathbf{j} - \mathbf{k}) \times (\mathbf{i} + \mathbf{k}) = \begin{vmatrix} \mathbf{i} & \mathbf{j} & \mathbf{k} \\ 1 & 3 & -1 \\ 1 & 0 & 1 \end{vmatrix} = [(3)(1) - (-1)(0)]\,\mathbf{i} - [(1)(1) - (-1)(1)]\,\mathbf{j} + [(1)0 - (3)(1)]\,\mathbf{k}$

$$= 3\,\mathbf{i} - 2\,\mathbf{j} - 3\,\mathbf{k}$$

14. $(3\,\mathbf{i} - 2\,\mathbf{j} + \mathbf{k}) \times (\mathbf{i} - \mathbf{j} + \mathbf{k}) = \begin{vmatrix} \mathbf{i} & \mathbf{j} & \mathbf{k} \\ 3 & -2 & 1 \\ 1 & -1 & 1 \end{vmatrix} = -\mathbf{i} - 2\,\mathbf{j} - \mathbf{k}$

15. $(\mathbf{i} + \mathbf{j} + \mathbf{k}) \times (2\,\mathbf{i} + \mathbf{k}) = \begin{vmatrix} \mathbf{i} & \mathbf{j} & \mathbf{k} \\ 1 & 1 & 1 \\ 2 & 0 & 1 \end{vmatrix} = [(1)(1) - (1)(0)]\,\mathbf{i} - [(1)(1) - (1)(2)]\,\mathbf{j} + [(1)(0) - (1)(2)]\,\mathbf{k}$

$$= \mathbf{i} + \mathbf{j} - 2\,\mathbf{k}$$

16. $(2\,\mathbf{i} - \mathbf{k}) \times (\mathbf{i} - 2\,\mathbf{j} + 2\,\mathbf{k}) = \begin{vmatrix} \mathbf{i} & \mathbf{j} & \mathbf{k} \\ 2 & 0 & -1 \\ 1 & -2 & 2 \end{vmatrix} = -2\,\mathbf{i} - 5\,\mathbf{j} - 4\,\mathbf{k}$

17. $[2\mathbf{i} + \mathbf{j}] \cdot [(\mathbf{i} - 3\mathbf{j} + \mathbf{k}) \times (4\mathbf{i} + \mathbf{k})] = \begin{vmatrix} 1 & -3 & 1 \\ 4 & 0 & 1 \\ 2 & 1 & 0 \end{vmatrix} =$

$$[(0)(0) - (1)(1)] - (-3)[(4)(0) - (1)(2)] + [(4)(1) - (0)(2)] = -3$$

18. $[(-2\mathbf{i} + \mathbf{j} - 3\mathbf{k}) \times \mathbf{i}] \times [\mathbf{i} + \mathbf{j}] = (-3\mathbf{j} - \mathbf{k}) \times (\mathbf{i} + \mathbf{j}) = \begin{vmatrix} \mathbf{i} & \mathbf{j} & \mathbf{k} \\ 0 & -3 & -1 \\ 1 & 1 & 0 \end{vmatrix} = \mathbf{i} - \mathbf{j} + 3\mathbf{k}$

19.

$$[(\mathbf{i} - \mathbf{j}) \times (\mathbf{j} - \mathbf{k})] \times [\mathbf{i} + 5\mathbf{k}] = \{[\mathbf{i} \times (\mathbf{j} - \mathbf{k})] - [\mathbf{j} \times (\mathbf{j} - \mathbf{k})]\} \times [\mathbf{i} + 5\mathbf{k}]$$

$$= [(\mathbf{k} + \mathbf{j}) - (-\mathbf{i})] \times [\mathbf{i} + 5\mathbf{k}]$$

$$= (\mathbf{i} + \mathbf{j} + \mathbf{k}) \times (\mathbf{i} + 5\mathbf{k})$$

$$= [(\mathbf{i} + \mathbf{j} + \mathbf{k}) \times \mathbf{i}] + [(\mathbf{i} + \mathbf{j} + \mathbf{k}) \times 5\mathbf{k}]$$

$$= (-\mathbf{k} + \mathbf{j}) + (-5\mathbf{j} + 5\mathbf{i})$$

$$= 5\mathbf{i} - 4\mathbf{j} - \mathbf{k}$$

20. $(\mathbf{i} - \mathbf{j}) \times [(\mathbf{j} - \mathbf{k}) \times (\mathbf{j} + 5\mathbf{k})] = (\mathbf{i} - \mathbf{j}) \times \begin{vmatrix} \mathbf{i} & \mathbf{j} & \mathbf{k} \\ 0 & 1 & -1 \\ 0 & 1 & 5 \end{vmatrix} = (\mathbf{i} - \mathbf{j}) \times 6\mathbf{i} = 6\mathbf{k}$

21. $\mathbf{a} \times \mathbf{b} = \begin{vmatrix} \mathbf{i} & \mathbf{j} & \mathbf{k} \\ 1 & 3 & -1 \\ 2 & 0 & 1 \end{vmatrix} = 3\mathbf{i} - 3\mathbf{j} - 6\mathbf{k}$

$$\frac{\mathbf{a} \times \mathbf{b}}{\|\mathbf{a} \times \mathbf{b}\|} = \frac{1}{\sqrt{6}}\mathbf{i} - \frac{1}{\sqrt{6}}\mathbf{j} - \frac{2}{\sqrt{6}}\mathbf{k}; \qquad \frac{\mathbf{b} \times \mathbf{a}}{\|\mathbf{b} \times \mathbf{a}\|} = -\frac{1}{\sqrt{6}}\mathbf{i} + \frac{1}{\sqrt{6}}\mathbf{j} + \frac{2}{\sqrt{6}}\mathbf{k}$$

22. $\mathbf{a} \times \mathbf{b} = \begin{vmatrix} \mathbf{i} & \mathbf{j} & \mathbf{k} \\ 1 & 2 & 3 \\ 2 & 1 & 1 \end{vmatrix} = -\mathbf{i} + 5\mathbf{j} - 3\mathbf{k}, \quad \text{so take} \quad \pm\frac{\sqrt{35}}{35}(-\mathbf{i} + 5\mathbf{j} - 3\mathbf{k})$

23. Set $\mathbf{a} = \overrightarrow{PQ} = -\mathbf{i} + 2\mathbf{k}$ and $\mathbf{b} = \overrightarrow{PR} = 2\mathbf{i} - \mathbf{k}$. Then

$$\mathbf{a} \times \mathbf{b} = \begin{vmatrix} \mathbf{i} & \mathbf{j} & \mathbf{k} \\ -1 & 0 & 2 \\ 2 & 0 & -1 \end{vmatrix} = 3\mathbf{j}; \qquad \frac{\mathbf{a} \times \mathbf{b}}{\|\mathbf{a} \times \mathbf{b}\|} = \mathbf{j}$$

and $A = \frac{1}{2}\|\mathbf{a} \times \mathbf{b}\| = \frac{1}{2}\|3\mathbf{j}\| = \frac{3}{2}$.

24. $\mathbf{N} = \vec{PQ} \times \vec{PR} = \begin{vmatrix} \mathbf{i} & \mathbf{j} & \mathbf{k} \\ -2 & 1 & -1 \\ 2 & -3 & -1 \end{vmatrix} = -4\,\mathbf{i} - 4\,\mathbf{j} + 4\,\mathbf{k}; \qquad \dfrac{\mathbf{N}}{\|\mathbf{N}\|} = -\dfrac{1}{\sqrt{3}}\mathbf{i} - \dfrac{1}{\sqrt{3}}\mathbf{j} + \dfrac{1}{\sqrt{3}}\mathbf{k}$

$A = \dfrac{1}{2}\|\mathbf{N}\| = 2\sqrt{3}$

25. Set $\quad \mathbf{a} = \vec{PQ} = \mathbf{i} + \mathbf{j} - 3\,\mathbf{k} \quad$ and $\quad \mathbf{b} = \vec{PR} = -\mathbf{i} + 3\,\mathbf{j} - \mathbf{k}. \quad$ Then

$$\mathbf{a} \times \mathbf{b} = \begin{vmatrix} \mathbf{i} & \mathbf{j} & \mathbf{k} \\ 1 & 1 & -3 \\ -1 & 3 & -1 \end{vmatrix} = 8\,\mathbf{j} + 4\,\mathbf{j} + 4\,\mathbf{k}; \qquad \dfrac{\mathbf{a} \times \mathbf{b}}{\|\mathbf{a} \times \mathbf{b}\|} = \dfrac{2}{\sqrt{6}}\mathbf{i} + \dfrac{1}{\sqrt{6}}\mathbf{j} + \dfrac{1}{\sqrt{6}}\mathbf{k}$$

and $A = \tfrac{1}{2}\|\mathbf{a} \times \mathbf{b}\| = \tfrac{1}{2}\|8\,\mathbf{i} + 4\,\mathbf{j} + 4\,\mathbf{k}\| = \tfrac{1}{2}\sqrt{8^2 + 4^2 + 4^2} = 2\sqrt{6}.$

26. $\mathbf{N} = \vec{PQ} \times \vec{PR} = \begin{vmatrix} \mathbf{i} & \mathbf{j} & \mathbf{k} \\ 2 & 2 & -4 \\ -5 & 1 & 2 \end{vmatrix} = 8\,\mathbf{i} + 16\,\mathbf{j} + 12\,\mathbf{k} \qquad \dfrac{\mathbf{N}}{\|\mathbf{N}\|} = \dfrac{2}{\sqrt{29}}\mathbf{i} + \dfrac{4}{\sqrt{29}}\mathbf{j} + \dfrac{3}{\sqrt{29}}\mathbf{k}$

Area $= \dfrac{1}{2}\|\mathbf{N}\| = 2\sqrt{29}$

27. $V = \left| [(\mathbf{i} + \mathbf{j}) \times (2\,\mathbf{i} - \mathbf{k})] \cdot (3\,\mathbf{j} + \mathbf{k}) \right| = \left| (-\mathbf{i} + \mathbf{j} - 2\,\mathbf{k}) \cdot (3\,\mathbf{j} + \mathbf{k}) \right| = 1$

28. $V = (\mathbf{i} - 3\,\mathbf{j} + \mathbf{k}) \times (2\,\mathbf{j} - \mathbf{k})] \cdot (\mathbf{i} + \mathbf{j} - 2\,\mathbf{k}) = \begin{vmatrix} 1 & -3 & 1 \\ 0 & 2 & -1 \\ 1 & 1 & -2 \end{vmatrix} = -2; \qquad V = |-2| = 2$

29. $V = \left| \vec{OP} \cdot \left(\vec{OQ} \times \vec{OR} \right) \right| = \left\| \begin{vmatrix} 1 & 2 & 3 \\ 1 & 1 & 2 \\ 2 & 1 & 1 \end{vmatrix} \right\| = 2$

30. $\vec{PQ} \cdot \left(\vec{PR} \times \vec{PS} \right) = \begin{vmatrix} 1 & 1 & -3 \\ -1 & 3 & -1 \\ 2 & 6 & 3 \end{vmatrix} = 52; \qquad V = 52$

31. $$(\mathbf{a} + \mathbf{b}) \times (\mathbf{a} - \mathbf{b}) = [\mathbf{a} \times (\mathbf{a} - \mathbf{b})] + [\mathbf{b} \times (\mathbf{a} - \mathbf{b})]$$
$$= [\mathbf{a} \times (-\mathbf{b})] + [\mathbf{b} \times \mathbf{a}]$$
$$= -(\mathbf{a} \times \mathbf{b}) - (\mathbf{a} \times \mathbf{b}) = -2(\mathbf{a} \times \mathbf{b})$$

32. $(\mathbf{a} + \mathbf{b}) \times \mathbf{c} = -[\mathbf{c} \times (\mathbf{a} + \mathbf{b})] = -(\mathbf{c} \times \mathbf{a}) - (\mathbf{c} \times \mathbf{b}) = (\mathbf{a} \times \mathbf{c}) + (\mathbf{b} \times \mathbf{c}).$

33. $\mathbf{a} \times \mathbf{i} = \mathbf{0}, \quad \mathbf{a} \times \mathbf{j} = \mathbf{0} \quad \Longrightarrow \quad \mathbf{a} \| \mathbf{i} \quad$ and $\quad \mathbf{a} \| \mathbf{j} \quad \Longrightarrow \quad \mathbf{a} = \mathbf{0}$

34. $\mathbf{a} \times \mathbf{b} = (a_1 b_2 - b_1 a_2)\,\mathbf{k}$

35. By (13.4.4) $\alpha \mathbf{a} \times \beta \mathbf{b} = (\alpha\beta)\mathbf{a} \times \mathbf{b}$. Therefore, $\|\alpha \mathbf{a} \times \beta \mathbf{b}\| = \|(\alpha\beta)\mathbf{a} \times \mathbf{b}\|$.

36. (a) The following statements are equivalent: (b)

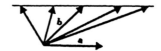

$$\mathbf{a} \times \mathbf{b} = \mathbf{a} \times \mathbf{c}$$
$$(\mathbf{a} \times \mathbf{b}) - (\mathbf{a} \times \mathbf{c}) = 0$$
$$(\mathbf{a} \times \mathbf{b}) + [\mathbf{a} \times (-\mathbf{c})] = 0$$
$$\mathbf{a} \times (\mathbf{b} - \mathbf{c}) = 0$$

The last equation holds iff $\mathbf{a}$ and $\mathbf{b} - \mathbf{c}$ are parallel. The tip of $\mathbf{c}$ must lie on the line which passes through the tip of $\mathbf{b}$ and is parallel to $\mathbf{a}$.

37. (a) $\mathbf{a} \cdot (\mathbf{b} \times \mathbf{c})$: makes sense – this is the dot product of two vectors.

(b) $\mathbf{a} \times (\mathbf{b} \cdot \mathbf{c})$: does not make sense – this is the cross product of a vector with a number.

(c) $\mathbf{a} \cdot (\mathbf{b} \cdot \mathbf{c})$: does not make sense – this is the dot product of a vector with a number.

(d) $\mathbf{a} \times (\mathbf{b} \times \mathbf{c})$: makes sense – this is the cross product of two vectors.

38. Given a vector $\mathbf{a}$. Since $\mathbf{a} \times \mathbf{b} \perp \mathbf{b}$, $(\mathbf{a} \times \mathbf{b}) \cdot \mathbf{b} = 0$ for all vectors $\mathbf{b}$. That is, there are no vectors $\mathbf{b}$ such that $(\mathbf{a} \times \mathbf{b}) \cdot \mathbf{b} \neq 0$.

39. $\mathbf{d} \cdot \mathbf{a} = \mathbf{d} \cdot \mathbf{b} \Longrightarrow \mathbf{d} \perp (\mathbf{a} - \mathbf{b})$; $\mathbf{d} \cdot \mathbf{a} = \mathbf{d} \cdot \mathbf{c} \Longrightarrow \mathbf{d} \perp (\mathbf{a} - \mathbf{c})$

Therefore, $\mathbf{d} = \lambda[(\mathbf{a} - \mathbf{b}) \times (\mathbf{a} - \mathbf{c})]$ for some number λ.

40. $\mathbf{b} \times \mathbf{c} \perp \mathbf{b}$ and $\mathbf{b} \times \mathbf{c} \perp \mathbf{c}$, so $\mathbf{b} \times \mathbf{c}$ must be parallel to $\mathbf{a}$. Hence $\mathbf{a} \times (\mathbf{b} \times \mathbf{c}) = 0$.

41. $\mathbf{a} \cdot \mathbf{b} = \mathbf{a} \cdot \mathbf{c} \implies \mathbf{a} \cdot (\mathbf{b} - \mathbf{c}) = 0$; $\mathbf{a}$ is perpendicular to $\mathbf{b} - \mathbf{c}$.

$\mathbf{a} \times \mathbf{b} = \mathbf{a} \times \mathbf{c} \implies \mathbf{a} \times (\mathbf{b} - \mathbf{c}) = 0$; $\mathbf{a}$ is parallel to $\mathbf{b} - \mathbf{c}$.

Since $\mathbf{a} \neq 0$ it follows that $\mathbf{b} - \mathbf{c} = 0$ or $\mathbf{b} = \mathbf{c}$.

42. (a) $\mathbf{i} -$ component of $\mathbf{a} \times (\mathbf{b} \times \mathbf{c}) = a_2(\mathbf{b} \times \mathbf{c})_3 - a_3(\mathbf{b} \times \mathbf{c})_2$

$$= a_2(b_1c_2 - b_2c_1) - a_3(b_3c_1 - b_1c_3) = (a_2c_2 + a_3c_3)b_1 - (a_2b_2 + a_3b_3)c_1$$
$$= (a_1c_1 + a_2c_2 + a_3c_3)b_1 - (a_1b_1 + a_2b_2 + a_3b_3)c_1$$
$$= (\mathbf{a} \cdot \mathbf{c})b_1 - (\mathbf{a} \cdot \mathbf{c})c_1 = \mathbf{i}\text{-component of } (\mathbf{a} \cdot \mathbf{c})\mathbf{b} - (\mathbf{a} \cdot \mathbf{c})\mathbf{c}$$

(b) $(\mathbf{a} \times \mathbf{b}) \times \mathbf{c} = -\mathbf{c} \times (\mathbf{a} \times \mathbf{b}) = -[(\mathbf{c} \cdot \mathbf{b})\mathbf{a} - (\mathbf{c} \cdot \mathbf{a})\mathbf{b}] = (\mathbf{c} \cdot \mathbf{a})\mathbf{b} - (\mathbf{c} \cdot \mathbf{b})\mathbf{a}$

(c) with $\mathbf{r} = \mathbf{c} \times \mathbf{d}$

$$(\mathbf{a} \times \mathbf{b}) \cdot (\mathbf{c} \times \mathbf{d}) = (\mathbf{a} \times \mathbf{b}) \cdot \mathbf{r} = (\mathbf{r} \times \mathbf{a}) \cdot \mathbf{b}$$
$$= [(\mathbf{c} \times \mathbf{d}) \times \mathbf{a}] \cdot \mathbf{b}$$
$$= [(\mathbf{a} \cdot \mathbf{c})\mathbf{d} - (\mathbf{a} \cdot \mathbf{d})\mathbf{c}] \cdot \mathbf{b}$$
$$= (\mathbf{a} \cdot \mathbf{c})(\mathbf{d} \cdot \mathbf{b}) - (\mathbf{a} \cdot \mathbf{d})(\mathbf{c} \cdot \mathbf{b})$$
$$= (\mathbf{a} \cdot \mathbf{c})(\mathbf{b} \cdot \mathbf{d}) - (\mathbf{a} \cdot \mathbf{d})(\mathbf{b} \cdot \mathbf{c})$$

43. $c \times a = (a \times b) \times a = (a \cdot a)b - (a \cdot b)a = (a \cdot a)b = \|a\|^2 b$

Exercise 42(a) $a \cdot b = 0$

44.

$$(a \cdot u)u + (u \times a) \times u = (a \cdot u)u + [(u \cdot u)a - (u \cdot a)u]$$

Exercise 42(b)

$$= (a \cdot u)u + a - (a \cdot u)u = a$$

45. Suppose $a \neq 0$. Then

$$a \cdot b = 0 \Longrightarrow b \perp a; \quad a \times b = 0 \Longrightarrow b \| a$$

Thus b is simultaneously perpendicular to, and parallel to a. It follows that $b = 0$.

46. $D = \frac{1}{2} | \overrightarrow{PQ} \times \overrightarrow{PR} | = \frac{1}{2}\sqrt{a^2 b^2 + a^2 c^2 + b^2 c^2}$

47. The result is an immediate consequence of Exercise 46.

48. $a \cdot (b \times c) = (a \times b) \cdot c = (c \times a) \cdot b = (b \times c) \cdot a = (a \times - c) \cdot b$
$a \cdot (c \times b) = c \cdot (b \times a) = (-a \times b) \cdot c$

SECTION 13.5

1. P (when $t = 0$) and Q (when $t = -1$)

2. l_1 and l_3 are parallel.

3. Take $r_0 = \overrightarrow{OP} = 3\,i + j$ and $d = k$. Then, $r(t) = (3\,i + j) + t\,k$.

4. $r(t) = i - j + 2\,k + t(3\,i - j + k)$

5. Take $r_0 = 0$ and $d = \overrightarrow{OQ}$. Then, $r(t) = t(x_1\,i + y_1\,j + z_1\,k)$.

6. $r(t) = x_0\,i + y_0\,j + z_0\,k + t\,[(x_1 - x_0)\,i + (y_1 - y_0)\,j + (z_1 - z_0)\,k]$

7. $\overrightarrow{PQ} = i - j + k$ so direction numbers are $1, -1, 1$. Using P as a point on the line, we have

$$x(t) = 1 + t, \quad y(t) = -t, \quad z(t) = 3 + t.$$

8. $x(t) = x_0 + t(x_1 - x_0), \quad y(t) = y_0 + t(y_1 - y_0), \quad z(t) = z_0 + t(z_1 - z_0).$

9. The line is parallel to the y-axis so we can take $0, 1, 0$ as direction numbers. Therefore

$$x(t) = 2, \quad y(t) = -2 + t, \quad z(t) = 3.$$

10. $x(t) = 1 + t, \quad y(t) = 4, \quad z(t) = -3$

11. Since the line $\quad 2(x+1) = 4(y-3) = z \quad$ can be written

$$\frac{x+1}{2} = \frac{y-3}{1} = \frac{z}{4},$$

it has direction numbers $2, 1, 4$. The line through $P(-1, 2, -3)$ with direction vector $2\mathbf{i} + \mathbf{j} + 4\mathbf{k}$ can be parameterized

$$\mathbf{r}(t) = (-\mathbf{i} + 2\mathbf{j} - 3\mathbf{k}) + t(2\mathbf{i} + \mathbf{j} + 4\mathbf{k}).$$

12. $\dfrac{x}{x_0} = \dfrac{y}{y_0} = \dfrac{z}{z_0}, \quad$ so $\quad \dfrac{x - x_0}{x_0} = \dfrac{y - y_0}{y_0} = \dfrac{z - z_0}{z_0} \quad$ provided $x_0 y_0 z_0 \neq 0$

13. $\mathbf{r}(t) = (3\mathbf{i} + \mathbf{j} + 5\mathbf{k}) + t(\mathbf{i} - \mathbf{j} + 2\mathbf{k}) = (3 + t)\mathbf{i} + (1 - t)\mathbf{j} + (5 + 2t)\mathbf{k}$

$\mathbf{R}(u) = (\mathbf{i} + 4\mathbf{j} + 2\mathbf{k}) + u(\mathbf{j} + \mathbf{k}) = \mathbf{i} + (4 + u)\mathbf{j} + (2 + u)\mathbf{k}$

$\mathbf{d} = \mathbf{i} - \mathbf{j} + 2\mathbf{k}$ is a direction vector for l_1; $\quad \mathbf{D} = \mathbf{j} + \mathbf{k}$ is a direction vector for l_2. Since $\mathbf{d}$ is not a multiple of $\mathbf{D}$, the lines either intersect or are skew. Setting $\mathbf{r}(t) = \mathbf{R}(u)$ we get the system of equations:

$$3 + t = 1, \quad 1 - t = 4 + u, \quad 5 + 2t = 2 + u$$

This system has the solution $t = -2, \ u = -1$. The point of intersection is: $(1, 3, 1)$.

14. $\mathbf{d}_2 = -2\mathbf{i} + 6\mathbf{j} - 4\mathbf{k} = -2(\mathbf{i} - 3\mathbf{j} + 2\mathbf{k})$. Therefore, the lines are either parallel or coincident. Since the point $(-1, 2, 1)$ on l_1 does not lie on l_2, the lines are parallel.

15. $\mathbf{d} = 2\mathbf{i} + 4\mathbf{j} - \mathbf{k}$ is a direction vector for l_1; $\quad \mathbf{D} = 2\mathbf{i} + \mathbf{j} + 2\mathbf{k}$ is a direction vector for l_2. Since $\mathbf{d}$ is not a multiple of $\mathbf{D}$, the lines either intersect or are skew. Equating coordinates, we get the system of equations:

$$3 + 2t = 3 + 2u, \quad -1 + 4t = 2 + u, \quad 2 - t = -2 + 2u$$

From the first two equations, we get $t = u = 1$. Since these values of t and u do not satisfy the third equation, the lines are skew.

16. The lines are skew.

17. $\mathbf{d} = -6\mathbf{i} + 9\mathbf{j} - 3\mathbf{k}$ is a direction vector for l_1; $\quad \mathbf{D} = 2\mathbf{i} - 3\mathbf{j} + \mathbf{k}$ is a direction vector for l_2. Since $\mathbf{d} = -3\mathbf{D}$, we conclude that l_1 and l_2 are either parallel or coincident. The point $(1, 2, 0)$ lies on l_1 but does not line on l_2. Therefore, the lines are parallel.

18. $\mathbf{d} = \mathbf{i} + 2\mathbf{j} + 3\mathbf{k}$ is a direction vector for l_1; $\quad \mathbf{D} = 3\mathbf{i} + 2\mathbf{j} + 1\mathbf{k}$ is a direction vector for l_2. Since $\mathbf{d}$ is not a multiple of $\mathbf{D}$, the lines either intersect or are skew. The system of equations

$$2 + t = 5 + 3u, \quad -1 + 2t = 1 + 2u, \quad 1 + 3t = 4 + u$$

does not have a solution. Therefore the lines are skew.

19. $\mathbf{d} = 2\mathbf{i} + 4\mathbf{j} + 3\mathbf{k}$ is a direction vector for l_1; $\mathbf{D} = \mathbf{i} + 3\mathbf{j} + 2\mathbf{k}$ is a direction vector for l_2. Since $\mathbf{d}$ is not a multiple of $\mathbf{D}$, the lines either intersect or are skew. The system of equations

$$4 + 2t = 2 + u, \quad -5 + 4t = -1 + 3u, \quad 1 + 3t = 2u$$

does not have a solution. Therefore the lines are skew.

20.   The lines intersect at the point $\left(\frac{3}{2}, 1, \frac{5}{2}\right)$.

21.   We set $\mathbf{r}_1(t) = \mathbf{r}_2(u)$ and solve for t and u:

$$\mathbf{i} + t\mathbf{j} = \mathbf{j} + u(\mathbf{i} + \mathbf{j}),$$

$$(1 - u)\mathbf{i} + (-1 - u + t)\mathbf{j} = \mathbf{0}.$$

Thus,

$$1 - u = 0 \quad \text{and} \quad -1 - u + t = 0.$$

These equations give $u = 1$, $t = 2$.   The point of intersection is $P(1, 2, 0)$. As direction vectors for the lines we can take $\mathbf{u} = \mathbf{j}$ and $\mathbf{v} = \mathbf{i} + \mathbf{j}$.   Thus

$$\cos\theta = \frac{\mathbf{u} \cdot \mathbf{v}}{\|\mathbf{u}\|\,\|\mathbf{v}\|} = \frac{1}{(1)(\sqrt{2})} = \frac{1}{2}\sqrt{2}.$$

The angle of intersection is $\frac{1}{4}\pi$ radians.

22.   Set $\mathbf{r}_1(t) = \mathbf{r}_2(u)$:

$$\mathbf{i} - 4\sqrt{3}\,\mathbf{j} + t(\mathbf{i} + \sqrt{3}\,\mathbf{j}) = 4\,\mathbf{i} + 3\sqrt{3}\,\mathbf{j} + u(\mathbf{i} - \sqrt{3}\,\mathbf{j})$$

$$(-3 + t - u)\mathbf{i} + (-7\sqrt{3} + t\sqrt{3} + u\sqrt{3})\mathbf{j} = \mathbf{0}$$

$$\implies t - u = 3, \quad t + u = 7, \quad \implies \quad t = 5, \quad u = 2$$

$$\mathbf{r}_1(5) = \mathbf{r}_2(2) = 6\,\mathbf{i} + \sqrt{3}\,\mathbf{j} : \quad \text{the point of intersection is: } P(6, \sqrt{3}, 0)$$

$$\cos\theta = \frac{|(\mathbf{i} + \sqrt{3}\,\mathbf{j}) \cdot (\mathbf{i} - \sqrt{3}\,\mathbf{j})|}{\|\mathbf{i} + \sqrt{3}\,\mathbf{j}\|\,\|\mathbf{i} - \sqrt{3}\,\mathbf{j}\|} = \frac{1}{2} \quad \implies \theta = \frac{\pi}{3} \text{ radians.}$$

23. $\left(x_0 - \dfrac{d_1}{d_3}z_0, \;\; y_0 - \dfrac{d_2}{d_3}z_0, \;\; 0\right)$

24.   The lines meet at (x_0, y_0, z_0), and since $\mathbf{d} \cdot \mathbf{D} = 0$, they are perpendicular.

25.   The lines are parallel.

26.   Note that $\mathbf{r}(0) = \mathbf{r}_0$ and $\mathbf{r}(1) = \mathbf{r}_1$, so we need $0 \leq t \leq 1$

27. $\mathbf{r}(t) = (2\,\mathbf{i} + 7\,\mathbf{j} - \mathbf{k}) + t(2\,\mathbf{i} - 5\,\mathbf{j} + 4\,\mathbf{k}), \quad 0 \leq t \leq 1$

28. $-1 \leq t \leq 2$.

29. Set
$$\mathbf{u} = \frac{\overrightarrow{PQ}}{\|\overrightarrow{PQ}\|} = \frac{-4\mathbf{i} + 2\mathbf{j} + 4\mathbf{k}}{\|-4\mathbf{i} + 2\mathbf{j} + 4\mathbf{k}\|} = -\frac{2}{3}\mathbf{i} + \frac{1}{3}\mathbf{j} + \frac{2}{3}\mathbf{k}.$$

Then $\mathbf{r}(t) = (6\mathbf{i} - 5\mathbf{j} + \mathbf{k}) + t\mathbf{u}$ is $\overrightarrow{OP}$ at $t = 9$ and it is $\overrightarrow{OQ}$ at $t = 15$. (Check this.)

Answer: $\mathbf{u} = -\frac{2}{3}\mathbf{i} + \frac{1}{3}\mathbf{j} + \frac{2}{3}\mathbf{k}, \quad 9 \le t \le 15.$

30. Since the two lines intersect, there exist numbers t_0 and u_0 such that
$$\mathbf{r}(t_0) = \mathbf{R}(u_0)$$

Suppose first that
$$\mathbf{r}(t_0) = \mathbf{R}(u_0) = 0.$$

Then
$$\mathbf{r}(t) = \mathbf{r}(t) - \mathbf{r}(t_0) = (\mathbf{r}_0 + t\mathbf{d}) - (\mathbf{r}_0 + t_0\mathbf{d}) = (t - t_0)\mathbf{d}.$$

Similarly $\mathbf{R}(u) = (u - u_0)\mathbf{D}.$ Since $l_1 \perp l_2,$ we have $\mathbf{d} \cdot \mathbf{D} = 0$ and thus
$$\mathbf{r}(t) \cdot \mathbf{R}(u) = (t - t_0)(u - u_0)(\mathbf{d} \cdot \mathbf{D}) = 0 \quad \text{for all } t, u$$

Suppose now that
$$\mathbf{r}(t_0) = \mathbf{R}(u_0) \ne \mathbf{0}$$

Then
$$\mathbf{r}(t_0) \cdot \mathbf{R}(u_0) = \mathbf{r}(t_0) \cdot \mathbf{r}(t_0) = \|\mathbf{r}(t_0)\|^2 \ne 0$$

and it is therefore not true that
$$\mathbf{R}(t) \cdot \mathbf{R}(u) = 0, \quad \text{for all } t, u.$$

31. The given line, call it l, has direction vector $2\mathbf{i} - 4\mathbf{j} + 6\mathbf{k}.$

If $a\mathbf{i} + b\mathbf{j} + c\mathbf{k}$ is a direction vector for a line perpendicular to l, then
$$(2\mathbf{i} - 4\mathbf{j} + 6\mathbf{k}) \cdot (a\mathbf{i} + b\mathbf{j} + c\mathbf{k}) = 2a - 4b + 6c = 0.$$

The lines through $P(3, -1, 8)$ perpendicular to l can be parameterized
$$X(u) = 3 + au, \quad Y(u) = -1 + bu, \quad Z(u) = 8 + cu$$

with $2a - 4b + 6c = 0.$

32. (a) Since $\mathbf{R}(0) = \mathbf{R}_0$ is on the line, there exists a number t_0 such that $\mathbf{r}(t_0) = \mathbf{r}_0 + t_0\mathbf{d} = \mathbf{R}_0.$

(b) Since $\mathbf{d}$ and $\mathbf{D}$ are both direction vectors for the same line, they are parallel. Since $\mathbf{d} \ne \mathbf{0},$ there exists a scalar α such that $\mathbf{D} = \alpha\mathbf{d}.$ It follows that, for all real $u,$
$$\mathbf{R}(u) = \mathbf{R}_0 + u\mathbf{D} = (\mathbf{r}_0 + t_0\mathbf{d}) + u(\alpha\mathbf{d}) = r_0 + (t_0 + \alpha u)\mathbf{d}.$$

33. $d(P, l) = \dfrac{\|(\mathbf{i} + 2\,\mathbf{k}) \times (2\,\mathbf{i} - \mathbf{j} + 2\,\mathbf{k})\|}{\|2\,\mathbf{i} - \mathbf{j} + 2\,\mathbf{k}\|} = 1$

34. $d(P, l) = \dfrac{\|(\mathbf{j} + \mathbf{k}) \times (\mathbf{i} - 2\,\mathbf{j} - 2\,\mathbf{k})\|}{\|\mathbf{i} - 2\,\mathbf{j} - 2\,\mathbf{k}\|} = \dfrac{1}{3}\sqrt{2} \cong 0.47$

35. The line contains the point $P_0(1, 0, 2)$. Therefore

$$d(P, l) = \dfrac{\|(2\,\mathbf{j} + \mathbf{k}) \times (\mathbf{i} - 2\,\mathbf{j} + 3\,\mathbf{k})\|}{\|\mathbf{i} - 2\,\mathbf{j} + 3\,\mathbf{k}\|} = \sqrt{\dfrac{69}{14}} \cong 2.22$$

36. $P_0 = (1, 0, 0), \quad \mathbf{d} = \mathbf{j}, \quad d(P, l) = \dfrac{\| - \mathbf{i} \times \mathbf{j}\|}{\|\mathbf{j}\|} = 1.$

37. The line contains the point $P_0(2, -1, 0)$. Therefore

$$d(P, l) = \dfrac{\|(\mathbf{i} - \mathbf{j} - \mathbf{k}) \times (\mathbf{i} + \mathbf{j})\|}{\|\mathbf{i} + \mathbf{j}\|} = \sqrt{3} \cong 1.73.$$

38. The line can be parameterized $\mathbf{r}(t) = b\,\mathbf{j} + t(\mathbf{i} + m\,\mathbf{j})$. Since the line contains the point $(0, b, 0)$

$$d(P, l) = \dfrac{\|[x_0\,\mathbf{i} + (y_0 - b)\,\mathbf{j} + z_0\,\mathbf{k}] \times (\mathbf{i} + m\,\mathbf{j})\|}{\|\mathbf{i} + m\,\mathbf{j}\|}$$

$$= \sqrt{\dfrac{(1 + m^2)z_0{}^2 + [y_0 - (mx_0 + b)]^2}{1 + m^2}}$$

If $P(x_0, y_0, z_0)$ lies directly above or below the line, then $y_0 = mx_0 + b$ and $d(P, l)$ reduces to $\sqrt{z_0{}^2} = |z_0|$. This is evident geometrically.

39. (a) The line passes through $P(1, 1, 1)$ with direction vector $\mathbf{i} + \mathbf{j}$. Therefore

$$d(0, l) = \dfrac{\|(\mathbf{i} + \mathbf{j} + \mathbf{k}) \times (\mathbf{i} + \mathbf{j})\|}{\|\mathbf{i} + \mathbf{j}\|} = 1.$$

(b) The distance from the origin to the line segment is $\sqrt{3}$.

Solution. The line segment can be parameterized

$$\mathbf{r}(t) = \mathbf{i} + \mathbf{j} + \mathbf{k} + t(\mathbf{i} + \mathbf{j}), \quad t \in [0, 1].$$

This is the set of all points $P(1 + t, 1 + t, 1)$ with $t \in [0, 1]$.

The distance from the origin to such a point is

$$f(t) = \sqrt{2\,(1 + t)^2 + 1}.$$

The minimum value of this function is $f(0) = \sqrt{3}$.

Explanation. The point on the line through P and Q closest to the origin is not on the line segment $\overline{PQ}$.

40. (a) We want $(\mathbf{r}_0 + t_0\mathbf{d}) \cdot \mathbf{d} = 0, \quad$ so $\quad \mathbf{r}_0 \cdot \mathbf{d} + t_0\|\mathbf{d}\|^2 = 0 \implies t_0 = -\dfrac{\mathbf{r}_0 \cdot \mathbf{d}}{\|\mathbf{d}\|^2}$

(b) $\mathbf{R}(t) = \mathbf{r}(t_0) \pm t\dfrac{\mathbf{d}}{\|\mathbf{d}\|}, \quad$ where t_0 is as in part (a).

41. We begin with $\mathbf{r}(t) = \mathbf{j} - 2\mathbf{k} + t(\mathbf{i} - \mathbf{j} + 3\mathbf{k})$. The scalar t_0 for which $\mathbf{r}(t_0) \perp l$ can be found by solving the equation

$$[\mathbf{j} - 2\mathbf{k} + t_0(\mathbf{i} - \mathbf{j} + 3\mathbf{k})] \cdot [\mathbf{i} - \mathbf{j} + 3\mathbf{k}] = 0.$$

This equation gives $-7 + 11t_0 = 0$ and thus $t_0 = 7/11$. Therefore

$$\mathbf{r}(t_0) = \mathbf{j} - 2\mathbf{k} + \tfrac{7}{11}(\mathbf{i} - \mathbf{j} + 3\mathbf{k}) = \tfrac{7}{11}\mathbf{i} + \tfrac{4}{11}\mathbf{j} - \tfrac{1}{11}\mathbf{k}.$$

The vectors of norm 1 parallel to $\mathbf{i} - \mathbf{j} + 3\mathbf{k}$ are

$$\pm\frac{1}{\sqrt{11}}(\mathbf{i} - \mathbf{j} + 3\mathbf{k}).$$

The standard parameterizations are

$$\mathbf{R}(t) = \frac{7}{11}\mathbf{i} + \frac{4}{11}\mathbf{j} - \frac{1}{11}\mathbf{k} \pm \frac{t}{\sqrt{11}}(\mathbf{i} - \mathbf{j} + 3\mathbf{k})$$

$$= \frac{1}{11}(7\mathbf{i} + 4\mathbf{j} - \mathbf{k}) \pm t\left[\frac{\sqrt{11}}{11}(\mathbf{i} - \mathbf{j} + 3\mathbf{k})\right].$$

42. Start with $\mathbf{r}(t) = \sqrt{3}\,\mathbf{i} + t(\mathbf{i} + \mathbf{j} + \mathbf{k})$ to get $t_0 = -\dfrac{\sqrt{3}}{3}$, so

$$\mathbf{R}(t) = \frac{\sqrt{3}}{3}(2\mathbf{i} - \mathbf{j} - \mathbf{k}) \pm t\frac{\sqrt{3}}{3}(\mathbf{i} + \mathbf{j} + \mathbf{k}).$$

43. $0 < t < s$

By similar triangles, if $0 < s < 1$, the tip of $\overrightarrow{OA} +$ $s\,\overrightarrow{AB} + s\,\overrightarrow{BC}$ falls on $\overline{AC}$. If $0 < t < s$, then the tip of $\overrightarrow{OA} + s\,\overrightarrow{AB} + t\,\overrightarrow{BC}$ falls short of $\overline{AC}$ and stays within the triangle. Clearly all points in the interior of the triangle can be reached in this manner.

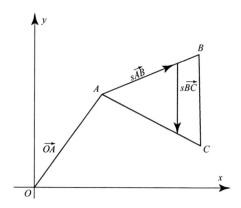

44. $\mathbf{d}_1 \times \mathbf{d}_2$ is a direction vector for the line that is perpendicular to both l_1 and l_2.

$$\frac{\left|\overrightarrow{PQ} \cdot (\mathbf{d}_1 \times \mathbf{d}_2)\right|}{|\mathbf{d}_1 \times \mathbf{d}_2|}$$

is the magnitude of the projection of the vector $\overrightarrow{PQ}$ onto the vector $\mathbf{d}_1 \times \mathbf{d}_2$.

45. $\mathbf{d} = \mathbf{i} + 3\mathbf{j} - 2\mathbf{k}$ is a direction vector for l_1; $\mathbf{D} = 4\mathbf{i} - \mathbf{j} + 2\mathbf{k}$ is a direction vector for l_2. Since $\mathbf{d}$ is not a multiple of $\mathbf{D}$, the lines either intersect or are skew. Equating coordinates, we get the system of equations:

$$2 + t = -1 + 4u, \quad -1 + 3t = 2 - u, \quad 1 - 2t = -3 + 2u$$

This system does not have a solution. Therefore the lines are skew. The point $P(2, -1, 1)$ is on l_1 and the point $Q(-1, 2, -3)$ is on l_2, and $\overrightarrow{PQ} = -3\,\mathbf{i} + 3\,\mathbf{j} - 4\,\mathbf{k}$. By Exercise 44, the distance between l_1 and l_2 is:

$$\frac{\left|\overrightarrow{PQ} \cdot (\mathbf{d} \times \mathbf{D})\right|}{\|\mathbf{d} \times \mathbf{D}\|} = \frac{|(-3\,\mathbf{i} + 3\,\mathbf{j} - 4\,\mathbf{k}) \cdot (4\,\mathbf{i} - 10\,\mathbf{j} - 13\,\mathbf{k})|}{\sqrt{285}} = \frac{10}{\sqrt{285}}$$

46. $\mathbf{D} = 2\,\mathbf{i} + \mathbf{j} + 4\mathbf{k}$ is not a multiple of $\mathbf{d} = \mathbf{i} + 3\,\mathbf{j} - 2\,\mathbf{k}$ and the system of equations

$$1 + t = 2u, \quad -2 + 3t = 3 + u, \quad 4 - 2t = -3 + 4u$$

does not have a solution. Therefore the lines are skew.

$$\overrightarrow{PQ} = -\mathbf{i} + 5\,\mathbf{j} - 7\,\mathbf{k} \quad \text{and} \quad \mathbf{d} \times \mathbf{D} = 14\,\mathbf{i} - 8\,\mathbf{j} - 5\,\mathbf{k}$$

The distance between the lines is: $\dfrac{19}{\sqrt{285}}$

SECTION 13.6

1. Q

2. An equation for the plane is: $(x - 4) - 3(y - 1) + (z + 1) = 0$; R and S lie on the plane.

3. Since $\mathbf{i} - 4\,\mathbf{j} + 3\,\mathbf{k}$ is normal to the plane, we have

$$(x - 2) - 4(y - 3) + 3(z - 4) = 0 \quad \text{and thus} \quad x - 4y + 3z - 2 = 0.$$

4. $\mathbf{N} = \mathbf{j} + 2\,\mathbf{k}, \quad P(1, -2, 3) \implies (y + 2) + 2(z - 3) = 0 \text{ or } y + 2z - 4 = 0.$

5. The vector $3\,\mathbf{i} - 2\,\mathbf{j} + 5\,\mathbf{k}$ is normal to the given plane and thus to every parallel plane: the equation we want can be written

$$3(x - 2) - 2(y - 1) + 5(z - 1) = 0, \quad 3x - 2y + 5z - 9 = 0.$$

6. $\mathbf{N} = 4\,\mathbf{i} + 2\,\mathbf{j} - 7\,\mathbf{k}, \quad P(3, -1, 5) \implies 4(x - 3) + 2(y + 1) - 7(z - 5) = 0$

7. The point $Q(0, 0, -2)$ lies on the line l; and $\mathbf{d} = \mathbf{i} + \mathbf{j} + \mathbf{k}$ is a direction vector for l. We want an equation for the plane which has the vector

$$\mathbf{N} = \overrightarrow{PQ} \times \mathbf{d} = (\mathbf{i} + 3\,\mathbf{j} + 3\,\mathbf{k}) \times (\mathbf{i} + \mathbf{j} + \mathbf{k})$$

as a normal vector:

$$\mathbf{N} = \begin{vmatrix} \mathbf{i} & \mathbf{j} & \mathbf{k} \\ -1 & -3 & -3 \\ 1 & 1 & 1 \end{vmatrix} = -2\,\mathbf{j} + 2\,\mathbf{k}$$

An equation for the plane is: $-2(y - 3) + 2(z - 1) = 0$ or $y - z - 2 = 0$

8. Another point is $Q(1, 1, 2)$, and the plane is parallel to the vectors $\mathbf{d} = -2\,\mathbf{i} + 4\,\mathbf{j} + \mathbf{k}$ and $\overrightarrow{PQ} = -\mathbf{i} + \mathbf{j} + \mathbf{k}$. Thus, $\mathbf{N} = \mathbf{d} \times \overrightarrow{PQ} = 3\,\mathbf{i} + \mathbf{j} + 2\,\mathbf{k}$ is a normal to the plane.
An equation for the plane is: $3(x - 2) + y + 2(z - 1) = 0$ or $3x + y + 2z - 8 = 0$

9. $\overrightarrow{OP_0} = x_0\,\mathbf{i} + y_0\,\mathbf{j} + z_0\,\mathbf{k}$ An equation for the plane is:

$$x_0\,(x - x_0) + y_0\,(y - y_0) + z_0\,(z - z_0) = 0$$

10. $\mathbf{N} = 2\,\mathbf{i} - 3\,\mathbf{j} + 7\,\mathbf{k}$,   unit normals: $\mathbf{u}_N = \pm \dfrac{1}{\sqrt{62}}(2\,\mathbf{i} - 3\,\mathbf{j} + 7\,\mathbf{k})$

11. The vector $\mathbf{N} = 2\,\mathbf{i} - \mathbf{j} + 5\,\mathbf{k}$ is normal to the plane $2x - y + 5z - 10 = 0$. The unit normals are:

$$\frac{\mathbf{N}}{\|\mathbf{N}\|} = \frac{1}{\sqrt{30}}(2\,\mathbf{i} - \mathbf{j} + 5\,\mathbf{k}) \quad \text{and} \quad -\frac{\mathbf{N}}{\|\mathbf{N}\|} = -\frac{1}{\sqrt{30}}(2\,\mathbf{i} - \mathbf{j} + 5\,\mathbf{k})$$

12. $(a, 0, 0)$, $(0, b, 0)$,   and $(0, 0, c)$ satisfy the equation.

13. Intercept form: $\dfrac{x}{15} + \dfrac{y}{12} - \dfrac{z}{10} = 1$ x-intercept: $(15, 0, 0)$
    y-intercept: $(0, 12, 0)$
    z-intercept: $(0, 0, -10)$

14. $\dfrac{x}{-2/3} + \dfrac{y}{2} + \dfrac{z}{-1/2} = 1$; $\left(-\dfrac{2}{3}, 0, 0\right)$, $(0, 2, 0)$, $\left(0, 0, -\dfrac{1}{2}\right)$.

15. $\mathbf{u}_{N_1} = \dfrac{\sqrt{38}}{38}(5\,\mathbf{i} - 3\,\mathbf{j} + 2\,\mathbf{k})$, $\mathbf{u}_{N_2} = \dfrac{\sqrt{14}}{14}(\mathbf{i} + 3\,\mathbf{j} + 2\,\mathbf{k})$, $\cos\theta = \left|\mathbf{u}_{N_1} \cdot \mathbf{u}_{N_2}\right| = 0$.
Therefore $\theta = \pi/2$ radians.

16. $\cos\theta = \dfrac{|(2\,\mathbf{i} - \mathbf{j} + 3\,\mathbf{k}) \cdot (5\,\mathbf{i} + 5\,\mathbf{j} - \mathbf{k})|}{\|2\,\mathbf{i} - \mathbf{j} + 3\,\mathbf{k}\|\|5\,\mathbf{i} + 5\,\mathbf{j} - \mathbf{k}\|} = \dfrac{2}{\sqrt{14}\sqrt{51}}$; $\theta \cong 1.50$ radians.

17. $\mathbf{u}_{N_1} = \dfrac{\sqrt{3}}{3}(\mathbf{i} - \mathbf{j} + \mathbf{k})$, $\mathbf{u}_{N_2} = \dfrac{\sqrt{14}}{14}(2\,\mathbf{i} + \mathbf{j} + 3\,\mathbf{k})$, $\cos\theta = \left|\mathbf{u}_{N_1} \cdot \mathbf{u}_{N_2}\right| = \dfrac{2}{21}\sqrt{42} \cong 0.617$.
Therefore $\theta \cong 0.91$ radians.

18. $\cos\theta = \dfrac{|(4\,\mathbf{i} + 4\,\mathbf{j} - 2\,\mathbf{k}) \cdot (2\,\mathbf{i} + \mathbf{j} + \mathbf{k})|}{\|4\,\mathbf{i} + 4\,\mathbf{j} - 2\,\mathbf{k}\|\|2\,\mathbf{i} + \mathbf{j} + \mathbf{k}\|} = \dfrac{10}{6\sqrt{6}}$; $\theta \cong 0.82$ radian.

19. coplanar since $0(4\,\mathbf{j} - \mathbf{k}) + 0(3\,\mathbf{i} + \mathbf{j} + 2\,\mathbf{k}) + 1(0) = 0$

20. $s\,\mathbf{i} + t(\mathbf{i} - 2\,\mathbf{j}) + u(3\,\mathbf{j} + \mathbf{k}) = (s + t)\,\mathbf{i} + (-2t + 3u)\,\mathbf{j} + u\,\mathbf{k} = \mathbf{0}$ only if $s = t = u = 0$, so not coplanar.

21. We need to determine whether there exist scalars s, t, u not all zero such that

$$s(\mathbf{i} + \mathbf{j} + \mathbf{k}) + t(2\,\mathbf{i} - \mathbf{j}) + u(3\,\mathbf{i} - \mathbf{j} - \mathbf{k}) = \mathbf{0}$$

$$(s + 2t + 3u)\,\mathbf{i} + (s - t - u)\,\mathbf{j} + (s - u)\,\mathbf{k} = \mathbf{0}.$$

The only solution of the system

$$s + 2t + 3u = 0, \quad s - t - u = 0, \quad s - u = 0$$

is $\quad s = t = u = 0.$ Thus, the vectors are not coplanar.

22. coplanar since $\quad 1(\mathbf{j} - \mathbf{k}) - 1(3\mathbf{i} - \mathbf{j} + 2\mathbf{k}) + 1(3\mathbf{i} - 2\mathbf{j} + 3\mathbf{k}) = 0$

23. By (13.6.5), $\quad d(P, p) = \dfrac{|2(2) + 4(-1) - (3) + 1|}{\sqrt{4 + 16 + 1}} = \dfrac{2}{\sqrt{21}} = \dfrac{2}{21}\sqrt{21}.$

24. $d = \dfrac{|8(3) - 2(-5) + 2 - 5|}{\sqrt{8^2 + (-2)^2 + 1^2}} = \dfrac{31}{\sqrt{69}}$

25. By (13.6.5), $\quad d(P, p) = \dfrac{|(-3)(1) + 0(-3) + 4(5) + 5|}{\sqrt{9 + 16}} = \dfrac{22}{5}.$

26. $d = \dfrac{|1 + 3 - 2 \cdot 4|}{\sqrt{1^2 + 1^2 + (-2)^2}} = \dfrac{2}{3}\sqrt{6}$

27. $\overrightarrow{P_1P} = (x - 1)\mathbf{i} + y\mathbf{j} + (z - 1)\mathbf{k}, \quad \overrightarrow{P_1P_2} = \mathbf{i} + \mathbf{j} - \mathbf{k}, \quad \overrightarrow{P_1P_3} = \mathbf{j}.$
Therefore

$$(\overrightarrow{P_1P_2} \times \overrightarrow{P_1P_3}) = (\mathbf{i} + \mathbf{j} - \mathbf{k}) \times \mathbf{j} = \mathbf{i} + \mathbf{k}$$

and

$$\overrightarrow{P_1P} \cdot (\overrightarrow{P_1P_2} \times \overrightarrow{P_1P_3}) = [(x - 1)\mathbf{i} + y\mathbf{j} + (z - 1)\mathbf{k}] \cdot [\mathbf{i} + \mathbf{k}] = x - 1 + z - 1.$$

An equation for the plane can be written $\quad x + z = 2.$

28. $\overrightarrow{P_1P_2} = (1, -3, -2), \quad \overrightarrow{P_1P_3} = (-1, 1, 0), \quad \mathbf{N} = \overrightarrow{P_1P_2} \times \overrightarrow{P_1P_3} = (2, 2, -2)$
$\implies 2(x - 1) + 2(y - 1) - 2(z - 1) = 0 \quad$ or $\quad x + y - z - 1 = 0$

29. $\overrightarrow{P_1P} = (x - 3)\mathbf{i} + (y + 4)\mathbf{j} + (z - 1)\mathbf{k}, \quad \overrightarrow{P_1P_2} = 6\mathbf{j}, \quad \overrightarrow{P_1P_3} = -2\mathbf{i} + 5\mathbf{j} - 3\mathbf{k}.$
Therefore

$$(\overrightarrow{P_1P_2} \times \overrightarrow{P_1P_3}) = 6\mathbf{j} \times (-4\mathbf{i} + 5\mathbf{j} - 3\mathbf{k}) = -18\mathbf{i} + 12\mathbf{k}$$

and

$$\overrightarrow{P_1P} \cdot (\overrightarrow{P_1P_2} \times \overrightarrow{P_1P_3}) = [(x - 3)\mathbf{i} + (y + 4)\mathbf{j} + (z - 1)\mathbf{k}] \cdot [-18\mathbf{i} + 12\mathbf{k}]$$
$$= -18(x - 3) + 12(z - 1)$$

An equation for the plane can be written $\quad -18(x - 3) + 12(z - 1) = 0 \quad$ or $\quad 3x - 2z - 7 = 0.$

30. $\overrightarrow{P_1P_2} = (0, -4, 5), \quad \overrightarrow{P_1P_3} = (-2, -3, 4) \quad \mathbf{N} = \overrightarrow{P_1P_2} \times \overrightarrow{P_1P_3} = (-1, -10, -8)$
$$\implies (x - 3) + 10(y - 2) + 8(z + 1) = 0$$

31. The line passes through the point $P_0(x_0, y_0, z_0)$ with direction numbers: A, B, C.
Equations for the line written in symmetric form are:

$$\frac{x - x_0}{A} = \frac{y - y_0}{B} = \frac{z - z_0}{C}, \quad \text{provided} \quad A \neq 0, \ B \neq 0, \ C \neq 0.$$

32. Take a point $P_1(x_1, y_1, z_1)$ on plane 1 (so $Ax_1 + By_1 + Cz_1 + D_1 = 0$) and find the distance to plane 2 :

$$d = \frac{|Ax_1 + By_1 + Cz_1 + D_2|}{\sqrt{A^2 + B^2 + C^2}} = \frac{|D_2 - D_1|}{\sqrt{A^2 + B^2 + C^2}}$$

33. $\dfrac{x - x_0}{d_1} = \dfrac{y - y_0}{d_2}, \qquad \dfrac{y - y_0}{d_2} = \dfrac{z - z_0}{d_3}$

34. (a) $x = t, \ y = y_0, \ z = z_0$ \qquad (b) $x = x_0, \ y = t, \ z = z_0$

35. We set $x = 0$ and find that $P_0(0, 0, 0)$ lies on the line of intersection. As normals to the plane we use

$$\mathbf{N}_1 = \mathbf{i} + 2\mathbf{j} + 3\mathbf{k} \quad \text{and} \quad \mathbf{N}_2 = -3\mathbf{i} + 4\mathbf{j} + \mathbf{k}.$$

Note that

$$\mathbf{N}_1 \times \mathbf{N}_2 = (\mathbf{i} + 2\mathbf{j} + 3\mathbf{k}) \times (-3\mathbf{i} + 4\mathbf{j} + \mathbf{k}) = -10\mathbf{i} - 10\mathbf{j} + 10\mathbf{k}.$$

We take $-\frac{1}{10}(\mathbf{N}_1 \times \mathbf{N}_2) = \mathbf{i} + \mathbf{j} - \mathbf{k}$ as a direction vector for the line through $P_0(0, 0, 0)$. Then

$$x(t) = t, \quad y(t) = t, \quad z(t) = -t.$$

36. Set $x = 0$ to find that $P(0, \frac{1}{2}, -\frac{3}{2})$ on the line of intersection.
For the direction vector, consider $\mathbf{N}_1 \times \mathbf{N}_2 = (\mathbf{i} + \mathbf{j} + \mathbf{k}) \times (\mathbf{i} - \mathbf{j} + \mathbf{k}) = 2\mathbf{i} - 2\mathbf{k}, \quad$ so we can use $\mathbf{d} = \mathbf{i} - \mathbf{k}$. Thus

$$x(t) = t, \quad y(t) = \frac{1}{2}, \quad z(t) = -\frac{3}{2} - t.$$

37. Straightforward computations give us

$$l : x(t) = 1 - 3t, \quad y(t) = -1 + 4t, \quad z(t) = 2 - t$$

and

$$p : x + 4y - z = 6.$$

Substitution of the scalar parametric equations for l in the equation for p gives

$$(1 - 3t) + 4(-1 + 4t) - (2 - t) = 6 \quad \text{and thus} \quad t = 11/14.$$

Using $t = 11/14$, we get $\quad x = -19/14, \quad y = 15/7, \quad z = 17/14.$

38. $l : x(t) = 4 - 2t, \quad y(t) = -3 + t, \quad z(t) = 1 + 2t \quad P : x + 4y - z = 6$

Note that $\mathbf{d} \cdot \mathbf{N} = (-2\mathbf{i} + \mathbf{j} + 2\mathbf{k}) \cdot (\mathbf{i} + 4\mathbf{j} - \mathbf{k}) = 0$, so the line is parallel to the plane, and since P_1 does not lie in the plane, l and P do not intersect.

39. Let $\mathbf{N} = A\mathbf{i} + B\mathbf{j} + C\mathbf{k}$ be normal to the plane. Then

$$\mathbf{N} \cdot \mathbf{d} = (\mathbf{i} + B\mathbf{j} + C\mathbf{k}) \cdot (\mathbf{i} + 2\mathbf{j} + 4\mathbf{k}) = 1 + 2B + 4C = 0$$

and

$$\mathbf{N} \cdot \mathbf{D} = (\mathbf{i} + B\mathbf{j} + C\mathbf{k}) \cdot (-\mathbf{i} - \mathbf{j} + 3\mathbf{k}) = -1 - B + 3C = 0.$$

This gives $B = -7/10$ and $C = 1/10$. The equation for the plane can be written

$$1(x - 0) - \tfrac{7}{10}(y - 0) + \tfrac{1}{10}(z - 0) = 0, \quad \text{which simplifies to} \quad 10x - 7y + z = 0.$$

40. If the two lines intersect, then there exist numbers t_0 and u_0 such that

$$\mathbf{r}_0 + t_0\mathbf{d} = \mathbf{R}_0 + u_0\mathbf{D}.$$

It follows then that

$$1(\mathbf{r}_0 - \mathbf{R}_0) + t_0\mathbf{d} - u_0\mathbf{D} = 0$$

and therefore the vectors $\mathbf{r}_0 - \mathbf{R}_0, \mathbf{d}$, and $\mathbf{D}$ are coplanar. Conversely, if $\mathbf{r}_0 - \mathbf{R}_0, \mathbf{d}, \mathbf{D}$ are coplanar, then there exist numbers α, β, γ not all zero such that

$$(*) \qquad \alpha(\mathbf{r}_0 - \mathbf{R}_0) + \beta\mathbf{d} + \gamma\mathbf{D} = 0.$$

We assert now that $\alpha \neq 0$. [If α were 0, then we would have $\beta\mathbf{d} + \gamma\mathbf{D} = 0$ so that $\mathbf{d}$ and $\mathbf{D}$ would have to be parallel, which (by assumption) they are not.] With $\alpha \neq 0$ we can divide equation $(*)$ by α and obtain

$$\mathbf{r}_0 - \mathbf{R}_0 + \frac{\beta}{\alpha}\mathbf{d} + \frac{\gamma}{\alpha}\mathbf{D} = 0.$$

This gives

$$\mathbf{r}_0 + \frac{\beta}{\alpha}\mathbf{d} = \mathbf{R}_0 - \frac{\gamma}{\alpha}\mathbf{D},$$

which means that

$$\mathbf{r}\left(\frac{\beta}{\alpha}\right) = \mathbf{R}\left(-\frac{\gamma}{\alpha}\right)$$

and the two lines intersect.

41. $\mathbf{N} + \overrightarrow{PQ}$ and $\mathbf{N} - \overrightarrow{PQ}$ are the diagonals of a rectangle with sides $\mathbf{N}$ and $\overrightarrow{PQ}$. Since the diagonals are perpendicular, the rectangle is a square; that is $\|\mathbf{N}\| = \| \overrightarrow{PQ} \|$. Thus, the points Q form a circle centered at P with radius $\|\mathbf{N}\|$.

42. Given that $\|\mathbf{a}\|, \|\mathbf{b}\|, \|\mathbf{c}\|$ are nonzero and $\mathbf{a} \cdot \mathbf{b} = \mathbf{a} \cdot \mathbf{c} = \mathbf{b} \cdot \mathbf{c} = 0$. The vectors $\mathbf{a}, \mathbf{b}, \mathbf{c}$ are coplanar iff there exist scalars s, t, u not all zero such that $s\mathbf{a} + t\mathbf{b} + u\mathbf{c} = 0$. Now assume that

$$s\mathbf{a} + t\mathbf{b} + u\mathbf{c} = 0.$$

The relation

$$0 = \mathbf{a} \cdot (s\mathbf{a} + t\mathbf{b} + u\mathbf{c}) = s\|\mathbf{a}\|^2 + t(\mathbf{a} \cdot \mathbf{b}) + u(\mathbf{a} \cdot \mathbf{c}) = s\|\mathbf{a}\|^2$$

gives $s = 0$. Similarly, we can show that $t = 0$ and $u = 0$. The vectors $\mathbf{a}, \mathbf{b}.\mathbf{c}$ can not be linearly dependent.

43. If $\alpha > 0$, then P_0 lies on the same side of the plane as the tip of $\mathbf{N}$; if $\alpha < 0$, then P_0 and the tip of $\mathbf{N}$ lie on opposite sides of the plane.

To see this, suppose that $\mathbf{N}$ emanates from the point $P_1(x_1, y_1, z_1)$ on the plane. Then

$$\mathbf{N} \cdot \overrightarrow{P_1 P_0} = A(x_0 - x_1) + B(y_0 - y_1) + C(z_0 - z_1) = Ax_0 + By_0 + Cz_0 + D = \alpha.$$

If $\alpha > 0$, $0 \leq \sphericalangle \left(\mathbf{N}, \overrightarrow{P_0 P_1}\right) < \pi/2$; if $\alpha < 0$, then $\pi/2 < \sphericalangle \left(\mathbf{N}, \overrightarrow{P_0 P_1}\right) < \pi$. Since $\mathbf{N}$ is perpendicular to the plane, the result follows.

45. (a) intercepts:

$(4, 0, 0), \ (0, 5, 0), \ (0, 0, 2)$

(b) traces:

in the x, y-plane: $5x + 4y = 20$

in the x, z-plane: $x + 2z = 4$

in the y, z-plane: $2y + 5z = 10$

(c) unit normals: $\pm \dfrac{1}{\sqrt{141}} (5\mathbf{i} + 4\mathbf{j} + 10\mathbf{k})$

(d)

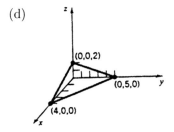

46. (a) intercepts:

$(6, 0, 0), \ (0, 3, 0), \ (0, 0, 2)$

(b) traces:

in the x, y-plane: $x + 2y = 6$

in the x, z-plane: $x + 3z = 6$

in the y, z-plane: $2y + 3z = 6$

(c) unit normals: $\pm \dfrac{1}{\sqrt{14}} (\mathbf{i} + 2\mathbf{j} + 3\mathbf{k})$

(d)

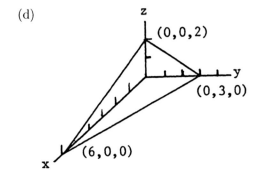

47. (a) intercepts:

$(4, 0, 0)$, no y-intercept, $(0, 0, 6)$

(b) traces:

in the x, y-plane: $x = 4$

in the x, z-plane: $3x + 2z = 12$

in the y, z-plane: $z = 6$

(c) unit normals: $\pm \dfrac{1}{\sqrt{13}} (3\mathbf{i} + 2\mathbf{k})$

(d)

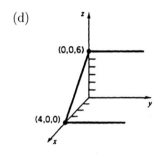

48. (a) intercepts:

$(2, 0, 0)$, $(0, 3, 0)$, no z-intercept

(b) traces:

in the x, y-plane: $3x + 2y = 6$

in the x, z-plane: $x = 2$

in the y, z-plane: $y = 3$

(c) unit normals: $\pm \dfrac{1}{\sqrt{13}} (3\mathbf{i} + 2\mathbf{j})$

(d)

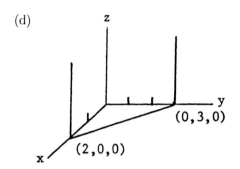

49. The normal vectors to the planes are: $\mathbf{N}_1 = 2\mathbf{i} + \mathbf{j} + 3\mathbf{k}$, $\mathbf{N}_2 = \mathbf{i} + 5\mathbf{j} - 2\mathbf{k}$.

The cosine of the angle θ between the planes is: $\cos\theta = \dfrac{|\mathbf{N}_1 \cdot \mathbf{N}_2|}{\|\mathbf{N}_1\| \, \|\mathbf{N}_2\|} = \dfrac{1}{2\sqrt{105}}$;

$\theta \cong 1.52$ radians $\cong 87.2°$.

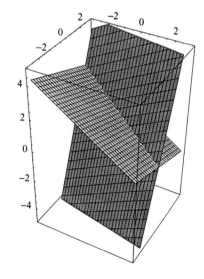

50. The normal vectors to the planes are: $\mathbf{N}_1 = -\mathbf{i} - 2\mathbf{j} + 4\mathbf{k}, \; \mathbf{N}_2 = 2\mathbf{i} - \mathbf{j} - 3\mathbf{k}$.

The cosine of the angle θ between the planes is: $\cos\theta = \dfrac{|\mathbf{N}_1 \cdot \mathbf{N}_2|}{\|\mathbf{N}_1\| \, \|\mathbf{N}_2\|} = \dfrac{12}{7\sqrt{6}}$;

$\theta \cong 0.796$ radians $\cong 45.6°$.

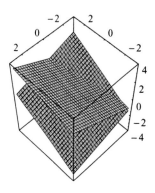

51. An equation of the plane that passes through $(2, 7, -3)$ with normal vector $\mathbf{N} = 3\mathbf{i} + \mathbf{j} + 4\mathbf{k}$ is:

$3(x - 2) + (y - 7) + 4(z + 3) = 0$ or $3x + y + 4z = 1$.

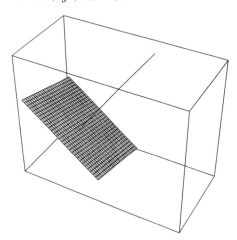

52. An equation of the plane that passes through $(5, -3, -4)$ with normal vector $\mathbf{N} = -\mathbf{i} + 2\mathbf{j} - 3\mathbf{k}$ is:

$-(x - 5) + 2(y + 3) - 3(z + 4) = 0$ or $x - 2y + 3z = -1$.

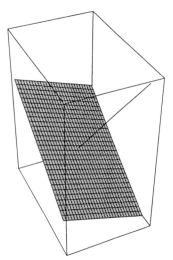

53. $\dfrac{x}{2} + \dfrac{y}{5} + \dfrac{z}{4} = 1$

    $10x + 4y + 5z = 20$

54. $4x + 3y + 6z - 12 = 0$

55. $\dfrac{x}{3} + \dfrac{y}{5} = 1$

    $5x + 3y = 15$

56. $3y + 4z - 12 = 0$

REVIEW EXERCISES

1. (a) $\overline{PQ} = \sqrt{(7-3)^2 + (-5-2)^2 + (-4-[-1])^2} = \sqrt{16 + 49 + 25} = 3\sqrt{10}$

(b) Midpoint of $\overline{QR}$: $\left(\dfrac{7+5}{2}, \dfrac{-5+6}{2}, \dfrac{4-3}{2}\right) = \left(6, \dfrac{1}{2}, \dfrac{1}{2}\right)$

(c) Let $X = (x, y, z)$. Then

$$(7, -5, 4) = \left(\frac{3+x}{2}, \frac{2+y}{2}, \frac{-1+z}{2}\right) \implies (x, y, z) = (11, -12, 9).$$

(d) Midpoint of $\overline{PR}$: $(4, 4, -2)$; radius of the sphere:

$$r = \frac{1}{2}|\overline{PR}| = \frac{1}{2}\sqrt{(5-3)^2 + (6-2)^2 + (-3+1)^2} = \sqrt{6}.$$

The equation of the sphere is: $(x-4)^2 + (y-4)^2 + (z+2)^2 = 6$.

2. (a) $\overline{PQ} = \sqrt{6^2 + 1^2 + (-7)^2} = \sqrt{36 + 1 + 49} = \sqrt{86}$

(b) Midpoint of $\overline{QR}$: $(\dfrac{-2+1}{2}, \dfrac{1-1}{2}, \dfrac{4-6}{2}) = (-\frac{1}{2}, 0, -1)$

(c) Let $X = (x, y, z)$. Then

$$(-2, 1, 4) = \left(\frac{4+x}{2}, \frac{2+y}{2}, \frac{-3+z}{2}\right) \implies (x, y, z) = (-8, 0, 11).$$

(d) Midpoint of $\overline{PR}$: $(\frac{5}{2}, \frac{1}{2}, -\frac{9}{2})$; radius of sphere:.

$$r = \frac{1}{2}|\overline{PR}| = \frac{1}{2}\sqrt{(4-1)^2 + (2+1)^2 + (-3+6)^2} = \frac{1}{2}\sqrt{27}.$$

The equation of the sphere is: $(x - \frac{5}{2})^2 + (y - \frac{1}{2})^2 + (z + \frac{9}{2})^2 = \frac{27}{4}$.

3. radius: $\sqrt{2^2 + (-3)^2 + 1^2} = \sqrt{14}$

equation: $(x-2)^2 + (y+3)^2 + (z-1)^2 = 14$

4. radius: $\frac{1}{2}\sqrt{4^2 + 6^2 + 4^2} = \sqrt{17}$

midpoint: $\left(\dfrac{-1+3}{2}, \dfrac{4-2}{2}, \dfrac{2+6}{2}\right) = (1, 1, 4)$

equation: $(x-1)^2 + (y-1)^2 + (z-4)^2 = 17$

5. By completing the square, the equation can be written as

$$(x+1)^2 + (y+2)^2 + (z-4)^2 = 4 = 2^2.$$

center: $(-1, -2, 4)$. radius: 2

6. By completing the square, the equation can be written as

$$(x-3)^2 + (y+5)^2 + (z-1)^2 = 33.$$

center: $(3, -5, 1)$ radius: $\sqrt{33}$

7. $\frac{3}{2}\mathbf{i} + \mathbf{j} - \frac{1}{2}\mathbf{k}$
$\qquad\qquad\qquad\qquad\qquad$ **8.** $29\mathbf{i} + 2\mathbf{j} - 4\mathbf{k}$

9. $\mathbf{b} + \mathbf{c} = 3\mathbf{i} + 7\mathbf{j} + \mathbf{k}$, $\quad \mathbf{a} \cdot (\mathbf{b} + \mathbf{c}) = (3\mathbf{i} + 2\mathbf{j} - \mathbf{k}) \cdot (3\mathbf{i} + 7\mathbf{j} + \mathbf{k}) = 22$

10. $\mathbf{a} + \mathbf{b} = 8\mathbf{i} + 5\mathbf{j} - \mathbf{k})$; $\quad \|\mathbf{a} + \mathbf{b}\| = \sqrt{64 + 25 + 1} = 3\sqrt{10}$

11. $\|\mathbf{c}\|^2 = (-2)^2 + 4^2 + 1^2 = 21$
$\qquad\qquad$ **12.** $\mathbf{b} \times \mathbf{b} = 0$. Therefore $\|\mathbf{b} \times \mathbf{b}\| = 0$

13. $2\mathbf{a} - \mathbf{b} = \mathbf{i} + \mathbf{j} - 2\mathbf{k}$; $\quad (2\mathbf{a} - \mathbf{b}) \cdot \mathbf{c} = (\mathbf{i} + \mathbf{j} - 2\mathbf{k}) \cdot (-2\mathbf{i} + 4\mathbf{j} + \mathbf{k}) = 0$

14. $\mathbf{b} + \mathbf{c} = 3\mathbf{i} + 7\mathbf{j} + \mathbf{k}$; $\quad \mathbf{a} \times (\mathbf{b} + \mathbf{c}) = \begin{vmatrix} \mathbf{i} & \mathbf{j} & \mathbf{k} \\ 3 & 2 & -1 \\ 3 & 7 & 1 \end{vmatrix} = 9\mathbf{i} - 6\mathbf{j} + 15\mathbf{k}$

15. $\|\mathbf{a}\| = \sqrt{14}$; $\quad \mathbf{u_a} = \dfrac{1}{\sqrt{14}}(3\mathbf{i} + 2\mathbf{j} - \mathbf{k})$
$\qquad$ **16.** $\|\mathbf{c}\| = \sqrt{21}$; $\quad -\mathbf{u_c} = -\dfrac{1}{\sqrt{21}}(-2\mathbf{i} + 4\mathbf{j} + \mathbf{k})$

17. $\cos\theta = \dfrac{\mathbf{a} \cdot \mathbf{c}}{\|\mathbf{a}\|\|\mathbf{c}\|} = \dfrac{1}{\sqrt{14}\sqrt{21}} = \dfrac{1}{7\sqrt{6}}$; $\quad \theta \cong 1.51$ radians

18. $\cos\theta = \dfrac{\mathbf{b} \cdot \mathbf{c}}{\|\mathbf{b}\|\|\mathbf{c}\|} = \dfrac{2}{\sqrt{34}\sqrt{21}}$; $\quad \theta \cong 85.7°$

19. $\|\mathbf{a}\| = \sqrt{14}$; $\quad \cos\alpha = \dfrac{3}{\sqrt{14}}$, $\alpha \cong 0.64$ radians, $\cos\beta = \dfrac{2}{\sqrt{14}}$, $\beta \cong 1.01$, $\cos\gamma = \dfrac{-1}{\sqrt{14}}$, $\gamma \approx 1.84$

20. $\text{comp}_\mathbf{a}\mathbf{b} = \mathbf{b} \cdot \mathbf{u_a} = (5\mathbf{i} + 3\mathbf{j}) \cdot \dfrac{1}{\sqrt{14}}(3\mathbf{i} + 2\mathbf{j} - \mathbf{k}) = \dfrac{21}{\sqrt{14}}$

21. $\mathbf{b} \times \mathbf{c} = \begin{vmatrix} \mathbf{i} & \mathbf{j} & \mathbf{k} \\ 5 & 3 & 0 \\ -2 & 4 & 1 \end{vmatrix} = 3\mathbf{i} - 5\mathbf{j} + 26\mathbf{k}$;

$\quad \text{comp}_\mathbf{a}(\mathbf{b} \times \mathbf{c}) = (\mathbf{b} \times \mathbf{c}) \cdot \mathbf{u_a} = (3\mathbf{i} - 5\mathbf{j} + 26\mathbf{k}) \cdot \dfrac{1}{\sqrt{14}}(3\mathbf{i} + 2\mathbf{j} - \mathbf{k}) = -\dfrac{27}{\sqrt{14}}$

22. $\mathbf{a} \times \mathbf{c} = 6\mathbf{i} - \mathbf{j} + 16\mathbf{k}, \ \|\mathbf{a} \times \mathbf{c}\| = \sqrt{293}; \quad \mathbf{u} = \pm \dfrac{6\mathbf{i} - \mathbf{j} + 16\mathbf{k}}{\sqrt{293}}$

23. $V = |(\mathbf{a} \times \mathbf{b}) \cdot \mathbf{c}|; \quad (\mathbf{a} \times \mathbf{b}) \cdot \mathbf{c} = \begin{vmatrix} 3 & 2 & -1 \\ 5 & 3 & 0 \\ -2 & 4 & 1 \end{vmatrix} = -27; \quad V = |-27| = 27$

24. $A = \dfrac{1}{2}\|\mathbf{a} \times \mathbf{b}\|; \quad \mathbf{a} \times \mathbf{b} = \begin{vmatrix} \mathbf{i} & \mathbf{j} & \mathbf{k} \\ 3 & 2 & -1 \\ 5 & 3 & 0 \end{vmatrix} = 3\mathbf{i} - 5\mathbf{j} - \mathbf{k}; \quad A = \dfrac{1}{2}\sqrt{35}$

25. (a) Direction vector: $\overrightarrow{QR} = (6, -3, 3);$ scalar parametric equations for the line:

$$x = 1 + 6t, \quad y = 1 - 3t, \quad z = 1 + 3t.$$

(b) Normal vector: $\overrightarrow{PR} = (3, -3, 2);$ equation of the plane:

$$3(x - 1) + (-3)(y - 1) + 2(z - 1) = 0.$$

(c) A normal vector for the plane is: $\overrightarrow{QR} \times \overrightarrow{PR} = (3, -3, -9)$ or $\mathbf{N} = (1, -1, -3);$
an equation for the plane: $(x - 1) - (y - 1) - 3(z - 1) = 0$

26. (a) Direction vector: $\overrightarrow{PQ} = (4, -7, 5);$ scalar parametric equations for the line:

$$x = 5 + 4t, \quad y = 6 - 7t, \quad z = -3 + 5t.$$

(b) Scalar parametric equations for the line l through P and Q are:

$$x = 3 + 4t, \quad y = 2 - 7t, \quad z = -1 + 5t.$$

For any point S on $l,$ the vector $\overrightarrow{RS}$ is $(4t - 2, -7t - 4, 5t + 2).$ We want $\overrightarrow{RS} \perp \overrightarrow{PQ}:$

$$(4t - 2, -7t - 4, 5t + 2) \cdot (4, -7, 5) = 0 \Longrightarrow t = -\frac{1}{3}.$$

Therefore $\left(-\frac{10}{3}, -\frac{5}{3}, \frac{1}{3}\right)$ is a direction vector for the line that passes through R perpendicular to $PQ.$ Scalar parametric equations for this line are:

$$x = 5 - \frac{10}{3}t, \quad y = 6 - \frac{5}{3}t, \quad z = -3 + \frac{1}{3}t.$$

(c) This plane is determined by the points $P, \ Q, \ R.$ A normal vector for the plane is:

$$\overrightarrow{PQ} \times \overrightarrow{PR} = -6\mathbf{i} + 18\mathbf{j} + 30\mathbf{k} \quad \text{or} \quad \mathbf{i} - 3\mathbf{j} - 5\mathbf{k}.$$

An equation for the plane is: $(x - 3) - 3(y - 2) - 5(z + 1) = 0$ or $x - 3y - 5z = 2.$

27. Solve, if possible, the system of equations: $t = 1 - u, \ -t = 1 + 3u, \ -6 + 2t = 2u.$ In this case, the solution is $t = 2, u = -1.$ The lines intersect at the point $(2, -2, -2).$

28. Solve, if possible, the system of equations: $1 - 2t = 3 + 2u, \ 3 + 3t = 1 - u, \ 5t = 6 + 3u.$ In this case there is no solution; the lines are skew.

29. The lines l_1 and l_2 written in scalar parametric form are:

$$l_1 : x = 1 + 2t, \quad y = -2 - t, \quad z = 3 + 4t; \quad l_2 : x = -2 + u, \quad y = 3 + 3u, \quad z = u.$$

Solve, if possible, the system of equations: $1 + 2t = -2 + u, \quad -2 - t = 3 + 3u, \quad 3 + 4t = u$. In this case there is no solution; the lines are skew.

30. Scalar parametric equations for the line l are: $x = -1 + t, \ y = -2 + t, \ -1 + t$ and a direction vector for l is $\mathbf{d} = (1, 1, 1)$. The point $Q \ (-1 + t, -2 + t, -1 + t)$ is a point on l and $\overrightarrow{PQ} = (-4 + t, -3 + t, 1 + t)$ is the vector from P to Q. We want $\mathbf{d} \cdot \overrightarrow{PQ} = 0$:

$$(-4 + t, -3 + t, 1 + t) \cdot (1, 1, 1) = 0 \Longrightarrow 3t = 6 \quad \text{and} \quad t = 2.$$

Therefore $Q = (1, 0, 1)$.

31. (a) No. $\overrightarrow{PQ} = (4, -7, 5), \quad \overrightarrow{PR} = (2, -3, 2);$ the vectors are not parallel; the points are not collinear.

(b) $\overrightarrow{PQ} = (4, -7, 5), \quad \overrightarrow{PR} = (2, -3, 2), \quad \overrightarrow{PS} = (-2, 0, 1)$

$$(\overrightarrow{PQ} \times \overrightarrow{PR}) \cdot \overrightarrow{PS} = \begin{vmatrix} 4 & -7 & 5 \\ 2 & -3 & 2 \\ -2 & 0 & 1 \end{vmatrix} = 0.$$

The points are coplanar.

32. The scalar parametric equations for the line are:

$$x = -2 + 3t, \quad y = 1 + 2t, \quad z = -6 + t.$$

Substituting x, y, z into the equation for the plane gives

$$2(-2 + 3t) + (1 + 2t) - 3(-6 + t) + 6 = 0 \Longrightarrow t = -\frac{21}{5}.$$

Therefore the line intersects the plane at the point $\left(-\frac{73}{5}, -\frac{37}{5}, -\frac{51}{5}\right)$.

33. $\overrightarrow{PQ} \times \overrightarrow{PR} = -10\,\mathbf{i} + 5\,\mathbf{k}$ is a normal vector for the plane; so is $\mathbf{N} = 2\,\mathbf{i} - \mathbf{k}$.

An equation for the plane is: $2(x - 1) - (z - 1) = 0 \quad$ or $\quad 2x - z = 1$

34. $\mathbf{N} = 2\,\mathbf{i} + 3\,\mathbf{j} - 4\,\mathbf{k}$ is a normal vector for the plane. An equation for the plane is:

$$2(x - 2) + 3(y - 1) - 4(z + 3) = 0 \quad \text{or} \quad 2x + 3y - 4z = 19.$$

35. $\mathbf{N} = 3\,\mathbf{i} + 2\,\mathbf{j} - 1\,\mathbf{k}$ is a normal vector for the plane. An equation for the plane is:

$$3(x - 1) + 2(y + 2) - (z + 1) = 0 \quad \text{or} \quad 3x + 2y - z = 0.$$

36. The point $Q \ (2, -1, 0)$ is in the plane since it is on the line. $\mathbf{d} = 2\,\mathbf{i} + 3\,\mathbf{j} - 2\,\mathbf{k}$ is a direction vector for the line. The vector $\overrightarrow{PQ} \times \mathbf{d}$ is a normal vector for the plane.

$$\overrightarrow{PQ} \times \mathbf{d} = \begin{vmatrix} \mathbf{i} & \mathbf{j} & \mathbf{k} \\ -1 & 0 & -2 \\ 2 & 3 & -2 \end{vmatrix} = 6\,\mathbf{i} - 6\,\mathbf{j} - 3\,\mathbf{k}.$$

Take $\mathbf{N} = 2\mathbf{i} - 2\mathbf{j} - \mathbf{k}$ as a normal vector for the plane. An equation for the plane is:

$$2(x - 3) - 2(y + 1) - (z - 2) = 0 \quad \text{or} \quad 2x - 2y - z = 6.$$

37. Let P be the plane that satisfies the conditions. A direction vector for the given line is $\mathbf{d} = (3, 2, 4)$; a normal vector for the given plane is $\mathbf{N} = (2, 1, -3)$. The cross product $\mathbf{d} \times \mathbf{N}$ is a normal vector for P.

$$\mathbf{d} \times \mathbf{N} = \begin{vmatrix} \mathbf{i} & \mathbf{j} & \mathbf{k} \\ 3 & 2 & 4 \\ 2 & 1 & -3 \end{vmatrix} = -10\,\mathbf{i} + 17\,\mathbf{j} - \mathbf{k}.$$

The point $Q(-1, 1, 2)$ is on the plane. An equation for P is:

$$-10(x + 1) + 17(y - 1) - (z - 2) = 0 \quad \text{or} \quad 10x - 17y + z + 25 = 0.$$

38. Let P be the plane that satisfies the given conditions. The vectors $\mathbf{N}_1 = 3\mathbf{i} + \mathbf{j} - \mathbf{k}$ and $\mathbf{N}_2 = 2\mathbf{i} + \mathbf{j} + 4\mathbf{k}$ are normal vectors for the given planes. The vector $\mathbf{d} = \mathbf{N}_1 \times \mathbf{N}_2 = 5\mathbf{i} - 14\mathbf{j} + \mathbf{k}$ is a direction vector for the line of intersection. Solving the equations $3x + y - z = 2$, $2x + y + 4z = 1$ simultaneously, we find that $Q(1, -1, 0)$ lies on the line of intersection (set $z = 0$ and solve for x and y). Now, the vector

$$\mathbf{N} = \overrightarrow{PQ} \times \mathbf{d} = \begin{vmatrix} \mathbf{i} & \mathbf{j} & \mathbf{k} \\ -1 & -2 & 3 \\ 5 & -14 & 1 \end{vmatrix} = 40\,\mathbf{i} + 16\,\mathbf{j} + 24\,\mathbf{k} \quad \text{or} \quad 5\mathbf{i} + 2\mathbf{j} + 3\mathbf{k}$$

is a normal vector for P. An equation for P is: $5(x - 2) + 2(y - 1) + 3(z + 3) = 0$.

39. The line l which passes through Q and R has direction vector $\mathbf{d} = \overrightarrow{QR} = (2, 1, -2)$. By (13.5.6), the distance from P to l is given by

$$d(P, l) = \frac{\| \overrightarrow{QP} \times \mathbf{d} \|}{\|\mathbf{d}\|} = \frac{\|(2, 4, -5) \times (2, 1, -2)\|}{3} = \frac{9}{3} = 3.$$

40. By (13.6.5), the distance from P to the plane is given by

$$d = \frac{|1(2) - 2(1) + 2(-1) + 5|}{\sqrt{1 + 4 + 4}} = \frac{3}{3} = 1.$$

41. The normals are: $\mathbf{N}_1 = (2, 1, 1)$, $\mathbf{N}_2 = (2, 2, -1)$. The cosine of the angle between the planes is:

$$\cos\theta = \frac{|\mathbf{N}_1 \cdot \mathbf{N}_2|}{\|\mathbf{N}_1\|\|\mathbf{N}_2\|} = \frac{5}{\sqrt{54}} \quad \text{and} \quad \theta \cong 0.822 \text{ radians.}$$

42. The normals are: $\mathbf{N}_1 = (2, -3, 1)$, $\mathbf{N}_2 = (1, 4, -5)$. The cosine of the angle between the planes is:

$$\cos\theta = \frac{|\mathbf{N}_1 \cdot \mathbf{N}_2|}{\|\mathbf{N}_1\|\|\mathbf{N}_2\|} = \frac{15}{7\sqrt{12}} \quad \text{and} \quad \theta \cong 0.904 \text{ radians.}$$

43. The normal vectors to the two planes are: $\mathbf{N}_1 = 3\mathbf{i} + 5\mathbf{j} + 2\mathbf{k}$, $\mathbf{N}_2 = \mathbf{i} + 2\mathbf{j} - \mathbf{k}$. A direction vector for the line of intersection is:

$$\mathbf{N}_1 \times \mathbf{N}_2 = \begin{vmatrix} \mathbf{i} & \mathbf{j} & \mathbf{k} \\ 3 & 5 & 2 \\ 1 & 2 & -1 \end{vmatrix} = -9\mathbf{i} + 5\mathbf{j} + \mathbf{k}.$$

A solution of the pair of equations $3x + 5y + 2z - 4 = 0$ $x + 2y - z - 2 = 0$ is $x = -2$, $y = 2$, $z = 0$ (set $z = 0$ and solve for x and y). Scalar parametric equations for the line of intersection are:

$$x = -2 - 9t, \quad y = 2 + 5t, \quad z = t.$$

44. The normal vectors to the two planes are: $\mathbf{N}_1 = \mathbf{i} - 2\mathbf{j} + 2\mathbf{k}$, $\mathbf{N}_2 = 3\mathbf{i} - \mathbf{j} - \mathbf{k}$. A direction vector for the line of intersection is:

$$\mathbf{N}_1 \times \mathbf{N}_2 = \begin{vmatrix} \mathbf{i} & \mathbf{j} & \mathbf{k} \\ 1 & -2 & 2 \\ 3 & -1 & -1 \end{vmatrix} = 4\mathbf{i} + 7\mathbf{j} + 5\mathbf{k}.$$

A solution of the pair of equations $x - 2y + 2z = 1$ $3x - y - z = 2$ is $x = 0$, $y = -5/4$, $z = -3/4$ (set $x = 0$ and solve for y and z). Scalar parametric equations for the line of intersection are:

$$x = 4t, \quad y = -\tfrac{5}{4} + 7t, \quad z = -\tfrac{3}{4} + 5t.$$

45. $\mathbf{a} \times \mathbf{b} = -5\mathbf{i} + 11\mathbf{j} + 7\mathbf{k}$ is perpendicular to both $\mathbf{a}$ and $\mathbf{b}$; $\|\mathbf{a} \times \mathbf{b}\| = \sqrt{195}$. The vectors are:

$$\pm \frac{4}{\sqrt{195}}(-5\mathbf{i} + 11\mathbf{j} + 7\mathbf{k}).$$

46. Since $\mathbf{a} \times (\mathbf{b} \times \mathbf{c}) = (\mathbf{a} \cdot \mathbf{c})\mathbf{b} - (\mathbf{a} \cdot \mathbf{b})\mathbf{c}$ (13.4.11),

$$(\mathbf{a} \times \mathbf{b}) \times (\mathbf{c} \times \mathbf{d}) = [(\mathbf{a} \times \mathbf{b}) \cdot \mathbf{d}]\mathbf{c} - [(\mathbf{a} \times \mathbf{b}) \cdot \mathbf{c}]\mathbf{d}.$$

47. $(\|\mathbf{b}\|\mathbf{a} - \|\mathbf{a}\|\mathbf{b}) \cdot (\|\mathbf{b}\|\mathbf{a} + \|\mathbf{a}\|\mathbf{b}) = \|\mathbf{a}\|^2\|\mathbf{b}\|^2 + \|\mathbf{a}\|\|\mathbf{b}\|\mathbf{a} \cdot \mathbf{b} - \|\mathbf{a}\|\|\mathbf{b}\|\mathbf{a} \cdot \mathbf{b} - \|\mathbf{a}\|^2\|\mathbf{b}\|^2 = 0.$
Therefore, $(\|\mathbf{b}\|\mathbf{a} - \|\mathbf{a}\|\mathbf{b}) \perp (\|\mathbf{b}\|\mathbf{a} + \|\mathbf{a}\|\mathbf{b})$

48. $\|\mathbf{a} + \mathbf{b}\|^2 - \|\mathbf{a} - \mathbf{b}\|^2 = (\mathbf{a} + \mathbf{b}) \cdot (\mathbf{a} + \mathbf{b}) - (\mathbf{a} - \mathbf{b}) \cdot (\mathbf{a} - \mathbf{b}) = \|\mathbf{a}\|^2 + \|\mathbf{b}\|^2 + 2\mathbf{a} \cdot \mathbf{b} - \|\mathbf{a}\|^2 - \|\mathbf{b}\|^2 + 2\mathbf{a} \cdot \mathbf{b} = 4\mathbf{a} \cdot \mathbf{b}$

49. Let $\mathbf{a}$ and $\mathbf{b}$ be adjacent sides of a parallelogram. Then the diagonals of the parallelogram are $\mathbf{a} + \mathbf{b}$ and $\mathbf{a} - \mathbf{b}$. By Exercise 48, the diagonals have equal length iff $\mathbf{a} \perp \mathbf{b}$, which means that the parallelogram is a rectangle.

50. Let A, B, C, D be the vertices of the quadrilateral, and let E, F, G, H be the midpoints of $\overline{AB}$, $\overline{BC}$, $\overline{CD}$, $\overline{DA}$, respectively. Then, $\overline{EF} \parallel \overline{AC} \parallel \overline{GH}$ and $\overline{FG} \parallel \overline{BD} \parallel \overline{EH}$. Therefore $EFGH$ is a parallelogram.

51. Let A, B, C be the vertices of a triangle. Without loss of generality, assume that $A(0,0)$, $B(x_1, y_1)$, $C(x_2, 0)$. Let D and E be the midpoints of $\overline{AB}$ and $\overline{BC}$, respectively. Then $D\left(\dfrac{x_1}{2}, \dfrac{y_1}{2}\right)$ and $E\left(\dfrac{x_1 + x_2}{2}, \dfrac{y_1}{2}\right)$. Now

$$\overrightarrow{DE} = \left(\frac{x_2}{2}, 0\right), \quad \text{and} \quad \overrightarrow{AC} = (x_2, 0).$$

Therefore $\overrightarrow{DE} \| \overrightarrow{AC}$ and $\| \overrightarrow{DE} \| = \frac{1}{2} \| \overrightarrow{AC} \|$.

52. (a) If $A \neq 0$, then $\left(-\frac{C}{A}, 0\right)$ and $\left(-\frac{B+C}{A}, 1\right)$ are two points on l. Therefore $\left(\frac{B}{A}, -1\right)$ or $(B, -A)$ are directional vectors for l. Thus

$$\mathbf{r}(t) = (-C/A)\,\mathbf{i} + (B\,\mathbf{i} - A\,\mathbf{j})t = (-C/A + Bt)\,\mathbf{i} + (-At)\,\mathbf{j}$$

is the parametrization of the line.

(b) $(A\,\mathbf{i} + B\,\mathbf{j}) \cdot (B\,\mathbf{i} - A\,\mathbf{j}) = 0$

(c) $P = (-C/A, 0)$ is a point on l and $\overrightarrow{OP} = -C/A\,\mathbf{i}$. By (13.5.6)

$$\frac{\| \overrightarrow{OP} \times (B\,\mathbf{i} - A\,\mathbf{j}) \|}{\|(B\,\mathbf{i} - A\,\mathbf{j})\|} = \frac{|C|}{\sqrt{A^2 + B^2}}.$$

CHAPTER 14

SECTION 14.1

1. $\mathbf{f}'(t) = 2\mathbf{i} - \mathbf{j} + 3\mathbf{k}$

2. $\mathbf{f}'(t) = \sin t \, \mathbf{k}$

3. $\mathbf{f}'(t) = -\dfrac{1}{2\sqrt{1-t}}\mathbf{i} \dfrac{1}{2\sqrt{1+t}}\mathbf{j} + \dfrac{1}{(1-t)^2}\mathbf{k}$

4. $\mathbf{f}'(t) = \dfrac{1}{1+t^2}(\mathbf{i} - \mathbf{j})$

5. $\mathbf{f}'(t) = \cos t \, \mathbf{i} - \sin t \, \mathbf{j} + \sec^2 t \, \mathbf{k}$

6. $\mathbf{f}'(t) = e^t\left[\mathbf{i} + (1+t)\mathbf{j} + (2t + t^2)\mathbf{k}\right]$

7. $\mathbf{f}'(t) = \dfrac{-1}{1-t}\mathbf{i} - \sin t \, \mathbf{j} + 2t \, \mathbf{k}$

8. $\mathbf{f}'(t) = e^t(\mathbf{i} - \mathbf{j}) - 2e^{-2t}(\mathbf{j} - \mathbf{k})$

9. $\mathbf{f}'(t) = 4\mathbf{i} + 6t^2\mathbf{j} + (2t + 2)\mathbf{k};$ $\mathbf{f}''(t) = 12t\mathbf{j} + 2\mathbf{k}$

10. $\mathbf{f}'(t) = (\sin t + t\cos t)\mathbf{i} + (\cos t - t\sin t)\mathbf{k}$

    $\mathbf{f}''(t) = (2\cos t - t\sin t)\mathbf{i} + (-2\sin t - t\cos t)\mathbf{k}$

11. $\mathbf{f}'(t) = -2\sin 2t\,\mathbf{i} + 2\cos 2t\,\mathbf{j} + 4t\,\mathbf{k};$ $\mathbf{f}''(t) = -4\cos 2t\,\mathbf{i} - 4\sin 2t\,\mathbf{j}$

12. $\mathbf{f}'(t) = \dfrac{1}{2}t^{-1/2}\mathbf{i} + \dfrac{3}{2}t^{1/2}\mathbf{j} + \dfrac{1}{t}\mathbf{k}$

    $\mathbf{f}''(t) = -\dfrac{1}{4}t^{-3/2}\mathbf{i} + \dfrac{3}{4}t^{-1/2}\mathbf{j} - \dfrac{1}{t^2}\mathbf{k}$

13. (a) $\mathbf{r}'(t) = -2te^{-t^2}\mathbf{i} - e^{-t}\mathbf{j};$ $\mathbf{r}'(0) = -\mathbf{j}$

    (b) $\mathbf{r}'(t) = \cot t\,\mathbf{i} - \tan t\,\mathbf{j} + (2\cos t + 3\sin t)\mathbf{k};$ $\mathbf{r}'(\pi/4) = \mathbf{i} - \mathbf{j} + \dfrac{5}{\sqrt{2}}\mathbf{k}$

14. (a) $\mathbf{r}''(t) = \left(2e^{-t} - 4te^{-t} + t^2e^{-t}\right)\mathbf{i} + (-2e^{-t} + te^{-t})\mathbf{j};$ $\mathbf{r}''(0) = 2\mathbf{i} - 2\mathbf{j}$

    (b) $\mathbf{r}''(t) = \dfrac{1}{t}\mathbf{i} + \dfrac{2 - 2\ln t}{t^2}\mathbf{j} - \dfrac{1 + 2\ln t}{4t^2(\ln t)^{3/2}}\mathbf{k};$ $\mathbf{r}''(e) = e^{-1}\mathbf{i} - \dfrac{3}{4}e^{-2}\mathbf{k}$

15. $\displaystyle\int_1^2 (\mathbf{i} + 2t\,\mathbf{j})\, dt = \left[t\mathbf{i} + t^2\mathbf{j}\right]_1^2 = \mathbf{i} + 3\mathbf{j}$

16. $\displaystyle\int_0^\pi (\sin t\,\mathbf{i} + \cos t\,\mathbf{j} + t\,\mathbf{k})\, dt = \left[-\cos t\,\mathbf{i} + \sin t\,\mathbf{j} + \dfrac{t^2}{2}\mathbf{k}\right]_0^\pi = 2\mathbf{i} + \dfrac{1}{2}\pi^2\mathbf{k}$

17. $\displaystyle\int_0^1 \left(e^t\,\mathbf{i} + e^{-t}\,\mathbf{k}\right)\, dt = \left[e^t\,\mathbf{i} - e^{-t}\,\mathbf{k}\right]_0^1 = (e-1)\mathbf{i} + \left(1 - \dfrac{1}{e}\right)\mathbf{k}$

18. $\displaystyle\int_0^1 (te^{-t}\,\mathbf{i} + 4e^{2t}\,\mathbf{j} + e^{-t}\,\mathbf{k})\, dt = \left[(-te^{-t} - e^{-t})\mathbf{i} + 2e^{2t}\,\mathbf{j} - e^{-t}\,\mathbf{k}\right]_0^1 = 2e^{-1}\mathbf{i} + 2e^2\mathbf{j} - e^{-1}\mathbf{k}$

19. $\displaystyle\int_0^1 \left(\dfrac{1}{1+t^2}\mathbf{i} + \sec^2 t\,\mathbf{j}\right)\, dt = \left[\tan^{-1} t\,\mathbf{i} + \tan t\,\mathbf{j}\right]_0^1 = \dfrac{\pi}{4}\mathbf{i} + \tan(1)\mathbf{j}$

20. $\int_1^3 \left(\frac{1}{t}\mathbf{i} + \frac{\ln t}{t}\mathbf{j} + e^{-2t}\mathbf{k}\right) dt = \left[\ln t\,\mathbf{i} + \frac{1}{2}(\ln t)^2\,\mathbf{j} - \frac{1}{2}e^{-2t}\,\mathbf{k}\right]_1^3 = \ln 3\,\mathbf{i} + \frac{1}{2}(\ln 3)^2\,\mathbf{j} + \frac{1}{2}(e^{-2} - e^{-6})\,\mathbf{k}$

21. $\lim_{t\to 0}\mathbf{f}(t) = \left(\lim_{t\to 0}\frac{\sin t}{2t}\right)\mathbf{i} + \left(\lim_{t\to 0}e^{2t}\right)\mathbf{j} + \left(\lim_{t\to 0}\frac{t^2}{e^t}\right)\mathbf{k} = \frac{1}{2}\mathbf{i} + \mathbf{j}$

22. Does not exist (because of $\frac{t}{|t|}\mathbf{k}$)

23. $\lim_{t\to 0}\mathbf{f}(t) = \left(\lim_{t\to 0}t^2\right)\mathbf{i} + \left(\lim_{t\to 0}\frac{1-\cos t}{3t}\right)\mathbf{j} + \left(\lim_{t\to 0}\frac{t}{t+1}\right)\mathbf{k} = 0\,\mathbf{i} + \frac{1}{3}\left(\lim_{t\to 0}\frac{1-\cos t}{t}\right)\mathbf{j} + 0\,\mathbf{k} = \mathbf{0}$

24. $\lim_{t\to 0}\mathbf{f}(t) = \mathbf{j} + \mathbf{k}$

25. (a) $\int_0^1 \left(te^t\,\mathbf{i} + te^{t^2}\,\mathbf{j}\right) dt = \mathbf{i} + \frac{e-1}{2}\,\mathbf{j}$

(b) $\int_3^8 \left(\frac{t}{t+1}\mathbf{i} + \frac{t}{(t+1)^2}\mathbf{j} + \frac{t}{(t+1)^3}\mathbf{k}\right) dt = \left[5 + \ln\left(\frac{4}{9}\right)\right]\mathbf{i} + \left[-\frac{5}{36} + \ln\left(\frac{4}{9}\right)\right]\mathbf{j} + \frac{295}{2592}\mathbf{k}$

26. (a) $\frac{3}{4}\mathbf{i} + \frac{1}{4}\mathbf{j} + \mathbf{k}$ (b) $2e^2\,\mathbf{i} + 2e^{-4}\,\mathbf{j} + 2\,\mathbf{k}$

27.
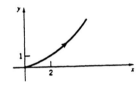

28.
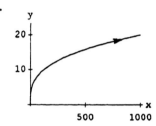

29.
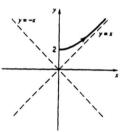

30.
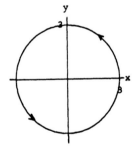

31.
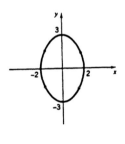

32.
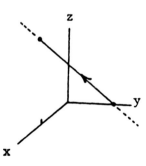

33.
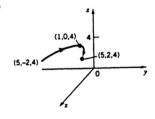

34.
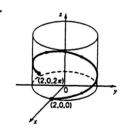

35.
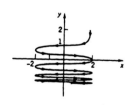

36.

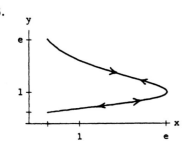

37.

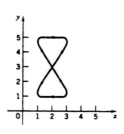

38.

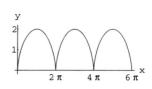

39. (a) $\mathbf{f}(t) = 3\cos t\,\mathbf{i} + 2\sin t\,\mathbf{j}$ (b) $\mathbf{f}(t) = 3\cos t\,\mathbf{i} - 2\sin t\,\mathbf{j}$

40. (a) $\mathbf{f}(t) = (1 + \cos t)\,\mathbf{i} + \sin t\,\mathbf{j}$ (b) $\mathbf{f}(t) = (1 + \cos t)\,\mathbf{i} - \sin t\,\mathbf{j}$

41. (a) $\mathbf{f}(t) = t\,\mathbf{i} + t^2\,\mathbf{j}$ (b) $\mathbf{f}(t) = -t\,\mathbf{i} + t^2\,\mathbf{j}$

42. (a) $\mathbf{f}(t) = t\,\mathbf{i} + t^3\,\mathbf{j}$ (b) $\mathbf{f}(t) = -t\,\mathbf{i} - t^3\,\mathbf{j}$

43. $\mathbf{f}(t) = (1 + 2t)\,\mathbf{i} + (4 + 5t)\,\mathbf{j} + (-2 + 8t)\,\mathbf{k}, \quad 0 \le t \le 1$

44. $\mathbf{f}(t) = (3 + 4t)\,\mathbf{i} + 2\,\mathbf{j} + (-5 + 14t)\,\mathbf{k}, \quad 0 \le t \le 1$

45. $\mathbf{f}'(t_0) = \mathbf{i} + m\,\mathbf{j},$

$$\int_a^b \mathbf{f}(t)\,dt = \left[\frac{1}{2}t^2\,\mathbf{i}\right]_a^b + \left[\int_a^b f(t)\,dt\right]\mathbf{j} = \frac{1}{2}\left(b^2 - a^2\right)\mathbf{i} + A\,\mathbf{j},$$

$$\int_a^b \mathbf{f}'(t)\,dt = [t\,\mathbf{i} + f(t)\,\mathbf{j}]_a^b = (b - a)\,\mathbf{i} + (d - c)\,\mathbf{j}$$

46. $\mathbf{f}(t) = \left(\frac{1}{2}t^2 + 1\right)\mathbf{i} + (\sqrt{1 + t^2} + 1)\,\mathbf{j} + (te^t - e^t + 4)\,\mathbf{k}$

47.

$$\mathbf{f}'(t) = \mathbf{i} + t^2\,\mathbf{j}$$
$$\mathbf{f}(t) = (t + C_1)\,\mathbf{i} + \left(\tfrac{1}{3}t^3 + C_2\right)\mathbf{j} + C_3\,\mathbf{k}$$
$$\mathbf{f}(0) = \mathbf{j} - \mathbf{k} \implies C_1 = 0, \quad C_2 = 1, \quad C_3 = -1$$
$$\mathbf{f}(t) = t\,\mathbf{i} + \left(\tfrac{1}{3}t^3 + 1\right)\mathbf{j} - \mathbf{k}$$

48. $\mathbf{f}(t) = e^{2t}\,\mathbf{i} - e^{2t}\,\mathbf{k} = e^{2t}(\mathbf{i} - \mathbf{k})$

49. $\mathbf{f}'(t) = \alpha\,\mathbf{f}(t) \implies \mathbf{f}(t) = e^{\alpha t}\,\mathbf{f}(0) = e^{\alpha t}\,\mathbf{c}$

50. For each $\epsilon > 0$ there exists $\delta > 0$ such that

$$\text{if} \quad 0 < |t - t_0| < \delta, \quad then \quad \|\mathbf{f}(t) - \mathbf{L}\| < \epsilon.$$

51. (a) If $\mathbf{f}'(t) = \mathbf{0}$ on an interval, then the derivative of each component is 0 on that interval, each component is constant on that interval, and therefore $\mathbf{f}$ itself is constant on that interval.

(b) Set $\mathbf{h}(t) = \mathbf{f}(t) - \mathbf{g}(t)$ and apply part (a).

52.
$$||[\mathbf{f}(t) \cdot \mathbf{g}(t)] - [\mathbf{L} \cdot \mathbf{M}]|| = ||[\mathbf{f}(t) \cdot \mathbf{g}(t)] - [\mathbf{L} \cdot \mathbf{g}(t)] + [\mathbf{L} \cdot \mathbf{g}(t)] - [\mathbf{L} \cdot \mathbf{M}]||$$
$$= ||[(\mathbf{f}(t) - \mathbf{L}) \cdot \mathbf{g}(t)] + [\mathbf{L} \cdot (\mathbf{g}(t) - \mathbf{M})]||$$
$$\leq ||(\mathbf{f}(t) - \mathbf{L}) \cdot \mathbf{g}(t)|| + ||\mathbf{L} \cdot (\mathbf{g}(t) - \mathbf{M})||$$
$$\leq ||\mathbf{f}(t) - \mathbf{L}||\, ||\mathbf{g}(t)|| + ||\mathbf{L}||\, ||\mathbf{g}(t) - \mathbf{M}||$$

by Schwarz's inequality

As $t \to t_0$, the right side tends to $(0)||\mathbf{M}|| + ||\mathbf{L}||(0) = 0$.

53. If $\mathbf{f}$ is differentiable at t, then each component is differentiable at t, each component is continuous at t, and therefore $\mathbf{f}$ is continuous at t.

54. Set $\mathbf{f}(t) = f_1(t)\,\mathbf{i} + f_2(t)\,\mathbf{j} + f_3(t)\,\mathbf{k}$ and apply the fundamental Theorem of Calculus to f_1, f_2, f_3.

55. no; as a counter-example, set $\mathbf{f}(t) = \mathbf{i} = \mathbf{g}(t)$.

56.
$$\int_a^b [\mathbf{c} \cdot \mathbf{f}(t)]\,dt = \int_a^b [c_1 f_1(t) + c_2 f_2(t) + c_3 f_3(t)]\,dt$$
$$= c_1 \int_a^b f_1(t)\,dt + c_2 \int_a^b f_2(t)\,dt + c_3 \int_a^b f_3(t)\,dt$$
$$= \mathbf{c} \cdot \int_a^b \mathbf{f}(t)\,dt$$

$\int_a^b [\mathbf{c} \times \mathbf{f}(t)]\,dt$ can be written

$$\int_a^b \{[c_2 f_3(t) - c_3 f_2(t)]\,\mathbf{i} - [c_1 f_3(t) - c_3 f_1(t)]\,\mathbf{j} + [c_1 f_2(t) - c_2 f_1(t)]\,\mathbf{k}\}\,dt.$$

This gives

$$\left[c_2 \int_a^b f_3(t)\,dt - c_3 \int_a^b f_2(t)\,dt \right]\mathbf{i} - \left[c_1 \int_a^b f_3(t)\,dt - c_3 \int_a^b f_1(t)\,dt \right]\mathbf{j}$$
$$+ \left[c_1 \int_a^b f_2(t)\,dt - c_2 \int_a^b f_1(t)\,dt \right]\mathbf{k}$$

which is $\mathbf{c} \times \int_a^b \mathbf{f}(t)\,dt$

57. Suppose $\mathbf{f}(t) = f_1(t)\,\mathbf{i} + f_2(t)\,\mathbf{j} + f_3(t)\,\mathbf{k}$. Then $||\mathbf{f}(t)|| = \sqrt{f_1^2(t) + f_2^2(t) + f_3^2(t)}$ and

$$\frac{d}{dt}(||\mathbf{f}||) = \frac{1}{2}\left[f_1^2 + f_2^2 + f_3^2 \right]^{-1/2}(2f_1 \cdot f_1' + 2f_2 \cdot f_2' + 2f_3 \cdot f_3') = \frac{\mathbf{f}(t) \cdot \mathbf{f}'(t)}{||\mathbf{f}(t)||}$$

The Answer Section of the text gives an alternative approach.

58.
$$\frac{d}{dt}\left(\frac{\mathbf{f}(t)}{||\mathbf{f}(t)||} \right) = \frac{d}{dt}\left(\frac{1}{||\mathbf{f}(t)||}\mathbf{f}(t) \right) = \frac{1}{||\mathbf{f}(t)||}\mathbf{f}'(t) + \mathbf{f}(t) \cdot \frac{-1}{||\mathbf{f}(t)||^2} \cdot \frac{d}{dt}[||\mathbf{f}(t)||]$$

$$= \frac{\mathbf{f}'(t)}{||\mathbf{f}(t)||} - \frac{\mathbf{f}(t) \cdot \mathbf{f}'(t)}{||\mathbf{f}(t)||^3}\mathbf{f}(t), \quad \text{using the result of Exercise 57.}$$

SECTION 14.2

1. $\mathbf{f}'t = \mathbf{b}, \quad \mathbf{f}''(t) = \mathbf{0}$

2. $\mathbf{f}'t = \mathbf{b} + 2t\,\mathbf{c}, \quad \mathbf{f}''(t) = 2\,\mathbf{c}$

3. $\mathbf{f}'(t) = 2e^{2t}\,\mathbf{i} - \cos t\,\mathbf{j}, \quad \mathbf{f}''(t) = 4e^{2t}\,\mathbf{i} + \sin t\,\mathbf{j}$

4. $\mathbf{f}(t) = 2t^2\,\mathbf{i} \implies \mathbf{f}'t = 4t\,\mathbf{i}, \quad \mathbf{f}''(t) = 4\,\mathbf{i}$

5. $\mathbf{f}'t = [(t^2\,\mathbf{i} - 2t\,\mathbf{j}) \cdot (\mathbf{i} + 3t^2\,\mathbf{j}) + (2t\,\mathbf{i} - 2\,\mathbf{j}) \cdot (t\,\mathbf{i} + t^3\,\mathbf{j}]\,\mathbf{j} = [3t^2 - 8t^3]\,\mathbf{j}$
   $\mathbf{f}''(t) = (6t - 24t^2)\,\mathbf{j}$

6. $\mathbf{f}(t) = (t - t^5)\,\mathbf{k} \implies \mathbf{f}'(t) = (1 - 5t^4)\,\mathbf{k}, \quad \mathbf{f}''(t) = -20t^3\,\mathbf{k}$

7.
$$\mathbf{f}'(t) = \left[(e^t\,\mathbf{i} + t\,\mathbf{k}) \times \frac{d}{dt}\,(t\,\mathbf{j} + e^{-t}\,\mathbf{k}) \right] + \left[\frac{d}{dt}\,(e^t\,\mathbf{i} + t\,\mathbf{k}) \times (t\,\mathbf{j} + e^{-t}\,\mathbf{k}) \right]$$
$$= \left[(e^t\,\mathbf{i} + t\,\mathbf{k}) \times (\mathbf{j} - e^{-t}\,\mathbf{k}) \right] + \left[(e^t\,\mathbf{i} + \mathbf{k}) \times (t\,\mathbf{j} + e^{-t}\,\mathbf{k}) \right]$$
$$= (-t\,\mathbf{i} + \mathbf{j} + e^t\,\mathbf{k}) + (-t\,\mathbf{i} - \mathbf{j} + te^t\,\mathbf{k})$$
$$= -2t\,\mathbf{i} + e^t(t + 1)\,\mathbf{k}$$
$$\mathbf{f}''(t) = -2\,\mathbf{i} + e^t(t + 2)\,\mathbf{k}$$

8. $\mathbf{f}'(t) = \mathbf{a} \times \mathbf{b} + 2t\,\mathbf{b}, \qquad \mathbf{f}''(t) = 2\,\mathbf{b}$

9. $\mathbf{f}'(t) = (\mathbf{a} \times t\,\mathbf{b}) \times 2t\,\mathbf{b} + (\mathbf{a} \times \mathbf{b}) \times (\mathbf{a} + t^2\,\mathbf{b}), \qquad \mathbf{f}''(t) = (\mathbf{a} \times t\,\mathbf{b}) \times 2\mathbf{b} + 4t(\mathbf{a} \times \mathbf{b}) \times \mathbf{b}$

10. $\mathbf{f}'(t) = \mathbf{g}(t^2) + t\,\mathbf{g}'(t^2)2t = \mathbf{g}(t^2) + 2t^2\mathbf{g}'(t^2)$
    $\mathbf{f}''(t) = \mathbf{g}'(t^2)2t + 4t\,\mathbf{g}'(t^2) + 2t^2\mathbf{g}''(t^2)2t = 6t\,\mathbf{g}'(t^2) + 4t^3\mathbf{g}''(t^2)$

11. $\mathbf{f}'(t) = \dfrac{1}{2}\,\sqrt{t}\,\mathbf{g}'\left(\sqrt{t}\right) + \mathbf{g}\left(\sqrt{t}\right), \quad \mathbf{f}''(t) = \dfrac{1}{4}\,\mathbf{g}''\left(\sqrt{t}\right) + \dfrac{3}{4\sqrt{t}}\,\mathbf{g}'\left(\sqrt{t}\right)$

12. $\mathbf{f}(t) = 2e^{-2t}\,\mathbf{i} - 2\,\mathbf{k} \implies \mathbf{f}'(t) = -4e^{-2t}\,\mathbf{i}, \quad \mathbf{f}''(t) = 8e^{-2t}\,\mathbf{i}.$

13. $-(\sin t)\,e^{\cos t}\,\mathbf{i} + (\cos t)\,e^{\sin t}\,\mathbf{j}$

14.
$$\frac{d^2}{dt^2}[e^t\,\cos t\,\mathbf{i} + e^t\,\sin t\,\mathbf{j}] = \frac{d}{dt}[e^t(\cos t - \sin t)\,\mathbf{i} + e^t(\sin t + \cos t)\,\mathbf{j}]$$
$$= -2e^t\,\sin t\,\mathbf{i} + 2e^t\,\cos t\,\mathbf{j}$$

15. $(e^t\,\mathbf{i} + e^{-t}\,\mathbf{j}) \cdot (e^t\,\mathbf{i} - e^{-t}\,\mathbf{j}) = e^{2t} - e^{-2t};$ therefore
$$\frac{d^2}{dt^2}\left[(e^t\,\mathbf{i} + e^{-t}\,\mathbf{j}) \cdot (e^t\,\mathbf{i} - e^{-t}\,\mathbf{j}) \right] = \frac{d^2}{dt^2}\left[e^{2t} - e^{-2t} \right] = \frac{d}{dt}\left[2e^{2t} + 2e^{-2t} \right] = 4e^{2t} - 4e^{-2t}$$

16. $(\ln t\,\mathbf{i} + t\,\mathbf{j}) \times (2t\,\mathbf{j} - \mathbf{k}) + \left(\dfrac{1}{t}\mathbf{i} + \mathbf{j}\right) \times (t^2\,\mathbf{j} - t\,\mathbf{k})$

17. $\dfrac{d}{dt}[(a + t\,\mathbf{b}) \times (c + t\,\mathbf{d})] = [(a + t\,\mathbf{b}) \times \mathbf{d}] + [\mathbf{b} \times (c + t\,\mathbf{d})] = (a \times \mathbf{d}) + (\mathbf{b} \times c) + 2t\,(\mathbf{b} \times \mathbf{d})$

18. $\mathbf{b} \times (a + t\,\mathbf{b} + t^2\,\mathbf{c}) + (a + t\,\mathbf{b}) \times (\mathbf{b} + 2t\,\mathbf{c}) = 2t(a \times \mathbf{c}) + 3t^2(\mathbf{b} \times \mathbf{c}).$

19. $\dfrac{d}{dt}[(a + t\,\mathbf{b}) \cdot (c + t\,\mathbf{d})] = [(a + t\,\mathbf{b}) \cdot \mathbf{d}] + [\mathbf{b} \cdot (c + t\,\mathbf{d})] = (a \cdot \mathbf{d}) + (\mathbf{b} \cdot c) + 2t\,(\mathbf{b} \cdot \mathbf{d})$

20. $\mathbf{b} \cdot (a + t\,\mathbf{b} + t^2\,\mathbf{c}) + (a + t\,\mathbf{b}) \cdot (\mathbf{b} + 2t\,\mathbf{c}) = 2(a \cdot \mathbf{b}) + 2t(a \cdot \mathbf{c}) + 2t\,\|\,\mathbf{b}\,\|^2 + 3t^2(\mathbf{b} \cdot \mathbf{c})$

21. $\mathbf{r}(t) = \mathbf{a} + t\,\mathbf{b}$ 　　　　　　　　　　　**22.** $\mathbf{r}(t) = \mathbf{a} + t\,\mathbf{b} + \dfrac{1}{2}t^2\,\mathbf{c}$

23. $\mathbf{r}(t) = \dfrac{1}{2}t^2\,\mathbf{a} + \dfrac{1}{6}t^3\,\mathbf{b} + t\,\mathbf{c} + \mathbf{d}$ 　　　**24.** $\mathbf{r}(t) = \left(1 + 2t - \dfrac{1}{4}\cos 2t\right)\mathbf{i} + \left(1 - \dfrac{1}{4}\sin 2t\right)\mathbf{j}$

25. $\mathbf{r}(t) = \sin t\,\mathbf{i} + \cos t\,\mathbf{j}$, 　　$\mathbf{r}'(t) = \cos t\,\mathbf{i} - \sin t\,\mathbf{j}$, 　　$\mathbf{r}''(t) = -\sin t\,\mathbf{i} - \cos t\,\mathbf{j} = -\mathbf{r}(t)$.

Thus $\mathbf{r}(t)$ and $\mathbf{r}''(t)$ are parallel, and they always point in opposite directions.

26. $\mathbf{r}''(t) = k^2 e^{kt}\,\mathbf{i} + k^2 e^{-kt}\,\mathbf{j} = k^2\,\mathbf{r}(t)$, 　　so $\mathbf{r}''(t)$ and $\mathbf{r}(t)$ are parallel.

27.
$$\mathbf{r}(t) \cdot \mathbf{r}'(t) = (\cos t\,\mathbf{i} + \sin t\,\mathbf{j}) \cdot (-\sin t\,\mathbf{i} + \cos t\,\mathbf{j}) = 0$$
$$\mathbf{r}(t) \times \mathbf{r}'(t) = (\cos t\,\mathbf{i} + \sin t\,\mathbf{j}) \times (-\sin t\,\mathbf{i} + \cos t\,\mathbf{j})$$
$$= \cos^2 t\,\mathbf{k} + \sin^2 t\,\mathbf{k} = \left(\cos^2 t + \sin^2 t\right)\mathbf{k} = \mathbf{k}$$

28.
$$(\mathbf{g} \times \mathbf{f})'(t) = [\mathbf{g}(t) \times \mathbf{f}'(t)] + [\mathbf{g}'(t) \times \mathbf{f}(t)]$$
$$= -[\mathbf{f}'(t) \times \mathbf{g}(t)] - [\mathbf{f}(t) \times \mathbf{g}'(t)]$$
$$= -\{[\mathbf{f}(t) \times \mathbf{g}'(t)] + [\mathbf{f}'(t) \times \mathbf{g}(t)]\} = -(\mathbf{f} \times \mathbf{g})'(t)$$

29. $\dfrac{d}{dt}[\mathbf{f}(t) \times \mathbf{f}'(t)] = [\mathbf{f}(t) \times \mathbf{f}''(t)] + \underbrace{[\mathbf{f}'(t) \times \mathbf{f}'(t)]}_{0} = \mathbf{f}(t) \times \mathbf{f}''(t).$

30.
$$\dfrac{d}{dt}[u_1(t)\mathbf{r}_1(t) \times u_2(t)\mathbf{r}_2(t)] = \dfrac{d}{dt}[(u_1(t)u_2(t))(\mathbf{r}_1(t) \times \mathbf{r}_2(t))]$$
$$= u_1(t)u_2(t)\dfrac{d}{dt}[\mathbf{r}_1(t) \times \mathbf{r}_2(t)] + [\mathbf{r}_1(t) \times \mathbf{r}_2(t)]\dfrac{d}{dt}[u_1(t)u_2(t)]$$

31. $[\mathbf{f} \cdot \mathbf{g} \times \mathbf{h}]' = \mathbf{f}' \cdot (\mathbf{g} \times \mathbf{h}) + \mathbf{f} \cdot (\mathbf{g} \times \mathbf{h})' = \mathbf{f}' \cdot (\mathbf{g} \times \mathbf{h}) + \mathbf{f} \cdot [\mathbf{g}' \times \mathbf{h} + \mathbf{g} \times \mathbf{h}']$
and the result follows.

32. $\dfrac{d}{dt}(\mathbf{f} \times \mathbf{f}') = \mathbf{f}(t) \times \mathbf{f}''(t)$ by Exercise 29. If $\mathbf{f}(t)$ and $\mathbf{f}''(t)$ are parallel, their cross product is zero,

so $\dfrac{d}{dt}(\mathbf{f} \times \mathbf{f}') = \mathbf{0}$, hence $\mathbf{f} \times \mathbf{f}'$ is constant.

33.

$$\| \mathbf{r}\,(t) \| \text{ is constant} \quad \Longleftrightarrow \quad \| \mathbf{r}\,(t) \|^2 = \mathbf{r}\,(t) \cdot \mathbf{r}\,(t) \text{ is constant}$$

$$\Longleftrightarrow \quad \frac{d}{dt}\,[\mathbf{r}\,(t) \cdot \mathbf{r}\,(t)] = 2\,[\mathbf{r}\,(t) \cdot \mathbf{r}'(t)] = 0 \text{ identically}$$

$$\Longleftrightarrow \quad \mathbf{r}\,(t) \cdot \mathbf{r}'(t) = 0 \text{ identically}$$

34. (a) Routine

(b) Write

$$\frac{[\mathbf{f}\,(t+h) \cdot \mathbf{g}(t+h)] - [\mathbf{f}\,(t) \cdot \mathbf{g}(t)]}{h}$$

so

$$\left(\mathbf{f}\,(t+h) \cdot \left[\frac{\mathbf{g}(t+h) - \mathbf{g}(t)}{h}\right]\right) + \left(\left[\frac{\mathbf{f}\,(t+h) - \mathbf{f}\,(t)}{h}\right] \cdot \mathbf{g}(t)\right)$$

and take the limit as $h \to 0$.   (Appeal to Theorem 13.1.3)

35. Write

$$\frac{[\mathbf{f}\,(t+h) \times \mathbf{g}\,(t+h)] - [\mathbf{f}\,(t) \times \mathbf{g}\,(t)]}{h}$$

as

$$\left(\mathbf{f}\,(t+h) \times \left[\frac{\mathbf{g}\,(t+h) - \mathbf{g}\,(t)}{h}\right]\right) + \left(\left[\frac{\mathbf{f}\,(t+h) - \mathbf{f}\,(t)}{h}\right] \times \mathbf{g}\,(t)\right)$$

and take the limit as $h \to 0$. (Appeal to Theorem 13.1.3.)

36. (a) and (b) can derived routinely by using components. An ϵ, δ derivation of (a) is also simple:
Let $\epsilon > 0$.   Since $\mathbf{f}$ is continuous at $u(t_0)$, there exists $\delta_1 > 0$ such that

$$\text{if } |z - u(t_0)| < \delta_1, \quad \text{then} \quad \| \mathbf{f}\,(z) - \mathbf{f}\,(u(t_0)) \| < \epsilon.$$

Since u is continuous at t_0, there exists $\delta > 0$ such that

$$\text{if } |t - t_0| < \delta, \quad \text{then} \quad |u(t) - u(t_0)| < \delta_1.$$

Thus

$$|t - t_0| < \delta \implies |u(t) - u(t_0)| < \delta_1 \implies \| \mathbf{f}\,(u(t)) - \mathbf{f}\,(u(t_0)) \| < \epsilon.$$

SECTION 14.3

1. $\mathbf{r}'\,(t) = -\pi \sin \pi t\,\mathbf{i} + \pi \cos \pi t\,\mathbf{j} + \mathbf{k}, \quad \mathbf{r}'(2) = \pi\,\mathbf{j} + \mathbf{k}$

$\mathbf{R}\,(u) = (\mathbf{i} + 2\,\mathbf{k}) + u(\pi\,\mathbf{j} + \mathbf{k})$

2. $\mathbf{r}'(t) = e^t\,\mathbf{i} - e^{-t}\,\mathbf{j} - \dfrac{1}{t}\,\mathbf{k}, \quad \mathbf{r}'(1) = e\,\mathbf{i} - e^{-1}\,\mathbf{j} - \mathbf{k}; \quad \mathbf{R}(u) = (e\,\mathbf{i} + e^{-1}\,\mathbf{j}) + u(e\,\mathbf{i} - e^{-1}\,\mathbf{j} - \mathbf{k})$

3. $\mathbf{r}'(t) = \mathbf{b} + 2t\,\mathbf{c}, \quad \mathbf{r}'(-1) = \mathbf{b} - 2\,\mathbf{c}, \quad \mathbf{R}\,(u) = (\mathbf{a} - \mathbf{b} + \mathbf{c}) + u(\mathbf{b} - 2\,\mathbf{c})$

4. $\mathbf{r}'(0) = \mathbf{i}; \quad \mathbf{R}(u) = (\mathbf{i} + \mathbf{j} + \mathbf{k}) + u\,\mathbf{i}.$

5. $\mathbf{r}'(t) = 4t\,\mathbf{i} - \mathbf{j} + 4t\,\mathbf{k}$, P is tip of $\mathbf{r}\,(1)$, $\mathbf{r}'(1) = 4\,\mathbf{i} - \mathbf{j} + 4\,\mathbf{k}$
    $\mathbf{R}\,(u) = (2\,\mathbf{i} + 5\,\mathbf{k}) + u\,(4\,\mathbf{i} - \mathbf{j} + 4\,\mathbf{k})$

6. $\mathbf{r}'(2) = 3\mathbf{a} - 4\,\mathbf{c}$; $\mathbf{R}(u) = (6\mathbf{a} + \mathbf{b} - 4\,\mathbf{c}) + u(3\mathbf{a} - 4\,\mathbf{c})$

7. $\mathbf{r}'(t) = -2\,\sin t\,\mathbf{i} + 3\,\cos t\,\mathbf{j} + \mathbf{k}$, $\mathbf{r}'(\pi/4) = -\sqrt{2}\,\mathbf{i} + \frac{3}{2}\,\sqrt{2}\,\mathbf{j} + \mathbf{k}$
    $\mathbf{R}\,(u) = \left(\sqrt{2}\,\mathbf{i} + \frac{3}{2}\,\sqrt{2}\,\mathbf{j} + \frac{\pi}{4}\,\mathbf{k}\right) + u\left(-\sqrt{2}\,\mathbf{i} + \frac{3}{2}\,\sqrt{2}\,\mathbf{j} + \mathbf{k}\right)$

8. $\mathbf{r}'(t) = (\sin t + t\,\cos t)\,\mathbf{i} + (\cos t - t\,\sin t)\,\mathbf{j} + 2\,\mathbf{k}$
    $\mathbf{r}'\left(\dfrac{\pi}{2}\right) = \mathbf{i} - \dfrac{\pi}{2}\,\mathbf{j} + 2\,\mathbf{k}$; $\mathbf{R}(u) = \left(\dfrac{\pi}{2}\,\mathbf{i} + \pi\,\mathbf{k}\right) + u\left(\mathbf{i} - \dfrac{\pi}{2}\,\mathbf{j} + 2\,\mathbf{k}\right)$

9. The scalar components $x(t) = at$ and $y(t) = bt^2$ satisfy the equation
    $$a^2 y(t) = a^2(bt^2) = b\,(a^2 t^2) = b\,[x(t)\,]^2$$
    and generate the parabola $a^2 y = bx^2$.

10. $x(t)^2 - y(t)^2 = \dfrac{a^2}{4}\left[(e^{wt} + e^{-wt})^2 - (e^{wt} - e^{-wt})^2\right] = a^2$,   with $x(t) > 0$;   the right branch of the
    hyperbola $x^2 - y^2 = a^2$.

11. $\mathbf{r}\,(t) = t\,\mathbf{i} + (1 + t^2)\,\mathbf{j}$, $\mathbf{r}'(t) = \mathbf{i} + 2t\,\mathbf{j}$
    (a) $\mathbf{r}\,(t) \perp \mathbf{r}'(t) \quad\Longrightarrow\quad \mathbf{r}\,(t) \cdot \mathbf{r}'(t) = \left[t\,\mathbf{i} + (1 + t^2)\,\mathbf{j}\right] \cdot (\mathbf{i} + 2t\,\mathbf{j})$
    $$= t(2t^2 + 3) = 0 \quad\Longrightarrow\quad t = 0$$
    $\mathbf{r}\,(t)$ and $\mathbf{r}'(t)$ are perpendicular at $(0, 1)$.
    (b) and (c) $\mathbf{r}\,(t) = \alpha\,\mathbf{r}'(t)$ with $\alpha \neq 0 \quad\Longrightarrow\quad t = \alpha$ and $1 + t^2 = 2t\alpha \quad\Longrightarrow\quad t = \pm 1$.
    If $\alpha > 0$, then $t = 1$. $\mathbf{r}\,(t)$ and $\mathbf{r}'(t)$ have the same direction at $(1, 2)$.
    If $\alpha < 0$, then $t = -1$. $\mathbf{r}\,(t)$ and $\mathbf{r}'(t)$ have opposite directions at $(-1, 2)$.

12. $\mathbf{r}(t) = e^{\alpha t}(\mathbf{i} + 2\,\mathbf{j} + 3\,\mathbf{k})$

13. The tangent line at $t = t_0$ has the form $\mathbf{R}\,(u) = \mathbf{r}\,(t_0) + u\,\mathbf{r}'(t_0)$. If $\mathbf{r}'(t_0) = \alpha\,\mathbf{r}\,(t_0)$, then
    $$\mathbf{R}\,(u) = \mathbf{r}\,(t_0) + u\,\alpha\mathbf{r}\,(t_0) = (1 + u\alpha)\,\mathbf{r}\,(t_0).$$
    The tangent line passes through the origin at $u = -1/\alpha$.

14. $\mathbf{r}_1'(0) = \mathbf{i}$, $\mathbf{r}_2'(0) = 2\,\mathbf{i} + \mathbf{j} + \mathbf{k}$
    $$\theta = \cos^{-1}\frac{\mathbf{r}_1'(0) \cdot \mathbf{r}_2'(0)}{\|\,\mathbf{r}_1'(0)\,\|\,\|\,\mathbf{r}_2'(0)\,\|} = \cos^{-1}\frac{2}{\sqrt{6}} \cong 0.62 \text{ radian, or } 35.3°$$

15. $\mathbf{r}_1(t)$ passes through $P\,(0, 0, 0)$ at $t = 0$; $\mathbf{r}_2(u)$ passes through $P\,(0, 0, 0)$ at $u = -1$.
    $\mathbf{r}_1'(t) = e^t\,\mathbf{i} + 2\,\cos t\,\mathbf{j} + \dfrac{1}{t + 1}\,\mathbf{k}$; $\mathbf{r}_1'(0) = \mathbf{i} + 2\,\mathbf{j} + \mathbf{k}$
    $\mathbf{r}_2'(u) = \mathbf{i} + 2u\,\mathbf{j} + 3u^2\,\mathbf{k}$; $\mathbf{r}_2'(-1) = \mathbf{i} - 2\,\mathbf{j} + 3\,\mathbf{k}$
    $\cos\theta = \dfrac{\mathbf{r}_1'(0) \cdot \mathbf{r}_2'(1)}{\|\mathbf{r}_1'(0)\|\,\|\mathbf{r}_2'(1)\|} = 0$; $\theta = \dfrac{\pi}{2} \cong 1.57$, or $90°$.

16. $\mathbf{r}'_1(0) = -\mathbf{i}, \quad \mathbf{r}'_2(-1) = \mathbf{i} - 4\mathbf{j} - 8\mathbf{k}$

$$\theta = \cos^{-1}\frac{\mathbf{r}'_1(0) \cdot \mathbf{r}'_2(-1)}{\|\mathbf{r}'_1(0)\| \|\mathbf{r}'_2(-1)\|} = \cos^{-1}\left(\frac{-1}{9}\right) \cong 1.68 \text{ radians, or } 96.4°$$

17. $\mathbf{r}_1(t) = \mathbf{r}_2(u)$ implies

$$\left.\begin{matrix} e^t = u \\ 2\sin\left(t + \frac{1}{2}\pi\right) = 2 \\ t^2 - 2 = u^2 - 3 \end{matrix}\right\} \quad \text{so that} \quad t = 0, \quad u = 1.$$

The point of intersection is $(1, 2, -2)$.

$$\mathbf{r}'_1(t) = e^t\mathbf{i} + 2\cos\left(t + \frac{\pi}{2}\right)\mathbf{j} + 2t\mathbf{k}, \quad \mathbf{r}'_1(0) = \mathbf{i}$$

$$\mathbf{r}'_2(u) = \mathbf{i} + 2u\mathbf{k}, \quad \mathbf{r}'_2(1) = \mathbf{i} + 2\mathbf{k}$$

$$\cos\theta = \frac{\mathbf{r}'_1(0) \cdot \mathbf{r}'_2(1)}{\|\mathbf{r}'_1(0)\| \|\mathbf{r}'_2(1)\|} = \frac{1}{5}\sqrt{5} \cong 0.447, \quad \theta \cong 1.11 \text{ radians}$$

18. (a) $\mathbf{R}(u) = \mathbf{r}(t_0) + u\mathbf{r}'(t_0) = (t_0\mathbf{i} + f(t_0)\mathbf{j}) + u(\mathbf{i} + f'(t_0)\mathbf{j})$

$\implies x(u) = t_0 + u, \quad y(u) = f(t_0) + uf'(t_0)$

(b)   From (a), we get $u = x(u) - t_0$ and $y(u) - f(t_0) = f'(t_0)u$.   so

$y - f(t_0) = f'(t_0)(x - t_0)$,   as expected.

19. (a) $\mathbf{r}(t) = a\cos t\,\mathbf{i} + b\sin t\,\mathbf{j}$ (b) $\mathbf{r}(t) = a\cos t\,\mathbf{i} - b\sin t\,\mathbf{j}$

(c) $\mathbf{r}(t) = a\cos 2t\,\mathbf{i} + b\sin 2t\,\mathbf{j}$ (d) $\mathbf{r}(t) = a\cos 3t\,\mathbf{i} - b\sin 3t\,\mathbf{j}$

20. (a) $\mathbf{r}(t) = -a\sin t\,\mathbf{i} + b\cos t\,\mathbf{j}$ (b) $\mathbf{r}(t) = a\sin t\,\mathbf{i} + b\cos t\,\mathbf{j}$

(c) $\mathbf{r}(t) = -a\sin 2t\,\mathbf{i} + b\cos 2t\,\mathbf{j}$ (d) $\mathbf{r}(t) = a\sin 3t\,\mathbf{i} + b\cos 3t\,\mathbf{j}$

21. $\mathbf{r}'(t) = t^3\mathbf{i} + 2t\mathbf{j}$ **22.** $\mathbf{r}'(t) = 2\mathbf{i} + 2t\mathbf{j}$ **23.** $\mathbf{r}'(t) = 2e^{2t}\mathbf{i} - 4e^{-4t}\mathbf{j}$

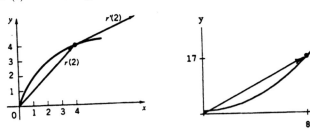

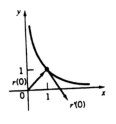

24. $\mathbf{r}'(t) = \cos t\,\mathbf{i} + 2\sin t\,\mathbf{j}$ **25.** $\mathbf{r}'(t) = -2\sin t\,\mathbf{i} + 3\cos t\,\mathbf{j}$ **26.** $\mathbf{r}'(t) = \sec t\tan t\,\mathbf{i} + \sec^2 t\,\mathbf{j}$

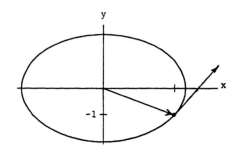

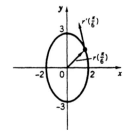

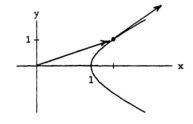

27. $\mathbf{r}(t) = (t^2 + 1)\mathbf{i} + t\mathbf{j}, \quad t \geq 1; \quad \text{or,} \quad \mathbf{r}(t) = \sec^2 t\,\mathbf{i} + \tan t\,\mathbf{j}, \quad t \in \left[\frac{1}{4}\pi, \frac{1}{2}\pi\right)$

28. $\mathbf{r}(t) = \cos t(1 - \cos t)\mathbf{i} + \sin t(1 - \cos t)\mathbf{j}, \quad t \in [0, 2\pi]$

29. $\mathbf{r}(t) = \cos t \sin 3t\,\mathbf{i} + \sin t \sin 3t\,\mathbf{j}, \quad t \in [0, \pi]$

30. $\mathbf{r}(t) = t^4\,\mathbf{i} + t^3\,\mathbf{j}, \quad t \leq 0$

31. $y^3 = x^2$

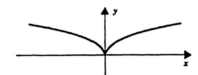

There is no tangent vector at the origin.

32.   (a) $\mathbf{r}(0) = \mathbf{i} + 2\mathbf{j} = \mathbf{r}(1)$

      (b) $\mathbf{T}(0) = \dfrac{\mathbf{r}'(0)}{\|\,\mathbf{r}'(0)\,\|} = \dfrac{1}{\sqrt{\pi^2 + 2}}(-\mathbf{i} - \mathbf{j} + \pi\,\mathbf{k}), \quad \mathbf{T}(1) = \dfrac{\mathbf{r}'(1)}{\|\,\mathbf{r}'(1)\,\|} = \dfrac{1}{\sqrt{\pi^2 + 5}}(\mathbf{i} + 2\mathbf{j} - \pi\,\mathbf{k})$

33.   We substitute $x = t, \; y = t^2, \; z = t^3$ in the plane equation to obtain

$$4t + 2t^2 + t^3 = 24, \quad (t - 2)\left(t^2 + 4t + 12\right) = 0, \quad t = 2.$$

The twisted cubic intersects the plane at the tip of $\mathbf{r}(2)$, the point $(2, 4, 8)$.

The angle between the curve and the normal line at the point of intersection is the angle between the tangent vector $\mathbf{r}'(2) = \mathbf{i} + 4\mathbf{j} + 12\mathbf{k}$ and the normal $\mathbf{N} = 4\mathbf{i} + 2\mathbf{j} + \mathbf{k}$:

$$\cos\theta = \frac{(\mathbf{i} + 4\mathbf{j} + 12\,\mathbf{k}) \cdot (4\mathbf{i} + 2\mathbf{j} + \mathbf{k})}{\|\mathbf{i} + 4\mathbf{j} + 12\,\mathbf{k}\|\,\|4\mathbf{i} + 2\mathbf{j} + \mathbf{k}\|} = \frac{24}{\sqrt{161}\,\sqrt{21}} \cong 0.412, \quad \theta \cong 1.15 \text{ radians.}$$

34.   (a) $T(t) = \dfrac{1}{\sqrt{a^2 \sin^2 t + b^2 \cos^2 t}}(-a \sin t\,\mathbf{i} + b \cos t\,\mathbf{j})$

         $N(t) = \dfrac{1}{\sqrt{a^2 \sin^2 t + b^2 \cos^2 t}}(-b \cos t\,\mathbf{i} + a \sin t\,\mathbf{j})$

      (b) $\mathbf{r}(u) = \dfrac{\sqrt{2}}{2}(a\,\mathbf{i} + b\,\mathbf{j}) + u(-a\,\mathbf{i} + b\,\mathbf{j}), \quad \mathbf{R}(u) = \dfrac{\sqrt{2}}{2}(a\,\mathbf{i} + b\,\mathbf{j}) + u(-b\,\mathbf{i} - a\,\mathbf{j})$

35. $\mathbf{r}'(t) = 2\mathbf{j} + 2t\,\mathbf{k}, \quad \|\mathbf{r}'(t)\| = 2\sqrt{1 + t^2}$

      $\mathbf{T}(t) = \dfrac{\mathbf{r}'(t)}{\|\mathbf{r}'(t)\|} = \dfrac{1}{\sqrt{1 + t^2}}(\mathbf{j} + t\,\mathbf{k})$,

      $\mathbf{T}'(t) = \dfrac{1}{(1 + t^2)^{3/2}}[-t\mathbf{j} + \mathbf{k}]$

      at $t = 1$:   tip of $\mathbf{r} = (1, 2, 1), \quad \mathbf{T} = \mathbf{T}(1) = \dfrac{1}{\sqrt{2}}\mathbf{j} + \dfrac{1}{\sqrt{2}}\mathbf{k}$;

$$\mathbf{T}'(1) = -\frac{1}{2\sqrt{2}}\,\mathbf{j} + \frac{1}{2\sqrt{2}}\,\mathbf{k}; \quad \|\mathbf{T}'(1)\| = \frac{1}{2}; \quad \mathbf{N} = \mathbf{N}(1) = \frac{\mathbf{T}'(1)}{\|\mathbf{T}'(1)\|} = -\frac{1}{\sqrt{2}}\,\mathbf{j} + \frac{1}{\sqrt{2}}\,\mathbf{k}$$

normal for osculating plane:

$$\mathbf{T} \times \mathbf{N} = \left(\frac{1}{\sqrt{2}}\,\mathbf{j} + \frac{1}{\sqrt{2}}\,\mathbf{k}\right) \times \left(-\frac{1}{\sqrt{2}}\,\mathbf{j} + \frac{1}{\sqrt{2}}\,\mathbf{k}\right) = \frac{1}{2}\,\mathbf{i}$$

equation for osculating plane:

$$\frac{1}{2}(x-1) + 0(y-2) + 0(z-1) = 0, \quad \text{which gives} \quad x - 1 = 0$$

36. $\mathbf{T}(t) = \dfrac{\mathbf{r}'(t)}{\|\,\mathbf{r}'(t)\,\|} = \dfrac{\mathbf{i} + 2t\,\mathbf{j} + 4t\,\mathbf{k}}{\sqrt{20t^2 + 1}}$ $\quad \mathbf{T}(1) = \dfrac{1}{\sqrt{21}}(\mathbf{i} + 2\,\mathbf{j} + 4\,\mathbf{k})$

$\mathbf{N}(t) = \dfrac{\mathbf{T}'(t)}{\|\,\mathbf{T}'(t)\,\|} = \dfrac{-20t\,\mathbf{i} + 2\,\mathbf{j} + 4\,\mathbf{k}}{\sqrt{400t^2 + 20}}$ $\quad \mathbf{N}(1) = \dfrac{1}{\sqrt{420}}(-20\,\mathbf{i} + 2\,\mathbf{j} + 4\,\mathbf{k}) = \dfrac{1}{\sqrt{105}}(-10\,\mathbf{i} + \mathbf{j} + 2\,\mathbf{k})$

$\mathbf{T}(1) \times \mathbf{N}(1) = \dfrac{1}{\sqrt{5}}(-2\,\mathbf{j} + \mathbf{k}).$

Osculating plane at $(1, 1, 2):$ $\quad -2(y - 1) + (z - 2) = 0 \implies -2y + z = 0$

37.
$$\mathbf{r}'(t) = -2\sin 2t\,\mathbf{i} + 2\cos 2t\,\mathbf{j} + \mathbf{k}, \quad \|\mathbf{r}'(t)\| = \sqrt{5}$$

$$\mathbf{T}(t) = \frac{\mathbf{r}'(t)}{\|\mathbf{r}'(t)\|} = \frac{1}{5}\sqrt{5}\,(-2\sin 2t\,\mathbf{i} + 2\cos 2t\,\mathbf{j} + \mathbf{k})$$

$$\mathbf{T}'(t) = -\frac{4}{5}\sqrt{5}\,(\cos 2t\,\mathbf{i} + \sin 2t\,\mathbf{j}), \quad \|\mathbf{T}'(t)\| = \frac{4}{5}\sqrt{5}$$

$$\mathbf{N}(t) = \frac{\mathbf{T}'(t)}{\|\mathbf{T}'(t)\|} = -(\cos 2t\,\mathbf{i} + \sin 2t\,\mathbf{j})$$

at $t = \pi/4$: tip of $\mathbf{r} = (0, 1, \pi/4)$, $\mathbf{T} = \dfrac{1}{5}\sqrt{5}\,(-2\,\mathbf{i} + \mathbf{k})$, $\mathbf{N} = -\mathbf{j}$

normal for osculating plane:

$$\mathbf{T} \times \mathbf{N} = \frac{1}{5}\sqrt{5}\,(-2\,\mathbf{i} + \mathbf{k}) \times (-\mathbf{j}) = \frac{1}{5}\sqrt{5}\,\mathbf{i} + \frac{2}{5}\sqrt{5}\,\mathbf{k}$$

equation for osculating plane:

$$\frac{1}{5}\sqrt{5}\,(x - 0) + \frac{2}{5}\sqrt{5}\left(z - \frac{\pi}{4}\right) = 0, \quad \text{which gives} \quad x + 2z = \frac{\pi}{2}$$

38. $\mathbf{T}(t) = \dfrac{\mathbf{r}'(t)}{\|\,\mathbf{r}'(t)\,\|} = \dfrac{\mathbf{i} + 2\,\mathbf{j} + 2t\,\mathbf{k}}{\sqrt{4t^2 + 5}}$ $\quad \mathbf{T}(2) = \dfrac{1}{\sqrt{21}}(\mathbf{i} + 2\,\mathbf{j} + 4\,\mathbf{k})$

$\mathbf{N}(t) = \dfrac{\mathbf{T}'(t)}{\|\,\mathbf{T}'(t)\,\|} = \dfrac{-2t\,\mathbf{i} - 4t\,\mathbf{j} + 5\,\mathbf{k}}{\sqrt{20t^2 + 25}}$ $\quad \mathbf{N}(2) = \dfrac{1}{\sqrt{105}}(-4\,\mathbf{i} - 8\,\mathbf{j} + 5\,\mathbf{k})$

$\mathbf{T}(2) \times \mathbf{N}(2) = \dfrac{1}{\sqrt{5}}(2\,\mathbf{i} - \mathbf{j})$

Osculating plane at $(2, 4, 2):$ $\quad 2(x - 2) - (y - 4) = 0 \implies 2x - y = 0$

39.

$$\mathbf{r}'(t) = \cosh t\,\mathbf{i} + \sinh t\,\mathbf{j} + \mathbf{k}, \quad \|\mathbf{r}'(t)\| = \sqrt{\cosh^2 t + \sinh^2 t + 1} = \sqrt{2}\,\cosh t$$

$$\mathbf{T}(t) = \frac{\mathbf{r}'(t)}{\|\mathbf{r}'(t)\|} = \frac{1}{\sqrt{2}}\left(\mathbf{i} + \tanh t\,\mathbf{j} + \operatorname{sech} t\,\mathbf{k}\right),$$

$$\mathbf{T}'(t) = \frac{1}{\sqrt{2}}\left(\operatorname{sech}^2 t\,\mathbf{j} - \operatorname{sech} t\,\tanh t\,\mathbf{k}\right)$$

at $t = 0$: tip of $\mathbf{r} = (0, 1, 0)$, $\quad \mathbf{T} = \frac{1}{\sqrt{2}}(\mathbf{i} + \mathbf{k})$, $\quad \mathbf{T}'(0) = \frac{1}{\sqrt{2}}\mathbf{j}$; $\quad \mathbf{N} = \mathbf{j}$

normal for osculating plane:

$$\mathbf{T} \times \mathbf{N} = \left(\frac{1}{\sqrt{2}}(-\mathbf{i} + \mathbf{k})\right) \times \mathbf{j} = \frac{1}{\sqrt{2}}(-\mathbf{i} + \mathbf{k}) \quad \text{or} \quad \mathbf{i} - \mathbf{k}$$

equation for osculating plane: $\quad x - z = 0$

40. $\mathbf{T}(t) = \dfrac{\mathbf{r}'(t)}{\|\,\mathbf{r}'(t)\,\|} = \dfrac{-3\sin 3t\,\mathbf{i} + \mathbf{j} - 3\cos 3t\,\mathbf{k}}{\sqrt{10}}$ $\qquad \mathbf{T}\left(\dfrac{\pi}{3}\right) = \dfrac{1}{\sqrt{10}}(\mathbf{j} + 3\,\mathbf{k})$

$\mathbf{N}(t) = \dfrac{\mathbf{T}'(t)}{\|\,\mathbf{T}'(t)\,\|} = \dfrac{-9\cos 3t\,\mathbf{i} + 9\sin 3t\,\mathbf{k}}{9}$ $\qquad \mathbf{N}\left(\dfrac{\pi}{3}\right) = \mathbf{i}$

$\mathbf{T}\left(\dfrac{\pi}{3}\right) \times \mathbf{N}\left(\dfrac{\pi}{3}\right) = 3\,\mathbf{j} - \mathbf{k}$

Osculating plane at $\left(-1, \frac{\pi}{3}, 0\right):\quad 3\left(y - \frac{\pi}{3}\right) - z = 0 \quad$ or $\quad 3y - z - \pi = 0$

41.

$$\mathbf{r}'(t) = e^t\left[(\sin t + \cos t)\,\mathbf{i} + (\cos t - \sin t)\,\mathbf{j} + \mathbf{k}\right], \quad \|\mathbf{r}'(t)\| = e^t\sqrt{3}$$

$$\mathbf{T}(t) = \frac{\mathbf{r}'(t)}{\|\mathbf{r}'(t)\|} = \frac{1}{\sqrt{3}}\left[(\sin t + \cos t)\,\mathbf{i} + (\cos t - \sin t)\,\mathbf{j} + \mathbf{k}\right],$$

$$\mathbf{T}'(t) = \frac{1}{\sqrt{3}}\left[(\cos t - \sin t)\,\mathbf{i} - (\sin t + \cos t)\,\mathbf{j}\right]$$

at $t = 0$: tip of $\mathbf{r} = (0, 1, 1)$, $\quad \mathbf{T} = \mathbf{T}(0) = \frac{1}{\sqrt{3}}(\mathbf{i} + \mathbf{j} + \mathbf{k})$;

$\mathbf{T}'(0) = \frac{1}{\sqrt{3}}(\mathbf{i} - \mathbf{j})$; $\quad \|\mathbf{T}'(0)\| = \frac{\sqrt{2}}{\sqrt{3}}$; $\quad \mathbf{N} = \mathbf{N}(0) = \frac{\mathbf{T}'(0)}{\|\mathbf{T}'(0)\|} = \frac{1}{\sqrt{2}}(\mathbf{i} - \mathbf{j})$

normal for osculating plane:

$$\mathbf{T} \times \mathbf{N} = \frac{1}{\sqrt{3}}(\mathbf{i} + \mathbf{j} + \mathbf{k}) \times \frac{1}{\sqrt{2}}(\mathbf{i} - \mathbf{j}) = \frac{1}{\sqrt{6}}(\mathbf{i} + \mathbf{j} - 2\,\mathbf{k})$$

equation for osculating plane:

$$\frac{1}{\sqrt{6}}(x - 0) + \frac{1}{\sqrt{6}}(y - 1) - \frac{2}{\sqrt{6}}(z - 1) = 0, \quad \text{or} \quad x + y - 2z + 1 = 0$$

42. $\mathbf{T}(t) = \dfrac{\mathbf{r}'(t)}{\|\,\mathbf{r}'(t)\,\|} = \dfrac{t\cos t\,\mathbf{i} + t\sin t\,\mathbf{j}}{t} = \cos t\,\mathbf{i} + \sin t\,\mathbf{j}$ $\mathbf{T}\!\left(\dfrac{\pi}{4}\right) = \dfrac{\sqrt{2}}{2}\,\mathbf{i} + \dfrac{\sqrt{2}}{2}\,\mathbf{j}$

$\mathbf{N}(t) = \dfrac{\mathbf{T}'(t)}{\|\,\mathbf{T}'(t)\,\|} = -\sin t\,\mathbf{i} + \cos t\,\mathbf{j}$ $\mathbf{N}\!\left(\dfrac{\pi}{4}\right) = -\dfrac{\sqrt{2}}{2}\,\mathbf{i} + \dfrac{\sqrt{2}}{2}\,\mathbf{j}$

$\mathbf{T}\!\left(\dfrac{\pi}{4}\right) \times \mathbf{N}\!\left(\dfrac{\pi}{4}\right) = \mathbf{k}$

Osculating plane at $\left(\dfrac{\sqrt{2}}{2}[1 + \tfrac{\pi}{4}],\ \dfrac{\sqrt{2}}{2}[1 - \tfrac{\pi}{4}],\ 2\right)$: $z - 2 = 0$ or $z = 2$

43. $\mathbf{T}_1 = \dfrac{\mathbf{R}'(u)}{\|\,\mathbf{R}'(u)\,\|} = -\dfrac{\mathbf{r}'(a + b - u)}{\|\,\mathbf{r}'(a + b - u)\,\|} = -\mathbf{T}.$

Therefore $\mathbf{T}_1'(u) = \mathbf{T}'(a + b - u)$ and $\mathbf{N}_1 = \mathbf{N}.$

44. (a) $\mathbf{r}'(t) = -\sqrt{2}\,\sin t\,\mathbf{i} + \sqrt{2}\,\cos t\,\mathbf{j} + \mathbf{k}$

tangent line $(t = \pi/4)$: $x = 1 - t,\quad y = 1 + t,\quad z = \tfrac{1}{4}\pi + t$

(c) Since $\tfrac{1}{4}\pi + t > 0$, $t \in [0, 2\pi]$, the tangent line is never parallel to the x, y-plane.

45. (a) $\mathbf{r}'(t) = -\sqrt{2}\,\sin t\,\mathbf{i} + \sqrt{2}\,\cos t\,\mathbf{j} + 5\cos t\,\mathbf{k}$

tangent line $(t = \pi/4)$: $x = 1 - t,\quad y = 1 + t,\quad z = -\dfrac{\sqrt{2}}{2} - \dfrac{5\sqrt{2}}{2}\,t$

(c) The tangent line is parallel to the x, y-plane at the points where $t = \dfrac{(2n + 1)\pi}{10}$, $n = 0, 1, 2, \ldots, 9$.

46. (a) $\mathbf{r}'(t) = -\sqrt{2}\,\sin t\,\mathbf{i} + \sqrt{2}\,\cos t\,\mathbf{j} + \dfrac{1}{t}\,\mathbf{k}$

tangent line $(t = \pi/4)$: $x = 1 - t,\quad y = 1 + t,\quad z = \ln(\pi/4) + \dfrac{4}{\pi} + t$

SECTION 14.4

1. $\mathbf{r}'(t) = \mathbf{i} + t^{1/2}\,\mathbf{j},\quad \|\mathbf{r}'(t)\| = \sqrt{1 + t}$

$L = \displaystyle\int_0^8 \sqrt{1 + t}\,dt = \left[\dfrac{2}{3}(1 + t)^{3/2}\right]_0^8 = \dfrac{52}{3}$

2. $\mathbf{r}'(t) = (t^2 - 1)\mathbf{i} + 2t\,\mathbf{j};\quad \|\,\mathbf{r}'(t)\,\| = t^2 + 1$

$L = \displaystyle\int_0^2 (t^2 + 1)\,dt = \dfrac{14}{3}$

3. $\mathbf{r}'(t) = -a\sin t\,\mathbf{i} + a\cos t\,\mathbf{j} + b\,\mathbf{k},\quad \|\mathbf{r}'(t)\| = \sqrt{a^2 + b^2}$

$L = \displaystyle\int_0^{2\pi} \sqrt{a^2 + b^2}\,dt = 2\pi\sqrt{a^2 + b^2}$

4. $\mathbf{r}'(t) = \mathbf{i} + \sqrt{2}\,t^{1/2}\,\mathbf{j} + t\,\mathbf{k};\quad \|\,\mathbf{r}'(t)\,\| = t + 1$

$L = \displaystyle\int_0^2 (t + 1)\,dt = 4$

5. $\mathbf{r}'(t) = \mathbf{i} + \tan t\,\mathbf{j},\quad \|\mathbf{r}'(t)\| = \sqrt{1 + \tan^2 t} = |\sec t|$

$L = \displaystyle\int_0^{\pi/4} |\sec t|\,dt = \int_0^{\pi/4} \sec t\,dt = \big[\ln|\sec t + \tan t|\big]_0^{\pi/4} = \ln\left(1 + \sqrt{2}\right)$

6. $\mathbf{r}'(t) = \dfrac{1}{1+t^2}\mathbf{i} + \dfrac{t}{1+t^2}\mathbf{j}; \quad \|\mathbf{r}'(t)\| = \dfrac{1}{\sqrt{1+t^2}}$

$L = \displaystyle\int_0^1 \dfrac{dt}{\sqrt{1+t^2}} = \left[\ln|t + \sqrt{t^2+1}|\right]_0^1 = \ln(1+\sqrt{2})$

7. $\mathbf{r}'(t) = 3t^2\,\mathbf{i} + 2t\,\mathbf{j}, \quad \|\mathbf{r}'(t)\| = \sqrt{9t^4 + 4t^2} = |t|\sqrt{4 + 9t^2}$

$L = \displaystyle\int_0^1 \left|t\sqrt{4+9t^2}\right|\,dt = \int_0^1 t\sqrt{4+9t^2}\,dt = \left[\dfrac{1}{27}\left(4 + 9t^2\right)^{3/2}\right]_0^1 = \dfrac{1}{27}\left(13\sqrt{13} - 8\right)$

8. $\mathbf{r}'(t) = \mathbf{i} + \left(\dfrac{1}{2}t^2 - \dfrac{1}{2}t^{-2}\right)\mathbf{k}; \quad \|\mathbf{r}'(t)\| = \dfrac{1}{2}t^2 + \dfrac{1}{2}t^{-2}$

$L = \displaystyle\int_1^3 \left(\dfrac{1}{2}t^2 + \dfrac{1}{2}t^{-2}\right)dt = \dfrac{14}{3}$

9. $\mathbf{r}'(t) = (\cos t - \sin t)e^t\,\mathbf{i} + (\sin t + \cos t)e^t\,\mathbf{j}, \quad \|\mathbf{r}'(t)\| = \sqrt{2}\,e^t$

$L = \displaystyle\int_0^\pi \sqrt{2}\,e^t\,dt = \sqrt{2}\left(e^\pi - 1\right)$

10. $\mathbf{r}'(t) = 2t\,\mathbf{i} + 2t\,\mathbf{j} - 2t\,\mathbf{k}; \quad \|\mathbf{r}'(t)\| = 2t\sqrt{3} \implies L = \displaystyle\int_0^2 2t\sqrt{3}\,dt = 4\sqrt{3}$

11. $\mathbf{r}'(t) = \dfrac{1}{t}\mathbf{i} + 2\,\mathbf{j} + 2t\,\mathbf{k}, \quad \|\mathbf{r}'(t)\| = \sqrt{\dfrac{1}{t^2} + 4 + 4t^2}$

$L = \displaystyle\int_1^e \sqrt{\dfrac{1}{t^2} + 4 + 4t^2}\,dt = \int_1^e \left(\dfrac{1}{t} + 2t\right)dt = \left[\ln|t| + t^2\right]_1^e = e^2$

12. $\mathbf{r}'(t) = t\cos t\,\mathbf{i} - t\sin t\,\mathbf{j}; \quad \|\mathbf{r}'(t)\| = t \implies L = \displaystyle\int_0^2 t\,dt = 2$

13. $\mathbf{r}'(t) = t\cos t\,\mathbf{i} + t\sin t\,\mathbf{j} + \sqrt{3}\,t\,\mathbf{k}, \quad \|\mathbf{r}'(t)\| = \sqrt{t^2\cos^2 t + t^2\sin^2 t + 3t^2} = \sqrt{4t^2} = 2t$

$L = \displaystyle\int_0^{2\pi} 2t\,dt = \left[t^2\right]_0^{2\pi} = 4\pi^2$

14. $\mathbf{r}'(t) = (1+t)^{1/2}\,\mathbf{i} + (1-t)^{1/2}\,\mathbf{j} + \sqrt{2}\,\mathbf{k}, \quad \|\mathbf{r}'(t)\| = \sqrt{(1+t) + (1-t) + 2} = \sqrt{4} = 2$

$L = \displaystyle\int_{-1/2}^{1/2} 2\,dt = 2$

15. $\mathbf{r}'(t) = 2\,\mathbf{i} + 2t\,\mathbf{j} - 2t\,\mathbf{k}, \quad \|\mathbf{r}'(t)\| = 2\sqrt{1 + 2t^2}$

$L = \displaystyle\int_0^2 2\sqrt{1+2t^2}\,dt = \sqrt{2}\int_0^{\tan^{-1}(2\sqrt{2})} \sec^3 u\,du$

$\quad\quad \underset{\displaystyle (t\sqrt{2} = \tan u)}{\Big\uparrow\!\!\!\rule{0pt}{0pt}}$

$= \tfrac{1}{2}\sqrt{2}\left[\sec u \tan u + \ln|\sec u + \tan u|\right]_0^{\tan^{-1}(2\sqrt{2})} = 6 + \tfrac{1}{2}\sqrt{2}\ln\left(3 + 2\sqrt{2}\right)$

16. $\mathbf{r}'(t) = (3\cos t - 3t\sin t)\mathbf{i} + (3\sin t + 3t\cos t)\mathbf{j} + 4\,\mathbf{k};$ $\|\,\mathbf{r}'(t)\,\| = \sqrt{9t^2 + 25}$

$$L = \int_0^4 3\sqrt{t^2 + \frac{25}{9}}\,dt = \left[\frac{3}{2}t\sqrt{t^2 + \frac{25}{9}} + \frac{3}{2}\cdot\frac{25}{9}\ln\left|t + \sqrt{t^2 + \frac{25}{9}}\right|\right]_0^4 = 26 + \frac{25}{6}\ln 5.$$

17.
$$s = s(t) = \int_a^t \|\mathbf{r}'(u)\|\,du$$

$$s'(t) = \|\mathbf{r}'(t)\| = \|x'(t)\,\mathbf{i} + y'(t)\,\mathbf{j} + z'(t)\,\mathbf{k}\|$$

$$= \sqrt{[x'(t)]^2 + [y'(t)]^2 + [z'(t)]^2}.$$

In the Leibniz notation this translates to

$$\frac{ds}{dt} = \sqrt{\left(\frac{dx}{dt}\right)^2 + \left(\frac{dy}{dt}\right)^2 + \left(\frac{dz}{dt}\right)^2}.$$

18. Parameterize the graph of f by setting $\mathbf{r}(x) = x\,\mathbf{i} + f(x)\,\mathbf{j}, \quad x \in [a, b]$.

19.
$$s = s(x) = \int_a^x \sqrt{1 + [f'(t)]^2}\,dt$$

$$s'(x) = \sqrt{1 + [f'(x)]^2}.$$

In the Leibniz notation this translates to

$$\frac{ds}{dx} = \sqrt{1 + \left(\frac{dy}{dx}\right)^2}.$$

20. $L_1 = \int_1^e \|\mathbf{r}'(t)\|\,dt = \int_1^e \sqrt{2 + \frac{2}{t^2}}\,dt;\quad L_2 = \int_0^1 \sqrt{1 + e^{2x}}\,dx$

Setting $t = e^x$, we get $L_1 = \sqrt{2}L_2$

21. $\mathbf{r}'(t) = -\sin t\,\mathbf{i} + \cos t\,\mathbf{j}.$ Since $\|\mathbf{r}'\| \equiv 1$, the parametrization is by arc length.

22. $\mathbf{r}'(t) = -\mathbf{a} + \mathbf{b};\quad \|\mathbf{r}'\| = \|\mathbf{b} - \mathbf{a}\|.$

$$s = \int_0^t \|\mathbf{b} - \mathbf{a}\|\,du = \|\mathbf{b} - \mathbf{a}\|\,t;\quad t = \frac{s}{\|\mathbf{b} - \mathbf{a}\|}.$$

$$\mathbf{R}(s) = \left(1 - \frac{s}{\|\mathbf{b} - \mathbf{a}\|}\right)\mathbf{a} + \frac{s}{\|\mathbf{b} - \mathbf{a}\|}\,\mathbf{b},\quad 0 \le s \le \|\mathbf{b} - \mathbf{a}\|.$$

23. $\mathbf{r}'(t) = t\sin t\,\mathbf{i} + t\cos t\,\mathbf{j} + t\,\mathbf{k};\quad \|\mathbf{r}'\| = t\sqrt{2}.$

$$s = \int_0^t u\sqrt{2}\,du = \frac{\sqrt{2}}{2}t^2;\quad t = 2^{1/4}\sqrt{s}.$$

$$\mathbf{R}(s) = \left(\sin 2^{1/4}\sqrt{s} - 2^{1/4}\sqrt{s}\cos 2^{1/4}\sqrt{s}\right)\mathbf{i} + \left(\cos 2^{1/4}\sqrt{s} + 2^{1/4}\sqrt{s}\sin 2^{1/4}\sqrt{s}\right)\mathbf{j} + \frac{1}{\sqrt{2}}s\,\mathbf{k}.$$

24. $\mathbf{r}'(t) = (e^t\cos t - e^t\sin t)\,\mathbf{i} + (e^t\sin t + e^t\cos t)\,\mathbf{j};\quad \|\mathbf{r}'\| = \sqrt{2}\,e^t.$

$$s = \int_0^t \sqrt{2}\,e^u\,du = \sqrt{2}(e^t - 1);\quad t = \ln(1 + s/\sqrt{2}).$$

$$\mathbf{R}(s) = e^v[\cos v\,\mathbf{i} + \sin v\,\mathbf{j}]\quad \text{where}\quad v = \ln(1 + s/\sqrt{2}),\ 0 \le s \le \sqrt{2}\,(e^\pi - 1).$$

25. $\mathbf{r}'(t) = t^{3/2}\,\mathbf{j} + \mathbf{k}, \quad \|\mathbf{r}'(t)\| = \sqrt{\left(t^{3/2}\right)^2 + 1} = \sqrt{t^3 + 1}$

$$s = \int_0^{1/2} \sqrt{t^3 + 1}\,dt \cong 0.5077$$

26. $\mathbf{r}'(t) = \mathbf{i} + t^2\,\mathbf{j}; \quad \|\mathbf{r}'(t)\| = \sqrt{1 + t^4} \quad L = \int_0^2 \sqrt{1 + t^4} \cong 3.6535$

27. $\mathbf{r}'(t) = -3\,\sin t\,\mathbf{i} + 4\,\cos t\,\mathbf{j}, \quad \|\mathbf{r}'(t)\| = \sqrt{9\,\sin^2 t + 16\,\cos^2 t}\,dt$

$$s = \int_0^{2\pi} \sqrt{9\,\sin^2 t + 16\,\cos^2 t}\,dt \cong 22.0939$$

28. $\mathbf{r}'(t) = \mathbf{i} + 2t\mathbf{j} + \dfrac{1}{t}\mathbf{k}; \quad \|\mathbf{r}'(t)\| = \sqrt{1 + 4t^2 + \dfrac{1}{t^2}} \quad L = \int_1^4 \sqrt{1 + 4t^2 + \dfrac{1}{t^2}}\,dt \cong 15.4480$

29. (a) $\qquad\qquad\qquad\qquad\qquad\qquad\qquad$ (b) $s = \displaystyle\int_0^{2\pi} \sqrt{1 + 16\,\cos^2 4t}\,dt \cong 17.6286$

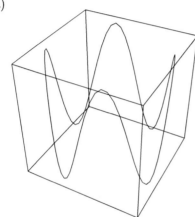

30. (a) $\qquad\qquad\qquad\qquad\qquad\qquad\qquad$ (b) $s = \displaystyle\int_0^{2\pi} \sqrt{1 + \left(\dfrac{1}{1+t}\right)^2}\,dt \cong 6.6818$

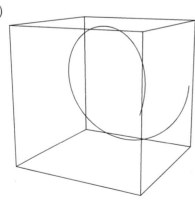

PROJECT 14.4

1. Given the differentiable curve $\mathbf{r} = \mathbf{r}(t)$, $t \in I$. Let $t = \phi(u)$ be a continuously differentiable one-to-one function that maps the interval J onto the interval I, and let $\mathbf{R}(u) = \mathbf{r}(\phi(u))$, $u \in J$. Suppose that $\phi'(u) > 0$ on J. Then, as u increases across J, $t = \phi(u)$ increases across I. As a result, $\mathbf{R}(u)$ takes on exactly the same values in exactly the same order as **r**. If $\phi'(u) < 0$ on J, then $\mathbf{R}(u)$ takes on exactly the same values as **r** but in the reverse order.

2. Let $\mathbf{r} = \mathbf{r}(t)$, $t \in I$ be a differentiable curve. Let $t = \phi(u)$ be a continuously differentiable one-to-one function that maps the interval J onto the interval I with $\phi'(u) > 0$. Set $\mathbf{R}(u) = \mathbf{r}(\phi(u))$, $u \in J$.

$$\mathbf{R}'(u) = [\mathbf{r}(\phi(u))]' = \mathbf{r}'(\phi(u))\phi'(u); \quad \frac{\mathbf{R}'(u)}{||\mathbf{R}'(u)||} = \frac{\mathbf{r}'(\phi(u))\phi'(u)}{||\mathbf{r}'(\phi(u))\phi'(u)||} = \frac{\mathbf{r}'(\phi(u))\phi'(u)}{\phi'(u)||\mathbf{r}'(\phi(u))||} = \frac{\mathbf{r}'(t)}{||\mathbf{r}'(t)||}.$$

$$\text{since } \phi'(u) > 0 \underline{\quad\uparrow}$$

Therefore, the unit tangent is left unchanged by a sense-preserving change of parameter. The invariance of the principal normal and the osculating plane follows from the invariance of the unit tangent.

3. Suppose that $\phi'(u) < 0$ on J. Then

$$\mathbf{R}'(u) = [\mathbf{r}(\phi(u))]' = \mathbf{r}'(\phi(u))\phi'(u); \quad \frac{\mathbf{R}'(u)}{\mathbf{R}'(u)} = \frac{\mathbf{r}'(\phi(u))\phi'(u)}{||\mathbf{r}'(\phi(u))\phi'(u)||} = -\frac{\mathbf{r}'(\phi(u))\phi'(u)}{\phi'(u)||\mathbf{r}'(\phi(u))||} = -\frac{\mathbf{r}'(t)}{||\mathbf{r}'(t)||}.$$

$$\text{since } \phi'(u) < 0 \underline{\quad\uparrow}$$

Thus, the unit tangent is reversed by a sense-reversing change of parameter. That is, if $T_\mathbf{R}$ and $T_\mathbf{r}$ are the respective unit tangents, then $T_\mathbf{R} = -T_\mathbf{r}$.

Now consider the principal normals:

$$\frac{T_\mathbf{R}'}{||T_\mathbf{R}'||} = -\frac{T_\mathbf{r}'(\phi(u)\phi'(u)}{||T_\mathbf{r}'(\phi(u)\phi'(u)||} = \frac{T_\mathbf{r}'(\phi(u)\phi'(u)}{\phi'(u)||T_\mathbf{r}'(\phi(u)\phi'(u)||} = \frac{T_\mathbf{r}'(t)}{||T_\mathbf{r}'(t)||}.$$

$$\text{since } \phi'(u) < 0 \underline{\quad\uparrow}$$

Thus the principal normal is unchanged by a sense-reversing change of parameter. The osculating plane is also unchanged since $\mathbf{T} \times \mathbf{N}$ and $-\mathbf{T} \times \mathbf{N}$ are each normal to the osculating plane.

4. $s = \displaystyle\int_0^t \{\mathbf{r}(v)|| \, dv = \psi(t)$. ψ is an continuously differentiable increasing function on I; $t = \psi^{-1}(s)$.

5. Let L be the length as computed from **r** and L^* the length as computed from **R**. Then

$$L^* = \int_c^d ||\mathbf{R}'(u)|| \, du = \int_c^d ||\mathbf{r}'(\phi(u))|| \, \phi'(u) \, du = \int_a^b ||\mathbf{r}'(t)|| \, dt = L.$$

$$t = \phi(u) \underline{\quad\uparrow}$$

SECTION 14.5

1. $\mathbf{r}(t) = a[\cos\theta(t)\,\mathbf{i} + \sin\theta(t)\,\mathbf{j}], \quad \mathbf{r}'(t) = a[-\sin\theta(t)\,\mathbf{i} + \cos\theta(t)\,\mathbf{j})]\theta'(t)$

$||\mathbf{r}'(t)|| = v \implies a|\theta'(t)| = v \implies |\theta'(t)| = v/a$

$\mathbf{r}''(t) = a[-\cos\theta(t)\,\mathbf{i} - \sin\theta(t)\,\mathbf{j}]\,[\theta'(t)]^2, \quad ||\mathbf{r}''(t)|| = a[\theta'(t)]^2 = v^2/a$

2. $\mathbf{r}'(t) = (-\pi a \sin \pi t + 2bt)\mathbf{i} + (\pi a \cos \pi t - 2bt)\mathbf{j}$

$\mathbf{r}''(t) = (-\pi^2 a \cos \pi t + 2b)\mathbf{i} + (-\pi^2 a \sin \pi t - 2b)\mathbf{j}$

At $t = 1$, $\mathbf{v} = 2b\,\mathbf{i} - (a\pi + 2b)\mathbf{j}$, $v = \|\,\mathbf{v}\,\| = \sqrt{4b^2 + (a\pi^2 + 2b)^2}$

$\mathbf{a} = (a\pi^2 + 2b)\mathbf{i} - 2b\,\mathbf{j}$, $\|\,\mathbf{a}\,\| = \sqrt{4b^2 + (a\pi^2 + 2b)^2}$

3. $\mathbf{r}(t) = at\,\mathbf{i} + b \sin at\,\mathbf{j}$, $= a\,\mathbf{i} + ab \cos at\,\mathbf{j}$

$\mathbf{r}'(t) = -a^2 b \sin at\,\mathbf{j}$, $\|\mathbf{r}'(t)\| = a^2 |b \sin at| = a^2 |y(t)|$

4. $\mathbf{r}'(t) = 2t\,\mathbf{j} + 2(t-1)\,\mathbf{k}$;   speed is minimum when $\|\,\mathbf{r}'(t)\,\|^2$ is minimum

$4t^2 + 4(t-1)^2 = 8t^2 - 8t + 4$ is minimum at $t = \dfrac{1}{2}$

5. $y = \cos \pi x$, $0 \le x \le 2$

6. $x = y^3$,   all real x

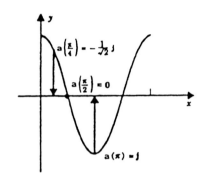

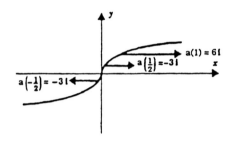

7. $x = \sqrt{1 + y^2}$, $y \ge -1$

8. $x = \sin \pi y$, $0 \le y \le 2$

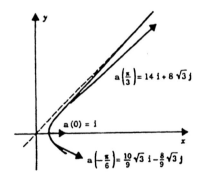

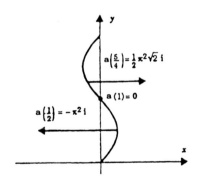

9. (a) initial position is tip of $\mathbf{r}(0) = x_0\,\mathbf{i} + y_0\,\mathbf{j} + z_0\,\mathbf{k}$

(b) $\mathbf{r}'(t) = (\alpha \cos \theta)\,\mathbf{j} + (\alpha \sin \theta - 32t)\,\mathbf{k}$, $\mathbf{r}'(0) = (\alpha \cos \theta)\,\mathbf{j} + (\alpha \sin \theta)\,\mathbf{k}$

(c) $|\mathbf{r}'(0)| = |\alpha|$ (d) $\mathbf{r}''(t) = -32\,\mathbf{k}$

(e) a parabolic arc from the parabola

$$z = z_0 + (\tan \theta)(y - y_0) - 16\,\frac{(y - y_0)^2}{\alpha^2 \cos^2 \theta}$$

in the plane $x = x_0$

10. (a) $x(t) = 2\cos 2t = 4\cos^2 t - 2$ and $y(t) = 3\cos t \implies x = \frac{4}{9}y^2 - 2$.   Since

$$-2 \le x(t) \le 2, \quad -3 \le y(t) \le 3,$$

the path consists only of the bounded arc

$$x = \frac{4}{9}y^2 - 2, \quad -3 \le y \le 3.$$

The motion traces out this arc twice on every t-interval of length 2π.

(b) included in part (d)

(c) $\mathbf{r}(t) = 2\cos 2t\,\mathbf{i} + 3\cos t\,\mathbf{j}, \quad \mathbf{r}'(t) = -4\sin 2t\,\mathbf{i} - 3\sin t\,\mathbf{j}, \quad \mathbf{r}''(t) = -8\cos 2t\,\mathbf{i} - 3\cos t\,\mathbf{j}$

The velocity $\mathbf{r}'(t)$ is $\mathbf{0}$ when $t = n\pi$.   At such points

$$\text{the acceleration} = \begin{cases} -8\mathbf{i} - 3\mathbf{j}, & \text{if } n \text{ is even} \\ -8\mathbf{i} + 3\mathbf{j}, & \text{if } n \text{ is odd} \end{cases}$$

(b) and (d)

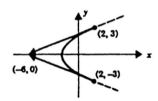

11.
$$\|\mathbf{r}(t)\| = C \quad \text{iff} \quad \|\mathbf{r}(t)\|^2 = \mathbf{r}(t) \cdot \mathbf{r}(t) = C$$

$$\text{iff} \quad \frac{d}{dt}\|\mathbf{r}(t)\| = 2\mathbf{r}(t) \cdot \mathbf{r}'(t) = 0$$

$$\text{iff} \quad \mathbf{r}(t) \perp \mathbf{r}'(t)$$

12. Problem 11 with $\mathbf{v}$ in place of $\mathbf{r}$.

13. $\kappa = \dfrac{e^{-x}}{\left(1 + e^{-2x}\right)^{3/2}}$

14. $y' = 3x^2, \quad y'' = 6x; \quad \kappa = \dfrac{6|x|}{(1 + 9x^4)^{3/2}}$

15. $y' = \dfrac{1}{2x^{1/2}}; \quad y'' = \dfrac{-1}{4x^{3/2}} \quad \kappa = \dfrac{\left|-1/4x^{3/2}\right|}{\left[1 + \left(1/2x^{1/2}\right)^2\right]^{3/2}} = \dfrac{2}{(1 + 4x)^{3/2}}$

16. $y' = 1 - 2x, \quad y'' = -2; \quad \kappa = \dfrac{2}{[1 + (1 - 2x)^2]^{3/2}} = \dfrac{\sqrt{2}}{2(1 - 2x + 2x^2)^{3/2}}$

17. $\kappa = \dfrac{\sec^2 x}{\left(1 + \tan^2 x\right)^{3/2}} = |\cos x|$

18. $y' = \sec^2 x, \quad y'' = 2\sec^2 x \tan x; \quad \kappa = \dfrac{\left|2\sec^2 x \tan x\right|}{(1 + \sec^4 x)^{3/2}}$

19. $\kappa = \dfrac{|\sin x|}{(1 + \cos^2 x)^{3/2}}$

20. $2x - 2yy' = 0 \implies y' = \dfrac{x}{y}, \quad y'' = \dfrac{y - x\left(\frac{x}{y}\right)}{y^2} = -\dfrac{a^2}{y^3}; \quad \kappa = \left| \dfrac{a^2/y^3}{[1 + (\frac{x}{y})^2]^{3/2}} \right| = \dfrac{a^2}{(x^2 + y^2)^{3/2}}$

21. $\kappa = \dfrac{|x|}{(1 + x^4/4)^{3/2}}; \quad$ at $\left(2, \dfrac{4}{3}\right), \quad \kappa = \dfrac{2}{5\sqrt{5}}$

22. $y' = x, \quad y'' = 1, \quad \kappa = \dfrac{1}{(1 + x^2)^{3/2}}; \quad$ at $(0,0), \quad \kappa = 1$

23. $\kappa = \dfrac{|-1/y^3|}{(1 + 1/y^2)^{3/2}} = \dfrac{1}{(1 + y^2)^{3/2}}; \quad$ at $(2,2), \quad \kappa = \dfrac{1}{5\sqrt{5}}$

24. $y' = 4\cos 2x, \quad y'' = -8\sin 2x; \quad$ at $x = \dfrac{\pi}{4}, \quad y' = 0, \quad y'' = -8 \implies \kappa = 8$

25. $y'(x) = \dfrac{1}{x+1}, \quad y'(2) = \dfrac{1}{3}; \quad y''(x) = \dfrac{-1}{(x+1)^2}, \quad y''(2) = -\dfrac{1}{9}.$

At $x = 2, \quad \kappa = \dfrac{\left| -\dfrac{1}{9} \right|}{\left[1 + \left(\dfrac{1}{3}\right)^2 \right]^{3/2}} = \dfrac{3}{10\sqrt{10}}$

26. At $x = \dfrac{\pi}{4}, \quad y' = \sec x \tan x = \sqrt{2}, \quad y'' = \sec x \tan^2 x + \sec^3 x = \sqrt{2} + 2^{3/2} = 3\sqrt{2}$

$\implies \kappa = \dfrac{3\sqrt{2}}{(1 + 2)^{3/2}} = \sqrt{\dfrac{2}{3}}$

27. $\kappa(x) = \dfrac{|-1/x^2|}{(1 + 1/x^2)^{3/2}} = \dfrac{x}{(x^2 + 1)^{3/2}}, \quad x > 0$

$\kappa'(x) = \dfrac{(1 - 2x^2)}{(x^2 + 1)^{5/2}}, \quad \kappa'(x) = 0 \implies x = \dfrac{1}{2}\sqrt{2}$

Since κ increases on $\left(0, \frac{1}{2}\sqrt{2}\right]$ and decreases on $\left[\frac{1}{2}\sqrt{2}, \infty\right), \quad \kappa$ is maximal at $\left(\frac{1}{2}\sqrt{2}, \frac{1}{2}\ln\frac{1}{2}\right).$

28. $y' = 3 - 3x^2, \quad y'' = -6x.$ local max at $x = 1: \quad y' = 0, \quad y'' = -6, \quad \kappa = \dfrac{6}{(1 + 0)^{3/2}} = 6$

29. $x(t) = t, \quad x'(t) = 1, \quad x''(t) = 0; \quad y(t) = \frac{1}{2}t^2, \quad y'(t) = t, \quad y''(t) = 1 \quad \kappa = \dfrac{1}{(1 + t^2)^{3/2}}$

30. $x' = e^t, \quad x'' = e^t, \quad y' = -e^{-t}, \quad y'' = e^{-t}$

$\kappa = \dfrac{|x'y'' - x''y'|}{[(x')^2 + (y')^2]^{3/2}} = \dfrac{2}{(e^{2t} + e^{-2t})^{3/2}}$

31. $x(t) = 2t,$ $x'(t) = 2,$ $x''(t) = 0;$ $\quad y(t) = t^3,$ $y'(t) = 3t^2,$ $y''(t) = 6t;$ $\quad \kappa = \dfrac{12|t|}{(4 + 9t^4)^{3/2}}$

32. $x' = 2t,$ $x'' = 2,$ $y' = 3t^2,$ $y'' = 6t;$ $\quad \kappa = \dfrac{|12t^2 - 6t^2|}{(4t^2 + 9t^4)^{3/2}} = \dfrac{6}{|t|(4 + 9t^2)^{3/2}}$

33. $x(t) = e^t \cos t,$ $\quad x'(t) = e^t(\cos t - \sin t),$ $\quad x''(t) = -2e^t \sin t$

$y(t) = e^t \sin t,$ $\quad y'(t) = e^t(\sin t + \cos t),$ $\quad y''(t) = 2e^t \cos t$

$$\kappa = \frac{\left|2e^{2t} \cos t \,(\cos t - \sin t) + 2e^{2t} \sin t \,(\cos t + \sin t)\right|}{\left[e^{2t}(\cos t - \sin t)^2 + e^{2t}(\cos t + \sin t)^2\right]^{3/2}} = \frac{2e^{2t}}{(2e^{2t})^{3/2}} = \frac{1}{2}\sqrt{2}\, e^{-t}$$

34. $x' = -2 \sin t,$ $x'' = -2 \cos t,$ $y' = 3 \cos t,$ $y'' = -3 \sin t;$ $\quad \kappa = \dfrac{6}{(4 \sin^2 t + 9 \cos^2 t)^{3/2}}$

35. $x(t) = t \cos t,$ $\quad x'(t) = \cos t - t \sin t,$ $\quad x''(t) = -2 \sin t - t \cos t$

$y(t) = t \sin t,$ $\quad y'(t) = \sin t + t \cos t,$ $\quad y''(t) = 2 \cos t - t \sin t$

$$\kappa = \frac{\left|(\cos t - t \sin t)(2 \cos t - t \sin t) - (\sin t + t \cos t)(-2 \sin t - t \cos t)\right|}{\left[(\cos t - t \sin t)^2 + (\sin t + t \cos t)^2\right]^{3/2}} = \frac{2 + t^2}{[1 + t^2]^{3/2}}$$

36. $x' = t \cos t,$ $\quad x'' = \cos t - t \sin t,$ $\quad y' = t \sin t,$ $\quad y'' = \sin t + t \cos t$

$$\kappa = \left|\frac{t \cos t(\sin t + t \cos t) - t \sin t(\cos t - t \sin t)}{(t^2 \cos^2 t + t^2 \sin^2 t)^{3/2}}\right| = \frac{t^2}{t^3} = \frac{1}{t}. \quad (t > 0)$$

37. $\kappa = \dfrac{|2/x^3|}{[1 + 1/x^4]^{3/2}} = \dfrac{2|x^3|}{(x^4 + 1)^{3/2}};$ $\quad$ at $x = \pm 1,$ $\quad \kappa = \dfrac{\sqrt{2}}{2}$

38. From Exercise 20, $\quad \kappa = \dfrac{1}{(x^2 + y^2)^{3/2}}.$ At $(\pm 1, 0),$ $\quad \kappa = 1$

39. We use (14.5.3) and the hint to obtain

$$\kappa = \frac{\left|ab \sinh^2 t - ab \cosh^2 t\right|}{\left[a^2 \sinh^2 t + b^2 \cosh^2 t\right]^{3/2}} = \frac{\left|\dfrac{a}{b} y^2 - \dfrac{b}{a} x^2\right|}{\left[\left(\dfrac{ay}{b}\right)^2 + \left(\dfrac{bx}{a}\right)^2\right]^{3/2}}$$

$$= \frac{a^3 b^3 \left|\dfrac{a}{b} y^2 - \dfrac{b}{a} x^2\right|}{\left[a^4 y^2 + b^4 x^2\right]^{3/2}} = \frac{a^4 b^4}{\left[a^4 y^2 + b^4 x^2\right]^{3/2}}.$$

40. $x' = r(1 - \cos t),$ $\quad x'' = r \sin t,$ $\quad y' = r \sin t,$ $\quad y'' = r \cos t$

Highest point when $t = \pi \implies x' = 2r,$ $x'' = 0,$ $y' = 0,$ $y'' = -r$

$\kappa = \dfrac{2r^2}{(4r^2)^{3/2}} = \dfrac{1}{4r}$

41. $\mathbf{r}'(t) = e^t(\cos t - \sin t)\,\mathbf{i} + e^t(\sin t + \cos t)\,\mathbf{j} + e^t\,\mathbf{k}$

$$\frac{ds}{dt} = \|\mathbf{r}'(t)\| = \sqrt{3}\,e^t, \qquad \frac{d^2 s}{dt^2} = \sqrt{3}\,e^t$$

$$\mathbf{T}(t) = \frac{\mathbf{r}'(t)}{\|\mathbf{r}'(t)\|} = \frac{1}{\sqrt{3}}\left[(\cos t - \sin t)\,\mathbf{i} + (\sin t + \cos t)\,\mathbf{j} + \mathbf{k}\right]$$

$$\mathbf{T}'(t) = \frac{1}{\sqrt{3}}\left[(-\sin t - \cos t)\,\mathbf{i} + (\cos t - \sin t)\,\mathbf{j}\right]; \quad \|\mathbf{T}'(t)\| = \sqrt{2/3}$$

$$\kappa = \frac{\|\mathbf{T}'(t)\|}{ds/dt} = \frac{\sqrt{2/3}}{\sqrt{3}\,e^t} = \frac{1}{3}\sqrt{2}\,e^{-t}; \qquad \mathbf{a_T} = \frac{d^2 s}{dt^2} = \sqrt{3}\,e^t, \quad \mathbf{a_N} = \kappa\left(\frac{ds}{dt}\right)^2 = \sqrt{2}\,e^t$$

42. $\mathbf{r}'(t) = \cosh t\,\mathbf{i} + \sinh t\,\mathbf{j} + \mathbf{k}; \qquad \dfrac{ds}{dt} = \|\mathbf{r}'\| = \sqrt{2}\,\cosh t; \qquad \dfrac{d^2 s}{dt^2} = \sqrt{2}\,\sinh t.$

$$\mathbf{T}(t) = \frac{\mathbf{r}'(t)}{\|\mathbf{r}'(t)\|} = \frac{1}{\sqrt{2}}\left(\mathbf{i} + \tanh t\,\mathbf{j} + \operatorname{sech} t\,\mathbf{k}\right).$$

$$\mathbf{T}'(t) = \frac{1}{\sqrt{2}}\left(\operatorname{sech}^2 t\,\mathbf{j} - \operatorname{sech} t\,\tanh t\,\mathbf{k}\right); \quad \|\mathbf{T}'(t)\| = \frac{1}{\sqrt{2}}\operatorname{sech} t$$

$$\kappa = \frac{\|\mathbf{T}'(t)\|}{ds/dt} = \frac{1}{2}\operatorname{sech}^2 t; \qquad \mathbf{a_T} = \frac{d^2 s}{dt^2} = \sqrt{2}\,\sinh t, \quad \mathbf{a_N} = \kappa\left(\frac{ds}{dt}\right)^2 = 1$$

43. $\mathbf{r}'(t) = -2\sin 2t\,\mathbf{i} + 2\cos 2t\,\mathbf{j}; \qquad \dfrac{ds}{dt} = \|\mathbf{r}'(t)\| = 2, \qquad \dfrac{d^2 s}{dt^2} = 0$

$$\mathbf{T}(t) = \frac{\mathbf{r}'(t)}{\|\mathbf{r}'(t)\|} = -\sin 2t\,\mathbf{i} + \cos 2t\,\mathbf{j}$$

$$\mathbf{T}'(t) = -2\left(\cos 2t\,\mathbf{i} + \sin 2t\,\mathbf{j}\right); \quad \|\mathbf{T}'(t)\| = 2$$

$$\kappa = \frac{\|\mathbf{T}'(t)\|}{ds/dt} = \frac{2}{2} = 1; \qquad \mathbf{a_T} = \frac{d^2 s}{dt^2} = 0, \quad \mathbf{a_N} = \kappa\left(\frac{ds}{dt}\right)^2 = 1 \cdot 4 = 4.$$

44. $\mathbf{r}'(t) = \mathbf{v}(t) = \mathbf{i} + 2t\,\mathbf{j} + \dfrac{1}{t}\,\mathbf{k}, \quad \mathbf{a}(t) = 2\,\mathbf{j} - \dfrac{1}{t^2}\,\mathbf{k}$

$$\frac{ds}{dt} = \|\mathbf{v}\| = \sqrt{1 + 4t^2 + 1/t^2} = \frac{\sqrt{4t^4 + t^2 + 1}}{t}; \quad \mathbf{v} \times \mathbf{a} = -\frac{4}{t}\,\mathbf{i} + \frac{1}{t^2}\,\mathbf{j} + 2\,\mathbf{k}$$

$$\kappa = \frac{\|\mathbf{v} \times \mathbf{a}\|}{(ds/dt)^3} = \frac{t\sqrt{4t^4 + 16t^2 + 1}}{(4t^4 + t^2 + 1)^{3/2}}$$

$$\mathbf{a_T} = \frac{d^2 s}{dt^2} = \frac{4t^4 - 1}{t^2\sqrt{4t^4 + t^2 + 1}}; \quad a_N = \kappa\left(\frac{ds}{dt}\right)^2 = \frac{1}{t}\sqrt{\frac{4t^4 + 16t^2 + 1}{4t^4 + t^2 + 1}}$$

45. $\mathbf{r}'(t) = -3\sin 3t\,\mathbf{i} + 4\,\mathbf{j} - 3\cos 3t\,\mathbf{k}; \qquad \dfrac{ds}{dt} = \|\mathbf{r}'\| = 5; \qquad \dfrac{d^2 s}{dt^2} = 0.$

$$\mathbf{T}(t) = \frac{\mathbf{r}'(t)}{\|\mathbf{r}'(t)\|} = -\frac{3}{5}\sin 3t\,\mathbf{i} + \frac{4}{5}\,\mathbf{j} - \frac{3}{5}\cos 3t\,\mathbf{k}$$

$$\mathbf{T}'(t) = -\frac{9}{5}\cos 3t\,\mathbf{i} + \frac{9}{5}\sin 3t\,\mathbf{k}; \quad \|\mathbf{T}'(t)\| = \frac{9}{5}$$

$$\kappa = \frac{\|\mathbf{T}'(t)\|}{ds/dt} = \frac{9/5}{5} = \frac{9}{25}; \qquad \mathbf{a_T} = \frac{d^2 s}{dt^2} = 0, \quad \mathbf{a_N} = \kappa\left(\frac{ds}{dt}\right)^2 = \frac{9}{25} \cdot 25 = 9.$$

46. $\mathbf{r}'(t) = t\cos t\,\mathbf{i} + t\sin t\,\mathbf{j} + t\sqrt{3}\,\mathbf{k}$, $\dfrac{ds}{dt} = \|\mathbf{r}'(t)\| = \sqrt{4t^2} = 2t$, $\dfrac{d^2s}{dt^2} = 2$

$\mathbf{T}(t) = \dfrac{\mathbf{r}'(t)}{\|\mathbf{r}'(t)\|} = \tfrac{1}{2}\cos t\,\mathbf{i} + \tfrac{1}{2}\sin t\,\mathbf{j} + \tfrac{1}{2}\sqrt{3}\,\mathbf{k}$

$\mathbf{T}'(t) = -\tfrac{1}{2}\sin t\,\mathbf{i} + \tfrac{1}{2}\cos t\,\mathbf{j}$, $\|\mathbf{T}'(t)\| = \dfrac{1}{2}$

Then, $\kappa = \dfrac{\|\mathbf{T}'(t)\|}{ds/dt} = \dfrac{1/2}{2t} = \dfrac{1}{4t}$

$\mathbf{a_T} = \dfrac{d^2s}{dt^2} = 2$, $\mathbf{a_N} = \kappa\left(\dfrac{ds}{dt}\right)^2 = \dfrac{1}{4t}(4t^2) = t.$

47. $\mathbf{r}'(t) = \sqrt{1+t}\,\mathbf{i} - \sqrt{1-t}\,\mathbf{j} + \sqrt{2}\,\mathbf{k}$, $\dfrac{ds}{dt} = \|\mathbf{r}'(t)\| = \sqrt{(1+t)+(1-t)+2} = 2$, $\dfrac{d^2s}{dt^2} = 0$

$\mathbf{T}(t) = \dfrac{\mathbf{r}'(t)}{\|\mathbf{r}'(t)\|} = \dfrac{\sqrt{1+t}}{2}\,\mathbf{i} - \dfrac{\sqrt{1-t}}{2}\,\mathbf{j} + \dfrac{\sqrt{2}}{2}\,\mathbf{k}$

$\mathbf{T}'(t) = \dfrac{1}{4\sqrt{1+t}}\,\mathbf{i} + \dfrac{1}{4\sqrt{1-t}}\,\mathbf{j}.$

$\|\mathbf{T}'(t)\| = \sqrt{\dfrac{1}{16(1+t)} + \dfrac{1}{16(1-t)}} = \dfrac{1}{4}\sqrt{\dfrac{2}{1-t^2}}$

Then, $\kappa = \dfrac{\|\mathbf{T}'(t)\|}{ds/dt} = \dfrac{1}{8}\sqrt{\dfrac{2}{1-t^2}}$

$\mathbf{a_T} = \dfrac{d^2s}{dt^2} = 0$, $\mathbf{a_N} = \kappa\left(\dfrac{ds}{dt}\right)^2 = \dfrac{1}{2}\sqrt{\dfrac{2}{1-t^2}}$.

48. (b) $\kappa(t) = \dfrac{\sqrt{1+4t^2+t^4}}{6\left(1+t^2+t^4\right)^{3/2}}$ maximum curvature occurs at $x \cong \pm 0.2715$

49.   tangential component: $\mathbf{a_T} = \dfrac{6t+12t^3}{\sqrt{1+t^2+t^4}}$;   normal component: $\mathbf{a_N} = 6\sqrt{\dfrac{1+4t^2+t^4}{1+t^2+t^4}}$

50.   Set $\mathbf{r}(\theta) = \cos\theta f(\theta)\mathbf{i} + \sin\theta f(\theta)\mathbf{j}$

51.   By Exercise 50

$$\kappa = \dfrac{\left|\left(e^{a\theta}\right)^2 + 2\left(ae^{a\theta}\right)^2 - \left(e^{a\theta}\right)\left(a^2 e^{a\theta}\right)\right|}{\left[\left(e^{a\theta}\right)^2 + \left(ae^{a\theta}\right)^2\right]^{3/2}} = \dfrac{e^{-a\theta}}{\sqrt{1+a^2}}.$$

52. $f(\theta) = a\theta$, $f'(\theta) = a$, $f''(\theta) = 0 \implies \kappa = \dfrac{a^2\theta^2 + 2a^2}{(a^2\theta^2 + a^2)^{3/2}} = \dfrac{\theta^2 + 2}{|a|(\theta^2+1)^{3/2}}$

53.   By Exercise 50,

$$\kappa = \dfrac{\left|a^2(1-\cos\theta)^2 + 2a^2\sin^2\theta - a^2(1-\cos\theta)(\cos\theta)\right|}{\left[a^2(1-\cos\theta)^2 + a^2\sin^2\theta\right]^{3/2}} = \dfrac{3a^2(1-\cos\theta)}{[2a^2(1-\cos\theta)]^{3/2}} = \dfrac{3ar}{[2ar]^{3/2}} = \dfrac{3}{2\sqrt{2ar}}.$$

54. By Exercise 50,

$$\kappa = \frac{\left|(a\sin 2\theta)^2 + 2(2a\cos 2\theta)^2 - (a\sin 2\theta)(2a\cos 2\theta)\right|}{[(a\sin 2\theta)^2 + (2a\cos 2\theta)^2]^{\frac{3}{2}}}$$

$$= \frac{\left|a^2\sin^2 2\theta + 8a^2\cos^2 2\theta - 3a^2\sin 2\theta\cos 2\theta\right|}{(a^2\sin^2 2\theta + 4a^2\cos^2 2\theta)^{\frac{3}{2}}}$$

$$= \frac{\left|r^2 + 8(a^2 - r^2) + 3r\sqrt{a^2 - r^2}\right|}{[r^2 + 4(a^2 - r^2)]^{\frac{3}{2}}} = \frac{\left|8a^2 - 7r^2 + 3r\sqrt{a^2 - r^2}\right|}{(4a^2 - 3r^2)^{\frac{3}{2}}}.$$

PROJECT 14.5A

1. The system of equations generated by the specified conditions is:

$a + b + c + d = 3$ $27a + 9b + 3c + d = 7$

$6a + 2b = 0$ $27\alpha + 9\beta + 3\gamma + \delta = 7$

$729\alpha + 81\beta + 9\gamma + \delta = -2$ $54\alpha + 2\beta = 0$

$27a + 6b + c = 27\alpha + 6\beta + \gamma$ $18a + 2b = 18\alpha + 2\beta$

$a \cong -0.1094$ $b \cong 0.3281$ $c \cong 2.1094$ $d \cong 0.6719$

$\alpha \cong 0.0365$ $\beta \cong -0.9844$ $\gamma \cong 6.0469$ $\delta \cong -3.2656$

2. Clearly p and q are continuous on their respective intervals. The conditions $p(3) = q(3)$, $p'(3) = q'(3)$ and $p''(3) = q''(3)$ imply that F, F', and F'' are continuous on $[1, 9]$.

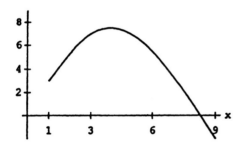

3. (a), (b) The system of equations generated by the specified conditions (and the derivative conditions of Problem 1) is:

$27\alpha + 9b + 3c + d = 10$ $64\alpha + 16b + 4c + d = 15$

$18a + 2b = 0$ $64\alpha + 64\beta + 4\gamma + \delta = 15$

$216\alpha + 36\beta + 6\gamma + \delta = 35$ $36\alpha + 2\beta = 0$

$48\alpha + 8b + c = 48\alpha + 8\beta + \gamma$ $24\alpha + 2b = 24\alpha + 2\beta$

(c) $a = b = 0$, $c = 5$, $d = -5$; $\alpha = 1.25$, $\beta = -15$, $\gamma = 65$, $\delta = -85$

4. $\quad \kappa = \dfrac{|y''|}{[1 + (y')^2]^{3/2}} = \dfrac{Cn(n-1)x^{n-2}}{[1 + (Cnx^{n-1})^2]^{3/2}}$

At $x = 0$, $\kappa = 0$ as desired. At $x = 1$, want $\kappa = \dfrac{Cn(n-1)}{(1+C^2n^2)^{3/2}} = \dfrac{96}{125}$

$y = \dfrac{1}{4}x^3$ works.

PROJECT 14.5B

1. $\dfrac{d\mathbf{T}}{dt} = \dfrac{d\mathbf{T}}{ds}\dfrac{ds}{dt} \implies \dfrac{d\mathbf{T}}{ds} = \dfrac{d\mathbf{T}/dt}{ds/dt} = \dfrac{\mathbf{T}'(t)}{\|\mathbf{T}'(t)\|}\dfrac{\|\mathbf{T}/(t)\|}{ds/dt} = \kappa\,\mathbf{N}.$

2. $\dfrac{d\mathbf{B}}{ds} = \dfrac{d}{ds}(\mathbf{T}\times\mathbf{N}) = \left(\mathbf{T}\times\dfrac{d\mathbf{N}}{ds}\right) + \left(\dfrac{d\mathbf{T}}{ds}\times\mathbf{N}\right) = \mathbf{T}\times\dfrac{d\mathbf{N}}{ds}.$ Therefore $\dfrac{d\mathbf{B}}{ds}\perp\mathbf{T}.$

   Since $\mathbf{B}$ has constant length, $\dfrac{d\mathbf{B}}{ds}\perp\mathbf{B}.$ Being perpendicular to both $\mathbf{T}$ and $\mathbf{B}$, $\dfrac{d\mathbf{B}}{ds}$ is parallel to $\mathbf{N}$ and is therefore a scalar multiple of $\mathbf{N}$.

3. $\dfrac{d\mathbf{N}}{ds} = \dfrac{d}{ds}(\mathbf{B}\times\mathbf{T}) = \left(\mathbf{B}\times\dfrac{d\mathbf{T}}{ds}\right) + \left(\dfrac{d\mathbf{B}}{ds}\times\mathbf{T}\right) = (\mathbf{B}\times\kappa\mathbf{N}) + \tau(\mathbf{N}\times\mathbf{T})$

   $\qquad = -\kappa(\mathbf{N}\times\mathbf{B}) - \tau(\mathbf{T}\times\mathbf{N}) = -\kappa\mathbf{T} - \tau\mathbf{B}$

4. We know that $d\mathbf{B}/ds = \tau\mathbf{N}$.   It follows that $\|\,d\mathbf{B}/ds\,\| = |\tau|\,\|\,\mathbf{N}\,\| = |\tau|$.   Thus $|\tau|$ is the magnitude of the change in direction of $\mathbf{B}$ per unit arc length, or equivalently, the rate per unit arc length at which the curve tends away from the osculating plane.

SECTION 14.6

1. (a) $\mathbf{r}'(t) = \dfrac{a\omega}{2}\left(e^{\omega t} - e^{-\omega t}\right)\mathbf{i} + \dfrac{b\omega}{2}\left(e^{\omega t} + e^{-\omega t}\right)\mathbf{j}$, $\mathbf{r}'(0) = b\omega\mathbf{j}$

   (b) $\mathbf{r}''(t) = \dfrac{a\omega^2}{2}\left(e^{\omega t} + e^{-\omega t}\right)\mathbf{i} + \dfrac{b\omega^2}{2}\left(e^{\omega t} - e^{-\omega t}\right)\mathbf{j} = \omega^2\mathbf{r}(t)$

   (c) The torque τ is $\mathbf{0}$: $\tau(t) = \mathbf{r}(t)\times m\mathbf{a}(t) = \mathbf{r}(t)\times m\omega^2\mathbf{r}(t) = \mathbf{0}.$

   The angular momentum $\mathbf{L}(t)$ is constant since $\mathbf{L}'(t) = \tau(t) = \mathbf{0}$.

2. (a) $\mathbf{F}(t) = m\mathbf{r}''(t) = mb^2\mathbf{r}(t)$, $mb^2 > 0$

   (b) $\mathbf{F}(t) = -m\mathbf{r}(t)$, $-m < 0$

   (c) $\mathbf{L}(t) = \mathbf{r}(t)\times m\mathbf{v}(t) = m(\mathbf{i}+\mathbf{j}-\mathbf{k})$

3. We begin with the force equation $\mathbf{F}(t) = \alpha\mathbf{k}$. In general, $\mathbf{F}(t) = m\mathbf{a}(t)$, so that here

$$\mathbf{a}(t) = \dfrac{\alpha}{m}\,\mathbf{k}.$$

Integration gives

$$\mathbf{v}(t) = C_1\mathbf{i} + C_2\mathbf{j} + \left(\dfrac{\alpha}{m}t + C_3\right)\mathbf{k}.$$

Since $\mathbf{v}(0) = 2\mathbf{j}$, we can conclude that $C_1 = 0$, $C_2 = 2$, $C_3 = 0$. Thus

$$\mathbf{v}(t) = 2\mathbf{j} + \dfrac{\alpha}{m}\,t\mathbf{k}.$$

Another integration gives

$$\mathbf{r}(t) = D_1\mathbf{i} + (2t + D_2)\mathbf{j} + \left(\frac{\alpha}{2m}t^2 + D_3\right)\mathbf{k}.$$

Since $\mathbf{r}(0) = y_0\mathbf{j} + z_0\mathbf{k}$, we have $D_1 = 0$, $D_2 = y_0$, $D_3 = z_0$, and therefore

$$\mathbf{r}(t) = (2t + y_0)\mathbf{j} + \left(\frac{\alpha}{2m}t^2 + z_0\right)\mathbf{k}.$$

The conditions of the problem require that t be restricted to nonnegative values.

To obtain an equation for the path in Cartesian coordinates, we write out the components

$$x(t) = 0, \quad y(t) = 2t + y_0, \quad z(t) = \frac{\alpha}{2m}t^2 + z_0. \qquad (t \geq 0)$$

From the second equation we have

$$t = \tfrac{1}{2}[y(t) - y_0]. \qquad (y(t) \geq y_0)$$

Substituting this into the third equation, we get

$$z(t) = \frac{\alpha}{8m}[y(t) - y_0]^2 + z_0. \qquad (y(t) \geq y_0)$$

Eliminating t altogether, we have

$$z = \frac{\alpha}{8m}(y - y_0)^2 + z_0. \qquad (y \geq y_0)$$

Since $x = 0$, the path of the object is a parabolic arc in the yz-plane.

Answers to (a) through (d):

(a) velocity: $\mathbf{v}(t) = 2\mathbf{j} + \dfrac{\alpha}{m}t\mathbf{k}.$ (b) speed: $v(t) = \dfrac{1}{m}\sqrt{4m^2 + \alpha^2 t^2}.$

(c) momentum: $\mathbf{p}(t) = 2m\mathbf{j} + \alpha t\mathbf{k}.$

(d) path in vector form: $\mathbf{r}(t) = (2t + y_0)\mathbf{j} + \left(\dfrac{\alpha}{2m}t^2 + z_0\right)\mathbf{k}, \quad t \geq 0.$

path in Cartesian coordinates: $z = \dfrac{\alpha}{8m}(y - y_0)^2 + z_0, \quad y \geq y_0, \quad x = 0.$

4. $\mathbf{F}(t) \cdot \mathbf{v}(t) = 0$ for all t

$\implies \mathbf{a}(t) \cdot \mathbf{v}(t) = \mathbf{v}'(t) \cdot \mathbf{v}(t) = \dfrac{1}{2}\dfrac{d}{dt}[\mathbf{v}(t) \cdot \mathbf{v}(t)] = 0$ for all t

$\implies \mathbf{v}(t) \cdot \mathbf{v}(t) = [v(t)]^2$ is constant $\implies v(t)$ is constant

5. $\mathbf{F}(t) = m\mathbf{a}(t) = m\mathbf{r}''(t) = 2m\mathbf{k}$

6. If $\mathbf{v}'(t) = \mathbf{0}$, then $\mathbf{L}'(t) = \mathbf{r}'(t) \times m\mathbf{v}(t) + \mathbf{r}(t) \times m\mathbf{v}'(t)$
$= \mathbf{v}(t) \times m\mathbf{v}(t) + \mathbf{r}(t) \times \mathbf{0} = \mathbf{0}$

7. From $\mathbf{F}(t) = m\mathbf{a}(t)$ we obtain

$$\mathbf{a}(t) = \pi^2[a\cos\pi t\,\mathbf{i} + b\sin\pi t\,\mathbf{j}].$$

By direct calculation using $\mathbf{v}(0) = -\pi b\mathbf{j} + \mathbf{k}$ and $\mathbf{r}(0) = b\mathbf{j}$ we obtain

$$\mathbf{v}(t) = a\pi\sin\pi t\,\mathbf{i} - b\pi\cos\pi t\,\mathbf{j} + \mathbf{k}$$
$$\mathbf{r}(t) = a(1 - \cos\pi t)\,\mathbf{i} + b(1 - \sin\pi t)\,\mathbf{j} + t\mathbf{k}.$$

(a) $\mathbf{v}\,(1) = b\pi\mathbf{j} + \mathbf{k}$

(b) $\|\mathbf{v}\,(1)\| = \sqrt{\pi^2 b^2 + 1}$

(c) $\mathbf{a}\,(1) = -\pi^2 a\mathbf{i}$

(d) $m\,\mathbf{v}(1) = m\,(\pi b\mathbf{j} + \mathbf{k})$

(e) $\mathbf{L}\,(1) = \mathbf{r}\,(1) \times m\,\mathbf{v}(1) = [2a\mathbf{i} + b\mathbf{j} + \mathbf{k}] \times [m\,(b\pi\mathbf{j} + \mathbf{k})]$
$= m\,[b(1 - \pi)\mathbf{i} - 2a\mathbf{j} + 2ab\pi\mathbf{k}]$

(f) $\tau\,(1) = \mathbf{r}\,(1) \times \mathbf{F}\,(1) = [2a\mathbf{i} + b\mathbf{j} + \mathbf{k}] \times \left[-m\pi^2 a\mathbf{i}\right] = -m\pi^2 a\,[\mathbf{j} - b\mathbf{k}]$

8.
$$\frac{d}{dt}\left(\frac{1}{2}m[v(t)]^2\right) = \frac{1}{2}m\frac{d}{dt}[\mathbf{v}(t) \cdot \mathbf{v}(t)] = \frac{1}{2}m[2\mathbf{v}'(t) \cdot \mathbf{v}(t)]$$
$$= m\mathbf{a}(t) \cdot \mathbf{v}(t) = \mathbf{F}\,(t) \cdot \mathbf{v}(t)$$

9. We have $\quad m\mathbf{v} = m\mathbf{v}_1 + m\mathbf{v}_2 \quad$ and $\quad \frac{1}{2}mv^2 = \frac{1}{2}mv_1{}^2 + \frac{1}{2}mv_2{}^2.$

Therefore $\quad \mathbf{v} = \mathbf{v}_1 + \mathbf{v}_2 \quad$ and $\quad v^2 = v_1{}^2 + v_2{}^2.$

Since $\quad v^2 = \mathbf{v} \cdot \mathbf{v} = (\mathbf{v}_1 + \mathbf{v}_2) \cdot (\mathbf{v}_1 + \mathbf{v}_2) = v_1{}^2 + v_2{}^2 + 2(\mathbf{v}_1 \cdot \mathbf{v}_2),$

we have $\quad \mathbf{v}_1 \cdot \mathbf{v}_2 = 0 \quad$ and $\quad \mathbf{v}_1 \perp \mathbf{v}_2.$

10. That the path is of the form $\mathbf{r}(t) = \cos \omega t\mathbf{A} + \sin \omega t\mathbf{B}$ can be seen by applying the hint to each component of the equation $\mathbf{F}\,(t) = -m\omega^2\mathbf{r}(t).$

$$\mathbf{A} = \mathbf{r}(0) = \text{ the initial position}$$

$$\mathbf{B} = \mathbf{r}'(0)/\omega = \text{ the initial velocity divided by } \omega$$

The path is circular if $\|\,\mathbf{A}\,\| = \|\,\mathbf{B}\,\|$ and $\mathbf{A} \perp \mathbf{B}.$

11. $\mathbf{r}''(t) = \mathbf{a}, \quad \mathbf{r}'(t) = \mathbf{v}\,(0) + t\mathbf{a}, \quad \mathbf{r}\,(t) = \mathbf{r}\,(0) + t\,\mathbf{v}(0) + \frac{1}{2}\,t^2\,\mathbf{a}.$

If neither $\mathbf{v}\,(0)$ nor $\mathbf{a}$ is zero, the displacement $\mathbf{r}\,(t) - \mathbf{r}\,(0)$ is a linear combination of $\mathbf{v}\,(0)$ and $\mathbf{a}$ and thus remains on the plane determined by these vectors. The equation of this plane can be written

$$[\mathbf{a} \times \mathbf{v}\,(0)] \cdot [\mathbf{r} - \mathbf{r}\,(0)] = 0.$$

(If either $\mathbf{v}\,(0)$ or $\mathbf{a}$ is zero, the motion is restricted to a straight line; if both of these vectors are zero, the particle remains at its initial position $\mathbf{r}\,(0).$)

12. Clearly we can take $\phi_1 \in [0, 2\pi).$ With $\omega = -1,$ the path takes the form

$$\mathbf{r}(t) = [A_1 \cos(-t + \phi_1) + D_1]\mathbf{i} - [A_1 \sin(-t + \phi_1) + D_2]\mathbf{j} + [Ct + D_3]\mathbf{k}.$$

Differentiation gives

$$\mathbf{v}(t) = A_1 \sin(-t + \phi_1)\mathbf{i} + A_1 \cos(-t + \phi_1)\mathbf{j} + C\mathbf{k}.$$

$\mathbf{r}(0) = a\,\mathbf{i} \implies A_1 \cos \phi_1 + D_1 = a, \quad A \sin \phi_1 + D_2 = 0, \quad D_3 = 0$

$\mathbf{v}(0) = a\,\mathbf{j} + b\,\mathbf{k} \implies A_1 \sin \phi_1 = 0, \quad A_1 \cos \phi_1 = a, \quad C = b$

From these equations we have

$$D_1 = D_2 = D_3 = 0, \quad c = b, \quad A_1 \sin \phi_1 = 0, \quad A_1 \cos \phi_1 = a.$$

The last two equations give

$$\phi_1 = 0 \quad \text{and } A_1 = a \quad \text{or} \quad \phi_1 = \pi \text{ and } A_1 = -a.$$

The first possibility gives

$$\mathbf{r}(t) = a \cos(-t)\,\mathbf{i} - a \sin(-t)\,\mathbf{j} + bt\,\mathbf{k}$$

$$= a \cos t\,\mathbf{i} + a \sin t\,\mathbf{j} + bt\,\mathbf{k}.$$

The second possibility gives the same path:

$$\mathbf{r}(t) = -a \cos(-t + \pi)\,\mathbf{i} - a \sin(-t + \pi)\,\mathbf{j} + bt\,\mathbf{k}$$

$$= a \cos t\,\mathbf{i} + a \sin t\,\mathbf{j} + bt\,\mathbf{k}$$

13. $\mathbf{r}(t) = \mathbf{i} + t\mathbf{j} + \left(\dfrac{qE_0}{2m}\right)t^2\mathbf{k}$

14. Since $\mathbf{v}$ has magnitude $w\,\|\,\mathbf{r}\,\|$, lies in the plane of the wheel, and makes an angle of $90°$ counterclockwise with $\mathbf{r}$, we have $\mathbf{v} = \boldsymbol{\omega} \times \mathbf{r}$

15. $\mathbf{r}(t) = \left(1 + \dfrac{t^3}{6m}\right)\mathbf{i} + \dfrac{t^4}{12m}\mathbf{j} + t\mathbf{k}$ 16. $\mathbf{r}(t) = \sin\left(\omega t + \dfrac{1}{2}\pi\right)\mathbf{k}$

17.
$$\frac{d}{dt}\left(\frac{1}{2}mv^2\right) = mv\frac{dv}{dt} = m\left(\mathbf{v} \cdot \frac{d\mathbf{v}}{dt}\right) = m\frac{d\mathbf{v}}{dt} \cdot \mathbf{v} = \mathbf{F} \cdot \frac{d\mathbf{r}}{dt}$$

$$= 4r^2\left(\mathbf{r} \cdot \frac{d\mathbf{r}}{dt}\right) = 4r^2\left(r\frac{dr}{dt}\right) = 4r^3\frac{dr}{dt} = \frac{d}{dt}\left(r^4\right).$$

Therefore $d/dt\left(\frac{1}{2}mv^2 - r^4\right) = 0$ and $\frac{1}{2}mv^2 - r^4$ is a constant E.   Evaluating E from $t = 0$,   we find that $E = 2m$.

Thus $\frac{1}{2}mv^2 - r^4 = 2m$ and $v = \sqrt{4 + (2/m)\,r^4}$.

SECTION 14.7

1. On Earth: year of length T, average distance from sun d.

On Venus: year of length αT, average distance from sun $0.72d$.

Therefore
$$\frac{(\alpha T)^2}{T^2} = \frac{(0.72d)^3}{d^3}.$$

This gives $\alpha^2 = (0.72)^3 \cong 0.372$ and $\alpha \cong 0.615$.   Answer: about 61.5% of an Earth year.

2. $\dfrac{dE}{dt} = \dfrac{1}{2}m\dfrac{d}{dt}(v^2) - m\rho\dfrac{d}{dt}(r^{-1})$

$\dfrac{d}{dt}(v^2) = \dfrac{d}{dt}(\mathbf{v}\cdot\mathbf{v}) = 2(\mathbf{a}\cdot\mathbf{v})$

$\dfrac{d}{dt}(r^{-1}) = -\dfrac{1}{r^2}\dfrac{dr}{dt} = -\dfrac{1}{r^3}\left(r\dfrac{dr}{dt}\right) = -\dfrac{1}{r^3}(\mathbf{r}\cdot\mathbf{v})$ (using 13.2.3)

$\dfrac{dE}{dt} = m(\mathbf{a}\cdot\mathbf{v}) + \dfrac{m\rho}{r^3}(\mathbf{r}\cdot\mathbf{v}) = (\mathbf{a}\cdot m\mathbf{v}) + \left(\dfrac{\rho\mathbf{r}}{r^3}\cdot m\mathbf{v}\right) = \left(\mathbf{a} + \dfrac{\rho\mathbf{r}}{r^3}\right)\cdot m\mathbf{v} = 0$

3.
$\left(\dfrac{dx}{dt}\right)^2 + \left(\dfrac{dy}{dt}\right)^2 = \left[\dfrac{d}{dt}(r\cos\theta)\right]^2 + \left[\dfrac{d}{dt}(r\sin\theta)\right]^2$

$= \left[r(-\sin\theta)\dfrac{d\theta}{dt} + \dfrac{dr}{dt}\cos\theta\right]^2 + \left[r\cos\theta\dfrac{d\theta}{dt} + \dfrac{dr}{dt}\sin\theta\right]^2$

$= r^2\sin^2\theta\left(\dfrac{d\theta}{dt}\right)^2 + \left(\dfrac{dr}{dt}\right)^2\cos^2\theta + r^2\cos^2\theta\left(\dfrac{d\theta}{dt}\right)^2 + \left(\dfrac{dr}{dt}\right)^2\sin^2\theta$

$= \left(\dfrac{dr}{dt}\right)^2 + r^2\left(\dfrac{d\theta}{dt}\right)^2$

4. Using $\dot{r} = \dfrac{dr}{d\theta}\dot{\theta}$ and $\dot{\theta} = \dfrac{L}{mr^2}$

we have

$$E + \dfrac{m\rho}{r} = \dfrac{1}{2}m(\dot{r}^2 + r^2\dot{\theta}^2) = \dfrac{1}{2}\left[\left(\dfrac{dr}{d\theta}\right)^2\dot{\theta}^2 + r^2\dot{\theta}^2\right]$$

$$= \dfrac{1}{2}m\left[\left(\dfrac{dr}{d\theta}\right)^2\dfrac{L^2}{m^2r^4} + \dfrac{L^2}{m^2r^2}\right] = \dfrac{L^2}{2m}\left[\dfrac{1}{r^4}\left(\dfrac{dr}{d\theta}\right)^2 + \dfrac{1}{r^2}\right]$$

and therefore

$$E = \dfrac{L^2}{2m}\left[\dfrac{1}{r^2} + \dfrac{1}{r^4}\left(\dfrac{dr}{d\theta}\right)^2\right] - \dfrac{m\rho}{r}$$

5. Substitute

$$r = \dfrac{a}{1 + e\cos\theta}, \quad \left(\dfrac{dr}{d\theta}\right)^2 = \left[\dfrac{-a}{(1 + e\cos\theta)^2}\cdot(-e\sin\theta)\right]^2 = \dfrac{(ae\sin\theta)^2}{(1 + e\cos\theta)^4}$$

into the right side of the equation and you will see that, with a and e^2 as given, the expression reduces to E.

REVIEW EXERCISES

1. $\mathbf{f}'(t) = 6t\,\mathbf{i} - 15t^2\,\mathbf{j}, \qquad \mathbf{f}''(t) = 6\,\mathbf{i} - 30t\,\mathbf{j}$

2. $\mathbf{f}'(t) = 2e^{2t}\,\mathbf{i} + \dfrac{2t}{1 + t^2}\,\mathbf{j}, \qquad \mathbf{f}''(t) = 4e^{2t}\,\mathbf{i} + \dfrac{2 - 2t^2}{(1 + t^2)^2}\,\mathbf{j}$

3. $\mathbf{f}'(t) = (e^t\cos t - e^t\sin t)\,\mathbf{i} + 2\sin 2t\,\mathbf{j}, \qquad \mathbf{f}''(t) = -2e^t\sin t\,\mathbf{i} + 4\cos 2t\,\mathbf{j}$

4. $\mathbf{f}'(t) = \cosh t\,\mathbf{i} - (2t - t^2)e^{-t}\,\mathbf{j} + \sinh t\mathbf{k}, \qquad \mathbf{f}''(t) = \sinh t\,\mathbf{i} + (t^2 - 4t + 2)e^{-t}\,\mathbf{j} + \cosh t\,\mathbf{k}$

5. $\displaystyle\int_0^2 \left[2t\,\mathbf{i} + (t^2 - 1)\,\mathbf{j}\right]\,dt = \left[t^2\,\mathbf{i} + \left(\frac{1}{3}t^3 - t\right)\mathbf{j}\right]_0^2 = 4\,\mathbf{i} + \frac{2}{3}\,\mathbf{j}$

6. $\displaystyle\int_0^\pi \left[\sin 2t\,\mathbf{i} + 2\cos t\,\mathbf{j} + \sqrt{t}\,\mathbf{k}\right] dt = \left[-\frac{1}{2}\cos 2t\,\mathbf{i} + 2\sin t\,\mathbf{j} + \frac{2}{3}t^{3/2}\,\mathbf{k}\right]_0^\pi = \frac{2}{3}\pi^{3/2}\,\mathbf{k}$

7.

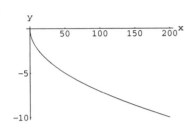

8.

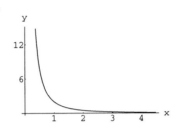

9.

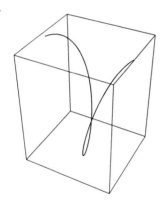

10.

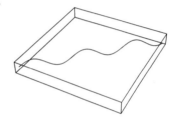

11. (a) $\mathbf{r}(t) = 2\cos\left(t + \dfrac{\pi}{2}\right)\mathbf{i} + 4\sin\left(t + \dfrac{\pi}{2}\right)\mathbf{j}$ (b) $\mathbf{r}(t) = -2\cos 2t\,\mathbf{i} + 4\sin 2t\,\mathbf{j}$

12. direction vector: $\mathbf{d} = (2, 4, 6);$ $\mathbf{r}(t) = (1 + 2t)\,\mathbf{i} + (1 + 4t)\,\mathbf{j} + (-2 + 6t)\,\mathbf{k}, \quad 0 \le t \le 1$

13. $\mathbf{f}(t) = \frac{1}{3}t^3\,\mathbf{i} + (\frac{1}{2}e^{2t} + t)\,\mathbf{j} + \frac{1}{3}(2t + 1)^{3/2}\mathbf{k} + \mathbf{C}.$

  $\mathbf{f}(0) = \mathbf{i} - 3\,\mathbf{j} + 3\mathbf{k} \Longrightarrow \mathbf{C} = \mathbf{i} - \frac{7}{2}\mathbf{j} + \frac{8}{3}\mathbf{k};$ $\mathbf{f}(t) = \left(\frac{1}{3}t^3 + 1\right)\mathbf{i} + \left(\frac{1}{2}e^{2t} + t - \frac{7}{2}\right)\mathbf{j} + \left(\frac{1}{3}(2t + 1)^{3/2} + \frac{8}{3}\right)\mathbf{k}$

14. $\mathbf{f}'(t) = -\mathbf{f}(t) \Longrightarrow \mathbf{f}(t) = \mathbf{f}_0 e^{-t}$

  $\mathbf{f}(0) = \mathbf{i} + 2\mathbf{k} \Longrightarrow \mathbf{f}_0 = \mathbf{i} + 2\mathbf{k}$ and so $\mathbf{f}(t) = e^{-t}\,\mathbf{i} + 2e^{-t}\mathbf{k}$

15. $\mathbf{f}'(t) = (6\,\mathbf{i} + 12t^3\,\mathbf{j}) + (8t\,\mathbf{i} - 12\,\mathbf{k}) = (6 + 8t)\,\mathbf{i} + 12t^3\,\mathbf{j} - 12\mathbf{k}$

16. Note: f is not a vector function. $f(t) = e^t + 1 \Longrightarrow f'(t) = e^t$

17. $\mathbf{f}(t) = (t^2 + 2t^3)\,\mathbf{i} - \left(2t^2 + \dfrac{1}{t^2}\right)\mathbf{j} + (t^4 - t)\,\mathbf{k}, \quad \mathbf{f}'(t) = (2t + 6t^2)\,\mathbf{i} - \left(4t - \dfrac{2}{t^3}\right)\mathbf{j} + (4t^3 - 1)\,\mathbf{k}$

18. Note: f is not a vector function. $f(t) = t^3 \cos t + t^2 \sin t + 3t \cos t = (t^3 + 3t)\cos t + t^2 \sin t$
$f'(t) = -(t^3 + 3t)\sin t + (3t^2 + 3)\cos t + 2t \sin t + t^2 \cos t = (4t^2 + 3)\cos t - (t^3 + t)\sin t$

19. $\mathbf{r}'(t) = 2\mathbf{r}(t) \Longrightarrow \mathbf{r}(t) = \mathbf{r}_0 e^{2t}$
$\mathbf{r}(0) = (1, 2, 1) \Longrightarrow \mathbf{r}_0 = (1, 2, 1) \quad \text{and} \quad \mathbf{r}(t) = (e^{2t}, 2e^{2t}, e^{2t})$

20. $\mathbf{F}(t) = e^{2t}\,\mathbf{i} + e^{-2t}\,\mathbf{j}, \quad \mathbf{F}'(t) = 2e^{2t}\,\mathbf{i} - 2e^{-2t}\,\mathbf{j}, \quad \mathbf{F}''(t) = 4e^{2t}\,\mathbf{i} + 4e^{-2t}\,\mathbf{j}$
Since $\mathbf{F}''(t) = 4\mathbf{F}(t)$ for all t, $\mathbf{F}$ and $\mathbf{F}''$ are parallel.
$\mathbf{F}$ and $\mathbf{F}'$ will have the same direction for some value of t iff there is a number $k > 0$ such that
$e^{2t}\,\mathbf{i} + e^{-2t}\,\mathbf{j} = k(2e^{2t}\,\mathbf{i} - 2e^{-2t}\,\mathbf{j})$. No such value of k exists.

21. The tip of $\mathbf{r}(t)$ is $P(1, 1, 1)$ when $t = 0$.
$\mathbf{r}'(t) = (2t + 2)\,\mathbf{i} + 3\,\mathbf{j} + (3t^2 + 1)\,\mathbf{k}, \quad \mathbf{r}'(0) = 2\,\mathbf{i} + 3\,\mathbf{j} + \mathbf{k}$
Scalar parametric equations for the tangent line are: $x = 1 + 2t, \quad y = 1 + 3t, \quad z = 1 + t$.

22. $\mathbf{r}(\pi/3) = (\sqrt{3}/2, -1/2, \pi/3)$
$\mathbf{r}'(t) = (2\cos 2t, -2\sin 2t, 1); \quad \mathbf{r}'(\pi/3) = (-1, -\sqrt{3}, 1)$
Scalar parametric equations for the tangent line are: $x = \frac{\sqrt{3}}{2} - t, \quad y = -\frac{1}{2} - \sqrt{3}\,t, \quad z = \frac{\pi}{3} + t$

23. $\mathbf{r}_1(t) = (2, 1, 1)$ at $t = 1$; $\mathbf{r}_2(u) = (2, 1, 1)$ at $u = -1$. Therefore the curves intersect at the point
$(2, 1, 1)$.
$\mathbf{r}_1'(t) = (2\mathbf{i} + 2t\,\mathbf{j} + \mathbf{k}, \quad \mathbf{r}_1'(1) = 2\,\mathbf{i} + 2\,\mathbf{j} + \mathbf{k}; \quad \mathbf{r}_2'(u) = -\mathbf{i} - 2u\,\mathbf{j} + 2u\,\mathbf{k}, \quad \mathbf{r}_2'(-1) = -\mathbf{i} + 2\,\mathbf{j} - 2\,\mathbf{k}$.
Since $\mathbf{r}_1'(1) \cdot \mathbf{r}_2'(-1) = 0$, the angle of intersection is $\pi/2$ radians

24. $\mathbf{r}(t) = 2t\,\mathbf{i} + (1 - t^2)\,\mathbf{j} - t^2\,\mathbf{k}, \quad \mathbf{r}'(t) = 2t\,\mathbf{i} - 2t\,\mathbf{j} - 2t\,\mathbf{k}$.
$(2t\,\mathbf{i} + (1 - t^2)\,\mathbf{j} - t^2\,\mathbf{k}) \cdot (2t\,\mathbf{i} - 2t\,w\mathbf{j} - 2t\,\mathbf{k}) = 4t^3; \quad 4t^3 = 0 \Longrightarrow t = 0$
The curve and the tangent line meet at right angles at the point where $t = 0$; $(0, 1, 0)$.

25. $\mathbf{r}(t) = t\,\mathbf{i} + e^{2t}\,\mathbf{j}, \quad \mathbf{r}(0) = \mathbf{j}$;
$\mathbf{r}'(t) = \mathbf{i} + 2e^{2t}\,\mathbf{j}, \quad \mathbf{r}'(0) = \mathbf{i} + 2\,\mathbf{j}$

26. $\mathbf{r}(t) = 2\sin t\,\mathbf{i} - 3\cos t\,\mathbf{j}, \quad \mathbf{r}(\pi/6) = \mathbf{i} - \frac{3}{2}\sqrt{3}\,\mathbf{j}$
$\mathbf{r}'(t) = 2\cos t\,\mathbf{i} + 3\sin t\,\mathbf{j} \quad \mathbf{r}'(\pi/6) = \sqrt{3}\,\mathbf{i} + \frac{3}{2}\,\mathbf{j}$

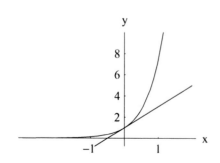

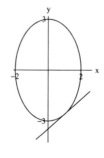

27. $\mathbf{r}'(t) = t\cos t\,\mathbf{i} + t\sin t\,\mathbf{j} + \sqrt{3}t\mathbf{k};\quad ||\mathbf{r}'|| = \dfrac{ds}{dt} = 2t$

unit tangent vector: $\mathbf{T} = \dfrac{\mathbf{r}'(t)}{||\mathbf{r}'(t)||} = \dfrac{1}{2}(\cos t\,\mathbf{i} + \sin t\,\mathbf{j} + \sqrt{3}\,\mathbf{k}).$

$\mathbf{T}'(t) = -\dfrac{1}{2}\sin t\,\mathbf{i} + \dfrac{1}{2}\cos t\,\mathbf{j};\quad ||\mathbf{T}'(t)|| = \dfrac{1}{2}.$

principal normal vector: $\mathbf{N} = \dfrac{\mathbf{T}'(t)}{||\mathbf{T}'(t)||} = -\sin t\,\mathbf{i} + \cos t\,\mathbf{j}$

28. $\mathbf{r}'(t) = (-a\sin t\,\mathbf{i} + a\cos t\,\mathbf{j} + b\,\mathbf{k}.$

The cosine of the angle θ between $\mathbf{r}'(t)$ and $\mathbf{k}$ is: $\dfrac{\mathbf{r}'\cdot\mathbf{k}}{||\mathbf{r}'||\,||\mathbf{k}||} = \dfrac{|b|}{\sqrt{a^2+b^2}} = \text{constant}.$

Therefore θ is a constant.

29. $\mathbf{r}'(t) = 2\,\mathbf{i} + \dfrac{1}{t}\,\mathbf{j} - 2t\mathbf{k};\quad ||\mathbf{r}'(t)|| = \dfrac{1+2t^2}{t}.$

$\mathbf{T}(t) = \dfrac{\mathbf{r}'(t)}{||\mathbf{r}'(t)||} = \dfrac{2t}{2t^2+1}\,\mathbf{i} + \dfrac{1}{2t^2+1}\,\mathbf{j} - \dfrac{2t^2}{2t^2+1}\mathbf{k};\quad \mathbf{T}(1) = \dfrac{2}{3}\,\mathbf{i} + \dfrac{1}{3}\,\mathbf{j} - \dfrac{2}{3}\mathbf{k}$

$\mathbf{T}'(t) = \dfrac{2-4t^2}{(2t^2+1)^2}\,\mathbf{i} - \dfrac{4t}{(2t^2+1)^2}\,\mathbf{j} - \dfrac{4t}{(2t^2+1)^2}\mathbf{k};\quad \mathbf{T}'(1) = -\dfrac{2}{9}\,\mathbf{i} - \dfrac{4}{9}\,\mathbf{j} - \dfrac{4}{9}\,\mathbf{k},\quad ||\mathbf{T}'(1)|| = \dfrac{2}{3}$

$\mathbf{N}(1) = \dfrac{\mathbf{T}'(1)}{||\mathbf{T}'(1)||} = -\dfrac{1}{3}\,\mathbf{i} - \dfrac{2}{3}\,\mathbf{j} - \dfrac{2}{3}\mathbf{k}$

A normal vector for the osculating plane is: $(2\,\mathbf{i} + \mathbf{j} - 2\,\mathbf{k})\times(\mathbf{i} + 2\,\mathbf{j} + 2\,\mathbf{k}) = 6\,\mathbf{i} - 6\,\mathbf{j} + 3\,\mathbf{k}.$

Since $\mathbf{r}(1) = 2\,\mathbf{i} - \mathbf{k},$ an equation for the osculating plane is

$$6(x-2) - 6y + 3(z+1) = 0 \ \text{ or } \ 2x - 2y + z = 3.$$

30. $\mathbf{r}'(t) = -\sin t\,\mathbf{i} - \sin t\,\mathbf{j} - \sqrt{2}\cos t\,\mathbf{k};\quad ||\mathbf{r}'(t)|| = \sqrt{2}.$

$\mathbf{T}(t) = \dfrac{\mathbf{r}'(t)}{||\mathbf{r}'(t)||} = \dfrac{1}{\sqrt{2}}(-\sin t\,\mathbf{i} - \sin t\,\mathbf{j} - \sqrt{2}\cos t\,\mathbf{k});\quad \mathbf{T}(\pi/4) = -\dfrac{1}{2}\,\mathbf{i} - \dfrac{1}{2}\,\mathbf{j} - \dfrac{1}{\sqrt{2}}\mathbf{k}$

$\mathbf{T}'(t) = \dfrac{1}{\sqrt{2}}(-\cos t\,\mathbf{i} - \cos t\,\mathbf{j} + \sqrt{2}\sin t\,\mathbf{k});\quad \mathbf{T}'(\pi/4) = -\dfrac{1}{2}\,\mathbf{i} - \dfrac{1}{2}\,\mathbf{j} + \dfrac{1}{\sqrt{2}}\mathbf{k},\quad ||\mathbf{T}'(\pi/4)|| = 1$

$\mathbf{N}(\pi/4) = \dfrac{\mathbf{T}'(\pi/4)}{||\mathbf{T}'(\pi/4)||} = -\dfrac{1}{2}\,\mathbf{i} - \dfrac{1}{2}\,\mathbf{j} + \dfrac{1}{\sqrt{2}}\mathbf{k}$

A normal vector for the osculating plane is: $(\mathbf{i} + \mathbf{j} + \sqrt{2}\,\mathbf{k})\times(\mathbf{i} + \mathbf{j} - \sqrt{2}\,\mathbf{k}) = -2\sqrt{2}\,\mathbf{i} + 2\sqrt{2}\,\mathbf{j}.$

Since $\mathbf{r}(\pi/4) = \dfrac{\sqrt{2}}{2}(\mathbf{i} + \mathbf{j} - 2\,\mathbf{k},$ an equation for the osculating plane is

$$-2\sqrt{2}\left(x - \dfrac{\sqrt{2}}{2}\right) + 2\sqrt{2}\left(y - \dfrac{\sqrt{2}}{2}\right) = 0 \ \text{ or } \ x - y = 0.$$

31. $\mathbf{r}'(t) = 2\,\mathbf{i} + t^{1/2}\,\mathbf{j};\quad L = \displaystyle\int_0^5 ||\mathbf{r}'(t)||\,dt = \int_0^5 \sqrt{4+t}\,dt = \dfrac{38}{3}$

32. $\mathbf{r}'(t) = e^t\,\mathbf{i} - e^{-t}\,\mathbf{j} - \sqrt{2}\mathbf{k};\quad ||\mathbf{r}'(t)|| = e^t + e^{-t};\quad L = \displaystyle\int_0^{\ln 3}(e^t + e^{-t})\,dt = \left[e^t - e^{-t}\right]_0^{\ln 3} = \dfrac{8}{3}$

33. $\mathbf{r}'(t) = \cosh t\,\mathbf{i} + \sinh t\,\mathbf{j} + \mathbf{k}$; $\|\mathbf{r}'(t)\| = \sqrt{\cosh^2 t + \sinh^2 t + 1} = \sqrt{2}\,\cosh t$;

$$L = \int_0^1 \sqrt{2}\,\cosh t\,dt = \left[\sqrt{2}\,\sinh t\right]_0^1 = \sqrt{2}\,\sinh 1.$$

34. $\mathbf{r}'(t) = -\sin t\,\mathbf{i} - \cos t\,\mathbf{j} + \sinh t\,\mathbf{k}$; $\|\mathbf{r}'\| = \sqrt{1 + \sinh^2 t} = \cosh t$;

$$\int_0^{\ln 2} \cosh t\,dt = \left[\sinh t\right]_0^{\ln 2} = \frac{3}{4}$$

35. (a) $\mathbf{r}'(t) = -\sin t\,\mathbf{i} + \cos t\,\mathbf{j} + t^{1/2}\,\mathbf{k}$; $\|\mathbf{r}'(t)\| = \sqrt{1+t}$.

$$s = \int_0^t \|\mathbf{r}'(u)\|\,du = \int_0^t \sqrt{1+t}\,dt = \left[\frac{2}{3}(1+u)^{3/2}\right]_0^t = \frac{2}{3}(1+t)^{3/2} - \frac{2}{3}$$

(b) $t = \left(\frac{3}{2}s + 1\right)^{2/3} - 1 = \phi(s)$; $\mathbf{R}(s) = \cos\,\phi(s)\,\mathbf{i} + \sin\,\phi(s)\,\mathbf{j} + \frac{2}{3}[\phi(s)]^{3/2}\,\mathbf{k}$

(c) $\mathbf{R}'(s) = \left[-\sin\,\phi(s)\,\mathbf{i} + \cos\,\phi(s)\,\mathbf{j} + \phi(s)^{1/2}\,\mathbf{k}\right]\phi'(s)$

$$\|\mathbf{R}'(s)\| = \phi'(s)\sqrt{1 + \phi(s)} = \frac{2}{3}\left[\frac{3}{2}s + 1\right]^{-1/3}\left(\frac{3}{2}\right)\sqrt{\left(\frac{3}{2}s + 1\right)^{2/3}} = 1$$

36. velocity: $\mathbf{v} = \mathbf{r}'(t) = -\sin t\,\mathbf{i} + \cos t\,\mathbf{j} + 2\sin 2t\,\mathbf{k}$ speed: $s = \|\mathbf{v}\| = \sqrt{1 + 4\sin^2 2t}$
acceleration: $\mathbf{a} = \mathbf{r}''(t) = -\cos t\,\mathbf{i} - \sin t\,\mathbf{j} + 4\cos 2t\,\mathbf{k}$; $\|\mathbf{a}\| = \sqrt{1 + 16\cos^2 2t}$

37. $\mathbf{r}''(t) = -\cos t\,\mathbf{i} - \sin t\,\mathbf{j}$ and $\mathbf{r}'(0) = \mathbf{k} \Longrightarrow \mathbf{r}'(t) = -\sin t\,\mathbf{i} + (\cos t - 1)\,\mathbf{j} + \mathbf{k}$.
Thus: velocity $\mathbf{v} = -\sin t\,\mathbf{i} + (\cos t - 1)\,\mathbf{j} + \mathbf{k}$ and speed $\|\mathbf{v}\| = \sqrt{3 - 2\cos t}$.
$\mathbf{r}'(t) = -\sin t\,\mathbf{i} + (\cos t - 1)\,\mathbf{j} + \mathbf{k}$ and $\mathbf{r}(0) = \mathbf{i} \Longrightarrow \mathbf{r}(t) = \cos t\,\mathbf{i} + (\sin t - t)\,\mathbf{j} + t\,\mathbf{k}$.

38. The acceleration vector remains perpendicular to the path means:

$$\mathbf{r}'' \cdot \mathbf{r}' = 0.$$

$\mathbf{r}'(t) = f'(t)\,\mathbf{i} + 2f(t)f'(t)\,\mathbf{j}$, $\mathbf{r}''(t) = f''(t)\,\mathbf{i} + [2f'(t)^2 + 2f(t)f''(t)]\,\mathbf{j}$
$\mathbf{r}'' \cdot \mathbf{r}' = 0 \Longrightarrow f'(t)f''(t) + 4f(t)[f'(t)]^3 + 4f^2(t)f'(t)f''(t) = 0$

39. $y' = \frac{3}{2}x^{1/2}$, $y'' = \frac{3}{4}x^{-1/2}$;

$$\kappa = \frac{|y''|}{[1 + (y')^2]^{3/2}} = \frac{\frac{3}{4}x^{-1/2}}{[1 + \frac{9}{4}x]^{3/2}} = \frac{6}{\sqrt{x}(4 + 9x)^{3/2}}$$

40. $y' = -2\sin 2x$, $y'' = -4\cos 2x$; $\kappa = \dfrac{4|\cos 2x|}{[1 + 4\sin^2 2x]^{3/2}}$

41. $x(t) = 2e^{-t}$, $y(t) = e^{-2t} \Longrightarrow x'(t) = -2e^{-t}$, $y'(t) = -2e^{-2t} \Longrightarrow x''(t) = 2e^{-t}$, $y''(t) = 4e^{-2t}$

$$\kappa = \frac{|(-2e^{-t})(4e^{-2t}) - (-2e^{-2t})(2e^{-t})|}{[4e^{-2t} + 4e^{-4t}]^{3/2}} = \frac{1}{2(1 + e^{-2t})^{3/2}} = \frac{e^{3t}}{2(e^{2t} + 1)^{3/2}}$$

42. $x(t) = \frac{1}{3}t^3, \quad y(t) = \frac{1}{2}t^2 \implies x'(t) = t^2, \quad y'(t) = t \implies x''(t) = 2t, \quad y''(t) = 1$

$$\kappa = \frac{|t^2 - 2t^2|}{(t^2 + t^4)^{\frac{3}{2}}} = \frac{1}{|t|(1 + t^2)^{\frac{3}{2}}}$$

43. $\mathbf{r}'(t) = -3\sin 3t\,\mathbf{i} - 4\mathbf{j} + 3\cos 3t\,\mathbf{k}, \quad \dfrac{ds}{dt} = |\mathbf{r}'(t)| = 5$

$\mathbf{T}(t) = -\dfrac{3}{5}\sin 3t\,\mathbf{i} - \dfrac{4}{5}\mathbf{j} + \dfrac{3}{5}\cos 3t\,\mathbf{k}, \quad \mathbf{T}'(t) = -\dfrac{9}{5}\cos 3t\,\mathbf{i} - \dfrac{9}{5}\sin 3t\,\mathbf{k}; \quad \|\mathbf{T}'(t)\| = 9/5$

$\kappa = \dfrac{\|\mathbf{T}'(t)\|}{ds/dt} = \dfrac{9}{25}$

44. $\mathbf{r}'(t) = 1\,\mathbf{i} + \sqrt{2}t^{1/2}\,\mathbf{j} + t\,\mathbf{k}, \quad \dfrac{ds}{dt} = \|\mathbf{r}'(t)\| = 1 + t$

$\mathbf{T} = \dfrac{1}{1+t}\,\mathbf{i} + \dfrac{\sqrt{2t}}{1+t}\,\mathbf{j} + \dfrac{t}{1+t}\,\mathbf{k} \quad \mathbf{T}'(t) = \dfrac{-1}{(1+t^2)}\,\mathbf{i} + \dfrac{\frac{\sqrt{2}}{2}(1/\sqrt{t} - \sqrt{t})}{(1+t)^2}\,\mathbf{j} + \dfrac{1}{(1+t)^2}\,\mathbf{k}$

$\|\mathbf{T}'(t)\| = \dfrac{1}{\sqrt{2t}(1+t)}; \qquad \kappa = \dfrac{\|\mathbf{T}'(t)\|}{ds/dt} = \dfrac{1}{\sqrt{2t}(1+t)^2}$

45. $y' = \sinh(x/a), \quad y'' = \dfrac{1}{a}\cosh(x/a); \qquad \kappa = \dfrac{|y''|}{[1 + (y')^2]^{\frac{3}{2}}} = \dfrac{1}{a\cosh^2(x/a)} = \dfrac{a}{y^2}$

46. (a) Suppose $\|\mathbf{v}\| = c$ constant. Then $\|\mathbf{v}\|^2 = c^2 \implies \mathbf{r}' \cdot \mathbf{r}' = c^2 \implies 2\mathbf{r}'\mathbf{r}'' = 0 \implies \mathbf{a} \perp \mathbf{v}$.

(b) $\mathbf{T} = \dfrac{\mathbf{v}}{\|\mathbf{v}\|}, \quad \mathbf{T}' = \dfrac{\mathbf{v}'}{\|\mathbf{v}\|} = \dfrac{\mathbf{a}}{\|\mathbf{v}\|}$ since $\|\mathbf{v}\|$ is constant. $\kappa = \dfrac{|\mathbf{T}'|}{|\mathbf{v}|} = \dfrac{|\mathbf{a}|}{|\mathbf{v}|^2}$

47. $\mathbf{r}'(t) = -\dfrac{4}{5}\sin t\,\mathbf{i} + \dfrac{3}{5}\sin t\,\mathbf{j} + \cos t\,\mathbf{k}; \quad \dfrac{ds}{dt} = \|\mathbf{r}'(t)\| = 1$

$\mathbf{T} = \mathbf{r}'(t) = -\dfrac{4}{5}\sin t\,\mathbf{i} + \dfrac{3}{5}\sin t\,\mathbf{j} + \cos t\,\mathbf{k}, \quad \mathbf{T}'(t) = -\dfrac{4}{5}\cos t\,\mathbf{i} + \dfrac{3}{5}\cos t\,\mathbf{j} - \sin t\,\mathbf{k}; \quad \|\mathbf{T}'(t)\| = 1$

$\kappa = 1; \qquad a_{\mathbf{T}} = \dfrac{d^2 s}{dt^2} = 0, \quad a_{\mathbf{N}} = \kappa\left(\dfrac{ds}{dt}\right)^2 = 1$

48. $\mathbf{r}'(t) = 2\,\mathbf{i} + 2t\,\mathbf{j} + t^2\,\mathbf{k}; \quad \dfrac{ds}{dt} = \|\mathbf{r}'(t)\| = 2 + t^2$

$\mathbf{T}(t) = \dfrac{1}{2+t^2}(2\,\mathbf{i} + 2t\,\mathbf{j} + t^2\,\mathbf{k}), \quad \mathbf{T}'(t) = \dfrac{1}{(2+t^2)^2}(-4t\,\mathbf{i} + (4 - 2t^2)\,\mathbf{j} + 4t\,\mathbf{k}); \quad \|\mathbf{T}'(t)\| = \dfrac{2}{2+t^2}$

$\kappa = \dfrac{2}{(2+t^2)^2}; \qquad a_{\mathbf{T}} = \dfrac{d^2 s}{dt^2} = 2t, \quad a_{\mathbf{N}} = \kappa\left(\dfrac{ds}{dt}\right)^2 = 2.$

CHAPTER 15

SECTION 15.1

1. dom (f) = the first and third quadrants, including the axes;   range $(f) = [0, \infty)$

2. dom (f) = the set of all points (x, y) with $xy \leq 1$; the two branches of the hyperbola $xy = 1$ and all points in between;   range $(f) = [0, \infty)$

3. dom (f) = the set of all points (x, y) except those on the line $y = -x$;   range $(f) = (-\infty, 0) \cup (0, \infty)$

4. dom (f) = the set of all points (x, y) other than the origin;   range $(f) = (0, \infty)$

5. dom (f) = the entire plane;   range $(f) = (-1, 1)$ since

$$\frac{e^x - e^y}{e^x + e^y} = \frac{e^x + e^y - 2e^y}{e^x + e^y} = 1 - \frac{2}{e^{x-y} + 1}$$

   and the last quotient takes on all values between 0 and 2.

6. dom (f) = the set of all points (x, y) other than the origin;   range $(f) = [0, 1]$

7. dom (f) = the first and third quadrants, excluding the axes;   range $(f) = (-\infty, \infty)$

8. dom (f) =the set of all points (x, y) between the branches of the hyperbola $xy = 1$;   range $(f) = (-\infty, \infty)$

9. dom (f) = the set of all points (x, y) with $x^2 < y$ —in other words, the set of all points of the plane above the parabola $y = x^2$;   range $(f) = (0, \infty)$

10. dom (f) = the set of all points (x, y) with $-3 \leq x \leq 3$, $-1 \leq y \leq 1$ (a rectangle); range $(f) = [0, 3]$

11. dom (f) = the set of all points (x, y) with $-3 \leq x \leq 3$, $-2 \leq y \leq 2$ (a rectangle); range $(f) = [-2, 3]$

12. dom (f) = all of space;   range $(f) = [-3, 3]$

13. dom (f) = the set of all points (x, y, z) not on the plane $x + y + z = 0$;   range $(f) = \{-1, 1\}$

14. dom (f) = the set of all points (x, y, z) with $x^2 \neq y^2$ —that is, all points of space except for those which lie on the plane $x - y = 0$ or on the plane $x + y = 0$;   range $(f) = (-\infty, \infty)$

15. dom (f) = the set of all points (x, y, z) with $|y| < |x|$;   range $(f) = (-\infty, 0]$

16. dom (f) =the set of all points (x, y, z) not on the plane $x - y = 0$;   range $(f) = (-\infty, \infty)$

17. dom (f) = the set of all points (x, y) with $x^2 + y^2 < 9$ —in other words, the set of all points of the plane inside the circle $x^2 + y^2 = 9$;   range $(f) = [2/3, \infty)$

18. dom (f) = all of space;   range $(f) = [0, \infty)$

19. dom (f) = the set of all points (x, y, z) with $x + 2y + 3z > 0$ — in other words, the set of all points in space that lie on the same side of the plane $x + 2y + 3z = 0$ as the point $(1, 1, 1)$;   range $(f) = (-\infty, \infty)$

20. dom (f) = the set of all points (x, y, z) with $x^2 + y^2 + z^2 \leq 4$ — in other words, the set of all points inside and on the sphere $x^2 + y^2 + z^2 = 4$;   range $(f) = [1, e^2]$

21. dom (f) = all of space;   range $(f) = (0, 1]$

22. dom (f) = the set of all points (x, y, z) with $-1 \leq x \leq 1$, $-2 \leq y \leq 2$, $-3 \leq z \leq 3$ (a rectangular solid);   range $(f) = [0, 3]$

23. dom $(f) = \{x : x \geq 0\}$; range $(f) = [0, \infty)$

dom $(g) = \{(x, y) : x \geq 0, \ y \text{ real}\}$; range $(g) = [0, \infty)$

dom $(h) = \{(x, y, z) : x \geq 0, \ y, z \text{ real}\}$; range $(h) = [0, \infty)$

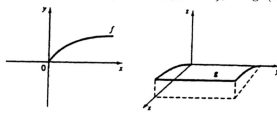

24. dom (f) = the entire plane,   range $(f) = [-1, 1]$

dom (g) = all of space,   range $(g) = [-1, 1]$

25. $\displaystyle\lim_{h \to 0} \frac{f(x + h, y) - f(x, y)}{h} = \lim_{h \to 0} \frac{2(x + h)^2 - y - (2x^2 - y)}{h} = \lim_{h \to 0} \frac{4xh + 2h^2}{h} = 4x$

$\displaystyle\lim_{h \to 0} \frac{f(x, y + h) - f(x, y)}{h} = \lim_{h \to 0} \frac{2x^2 - (y + h) - (2x^2 - y)}{h} = -1$

26. $\displaystyle\lim_{h \to 0} \frac{f(x + h, y) - f(x, y)}{h} = \lim_{h \to 0} \frac{xy + hy + 2y - (xy + 2y)}{h} = \lim_{h \to 0} y = y.$

$\displaystyle\lim_{h \to 0} \frac{f(x, y + h) - f(x, y)}{h} = \lim_{h \to 0} \frac{xy + xh + 2y + 2h - (xy + 2y)}{h} = \lim_{h \to 0} (x + 2) = x + 2$

27. $\displaystyle\lim_{h \to 0} \frac{f(x + h, y) - f(x, y)}{h} = \lim_{h \to 0} \frac{3(x + h) - (x + h)y + 2y^2 - (3x - xy + 2y^2)}{h} = \lim_{h \to 0} \frac{3h - hy}{h} = 3 - y$

$\displaystyle\lim_{h \to 0} \frac{f(x, y + h) - f(x, y)}{h} = \lim_{h \to 0} \frac{3x - x(y + h) + 2(y + h)^2 - (3x - xy + 2y^2)}{h}$

$= \displaystyle\lim_{h \to 0} \frac{-xh + 4yh + 2h^2}{h} = -x + 4y$

28. $\displaystyle\lim_{h \to 0} \frac{f(x + h, y) - f(x, y)}{h} = \lim_{h \to 0} \frac{x \sin y + h \sin y - x \sin y}{h} = \lim_{h \to 0} \sin y = \sin y.$

$\displaystyle\lim_{h \to 0} \frac{f(x, y + h) - f(x, y)}{h} = \lim_{h \to 0} \frac{x \sin(y + h) - x \sin y}{h} = x \lim_{h \to 0} \frac{\sin(y + h) - \sin y}{h} = x \cos y$

29. $\displaystyle\lim_{h \to 0} \frac{f(x + h, y) - f(x, y)}{h} = \lim_{h \to 0} \frac{\cos[(x + h)y] - \cos[xy]}{h}$

$= \displaystyle\lim_{h \to 0} \frac{\cos[xy] \cos[hy] - \sin[xy] \sin[hy] - \cos[xy]}{h}$

$= \cos[xy] \left(\displaystyle\lim_{h \to 0} \frac{\cos[hy] - 1}{h} \right) - \sin[xy] \lim_{h \to 0} \frac{\sin hy}{h}$

$= y \cos[xy] \left(\displaystyle\lim_{h \to 0} \frac{\cos[hy] - 1}{hy} \right) - y \sin[xy] \lim_{h \to 0} \frac{\sin hy}{hy}$

$= -y \sin[xy]$

and

$$\lim_{h\to 0}\frac{f(x,y+h)-f(x,y)}{h} = \lim_{h\to 0}\frac{\cos[x(y+h)]-\cos[xy]}{h}$$

$$= \lim_{h\to 0}\frac{\cos[xy]\cos[hx]-\sin[xy]\sin[hx]-\cos[xy]}{h}$$

$$= \cos[xy]\left(\lim_{h\to 0}\frac{\cos[hx]-1}{h}\right) - \sin[xy]\lim_{h\to 0}\frac{\sin hx}{h}$$

$$= x\cos[xy]\left(\lim_{h\to 0}\frac{\cos[hx]-1}{hx}\right) - x\sin[xy]\lim_{h\to 0}\frac{\sin hx}{hx}$$

$$= -x\sin[xy]$$

30. $\displaystyle \lim_{h\to 0}\frac{f(x+h,y)-f(x,y)}{h} = \lim_{h\to 0}\frac{(x^2+2xh+h^2)e^y - x^2 e^y}{h} = \lim_{h\to 0}(2x+h)e^y = 2xe^y.$

$\displaystyle \lim_{h\to 0}\frac{f(x,y+h)-f(x,y)}{h} = \lim_{h\to 0}\frac{x^2 e^{y+h}-x^2 e^y}{h} = x^2\lim_{h\to 0}\frac{e^{y+h}-e^y}{h} = x^2 e^y.$

31. (a) $f(x,y)=Ay$ (b) $f(x,y)=\pi x^2 y$ (b) $f(x,y)=|2\mathbf{i}\times(x\mathbf{i}+y\mathbf{j})|=2|y|$

32. (a) $f(x,y,z)=xy+2xz+2yz$

(b) $f(x,y,z)=\cos^{-1}\dfrac{(\mathbf{i}+\mathbf{j})\cdot(x\mathbf{i}+y\mathbf{j}+z\mathbf{k})}{\|\mathbf{i}+\mathbf{j}\|\|x\mathbf{i}+y\mathbf{j}+z\mathbf{k}\|} = \cos^{-1}\dfrac{x+y}{\sqrt{2}\sqrt{x^2+y^2+z^2}}$

(c) $f(x,y,z)=[\mathbf{i}\times(\mathbf{i}+\mathbf{j})]\cdot(x\mathbf{i}+y\mathbf{j}+z\mathbf{k})=z$

33. Surface area: $S=2lw+2lh+2hw=20 \Longrightarrow w=\dfrac{20-2lh}{2l+2h}=\dfrac{10-lh}{l+h}$

Volume: $V=lwh=\dfrac{lh(10-lh)}{l+h}$

34. $wlh=12 \quad\Longrightarrow\quad h=\dfrac{12}{wl}; \quad C=4wl+2(2wh+2lh)=4wl+\dfrac{48}{l}+\dfrac{48}{w}$

35. $V=\pi r^2 h+\dfrac{4}{3}\pi r^3$

36. $A=\dfrac{1}{2}[2(12-2x)+2x\cos\theta]\cdot x\sin\theta=(12-2x+x\cos\theta)\cdot x\sin\theta$

SECTION 15.2

1. an elliptic cone   2. an ellipsoid

3. a parabolic cylinder   4. a hyperbolic paraboloid

5. a hyperboloid of one sheet   6. an elliptic cylinder

7. a sphere   8. a hyperboloid of two sheets

9. an elliptic paraboloid   10. a hyperbolic cylinder

11. a hyperbolic paraboloid   12. an elliptic paraboloid

13.

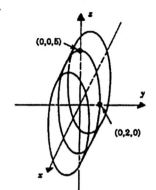

14.

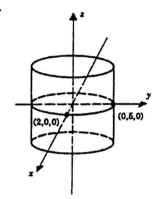

15.

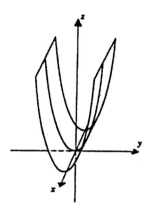

16.

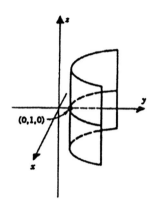

17.

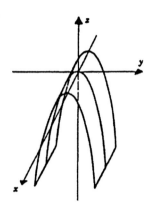

18.

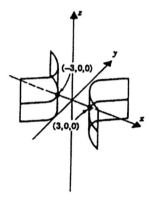

19.

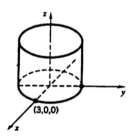

20.

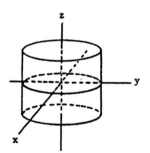

21.

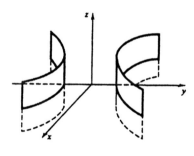

22.

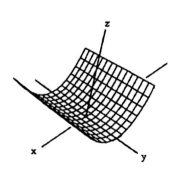

23.

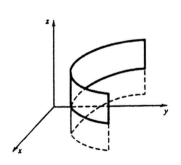

24.

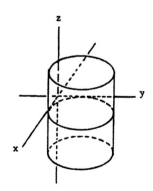

25. elliptic paraboloid
xy-trace: the origin
xz-trace: the parabola $x^2 = 4z$
yz-trace: the parabola $y^2 = 9z$
surface has the form of Figure 15.2.5

26. ellipsoid
xy-trace: the ellipse $9x^2 + 4y^2 = 36$
xz-trace: the ellipse $x^2 + 4z^2 = 4$
yz-trace: the ellipse $y^2 + 9z^2 = 9$
surface has the form of Figure 15.2.1

27. elliptic cone
xy-trace: the origin
xz-trace: the lines $x = \pm 2z$
yz-trace: the lines $y = \pm 3z$
surface has the form of Figure 15.2.4

28. hyperboloid of one sheet
xy-trace: the ellipse $9x^2 + 4y^2 = 36$
xz-trace: the hyperbola $x^2 - 4z^2 = 4$
yz-trace: the hyperbola $y^2 - 9z^2 = 9$
surface has the form of Figure 15.2.2

29. hyperboloid of two sheets
xy-trace: none
xz-trace: the hyperbola $4z^2 - x^2 = 4$
yz-trace: the hyperbola $9z^2 - y^2 = 9$
surface has the form of Figure 15.2.3

30. hyperbolic paraboloid
xy-trace: the lines $y = \pm \frac{3}{2}x$
xz-trace: the parabola $x^2 = 4z$
yz-trace: the parabola $y^2 = -9z$
surface has the form Figure 15.2.6

31. hyperboloid of two sheets
xy-trace: the hyperbola $9x^2 - 4y^2 = 36$
xz-trace: the hyperbola $x^2 - 4z^2 = 4$
yz-trace: none
see Figure 15.2.3

32. hyperboloid of one sheet
xy-trace: the hyperbola $x^2 - 9y^2 = 9$
xz-trace: the circle $x^2 + y^2 = 9$
yz-trace: the hyperbola $z^2 - 9y^2 = 9$
surface has the form of Figure 15.2.2,
rotated $90°$ about the x-axis.

33. elliptic paraboloid
xy-trace: the parabola $x^2 = 9y$
xz-trace: the origin
yz-trace: the parabola $z^2 = 4y$
surface has the form of Figure 15.2.5

34. elliptic cone
xy-trace: the lines $x = \pm 2y$
xz-trace: the origin
yz-trace: the lines $z = \pm 3y$
surface has the form of Figure 15.2.4,
rotated $90°$ about the x-axis.

35. hyperboloid of two sheets
xy-trace: the hyperbola $9y^2 - 4x^2 = 36$
xz-trace: none
yz-trace: the hyperbola $y^2 - 4z^2 = 4$
see Figure 15.2.3

36. elliptic paraboloid
xy-trace: the parabola $y^2 = 4x$
xz-trace: the parabola $z^2 = 9x$
yz-trace: the origin
surface has the form of Figure 15.2.5,
but opening along the positive x-axis.

37. paraboloid of revolution
xy-trace: the origin
xz-trace: the parabola $x^2 = 4z$
yz-trace: the parabola $y^2 = 4z$
surface has the form of Figure 15.2.5

38. ellipsoid
xy-trace: the ellipse $4x^2 + y^2 = 4$
xz-trace: the ellipse $9x^2 + z^2 = 9$
yz-trace: the ellipse $9y^2 + 4z^2 = 36$
the surface has the form of Figure 15.2.1,
rotated $90°$ about the x-axis.

39. (a) an elliptic paraboloid (vertex down if A and B are both positive, vertex up if A and B are both negative)

(b) a hyperbolic paraboloid

(c) the xy-plane if A and B are both zero; otherwise a parabolic cylinder

40. The xz-plane and all planes parallel to the xy-plane.

41. $x^2 + y^2 - 4z = 0$ (paraboloid of revolution)

42. $c^2 x^2 + c^2 y^2 - b^2 z^2 = b^2 c^2$ (hyperboloid of revolution, one sheet)

43. (a) a circle

(b) (i) $\sqrt{x^2 + y^2} = -3z$ (ii) $\sqrt{x^2 + z^2} = \frac{1}{3}y$

44. (a) the ellipse $b^2 x^2 + y^2 = b^2$

(b) ellipse approaches parallel lines $x = \pm 1$ in the plane $z = 1$

(c) paraboloid approaches parabolic cylinder $z = x^2$

45. $x + 2y + 3\left(\dfrac{x + y - 6}{2}\right) = 6$ or $5x + 7y = 30,$ a line

46. $3x + y - 2(4 - x + 2y) = 1,$ or $5x - 3y = 9,$ a line

47. $\left.\begin{array}{c} x^2 + y^2 + (z - 1)^2 = \frac{3}{2} \\ x^2 + y^2 - z^2 = 1 \end{array}\right\}$ $(z^2 + 1) + (z - 1)^2 = \dfrac{3}{2};$ $(2z - 1)^2 = 0,$ $z = \dfrac{1}{2}$ so that $x^2 + y^2 = \dfrac{5}{4}$

48. $z^2 + (z - 2)^2 = 2 \implies 2(z - 1)^2 = 0 \implies z = 1 \implies x^2 + y^2 = 1,$ a circle.

49. $x^2 + y^2 + \left(x^2 + 3y^2\right) = 4$ or $x^2 + 2y^2 = 2,$ an ellipse

50. $y^2 + (x^2 + 3y^2) = 4 \implies x^2 + 4y^2 = 4,$ an ellipse.

51. $x^2 + y^2 = (2 - y)^2$ or $x^2 = -4(y - 1),$ a parabola

52. $x^2 + y^2 = \left(\dfrac{2 - y}{2}\right)^2 \implies 4x^2 + 3y^2 + 4y = 4,$ an ellipse.

53. (a) Set $x = a \cos u \, \cos v,\ y = b \cos u \, \sin v,\ z = c \sin u.$ Then: $\dfrac{x^2}{a^2} + \dfrac{y^2}{b^2} + \dfrac{z^2}{c^2} = 1.$

(b)

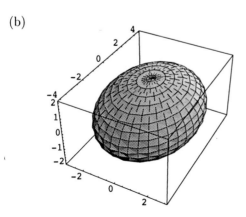

54. (a) Set $x = a \sec u \cos v$,
        $y = b \sec u \sin v$,
        $z = c \tan u$

(b)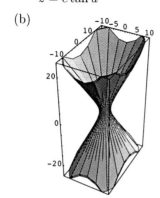

55. (a) Set $x = av \cos u$,
        $y = bv \sin u$,
        $z = cv$

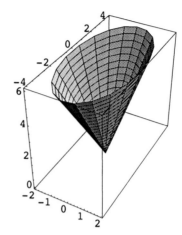

SECTION 15.3

1. lines of slope 1: $y = x - c$

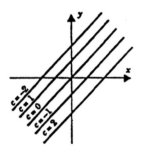

2. lines of slope 2 : $y = 2x - c$

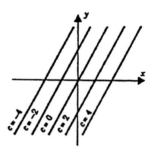

3. parabolas: $y = x^2 - c$

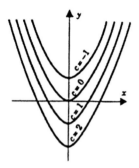

4. parabolas: $x - y^2 = \dfrac{1}{c}$

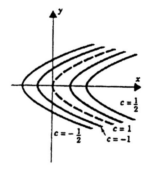

5. the y-axis and the lines $y = \left(\dfrac{1-c}{c}\right)x$ with the origin omitted throughout

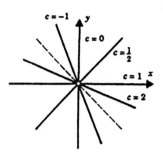

6. the x-axis and the parabolas $y = cx^2$ with the origin omitted throughout

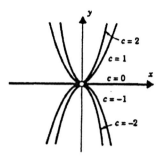

7. the cubics $y = x^3 - c$

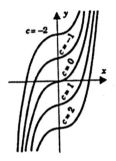

8. the coordinate axes and the hyperbolas $xy = \ln c$

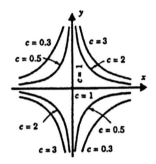

9. the lines $y = \pm x$ and the hyperbolas $x^2 - y^2 = c$

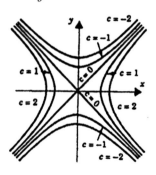

10. pairs of vertical lines $x = \pm\sqrt{c}$ and the y-axis

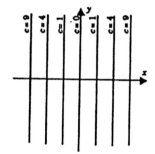

11. pairs of horizontal lines $y = \pm\sqrt{c}$ and the x-axis

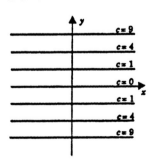

12. the lines $x = 0$, $y = 1$ and the hyperbolas $y = \dfrac{c}{x} + 1$

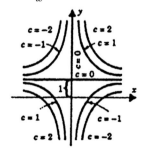

13. the circles $x^2 + y^2 = e^c$, c real

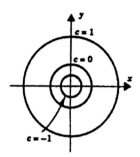

14. the parabolas $y = e^c x^2$ with the origin omitted throughout

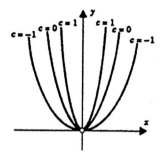

15. the curves $y = e^{cx^2}$ with the point $(0, 1)$ omitted

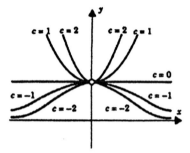

16. the coordinate axes and pairs of hyperbolas $xy = \pm\sqrt{c}$

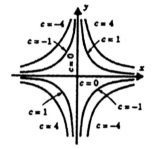

17. the coordinate axes and pairs of lines $y = \pm\dfrac{\sqrt{1-c}}{\sqrt{c}}\, x$, with the origin omitted throughout

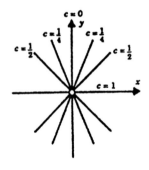

18. the curves $y = e^{cx}$ with the point $(0, 1)$ omitted

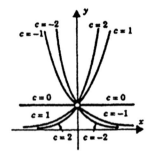

19. $x + 2y + 3z = 0$, plane through the origin

20. circular cylinder $x^2 + y^2 = 4$ (Figure 15.2.8)

21. $z = \sqrt{x^2 + y^2}$, the upper nappe of the circular cone $z^2 = x^2 + y^2$ (Figure 15.2.4)

22. ellipsoid $\dfrac{x^2}{4} + \dfrac{y^2}{6} + \dfrac{z^2}{9} = 1$ (Figure 15.2.1)

23. the elliptic paraboloid $\dfrac{x^2}{18} + \dfrac{y^2}{8} = z$ (Figure 15.2.5)

24. hyperboloid of two sheets $\dfrac{x^2}{(1/6)^2} + \dfrac{y^2}{(1/3)^2} - z^2 = -1$ (Figure 15.2.3)

25. (i) hyperboloid of two sheets   (Figure 15.2.3)
(ii) circular cone   (Figure 15.2.4)
(iii) hyperboloid of one sheet   (Figure 15.2.2)

26. (i) hyperboloid of two sheets
(ii) elliptic cone
(iii) hyperboloid of one sheet

27. The level curves of f are: $1 - 4x^2 - y^2 = c$. Substituting $P(0, 1)$ into this equation, we have
$$1 - 4(0)^2 - (1)^2 = c \quad \Longrightarrow \quad c = 0$$
The level curve that contains P is: $1 - 4x^2 - y^2 = 0$,   or $4x^2 + y^2 = 1$.

28. $(x^2 + y^2)e^{xy} = 1$

29. The level curves of f are: $y^2 \arctan x = c$. Substituting $P(1, 2)$ into this equation, we have
$$4 \arctan 1 = c \quad \Longrightarrow \quad c = \pi$$
The level curve that contains P is: $y^2 \tan^{-1} x = \pi$.

30. $(x^2 + y)\ln(2 - x + e^y) = 5$

31. The level surfaces of f are: $x^2 + 2y^2 - 2xyz = c$. Substituting $P(-1, 2, 1)$ into this equation, we have
$$(-1)^2 + 2(2)^2 - 2(-1)(2)(1) = c \quad \Longrightarrow \quad c = 13$$
The level surface that contains P is: $x^2 + 2y^2 - 2xyz = 13$.

32. $\sqrt{x^2 + y^2} - \ln z = 4$

33. (a)

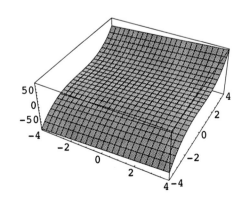
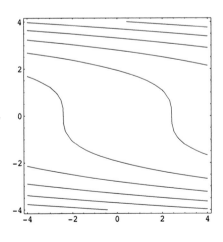

(b)

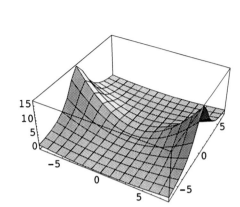
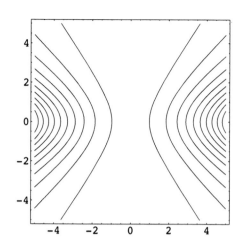

34. (a) the level surfaces are planes   (b)

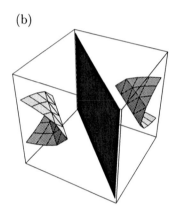

35. (a) $\dfrac{3x + 2y + 1}{4x^2 + 9} = \dfrac{3}{5}$ (b) $x^2 + 2y^2 - z^2 = 21$

36. (a)

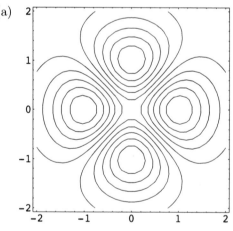

(b)

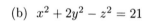

37. $\dfrac{GmM}{x^2 + y^2 + z^2} = c \implies x^2 + y^2 + z^2 = \dfrac{GmM}{c}$; the surfaces of constant gravitational force are concentric spheres.

38. Circular cylinders around the positive y-axis: $x^2 + z^2 = \dfrac{k^2}{c^2}$

39. (a) $T(x, y, z) = \dfrac{k}{\sqrt{x^2 + y^2 + z^2}}$,   where k is a constant.

(b) $\dfrac{k}{\sqrt{x^2 + y^2 + z^2}} = c \implies x^2 + y^2 + z^2 = \dfrac{k^2}{c^2}$;   the level surfaces are concentric spheres.

(c) $T(1, 2, 1) = \dfrac{k}{\sqrt{1^2 + 2^2 + 1^2}} = 50 \implies k = 50\sqrt{6} \implies T(x, y, z) = \dfrac{50\sqrt{6}}{\sqrt{x^2 + y^2 + z^2}}$

   Now, $T(4, 0, 3) = \dfrac{50\sqrt{6}}{\sqrt{3^2 + 4^2}} = 10\sqrt{6}$ degrees

40. $x^2 + y^2 = r^2 - \dfrac{k^2}{c^2}$;   circles about the origin, for $|c| > \dfrac{k}{r}$.

41. $f(x, y) = y^2 - y^3$; F   **42.** D.   **43.** $f(x, y) = \cos\sqrt{x^2 + y^2}$; A

44. B.   **45.** $f(x, y) = xye^{-(x^2 + y^2)/2}$; E   **46.** C.

PROJECT 15.3

1.

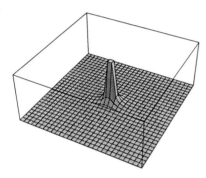

2.

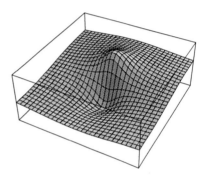

3.

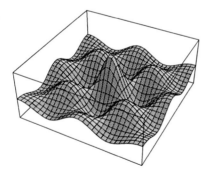

4.

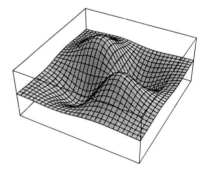

SECTION 15.4

1. $\dfrac{\partial f}{\partial x} = 6x - y$, $\dfrac{\partial f}{\partial y} = 1 - x$ **2.** $\dfrac{\partial g}{\partial x} = 2xe^{-y}$, $\dfrac{\partial g}{\partial y} = -x^2 e^{-y}$

3. $\dfrac{\partial \rho}{\partial \phi} = \cos\phi \cos\theta$, $\dfrac{\partial \rho}{\partial \theta} = -\sin\phi \sin\theta$

4. $\dfrac{\partial \rho}{\partial \theta} = 2\sin(\theta - \phi)\cos(\theta - \phi), \quad \dfrac{\partial \rho}{\partial \phi} = -2\sin(\theta - \phi)\cos(\theta - \phi)$

5. $\dfrac{\partial f}{\partial x} = e^{x-y} + e^{y-x}, \quad \dfrac{\partial f}{\partial y} = -e^{x-y} - e^{y-x}$
6. $\dfrac{\partial z}{\partial x} = \dfrac{x}{\sqrt{x^2 - 3y}}, \quad \dfrac{\partial z}{\partial y} = \dfrac{-3}{2\sqrt{x^2 - 3y}}$

7. $\dfrac{\partial g}{\partial x} = \dfrac{(AD - BC)y}{(Cx + Dy)^2}, \quad \dfrac{\partial g}{\partial y} = \dfrac{(BC - AD)x}{(Cx + Dy)^2}$
8. $\dfrac{\partial u}{\partial x} = \dfrac{-e^z}{x^2 y^2}, \quad \dfrac{\partial u}{\partial y} = \dfrac{-2e^z}{xy^3}, \quad \dfrac{\partial u}{\partial z} = \dfrac{e^z}{xy^2}$

9. $\dfrac{\partial u}{\partial x} = y + z, \quad \dfrac{\partial u}{\partial y} = x + z, \quad \dfrac{\partial u}{\partial z} = x + y$
10. $\dfrac{\partial z}{\partial x} = 2Ax + By, \quad \dfrac{\partial z}{\partial y} = Bx + 2Cy$

11. $\dfrac{\partial f}{\partial x} = z\cos(x - y), \quad \dfrac{\partial f}{\partial y} = -z\cos(x - y), \quad \dfrac{\partial f}{\partial z} = \sin(x - y)$

12. $\dfrac{\partial g}{\partial u} = \dfrac{2u}{u^2 + vw - w^2}, \quad \dfrac{\partial g}{\partial v} = \dfrac{w}{u^2 + vw - w^2}, \quad \dfrac{\partial g}{\partial w} = \dfrac{v - 2w}{u^2 + vw - w^2}$

13. $\dfrac{\partial \rho}{\partial \theta} = e^{\theta+\phi}\left[\cos(\theta - \phi) - \sin(\theta - \phi)\right], \quad \dfrac{\partial \rho}{\partial \phi} = e^{\theta+\phi}\left[\cos(\theta - \phi) + \sin(\theta - \phi)\right]$

14. $\dfrac{\partial f}{\partial x} = (x + y)\cos(x - y) + \sin(x - y), \quad \dfrac{\partial f}{\partial y} = -(x + y)\cos(x - y) + \sin(x - y)$

15. $\dfrac{\partial f}{\partial x} = 2xy\sec xy + x^2 y(\sec xy)(\tan xy)y = 2xy\sec xy + x^2 y^2 \sec xy \, \tan xy$

$\dfrac{\partial f}{\partial y} = x^2 \sec xy + x^2 y(\sec xy)(\tan xy)x = x^2 \sec xy + x^3 y \sec xy \, \tan xy$

16. $\dfrac{\partial g}{\partial x} = \dfrac{2}{1 + (2x + y)^2}, \quad \dfrac{\partial g}{\partial y} = \dfrac{1}{1 + (2x + y)^2}$

17. $\dfrac{\partial h}{\partial x} = \dfrac{x^2 + y^2 - x(2x)}{(x^2 + y^2)^2} = \dfrac{y^2 - x^2}{(x^2 + y^2)^2} \qquad \dfrac{\partial h}{\partial y} = \dfrac{-2xy}{(x^2 + y^2)^2}$

18. $\dfrac{\partial z}{\partial x} = \dfrac{x}{x^2 + y^2}, \quad \dfrac{\partial z}{\partial y} = \dfrac{y}{x^2 + y^2}$

19. $\dfrac{\partial f}{\partial x} = \dfrac{(y\cos x)\sin y - (x\sin y)(-y\sin x)}{(y\cos x)^2} = \dfrac{\sin y(\cos x + x\sin x)}{y\cos^2 x}$

$\dfrac{\partial f}{\partial y} = \dfrac{(y\cos x)(x\cos y) - (x\sin y)\cos x}{(y\cos x)^2} = \dfrac{x(y\cos y - \sin y)}{y^2 \cos x}$

20. $\dfrac{\partial f}{\partial x} = e^{xy}(y\sin xz + z\cos xz), \quad \dfrac{\partial f}{\partial y} = xe^{xy}\sin xz, \quad \dfrac{\partial f}{\partial z} = xe^{xy}\cos xz$

21. $\dfrac{\partial h}{\partial x} = 2f(x)f'(x)g(y), \quad \dfrac{\partial h}{\partial y} = [f(x)]^2 g'(y)$

22. $\dfrac{\partial h}{\partial x} = f'(x)g(y)e^{f(x)g(x)}, \quad \dfrac{\partial h}{\partial y} = f(x)g'(y)e^{f(x)g(y)}$

23. $\dfrac{\partial f}{\partial x} = (y^2 \ln z)z^{xy^2}, \quad \dfrac{\partial f}{\partial y} = (2xy \ln z)z^{xy^2}, \quad \dfrac{\partial f}{\partial z} = xy^2 z^{xy^2-1}$

24. $\dfrac{\partial h}{\partial x} = 2[f(x,y)]^3 g(x,z)\dfrac{\partial g}{\partial x} + 3[f(x,y)]^2[g(x,z)]^2 \dfrac{\partial f}{\partial x}$

$\dfrac{\partial h}{\partial y} = 3[f(x,y)]^2[g(x,z)]^2 \dfrac{\partial f}{\partial y}, \quad \dfrac{\partial h}{\partial z} = 2[f(x,y)]^3 g(x,y)\dfrac{\partial g}{\partial z}$

25. $\dfrac{\partial h}{\partial r} = 2re^{2t}\cos(\theta - t) \qquad\qquad \dfrac{\partial h}{\partial \theta} = -r^2 e^{2t}\sin(\theta - t)$

$\dfrac{\partial h}{\partial t} = 2r^2 e^{2t}\cos(\theta - t) + r^2 e^{2t}\sin(\theta - t) = r^2 e^{2t}[2\cos(\theta - t) + \sin(\theta - t)]$

26. $\dfrac{\partial u}{\partial x} = \dfrac{1}{x} - yze^{xz}, \quad \dfrac{\partial u}{\partial y} = -\dfrac{1}{y} - e^{xz}, \quad \dfrac{\partial u}{\partial z} = -xye^{xz}$

27. $\dfrac{\partial f}{\partial x} = z\dfrac{1}{1+(y/x)^2}\left(\dfrac{-y}{x^2}\right) = -\dfrac{yz}{x^2+y^2} \qquad \dfrac{\partial f}{\partial y} = z\dfrac{1}{1+(y/x)^2}\left(\dfrac{1}{x}\right) = \dfrac{xz}{x^2+y^2}$

$\dfrac{\partial f}{\partial x} = \arctan(y/x)$

28. $\dfrac{\partial w}{\partial x} = y\sin z - yz\cos x, \quad \dfrac{\partial w}{\partial y} = x\sin z - z\sin x, \quad \dfrac{\partial w}{\partial z} = xy\cos z - y\sin x$

29. $f_x(x,y) = e^x \ln y, \quad f_x(0,e) = 1; \quad f_y(x,y) = \dfrac{1}{y}e^x, \quad f_y(0,e) = e^{-1}$

30. $g_x = e^{-x}[-\sin(x+2y) + \cos(x+2y)], \quad g_x(0, \tfrac{1}{4}\pi) = -1$

$g_y = 2e^{-x}\cos(x+2y), \quad g_y(0, \tfrac{1}{4}\pi) = 0$

31. $f_x(x,y) = \dfrac{y}{(x+y)^2}, \quad f_x(1,2) = \dfrac{2}{9}; \quad f_y(x,y) = \dfrac{-x}{(x+y)^2}, \quad f_y(1,2) = -\dfrac{1}{9}$

32. $g_x(x,y) = \dfrac{y^2}{(x+y^2)^2}, \quad g_x(1,2) = \dfrac{4}{25} \qquad g_y(x,y) = \dfrac{-2xy}{(x+y^2)^2}, \quad g_y(1,2) = -\dfrac{4}{25}$

33. $f_x(x,y) = \lim\limits_{h\to 0}\dfrac{(x+h)^2\,y - x^2 y}{h} = \lim\limits_{h\to 0} y\left(\dfrac{2xh + h^2}{h}\right) = y\lim\limits_{h\to 0}(2x+h) = 2xy$

$f_x(x,y) = \lim\limits_{h\to 0}\dfrac{x^2(y+h) - x^2 y}{h} = \lim\limits_{h\to 0}\dfrac{x^2 h}{h} = \lim\limits_{h\to 0} x^2 = x^2$

34. $f_x(x,y) = 0, \quad f_y(x,y) = 2y$

35. $f_x(x,y) = \lim\limits_{h\to 0}\dfrac{\ln\left(y(x+h)^2\right) - \ln x^2 y}{h} = \lim\limits_{h\to 0}\dfrac{\ln y + 2\ln(x+h) - 2\ln x - \ln y}{h}$

$= 2\lim\limits_{h\to 0}\dfrac{\ln(x+h) - \ln x}{h} = 2\dfrac{d}{dx}(\ln x) = \dfrac{2}{x}$

$$f_y(x, y) = \lim_{h \to 0} \frac{\ln\left(x^2(y+h)\right) - \ln x^2 y}{h} = \lim_{h \to 0} \frac{\ln x^2 + \ln(y+h) - \ln x^2 - \ln y}{h}$$

$$= \lim_{h \to 0} \frac{\ln(y+h) - \ln y}{h} = \frac{d}{dy}(\ln y) = \frac{1}{y}$$

36. $f_x(x, y) = -(x + 4y)^{-2}, \quad f_y(x, y) = -4(x + 4y)^{-2}$

37. $f_y(x, y) = \lim_{h \to 0} \frac{1}{h}\left\{\frac{1}{(x+h) - y} - \frac{1}{x - y}\right\} = \lim_{h \to 0} \frac{1}{h}\left\{\frac{-h}{(x + h - y)(x - y)}\right\}$

$$= \lim_{h \to 0} \frac{-1}{(x + h - y)(x - y)} = \frac{-1}{(x - y)^2}$$

$$f_y(x, y) = \lim_{h \to 0} \frac{1}{h}\left\{\frac{1}{x - (y + h)} - \frac{1}{x - y}\right\} = \lim_{h \to 0} \frac{1}{h}\left\{\frac{h}{(x - y - h)(x - y)}\right\}$$

$$= \lim_{h \to 0} \frac{1}{(x - y - h)(x - y)} = \frac{1}{(x - y)^2}$$

38. $f_x(x, y) = 2e^{2x+3y}, \quad f_y(x, y) = 3e^{2x+3y}$

39. $f_x(x, y, z) = \lim_{h \to 0} \frac{(x + h)y^2 z - xy^2 z}{h} = \lim_{h \to 0} y^2 z = y^2 z$

$$f_y(x, y, z) = \lim_{h \to 0} \frac{x(y + h)^2 z - xy^2 z}{h} = \lim_{h \to 0} \frac{xz(2yh + h^2)}{h}$$

$$= \lim_{h \to 0} xz(2y + h) = 2xyz$$

$$f_z(x, y, z) = \lim_{h \to 0} \frac{xy^2 (z + h) - xy^2 z}{h} = \lim_{h \to 0} xy^2 = xy^2$$

40. $f_x(x, y, z) = \dfrac{2xy}{z}, \quad f_y(x, y, z) = \dfrac{x^2}{z}, \quad f_z(x, y, z) = -\dfrac{x^2 y}{z^2}$

41. (b) The slope of the tangent line to C at the point $P(x_0, y_0, f(x_0, y_0))$ is $f_y(x_0, y_0)$
Thus, equations for the tangent line are:

$$x = x_0, \quad z - z_0 = f_y(x_0, y_0)(y - y_0)$$

42. $f_x(x, y) = 2x, \quad f_x(1, 3) = 2,$ equations for the tangent line are: $\quad y = 3, \quad z - 10 = 2(x - 1).$

43. Let $z = f(x, y) = x^2 + y^2.$ Then $f(2, 1) = 5, \quad f_y(x, y) = 2y$ and $f_y(2, 1) = 2;$
equations for the tangent line are: $\quad x = 2, \quad z - 5 = 2(y - 1)$

44. $f(x, y) = \dfrac{x^2}{y^2 - 3}, \quad f_y(x, y) = \dfrac{-x^2 2y}{(y^2 - 3)^2}$

Tangent line: $x = x_0, \quad z - z_0 = f_y(x_0, y_0)(y - y_0) \implies x = 3, \quad z - 9 = -36(y - 2)$

45. Let $z = f(x, y) = \dfrac{x^2}{y^2 - 3}$. Then $f(3, 2) = 9$, $\quad f_x(x, y) = \dfrac{2x}{y^2 - 3}$ $\quad$ and $\quad f_x(3, 2) = 6$;

equations for the tangent line are: $\quad y = 2, \quad z - 9 = 6(x - 3)$

46. $f(x, y) = (4 - x^2 - y^2)^{1/2}, \quad f_x(x, y) = -x(4 - x^2 - y^2)^{-1/2}, \quad f_y(x, y) = -y(4 - x^2 - y^2)^{-1/2}$

(a) $f_y(1, 1) = -\dfrac{\sqrt{2}}{2} \implies x = 1, \; z - \sqrt{2} = -\dfrac{\sqrt{2}}{2}(y - 1)$

(b) $f_x(1, 1) = -\dfrac{\sqrt{2}}{2} \implies y = 1, \quad z - \sqrt{2} = -\dfrac{\sqrt{2}}{2}(x - 1)$

(c) l_1 and l_2 have direction vectors $\mathbf{j} - \frac{\sqrt{2}}{2}\mathbf{k}, \mathbf{i} - \frac{\sqrt{2}}{2}\mathbf{k}$ respectively. The normal to the plane is

$$\left(\mathbf{j} - \frac{\sqrt{2}}{2}\mathbf{k}\right) \times \left(\mathbf{i} - \frac{\sqrt{2}}{2}\mathbf{k}\right) = -\frac{\sqrt{2}}{2}\mathbf{i} - \frac{\sqrt{2}}{2}\mathbf{j} - \mathbf{k}, \quad \text{so the tangent plane is}$$

$$-\frac{\sqrt{2}}{2}(x - 1) - \frac{\sqrt{2}}{2}(y - 1) - (z - \sqrt{2}) = 0, \quad \text{or} \quad (x - 1) + (y - 1) + \sqrt{2}(z - \sqrt{2}) = 0$$

47. (a) $m_x = -6$; $\quad$ tangent line: $\quad y = 2, \quad z = -6x + 13$

(b) $m_y = 18$; $\quad$ tangent line: $\quad x = 1, \quad z = 18y - 29$

48. (a) $m_x = \frac{7}{25}$; $\quad$ tangent line: $\quad y = 2, \quad z = \frac{1}{25}(7x - 12)$

(b) $m_y = -\frac{1}{25}$; $\quad$ tangent line: $\quad x = 1, \quad z = -\frac{1}{25}(y + 3)$

49. $u_x(x, y) = 2x = v_y(x, y); \quad u_y(x, y) = -2y = -v_x(x, y)$

50. $u_x = e^x \cos y, \quad u_y = -e^x \sin y, \quad v_x = e^x \sin y = -u_y, \quad v_y = e^x \cos y = u_x$

51. $u_x(x, y) = \dfrac{1}{2}\dfrac{1}{x^2 + y^2}2x = \dfrac{x}{x^2 + y^2}; \quad v_y(x, y) = \dfrac{1}{1 + (y/x)^2}\left(\dfrac{1}{x}\right) = \dfrac{x}{x^2 + y^2}$

Thus, $u_x(x, y) = v_y(x, y)$.

$u_y(x, y) = \dfrac{1}{2}\dfrac{1}{x^2 + y^2}2y = \dfrac{y}{x^2 + y^2}; \quad v_x(x, y) = \dfrac{1}{1 + (y/x)^2}\left(\dfrac{-y}{x^2}\right) = \dfrac{-y}{x^2 + y^2}$

Thus, $u_y(x, y) = -v_x(x, y)$.

52. $u_x = \dfrac{y^2 - x^2}{(x^2 + y^2)^2}, \quad u_y = \dfrac{-2xy}{(x^2 + y^2)^2}, \quad v_x = \dfrac{2xy}{(x^2 + y^2)^2} = -u_y, \quad v_y = \dfrac{y^2 - x^2}{(x^2 + y^2)^2} = u_x$

53. (a) f depends only on y. $\qquad\qquad$ (b) f depends only on x.

54. (a) $a_0 = \left(b_0^2 + c_0^2 - 2b_0c_0 \cos \theta_0\right)^{1/2} = 5\sqrt{7}$

(b) $\dfrac{\partial a}{\partial b} = (2b - 2c \cos \theta)\left(\dfrac{1}{2}\right)(b^2 + c^2 - 2bc \cos \theta)^{-1/2} = \dfrac{\sqrt{7}}{14}$

(c) $a \cong a_0 + \dfrac{\sqrt{7}}{14}(b - b_0) = 5\sqrt{7} + \dfrac{\sqrt{7}}{14} \cdot (-1)$ decreases by about $\dfrac{\sqrt{7}}{14}$ inches.

(d) $\dfrac{\partial a}{\partial \theta} = 2bc \sin \theta(\dfrac{1}{2})(b^2 + c^2 - 2bc \cos \theta)^{-1/2} = \dfrac{15}{7}\sqrt{21}$

(e) Differentiate implicitly: $0 = 2c\dfrac{\partial c}{\partial \theta} - 2b\dfrac{\partial c}{\partial \theta}\cos\theta + 2bc\sin\theta$

$$\frac{\partial c}{\partial \theta} = \frac{bc\sin\theta}{b\cos\theta - c} = -\frac{15}{2}\sqrt{3}$$

55. (a) $\dfrac{75\sqrt{3}}{2}$ in.^{2}

(b) $\dfrac{\partial A}{\partial b} = \dfrac{1}{2}c\sin\theta;$ at time $t_0,$ $\dfrac{\partial A}{\partial b} = \dfrac{15\sqrt{3}}{4}$

(c) $\dfrac{\partial A}{\partial \theta} = \dfrac{1}{2}bc\cos\theta;$ at time $t_0,$ $\dfrac{\partial A}{\partial \theta} = \dfrac{75}{2}$

(d) with $h = \dfrac{\pi}{180},$ $A(b, c, \theta + h) - A(b, c, \theta) \cong h\dfrac{\partial A}{\partial \theta} = \dfrac{\pi}{180}\dfrac{75}{2} = \dfrac{5\pi}{24}$ in.^{2}

(e) $0 = \dfrac{1}{2}\sin\theta\left(b\dfrac{\partial c}{\partial b} + c\right);$ at time $t_0,$ $\dfrac{\partial c}{\partial b} = \dfrac{-c}{b} = -\dfrac{3}{2}$

56. By theorem 7.6.1, $f(x, y) = Ce^{kx}$ where C is independent of x. Since C may depend on y, we write $C = g(y)$.

57. (a) y_0-section: $\mathbf{r}(x) = x\mathbf{i} + y_0\mathbf{j} + f(x, y_0)\mathbf{k}$

tangent line: $\mathbf{R}(t) = [x_0\mathbf{i} + y_0\mathbf{j} + f(x_0, y_0)\mathbf{k}] + t\left[\mathbf{i} + \dfrac{\partial f}{\partial x}(x_0, y_0)\mathbf{k}\right]$

(b) x_0-section: $\mathbf{r}(y) = x_0\mathbf{i} + y\mathbf{j} + f(x_0, y)\mathbf{k}$

tangent line: $\mathbf{R}(t) = [x_0\mathbf{i} + y_0\mathbf{j} + f(x_0, y_0)\mathbf{k}] + t\left[\mathbf{j} + \dfrac{\partial f}{\partial y}(x_0, y_0)\mathbf{k}\right]$

(c) For (x, y, z) in the plane

$$[(x - x_0)\mathbf{i} + (y - y_0)\mathbf{j} + (z - f(x_0, y_0))\mathbf{k}] \cdot \left[\left(\mathbf{i} + \dfrac{\partial f}{\partial x}(x_0, y_0)\mathbf{k}\right) \times \left(\mathbf{j} + \dfrac{\partial f}{\partial y}(x_0, y_0)\mathbf{k}\right)\right] = 0.$$

From this it follows that

$$z - f(x_0, y_0) = (x - x_0)\frac{\partial f}{\partial x}(x_0, y_0) + (y - y_0)\frac{\partial f}{\partial y}(x_0, y_0).$$

58. Fix y and set $F(x) = f(x, y)$. Then, for that value of y, $h(x, y) = g(F(x))$ and thus

$$h_x(x, y) = \frac{d}{dx}[g(F(x))] = g'(F(x))F'(x) = g'(f(x, y))f_x(x, y).$$

The second formula can be derived in the same manner.

59. (a) Set $u = ax + by$. Then

$$b\frac{\partial w}{\partial x} - a\frac{\partial w}{\partial y} = b(a\,g'(u)) - a(b\,g'(u)) = 0.$$

(b) Set $u = x^m y^n$. Then

$$nx\frac{\partial w}{\partial x} - my\frac{\partial w}{\partial y} = nx\left[mx^{m-1}y^n g'(u)\right] - my\left[nx^m y^{n-1}g'(u)\right] = 0.$$

60. $\dfrac{\partial x}{\partial r}\dfrac{\partial y}{\partial \theta} - \dfrac{\partial x}{\partial \theta}\dfrac{\partial y}{\partial r} = (\cos\theta)(r\cos\theta) - (-r\sin\theta)(\sin\theta) = r$

61. $V\dfrac{\partial P}{\partial V} = V\left(-\dfrac{kT}{V^2}\right) = -k\dfrac{T}{V} = -P;$ $V\dfrac{\partial P}{\partial V} + T\dfrac{\partial P}{\partial T} = -k\dfrac{T}{V} + T\left(\dfrac{k}{V}\right) = 0$

62. $R = \dfrac{R_1 R_2 R_3}{R_1 R_2 + R_1 R_3 + R_2 R_3};$ $\dfrac{\partial R}{\partial R_1} = \left(\dfrac{R_2 R_3}{R_1 R_2 + R_1 R_3 + R_2 R_3}\right)^2$

SECTION 15.5

1. interior $= \{(x,y) : 2 < x < 4,\ \ 1 < y < 3\}$ (the inside of the rectangle), boundary = the union of the four boundary line segments; set is closed.

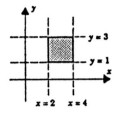

2. same interior and same boundary as in Exercise 1; set is open

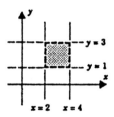

3. interior = the entire set (region between the two concentric circles), boundary = the two circles, one of radius 1, the other of radius 2; set is open.

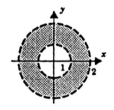

4. interior $= \{(x,y) : 1 < x^2 < 4\} = \{(x,y) : -2 < x < -1\} \cup \{(x,y) : 1 < x < 2\}$ (two vertical stripes without the boundary lines), boundary $= \{(x,y) : x = -2,\ x = -1,\ x = 1, \text{ or } x = 2\}$ (four vertical lines);   set is closed.

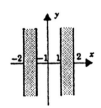

5. interior $= \{(x,y) : 1 < x^2 < 4\} = \{(x,y) : -2 < x < -1\} \cup \{(x,y) : 1 < x < 2\}$ (two vertical strips without the boundary lines), boundary $= \{(x,y) : x = -2,\ x = -1,\ x = 1, \text{ or } x = 2\}$ (four vertical lines); set is neither open nor closed.

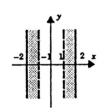

6. interior = the entire set (region below the parabola $y = x^2$), boundary = the parabola $y = x^2$;   the set is open.

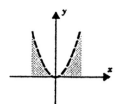

7. interior = region below the parabola $y = x^2$, boundary = the parabola $y = x^2$;   the set is closed.

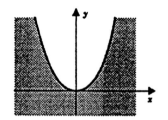

8. interior = the inside of the cube;   boundary = the faces of the cube; set is neither open nor closed (upper face of cube is omitted)

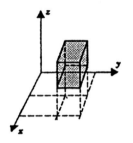

9. interior = $\{ (x, y, z) : x^2 + y^2 < 1, 0 < z \leq 4 \}$ (the inside of the cylinder), boundary = the total surface of the cylinder (the curved part, the top, and the bottom);   the set is closed.

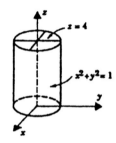

10. interior = the entire set (the inside of the ball of radius $\frac{1}{2}$, centered at $(1,1,1)$), boundary = the spherical surface; set is open.

11. (a) ϕ (b) S (c)   closed

12. interior = the entire set, boundary = $\{1, 3\}$;   set is open.

13. interior = $\{x : 1 < x < 3\}$, boundary = $\{1, 3\}$;   set is closed.

14. interior = $\{x : 1 < x < 3\}$,   boundary = $\{1, 3\}$;   set is neither open nor closed.

15. interior = the entire set, boundary = $\{1\}$;   set is open.

16. interior = $\{x : x < -1\}$,   boundary = $\{-1\}$;   set is closed.

17. interior $= \{x : |x| > 1\}$, boundary $= \{1, -1\}$; set is neither open nor closed.

18. interior $= \phi$, boundary = the entire set; set is closed.

19. interior $= \phi$, boundary = {the entire set} $\cup$ {0}; the set is neither open nor closed.

20. (a) ϕ is open because it contains no boundary points,

ϕ is closed because it contains its boundary (the boundary is empty).

(b) X is open because it contains a neighborhood of each of its points,

X is closed because it contains its boundary (the boundary is empty).

(c) Suppose that U is open. Let x be a boundary point of $X - U$. Then every neighborhood of x contains points from $X - U$. The point x can not be in U because U contains a neighborhood of each of its points. Thus $x \in X - U$. This shows that $X - U$ contains its boundary and is therefore closed.

Suppose now that $X - U$ is closed. Let x be a point of U. If no neighborhood of x lies entirely in U, then every neighborhood of x contains points from $X - U$. This makes x a boundary point of $X - U$ and, since $X - U$ is closed, places x in $X - U$. This contradiction shows that some neighborhood of x lies entirely in U. Thus U contains a neighborhood of each of its points and is therefore open.

(d) Set $U = X - F$ and note that $F = X - U$. By (c)

$$F = X - U \quad \text{is closed} \quad \text{iff} \quad X - F = U \quad \text{is open.}$$

SECTION 15.6

1. $\dfrac{\partial^2 f}{\partial x^2} = 2A, \quad \dfrac{\partial^2 f}{\partial y^2} = 2C, \quad \dfrac{\partial^2 f}{\partial y \partial x} = \dfrac{\partial^2 f}{\partial x \partial y} = 2B$

2. $\dfrac{\partial^2 f}{\partial x^2} = 6Ax + 2By, \quad \dfrac{\partial^2 f}{\partial y^2} = 2Cx, \quad \dfrac{\partial^2 f}{\partial y \partial x} = \dfrac{\partial^2 f}{\partial x \partial y} = 2Bx + 2Cy$

3. $\dfrac{\partial^2 f}{\partial x^2} = Cy^2 e^{xy}, \quad \dfrac{\partial^2 f}{\partial y^2} = Cx^2 e^{xy}, \quad \dfrac{\partial^2 f}{\partial y \partial x} = \dfrac{\partial^2 f}{\partial x \partial y} = Ce^{xy}(xy + 1)$

4. $\dfrac{\partial^2 f}{\partial x^2} = 2\cos y - y^2 \sin x, \quad \dfrac{\partial^2 f}{\partial y^2} = 2\sin x - x^2 \cos y, \quad \dfrac{\partial^2 f}{\partial y \partial x} = \dfrac{\partial^2 f}{\partial x \partial y} = 2(y\cos x - x\sin y)$

5. $\dfrac{\partial^2 f}{\partial x^2} = 2, \quad \dfrac{\partial^2 f}{\partial y^2} = 4(x + 3y^2 + z^3), \quad \dfrac{\partial^2 f}{\partial z^2} = 6z(2x + 2y^2 + 5z^3)$

$\dfrac{\partial^2 f}{\partial x \partial y} = \dfrac{\partial^2 f}{\partial y \partial x} = 4y, \quad \dfrac{\partial^2 f}{\partial z \partial x} = \dfrac{\partial^2 f}{\partial x \partial z} = 6z^2, \quad \dfrac{\partial^2 f}{\partial z \partial y} = \dfrac{\partial^2 f}{\partial y \partial z} = 12yz^2$

6. $\dfrac{\partial^2 f}{\partial x^2} = -\dfrac{1}{4(x + y^2)^{3/2}}, \quad \dfrac{\partial^2 f}{\partial y^2} = \dfrac{x}{(x + y^2)^{3/2}}, \quad \dfrac{\partial^2 f}{\partial x \partial y} = \dfrac{\partial^2 f}{\partial y \partial x} = -\dfrac{y}{2(x + y^2)^{3/2}}$

7. $\dfrac{\partial^2 f}{\partial x^2} = \dfrac{1}{(x+y)^2} - \dfrac{1}{x^2}, \quad \dfrac{\partial^2 f}{\partial y^2} = \dfrac{1}{(x+y)^2}, \quad \dfrac{\partial^2 f}{\partial y \partial x} = \dfrac{\partial^2 f}{\partial x \partial y} = \dfrac{1}{(x+y)^2}$

8. $\dfrac{\partial^2 f}{\partial x^2} = -\dfrac{2C(AD-BC)y}{(Cx+Dy)^3}, \quad \dfrac{\partial^2 f}{\partial y^2} = \dfrac{2D(AD-BC)x}{(Cx+Dy)^3}, \quad \dfrac{\partial^2 f}{\partial y \partial x} = \dfrac{\partial^2 f}{\partial x \partial y} = \dfrac{(AD-BC)(Cx-Dy)}{(Cx+Dy)^3}$

9. $\dfrac{\partial^2 f}{\partial x^2} = 2(y+z), \quad \dfrac{\partial^2 f}{\partial y^2} = 2(x+z), \quad \dfrac{\partial^2 f}{\partial z^2} = 2(x+y)$

   all the second mixed partials are $2(x+y+z)$

10. $\dfrac{\partial^2 f}{\partial x^2} = -\dfrac{2xy^3z^3}{(1+x^2y^2z^2)^2}, \quad \dfrac{\partial^2 f}{\partial y^2} = -\dfrac{2yx^3z^3}{(1+x^2y^2z^2)^2}$

   $\dfrac{\partial^2 f}{\partial z^2} = -\dfrac{2zx^3y^3}{(1+x^2y^2z^2)^2}, \quad \dfrac{\partial^2 f}{\partial y \partial x} = \dfrac{\partial^2 f}{\partial x \partial y} = \dfrac{z(1-x^2y^2z^2)}{(1+x^2y^2z^2)^2}$

   $\dfrac{\partial^2 f}{\partial z \partial x} = \dfrac{\partial^2 f}{\partial x \partial z} = \dfrac{y(1-x^2y^2z^2)}{(1+x^2y^2z^2)^2}, \quad \dfrac{\partial^2 f}{\partial y \partial z} = \dfrac{\partial^2 f}{\partial z \partial y} = \dfrac{x(1-x^2y^2z^2)}{(1+x^2y^2z^2)^2}$

11. $\dfrac{\partial^2 f}{\partial x^2} = y(y-1)x^{y-2}, \quad \dfrac{\partial^2 f}{\partial y^2} = (\ln x)^2 x^y, \quad \dfrac{\partial^2 f}{\partial y \partial x} = \dfrac{\partial^2 f}{\partial x \partial y} = x^{y-1}(1 + y \ln x)$

12. $\dfrac{\partial^2 f}{\partial x^2} = -\sin(x+z^y), \quad \dfrac{\partial^2 f}{\partial y^2} = z^y(\ln z)^2[\cos(x+z^y) - z^y \sin(x+z^y)]$

   $\dfrac{\partial^2 f}{\partial z^2} = y(y-1)z^{y-2}\cos(x+z^y) - y^2 z^{2y-2}\sin(x+z^y)$

   $\dfrac{\partial^2 f}{\partial y \partial x} = \dfrac{\partial^2 f}{\partial x \partial y} = -z^y \ln z \sin(x+z^y), \quad \dfrac{\partial^2 f}{\partial z \partial x} = \dfrac{\partial^2 f}{\partial x \partial z} = -yz^{y-1}\sin(x+z^y)$

   $\dfrac{\partial^2 f}{\partial z \partial y} = \dfrac{\partial^2 f}{\partial y \partial z} = z^{y-1}(1 + y \ln z)\cos(x+z^y) - yz^{2y-1}(\ln z)\sin(x+z^y)$

13. $\dfrac{\partial^2 f}{\partial x^2} = ye^x, \quad \dfrac{\partial^2 f}{\partial y^2} = xe^y, \quad \dfrac{\partial^2 f}{\partial y \partial x} = \dfrac{\partial^2 f}{\partial x \partial y} = e^y + e^x$

14. $\dfrac{\partial^2 f}{\partial x^2} = \dfrac{2xy}{(x^2+y^2)^2}, \quad \dfrac{\partial^2 f}{\partial y^2} = \dfrac{-2xy}{(x^2+y^2)^2}, \quad \dfrac{\partial^2 f}{\partial y \partial x} = \dfrac{\partial^2 f}{\partial x \partial y} = \dfrac{y^2-x^2}{(x^2+y^2)^2}$

15. $\dfrac{\partial^2 f}{\partial x^2} = \dfrac{y^2-x^2}{(x^2+y^2)^2}, \quad \dfrac{\partial^2 f}{\partial y^2} = \dfrac{x^2-y^2}{(x^2+y^2)^2}, \quad \dfrac{\partial^2 f}{\partial y \partial x} = \dfrac{\partial^2 f}{\partial x \partial y} = -\dfrac{2xy}{(x^2+y^2)^2}$

16. $\dfrac{\partial^2 f}{\partial x^2} = 6xy^2\cos(x^3y^2) - 9x^4y^4\sin(x^3y^2), \quad \dfrac{\partial^2 f}{\partial y^2} = 2x^3\cos(x^3y^2) - 4x^6y^2\sin(x^3y^2)$

   $\dfrac{\partial^2 f}{\partial y \partial x} = \dfrac{\partial^2 f}{\partial x \partial y} = 6x^2y\cos(x^3y^2) - 6x^5y^3\sin(x^3y^2)$

17. $\dfrac{\partial^2 f}{\partial x^2} = -2y^2\cos 2xy, \quad \dfrac{\partial^2 f}{\partial y^2} = -2x^2\cos 2xy, \quad \dfrac{\partial^2 f}{\partial y \partial x} = \dfrac{\partial^2 f}{\partial x \partial y} = -[\sin 2xy + 2xy \cos 2xy]$

18. $\dfrac{\partial^2 f}{\partial x^2} = y^4 e^{xy^2}, \quad \dfrac{\partial^2 f}{\partial y^2} = e^{xy^2}(2x + 4x^2 y^2), \quad \dfrac{\partial^2 f}{\partial y \partial x} = \dfrac{\partial^2 f}{\partial x \partial y} = e^{xy^2}(2y + 2xy^3)$

19. $\dfrac{\partial^2 f}{\partial x^2} = 0, \quad \dfrac{\partial^2 f}{\partial y^2} = xz \sin y, \quad \dfrac{\partial^2 f}{\partial z^2} = -xy \sin z,$

$\dfrac{\partial^2 f}{\partial y \partial x} = \dfrac{\partial^2 f}{\partial x \partial y} = \sin z - z \cos y, \quad \dfrac{\partial^2 f}{\partial x \partial z} = \dfrac{\partial^2 f}{\partial z \partial x} = y \cos z - \sin y, \quad \dfrac{\partial^2 f}{\partial y \partial z} = \dfrac{\partial^2 f}{\partial z \partial y} = x \cos z - x \cos y$

20. $\dfrac{\partial^2 f}{\partial x^2} = z e^x, \quad \dfrac{\partial^2 f}{\partial y^2} = x e^y, \quad \dfrac{\partial^2 f}{\partial z^2} = y e^z, \quad \dfrac{\partial^2 f}{\partial y \partial x} = \dfrac{\partial^2 f}{\partial x \partial y} = e^y,$

$\dfrac{\partial^2 f}{\partial z \partial x} = \dfrac{\partial^2 f}{\partial x \partial z} = e^x, \quad \dfrac{\partial^2 f}{\partial z \partial y} = \dfrac{\partial^2 f}{\partial y \partial z} = e^z$

21. $x^2 \dfrac{\partial^2 u}{\partial x^2} + 2xy \dfrac{\partial^2 u}{\partial x \partial y} + y^2 \dfrac{\partial^2 u}{\partial y^2} = x^2 \left(\dfrac{-2y^2}{(x+y)^3} \right) + 2xy \left(\dfrac{2xy}{(x+y)^3} \right) + y^2 \left(\dfrac{-2x^2}{(x+y)^3} \right) = 0$

22. (a) mixed partials are 0

(b) mixed partials are $g'(x)\, h'(y)$

(c) by the hint mixed partials for each term $x^m y^n$ are $mn x^{m-1} y^{n-1}$

23. (a) no, since $\dfrac{\partial^2 f}{\partial y \partial x} \neq \dfrac{\partial^2 f}{\partial x \partial y}$ (b) no, since $\dfrac{\partial^2 f}{\partial y \partial x} \neq \dfrac{\partial^2 f}{\partial x \partial y}$ for $x \neq y$

24. $\dfrac{\partial h}{\partial x} = g'(x + y) + g'(x - y), \quad \dfrac{\partial h}{\partial y} = g'(x + y) - g'(x - y)$

$\dfrac{\partial^2 h}{\partial x^2} = g''(x + y) + g''(x - y), \quad \dfrac{\partial^2 h}{\partial y^2} = g''(x + y) + g''(x - y) = \dfrac{\partial^2 h}{\partial x^2}$

25. $\dfrac{\partial^3 f}{\partial x^2 \partial y} = \dfrac{\partial}{\partial x}\left(\dfrac{\partial^2 f}{\partial x \partial y} \right) = \dfrac{\partial}{\partial x}\left(\dfrac{\partial^2 f}{\partial y \partial x} \right) = \dfrac{\partial^2}{\partial x \partial y}\left(\dfrac{\partial f}{\partial x} \right) = \dfrac{\partial^2}{\partial y \partial x}\left(\dfrac{\partial f}{\partial x} \right) = \dfrac{\partial}{\partial y}\left(\dfrac{\partial^2 f}{\partial x^2} \right) = \dfrac{\partial^3 f}{\partial y \partial x^2}$

by def.   (15.6.5)   by def.   (15.6.5)   by def.   by def.

26. (a) as (x, y) tends to $(0, 0)$ along the x-axis, $f(x, y) = f(x, 0) = 1$ tends to 1;

as (x, y) tends to $(0, 0)$ along the line $y = x$, $f(x, y) = f(x, x) = 0$ tends to 0;

(b) as (x, y) tends to $(0, 0)$ along the x-axis, $f(x, y) = f(x, 0) = 0$ tends to 0;

as (x, y) tends to $(0, 0)$ along the line $y = x$, $f(x, y) = f(x, x) = \frac{1}{2}$ tends to $\frac{1}{2}$;

27. (a) $\lim\limits_{x \to 0} \dfrac{(x)(0)}{x^2 + 0} = \lim\limits_{x \to 0} 0 = 0$ (b) $\lim\limits_{y \to 0} \dfrac{(0)(y)}{0 + y^2} = \lim\limits_{y \to 0} 0 = 0$

(c) $\lim\limits_{x \to 0} \dfrac{(x)(mx)}{x^2 + (mx)^2} = \lim\limits_{x \to 0} \dfrac{m}{1 + m^2} = \dfrac{m}{1 + m^2}$

(d) $\lim\limits_{\theta \to 0^+} \dfrac{(\theta \cos \theta)(\theta \sin \theta)}{(\theta \cos \theta)^2 + (\theta \sin \theta)^2} = \lim\limits_{\theta \to 0^+} \cos \theta \sin \theta = 0$

(e) By L'Hospital's rule $\quad \lim\limits_{x\to 0} \dfrac{f(x)}{x} = \lim\limits_{x\to 0} f'(x) = f'(0).$ Thus

$$\lim_{x\to 0} \frac{xf(x)}{x^2 + [f(x)]^2} = \lim_{x\to 0} \frac{f(x)/x}{1 + [f(x)/x]^2} = \frac{f'(0)}{1 + [f'(0)]^2}.$$

(f) $\displaystyle \lim_{\theta\to(\pi/3)^-} = \frac{(\cos\theta \sin 3\theta)(\sin\theta \sin 3\theta)}{(\cos\theta \sin 3\theta)^2 + (\sin\theta \sin 3\theta)^2} = \lim_{\theta\to(\pi/3)^-} \cos\theta \sin\theta = \frac{1}{4}\sqrt{3}$

(g) $\displaystyle \lim_{t\to\infty} \frac{(1/t)(\sin t)/t}{1/t^2 + (\sin^2 t)/t^2} = \lim_{t\to\infty} \frac{\sin t}{1 + \sin^2 t};$ does not exist

28. (a) $\displaystyle \lim_{x\to 0} \frac{x(0)^2}{(x^2 + 0^2)^{3/2}} = \lim_{x\to 0} 0 = 0$ (b) $\displaystyle \lim_{y\to 0} \frac{0\cdot y^2}{(0 + y^2)^{3/2}} = \lim_{y\to 0} 0 = 0$

(c) $\displaystyle \lim_{x\to 0} \frac{xm^2 x^2}{(x^2 + m^2 x^2)^{3/2}} = \lim_{x\to 0} \frac{m^2 x^3}{|x|^3(1 + m^2)^{3/2}} = \lim_{x\to 0} \frac{m^2 x}{|x|\,(1 + m^2)^{3/2}};$ does not exist

(d) $\displaystyle \lim_{\theta\to 0^+} \frac{(\theta\cos\theta)(\theta\sin\theta)^2}{[(\theta\cos\theta)^2 + (\theta\sin\theta)^2]^{3/2}} = \lim_{\theta\to 0^+} \cos\theta \sin^2\theta = 0$

(e) $\displaystyle \lim_{x\to 0} \frac{x[f(x)]^2}{(x^2 + [f(x)]^2)^{3/2}} = \lim_{x\to 0} \frac{[f(x)/x]^2}{(1 + [f(x)/x]^2)^{3/2}} = \lim_{x\to 0} \frac{x^3[f'(0)]^2}{|x|^3(1 + [f'(0)]^2)^{3/2}};$ does not exist

(f) $\displaystyle \lim_{\theta\to\frac{\pi}{3}^-} \frac{(\cos\theta \sin 3\theta)(\sin\theta \sin 3\theta)^2}{[(\cos\theta \sin 3\theta)^2 + (\sin\theta \sin 3\theta)^2]^{3/2}} = \lim_{\theta\to\frac{\pi}{3}^-} \cos\theta \sin^2\theta = \frac{3}{8}$

(g) $\displaystyle \lim_{t\to\infty} \frac{(1/t)(\sin t/t)^2}{\left[(1/t^2) + (\sin^2 t/t^2)\right]^{3/2}} = \lim_{t\to\infty} \frac{\sin^2 t}{(1 + \sin^2 t)^{3/2}};$ does not exist

29. (a) $\displaystyle \frac{\partial g}{\partial x}(0,0) = \lim_{h\to 0} \frac{g(h,0) - g(0,0)}{h} = \lim_{h\to 0} 0 = 0,$

$\displaystyle \frac{\partial g}{\partial y}(0,0) = \lim_{h\to 0} \frac{g(0,h) - g(0,0)}{h} = \lim_{h\to 0} 0 = 0$

(b) as (x,y) tends to $(0,0)$ along the x-axis, $g(x,y) = g(x,0) = 0$ tends to 0;

as (x,y) tends to $(0,0)$ along the line $y = x$, $g(x,y) = g(x,x) = \frac{1}{2}$ tends to $\frac{1}{2}$

30. No; as (x,y) tends to $(1,1)$ along the line $x = 1$, $f(x,y) = 1$ tends to 1; as (x,y) tends to $(1,1)$

along the line $y = 1$,

$$f(x,y) = \frac{x-1}{x^3-1} = \frac{1}{x^2+x+1} \quad \text{tends to} \quad \frac{1}{3}$$

31. For $y \neq 0$, $\qquad \displaystyle \frac{\partial f}{\partial x}(0,y) = \lim_{h\to 0} \frac{f(h,y) - f(0,y)}{h} = \lim_{h\to 0} \frac{y(y^2 - h^2)}{h^2 + y^2} = y.$

Since $\qquad \displaystyle \frac{\partial f}{\partial x}(0,0) = \lim_{h\to 0} \frac{f(h,0) - f(0,0)}{h} = \lim_{h\to 0} 0 = 0,$

we have $\dfrac{\partial f}{\partial x}(0, y) = y$ for all y.

For $x \neq 0$, $\quad \dfrac{\partial f}{\partial y}(x, 0) = \lim\limits_{h \to 0} \dfrac{f(x, h) - f(x, 0)}{h} = \lim\limits_{h \to 0} \dfrac{x(h^2 - x^2)}{x^2 + h^2} = -x.$

Since $\quad \dfrac{\partial f}{\partial y}(0, 0) = \lim\limits_{h \to 0} \dfrac{f(0, h) - f(0, 0)}{h} = \lim\limits_{h \to 0} 0 = 0,$

we have $\quad \dfrac{\partial f}{\partial y}(x, 0) = -x \quad$ for all x.

Therefore $\quad \dfrac{\partial^2 f}{\partial y \partial x}(0, y) = 1 \;$ for all $y \quad$ and $\quad \dfrac{\partial^2 f}{\partial x \partial y}(x, 0) = -1$ for all x.

In particular $\quad \dfrac{\partial^2 f}{\partial y \partial x}(0, 0) = 1 \quad$ while $\quad \dfrac{\partial^2 f}{\partial x \partial y}(0, 0) = -1.$

32.
$$\lim_{h \to 0} [f(x_0 + h, y_0) - f(x_0, y_0)] = \lim_{h \to 0} \left[(h) \left(\frac{f(x_0 + h, y_0) - f(x_0, y_0)}{h} \right) \right]$$
$$= \left(\lim_{h \to 0} h \right) \left(\lim_{h \to 0} \frac{f(x_0 + h, y_0) - f(x_0, y_0)}{h} \right)$$
$$= 0 \cdot \frac{\partial f}{\partial x}(x_0, y_0) = 0$$

33. Since $f_{xy}(x, y) = 0$, $f_x(x, y)$ must be a function of x alone, and $f_y(x, y)$ must be a function of y alone. Then f must be of the form

$f(x, y) = g(x) + h(y).$

34.

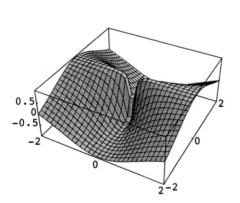

35.

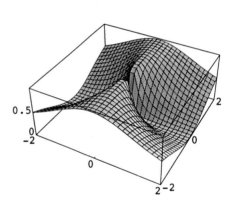

36.

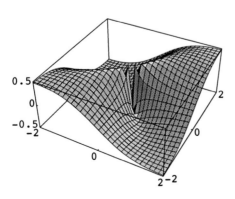

PROJECT 15.6

1. (a) $\dfrac{\partial u}{\partial x} = \dfrac{x^2 y^2 + 2xy^3}{(x+y)^2}, \quad \dfrac{\partial u}{\partial y} = \dfrac{x^2 y^2 + 2x^3 y}{(x+y)^2}$

$$x\frac{\partial u}{\partial x} + y\frac{\partial u}{\partial y} = \frac{3x^2 y^2 (x+y)}{(x+y)^2} = 3u$$

(b) $\dfrac{\partial u}{\partial x} = 2xy + z^2, \quad \dfrac{\partial u}{\partial y} = 2yz + x^2, \quad \dfrac{\partial u}{\partial z} = 2xz + y^2$

$$\frac{\partial u}{\partial x} + \frac{\partial u}{\partial y} + \frac{\partial u}{\partial z} = 2xy + z^2 + 2yz + x^2 + 2xz + y^2 = (x+y+z)^2$$

2. (a) (i) $\dfrac{\partial^2 f}{\partial x^2} + \dfrac{\partial^2 f}{\partial y^2} = 6x - 6x = 0$

(ii) $\dfrac{\partial^2 f}{\partial x^2} + \dfrac{\partial^2 f}{\partial y^2} = (-\cos x \sinh y - \sin x \cosh y) + (\cos x \sinh y + \sin x \cosh y) = 0$

(iii) $\dfrac{\partial^2 f}{\partial x^2} = \dfrac{y^2 - x^2}{(x^2 + y^2)^2}, \quad \dfrac{\partial^2 f}{\partial y^2} = \dfrac{x^2 - y^2}{(x^2 + y^2)^2}$

$$\frac{\partial^2 f}{\partial x^2} + \frac{\partial^2 f}{\partial y^2} = \frac{y^2 - x^2}{(x^2 + y^2)^2} + \frac{x^2 - y^2}{(x^2 + y^2)^2} = 0$$

(b) (i) $\dfrac{\partial^2 f}{\partial x^2} + \dfrac{\partial^2 f}{\partial y^2} + \dfrac{\partial^2 f}{\partial z^2} = \dfrac{2x^2 - y^2 - z^2}{(x^2 + y^2 + z^2)^{5/2}} + \dfrac{2y^2 - x^2 - z^2}{(x^2 + y^2 + z^2)^{5/2}} + \dfrac{2z^2 - x^2 - y^2}{(x^2 + y^2 + z^2)^{5/2}} = 0$

(ii) $\dfrac{\partial^2 f}{\partial x^2} = e^{x+y} \cos\left(\sqrt{2}\,z\right), \quad \dfrac{\partial^2 f}{\partial y^2} = e^{x+y} \cos\left(\sqrt{2}\,z\right), \quad \dfrac{\partial^2 f}{\partial z^2} = -2e^{x+y} \cos\left(\sqrt{2}\,z\right)$

$$\frac{\partial^2 f}{\partial x^2} + \frac{\partial^2 f}{\partial y^2} + \frac{\partial^2 f}{\partial z^2} = e^{x+y} \cos\left(\sqrt{2}\,z\right) + e^{x+y} \cos\left(\sqrt{2}\,z\right) + \left[-2e^{x+y} \cos\left(\sqrt{2}\,z\right)\right] = 0$$

3. (i) $\dfrac{\partial^2 f}{\partial t^2} = \dfrac{\partial^2 f}{\partial x^2} = 0 \implies \dfrac{\partial^2 f}{\partial t^2} - c^2 \dfrac{\partial^2 f}{\partial x^2} = 0$

(ii) $\dfrac{\partial^2 f}{\partial t^2} = -5c^2 \sin(x + ct) \cos(2x + 2ct) - 4c^2 \cos(x + ct) \sin(2x + 2ct)$

$$\frac{\partial^2 f}{\partial x^2} = -5 \sin(x + ct) \cos(2x + 2ct) - 4 \cos(x + ct) \sin(2x + 2ct)$$

It now follows that $\dfrac{\partial^2 f}{\partial t^2} - c^2 \dfrac{\partial^2 f}{\partial x^2} = 0$

(iii) $\dfrac{\partial^2 f}{\partial t^2} = -\dfrac{c^2}{(x+ct)^2}, \quad \dfrac{\partial^2 f}{\partial x^2} = -\dfrac{1}{(x+ct)^2} \implies \dfrac{\partial^2 f}{\partial t^2} - c^2 \dfrac{\partial^2 f}{\partial x^2} = 0$

(iv) $\dfrac{\partial^2 f}{\partial t^2} = c^2 k^2 \left(Ae^{kx} + Be^{-kx}\right)\left(Ce^{ckt} + De^{-ckt}\right), \quad \dfrac{\partial^2 f}{\partial x^2} = k^2 \left(Ae^{kx} + Be^{-kx}\right)\left(Ce^{ckt} + De^{-ckt}\right)$

It now follows that $\dfrac{\partial^2 f}{\partial t^2} - c^2 \dfrac{\partial^2 f}{\partial x^2} = 0$

4. $\dfrac{\partial^2 f}{\partial t^2} - c^2 \dfrac{\partial^2 f}{\partial x^2} = \left[c^2 g''(x + ct) + c^2 h''(x - ct)\right] - c^2 [g''(x + ct) + h''(x - ct)] = 0$

REVIEW EXERCISES

1. domain $\{(x, y) : y > x^2\}$,   range $(0, \infty)$

2. domain $\{(x, y) : x, y \in R\}$,   range $(0, \infty)$

3. domain $\{(x, y, x) : z \geq x^2 + y^2\}$,   dange $[0, +\infty)$

4. domain $\{(x, y, z) : x + 2y + z > 0\}$,   range R

5. (a) $f(x, y) = \dfrac{1}{3}\pi x^2 y$;

   (b) $f(x, y) = \dfrac{1}{2}y x^2$;

   (c) $\theta = \arccos \dfrac{x + 2y}{\sqrt{5}\sqrt{x^2 + y^2}}$

6. Assume one of the vertices is $(x, y, z), x > 0, y > 0, z > 0$.
   $$V = 8cxy\sqrt{1 - \frac{x^2}{a^2} - \frac{y^2}{b^2}}$$

7. ellipsoid
   xy−trace: ellipse $4x^2 + 9y^2 = 36$
   xz−trace: ellipse $4x^2 + 36z^2 = 36$
   yz−trace: ellipse $9y^2 + 36z^2 = 36$

8. hyperboloid of two sheets
   xy−trace: none
   xz−trace: hyperbola $4z^2 - x^2 = 4$
   yz−trace: hyperbola $4z^2 - y^2 = 4$

9. hyperbolic paraboloid
   xy−trace: lines $x = \pm y$
   xz−trace: parabola $z = -x^2$
   yz−trace: parabola $z = y^2$

10. elliptic paraboloid
   xy−trace: parabola $4x^2 = y$
   xz−trace: $(0, 0)$
   yz−trace: parabola $9z^2 = y$

11. cone
   xy−trace: lines $x = \pm y$
   xz−trace: lines $x = \pm z$
   yz−trace: $(0, 0)$

12. hyperboloid of one sheet
   xy−trace: ellipse $9x^2 + 4y^2 = 36$
   xz−trace: hyperbola $z^2 = 9x^2 - 36$
   yz−trace: hyperbola $z^2 = 4y^2 - 36$

13.

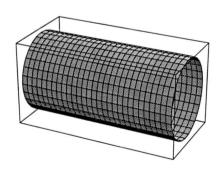

14.

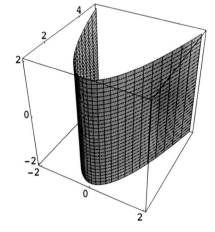

15.

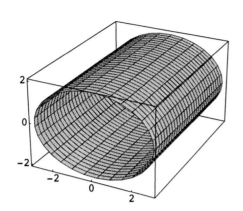

16.

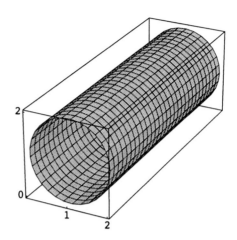

17. $c = 0, \Longrightarrow 0 = 2x^2 + 3y^2 \Longrightarrow (0, 0)$

$c = 6, \Longrightarrow 6 = 2x^2 + 3y^2$, ellipse

$c = 12, \Longrightarrow 6 = 2x^2 + 3y^2$, ellipse

18. $c = 0, \Longrightarrow 0 = x^2 + y^2 - 4$, circle

$c = 1, \Longrightarrow 5 = x^2 + y^2$, circle

$c = 2, \Longrightarrow 8 = x^2 + y^2$, circle

$c = \sqrt{5}, \Longrightarrow 9 = x^2 + y^2$, circle

19. $c = -4, \Longrightarrow x = -4y^2$, parabola

$c = -1, \Longrightarrow x = -y^2$, parabola

$c = 1, \Longrightarrow x = y^2$, parabola

$c = 4, \Longrightarrow x = 4y^2$, parabola

the origin is omitted

20. $c = 1, \Longrightarrow x^2 + y^2 = 1$ circle

$c = 4, \Longrightarrow x^2 + y^2 = 4$ circle

$c = 9, \Longrightarrow x^2 + y^2 = 9$ circle

21. $c = 6, 2x + y + 3z = 6$, plane

22. $c = 16, x^2 + y^2 + 4z^2 = 16$, ellipsoid

23. (a) $f(0, 0) = 1$, level curve: $f(x, y) = 1$

(b) $f(\ln 2, 1) = 4$, level curve: $f(x, y) = 4$

(c) $f(1, -1) = 2e$, level curve: $f(x, y) = 2e$

24. (a) $f(2, 0, 1) = 4$, level surface: $f(x, y, z) = 4$

(b) $f(1, \pi, -1) = -1$, level surface:

$f(x, y, z) = -1$

(c) $f(4, \pi, 1/2) = 0$, level surface: $f(x, y, z) = 0$

25.
$$f_x = \lim_{h \to 0} \frac{f(x + h, y) - f(x, y)}{h} = \lim_{h \to 0} \frac{(x + h)^2 + 2(x + h)y - x^2 - 2xy}{h}$$

$$= \lim_{h \to 0} (2x + h + 2y) = 2x + 2y$$

$$f_y = \lim_{h \to 0} \frac{f(x, y + h) - f(x, y)}{h} = \lim_{h \to 0} \frac{x^2 + 2x(y + h) - x^2 - 2xy}{h} = 2x$$

26.
$$f_x = \lim_{h \to 0} \frac{f(x + h, y) - f(x, y)}{h} = \lim_{h \to 0} \frac{y^2 \cos 2(x + h) - y^2 \cos 2x}{h}$$

$$= y^2 \lim_{h \to 0} \frac{\cos 2(x + h) - \cos 2x}{h}$$

$$= y^2 \cos' 2x = -2y^2 \sin 2x$$

$$f_y = \lim_{h \to 0} \frac{f(x, y+h) - f(x,y)}{h} = \lim_{h \to 0} \frac{(y+h)^2 \cos 2x - y^2 \cos 2x}{h}$$

$$= \cos 2x \lim_{h \to 0} \frac{(y+h)^2 - y^2}{h} = 2y \cos 2x$$

27. $f_x = 2xy - 2y^3; \qquad f_y = x^2 - 6xy^2$

28. $g_x = (x^2 + y^2)^{-1/2} - x^2(x^2 + y^2)^{-3/2}; \qquad g_y = -xy(x^2 + y^2)^{-3/2}$

29. $\dfrac{\partial z}{\partial x} = 2x \sin(xy^2) + x^2 y^2 \cos(xy^2); \qquad \dfrac{\partial z}{\partial y} = 2x^3 y \cos(xy^2).$

30. $f_x = y e^{xy} \ln(y/x) - \dfrac{1}{x} e^{xy} \qquad f_y = x e^{xy} \ln(y/x) + \dfrac{e^{xy}}{y}$

31. $h_x = -e^{-x} \cos(2x - y) - 2e^{-x} \sin(2x - y) \qquad h_y = e^{-x} \sin(2x - y)$

32. $u_x = y^2 \sec x \tan x + 2x \tan y \qquad u_y = 2y \sec x + x^2 \sec^2 y$

33. $f_x = \dfrac{2y^2 + 2yz}{(x + y + z)^2}; \qquad f_y = \dfrac{2x^2 + 2xz}{(x + y + z)^2}; \qquad f_z = \dfrac{-2xy}{(x + y + z)^2}$

34. $w_x = \arctan(y - z) \qquad w_y = \dfrac{x}{1 + (y - z)^2} \qquad w_z = \dfrac{-x}{1 + (y - z)^2}$

35. $\dfrac{\partial g}{\partial x} = \dfrac{x}{x^2 + y^2 + z^2}; \qquad \dfrac{\partial g}{\partial y} = \dfrac{y}{x^2 + y^2 + z^2}; \qquad \dfrac{\partial g}{\partial z} = \dfrac{z}{x^2 + y^2 + z^2}.$

36. $h_u = v e^{uv} \sin uw + w e^{uv} \cos uw; \qquad h_v = u e^{uv} \sin uw; \qquad h_w = u e^{uv} \cos uw$

37. $f_x = 3x^2 y^2 - 4y^3 + 2, \qquad f_y = 2x^3 y - 12xy^2 - 1;$

$f_{xx} = 6xy^2, \qquad f_{yy} = 2x^3 - 24xy, \qquad f_{yx} = f_{xy} = 6x^2 y - 12y^2$

38. $g_x = 2x \ln(y - x) - \dfrac{x^2}{y - x}, \qquad g_{xx} = 2 \ln(y - x) - \dfrac{4x}{y - x} - \dfrac{x^2}{(y - x)^2};$

$g_y = \dfrac{x^2}{y - x}, \quad g_{yy} = -\dfrac{x^2}{(y - x)^2}, \qquad g_{xy} = g_{yx} = \dfrac{2x}{y - x} + \dfrac{x^2}{(y - x)^2}$

39. $g_x = y \sin xy + xy^2 \cos xy, \quad g_{xx} = 2y^2 \cos xy - xy^3 \sin xy;$

$g_y = x \sin xy + x^2 y \cos xy, \qquad g_{yy} = 2x^2 \cos xy - yx^3 \sin xy,$

$g_{xy} = g_{yx} = \sin xy + 3xy \cos xy - x^2 y^2 \sin xy$

40. $f_x = 2x e^{x/y} + \dfrac{x^2}{y} e^{x/y}, \qquad f_{xx} = 2 e^{x/y} + \dfrac{4x}{y} e^{x/y} + \dfrac{x^2}{y^2} e^{x/y};$

$f_y = -\dfrac{x^3}{y^2} e^{x/y}, \qquad f_{yy} = \dfrac{2x^3}{y^3} e^{x/y} + \dfrac{x^4}{y^4} e^{x/y}, \qquad f_{xy} = f_{yx} = -\dfrac{3x^2}{y^2} e^{x/y} - \dfrac{x^3}{y^3} e^{x/y}$

41. $f_x = 2xe^{2y}\cos(2z+1)$, $f_y = 2x^2e^{2y}\cos(2z+1)$, $f_z = -2x^2e^{2y}\sin(2z+1)$;

$f_{xx} = 2e^{2y}\cos(2z+1)$, $f_{yy} = 4x^2e^{2y}\cos(2z+1)$, $f_{zz} = -4x^2e^{2y}\cos(2z+1)$;

$f_{xy} = f_{yx} = 4xe^{2y}\cos(2z+1)$, $f_{xz} = f_{zx} = -4xe^{2y}\sin(2z+1)$, $f_{yz} = f_{zy} = -4x^2e^{2y}\sin(2z+1)$

42. $g_x = 4xyz^3 + yze^{xyz}$, $\quad g_y = 2x^2z^3 + xze^{xyz}$, $\quad g_z = 6x^2yz^2 + xye^{xyz}$;

$g_{xx} = 4yz^3 + y^2z^2e^{xyz}$, $\quad g_{yy} = x^2z^2e^{xyz}$, $\quad g_{zz} = 12x^2yz + x^2y^2e^{xyz}$;

$g_{xy} = g_{yx} = 4xz^3 + ze^{xyz} + xyz^2e^{xyz}$, $\quad g_{xz} = g_{zx} = 12xyz^2 + ye^{xyz} + xy^2ze^{xyz}$

$g_{yz} = g_{zy} = 6x^2z^2 + xe^{xyz} + x^2yze^{xyz}$

43. $\dfrac{\partial z}{\partial x} = 4x + 6\big|_{(1,2,8)} = 10$

$x = 1+t, \quad y = 2, \quad z = 8 + 10t$

44. $\dfrac{\partial z}{\partial y} = 3x\big|_{(2,-1,2)} = 6$

$x = 2, \quad y = -1+t, \quad z = 2 + 6t$

45. (a) $z_y(2,1) = \dfrac{-6y}{2\sqrt{20 - 2x^2 - 3y^2}}(2,1) = -1$; the equation for l_1 is:

$x = 2; \quad y = 1 - t; \quad z = 3 + t$

(b) $z_x(2,1) = \dfrac{-4x}{2\sqrt{20 - 2x^2 - 3y^2}}(2,1) = -\dfrac{4}{3}$; the equation for l_2 is:

$x = 2 - \dfrac{3}{4}t; \quad y = 1; \quad z = 3 + t$

(c) The normal vector for this plane is: $-\mathbf{i} - \dfrac{3}{4}\mathbf{j} - \dfrac{3}{4}\mathbf{k}$ or $4\mathbf{i} + 3\mathbf{j} + 3\mathbf{k}$;
an equation for the plane is: $4(x-2) + 3(y-1) + 3(z-3) = 0$.

46. Neither.

interior: $\{(x,y) : 0 < x < 3, 2 < y < 5\}$

boundary: $\{(x,y) : x = 0 \text{ or } x = 3, \ 2 \le y \le 5\} \cup \{(x,y) : y = 2 \text{ or } y = 5, \ 0 \le x \le 3\}$

47. Open.

interior: $\{(x,y) : 0 < x^2 + y^2 < 4\}$

boundary: $\{(0,0)\} \cup \{(x,y) : x^2 + y^2 = 4\}$

48. Closed.

interior: $\{(x,y) : x + y > 4\}$

boundary: $\{(x,y) : x + y = 4\}$

49. Closed.

interior: $\{(x,y,z) : 0 < x < 2, \ 0 < y, \ 0 < z, \ y^2 + z^2 < 4\}$

boundary: the quarter disks $x = 0, \ y^2 + z^2 \le 4$; $\quad x = 2, \ y^2 + z^2 \le 4$;

the squares $z = 0, \ 0 \le x, y \le 2$; $\quad y = 0, 0 \le x, \ z \le 2$; and

the cylindrical surface $y^2 + z^2 = 4, \ 0 \le x \le 2, \ y, \ z \ge 0$

50. Neither.

interior: $\{(x, y, z) : 0 < x^2 + y^2 < z < 4\}$

boundary: the cone $z = x^2 + y^2$ and the disk $x^2 + y^2 \le 4$, $z = 4$

51. (a) $f_x = yg'(xy)$, $f_y = xg'(xy)$; $xf_x - yf_y = xyg' - xyg' = 0$

(b) $f_{xx} = y^2g''(xy)$, $f_{yy} = x^2g''(xy)$; $x^2f_{xx} - y^2f_{yy} = x^2y^2g'' - x^2y^2g'' = 0$

52. $f_x = -\dfrac{y}{x^2 + y^2}$, $f_y = \dfrac{x}{x^2 + y^2}$; $f_{xx} = \dfrac{2xy}{(x^2 + y^2)^2}$, $f_{yy} = \dfrac{-2xy}{(x^2 + y^2)^2}$

$f_{xx} + f_{yy} = 0$

53. No. $\dfrac{\partial^2 f}{\partial y \partial x} = x^2 e^{xy} \ne y^2 e^{xy} = \dfrac{\partial^2 f}{\partial x \partial y}$

54. (a) $\displaystyle\lim_{x \to 0} f(x, 0) = \lim_{x \to 0} 0 = 0$

(b) $\displaystyle\lim_{y \to 0} f(0, y) = \lim_{y \to 0} 0 = 0$

(c) $\displaystyle\lim_{x \to 0} \frac{2x^2 mx}{x^4 + m^2 x^2} = \lim_{x \to 0} \frac{2mx}{m^2 + x^2} = 0$

(d) $\displaystyle\lim_{x \to 0} \frac{2x^2 ax^2}{x^4 + a^2 x^4} = \frac{2a}{1 + a^2}$

$\lim_{(x,y) \to (0,0)} f(x, y)$ does not exist.

CHAPTER 16

SECTION 16.1

1. $\nabla f = (6x - y)\,\mathbf{i} + (1 - x)\,\mathbf{j}$

2. $\nabla f = (2Ax + By)\mathbf{i} + (Bx + 2Cy)\mathbf{j}$

3. $\nabla f = e^{xy}\big[\,(xy + 1)\,\mathbf{i} + x^2\mathbf{j}\,\big]$

4. $\nabla f = \dfrac{1}{(x^2 + y^2)^2}\big[(y^2 - x^2 + 2xy)\mathbf{i} + (y^2 - x^2 - 2xy)\mathbf{j}\big]$

5. $\nabla f = \big[2y^2 \sin(x^2 + 1) + 4x^2y^2 \cos(x^2 + 1)\big]\,\mathbf{i} + 4xy \sin(x^2 + 1)\,\mathbf{j}$

6. $\nabla f = \dfrac{2x}{x^2 + y^2}\,\mathbf{i} + \dfrac{2y}{x^2 + y^2}\,\mathbf{j}$

7. $\nabla f = (e^{x-y} + e^{y-x})\,\mathbf{i} + (-e^{x-y} - e^{y-x})\,\mathbf{j} = (e^{x-y} + e^{y-x})(\mathbf{i} - \mathbf{j})$

8. $\nabla f = \dfrac{AD - BC}{(Cx + Dy)^2}\,[\,y\mathbf{i} - x\mathbf{j}\,]$

9. $\nabla f = (z^2 + 2xy)\,\mathbf{i} + (x^2 + 2yz)\,\mathbf{j} + (y^2 + 2zx)\,\mathbf{k}$

10. $\nabla f = \dfrac{x}{\sqrt{x^2 + y^2 + z^2}}\,\mathbf{i} + \dfrac{y}{\sqrt{x^2 + y^2 + z^2}}\,\mathbf{j} + \dfrac{z}{\sqrt{x^2 + y^2 + z^2}}\,\mathbf{k}$

11. $\nabla f = e^{-z}(2xy\,\mathbf{i} + x^2\,\mathbf{j} - x^2y\,\mathbf{k})$

12. $\nabla f = \left[\dfrac{xyz}{x + y + z} + yz\,\ln(x + y + z)\right]\mathbf{i} + \left[\dfrac{xyz}{x + y + z} + xz\,\ln(x + y + z)\right]\mathbf{j}$
$\qquad + \left[\dfrac{xyz}{x + y + z} + xy\,\ln(x + y + z)\right]\mathbf{k}$

13. $\nabla f = e^{x+2y}\cos\left(z^2 + 1\right)\mathbf{i} + 2e^{x+2y}\cos\left(z^2 + 1\right)\mathbf{j} - 2ze^{x+2y}\sin\left(z^2 + 1\right)\mathbf{k}$

14. $\nabla f = e^{yz^2/x^3}\left(-\dfrac{3yz^2}{x^4}\mathbf{i} + \dfrac{z^2}{x^3}\mathbf{j} + \dfrac{2yz}{x^3}\mathbf{k}\right)$

15. $\nabla f = \left[2y\cos(2xy) + \dfrac{2}{x}\right]\mathbf{i} + 2x\cos(2xy)\,\mathbf{j} + \dfrac{1}{z}\,\mathbf{k}$

16. $\nabla f = \left(\dfrac{2xy}{z} - 3z^4\right)\mathbf{i} + \dfrac{x^2}{z}\,\mathbf{j} - \left(\dfrac{x^2y}{z^2} + 12xz^3\right)\mathbf{k}$

17. $\nabla f = (4x - 3y)\,\mathbf{i} + (8y - 3x)\,\mathbf{j}; \quad$ at $(2, 3),\ \nabla f = -\mathbf{i} + 18\mathbf{j}$

18. $\nabla f = \dfrac{1}{(x - y)^2}(-2y\mathbf{i} + 2x\mathbf{j}), \quad \nabla f(3, 1) = -\dfrac{1}{2}\mathbf{i} + \dfrac{3}{2}\mathbf{j}$

19. $\nabla f = \dfrac{2x}{x^2 + y^2}\,\mathbf{i} + \dfrac{2y}{x^2 + y^2}\,\mathbf{j}; \quad$ at $(2, 1),\ \nabla f = \dfrac{4}{5}\mathbf{i} + \dfrac{2}{5}\mathbf{j}$

20. $\nabla f = \left(\tan^{-1}(y/x) - \dfrac{xy}{x^2 + y^2} \right) \mathbf{i} + \left(\dfrac{x^2}{x^2 + y^2} \right) \mathbf{j}, \quad \nabla f(1,1) = \left(\dfrac{\pi}{4} - \dfrac{1}{2} \right) \mathbf{i} + \dfrac{1}{2} \mathbf{j}$

21. $\nabla f = (\sin xy + xy \cos xy) \mathbf{i} + x^2 \cos xy \, \mathbf{j}; \quad$ at $(1, \pi/2), \ \nabla f = \mathbf{i}$

22. $\nabla f = e^{-(x^2 + y^2)}[(y - 2x^2 y)\mathbf{i} + (x - 2xy^2)\mathbf{j}], \quad \nabla f(1, -1) = e^{-2}(\mathbf{i} - \mathbf{j})$

23. $\nabla f = -e^{-x} \sin(z + 2y) \mathbf{i} + 2e^{-x} \cos(z + 2y) \mathbf{j} + e^{-x} \cos(z + 2y) \mathbf{k};$

at $(0, \pi/4, \pi/4), \quad \nabla f = -\frac{1}{2}\sqrt{2}\,(\mathbf{i} + 2\mathbf{j} + \mathbf{k})$

24. $\nabla f = \cos \pi z \, \mathbf{i} - \cos \pi z \, \mathbf{j} - \pi(x - y) \sin \pi z \, \mathbf{k}, \quad \nabla f \left(1, 0, \dfrac{1}{2} \right) = -\pi \mathbf{k}$

25. $\nabla f = \mathbf{i} - \dfrac{y}{\sqrt{y^2 + z^2}}\mathbf{j} - \dfrac{z}{\sqrt{y^2 + z^2}}\mathbf{k}; \quad$ at $(2, -3, 4), \quad \nabla f = \mathbf{i} + \dfrac{3}{5}\mathbf{j} - \dfrac{4}{5}\mathbf{k}$

26. $\nabla f = -\sin(xyz^2)(yz^2\mathbf{i} + xz^2\mathbf{j} + 2xyz\mathbf{k}), \quad \nabla f \left(\pi, \dfrac{1}{4}, -1 \right) = -\dfrac{\sqrt{2}}{2}\left(\dfrac{1}{4}\mathbf{i} + \pi\mathbf{j} - \dfrac{\pi}{2}\mathbf{k} \right)$

27. (a) $\nabla f(0, 2) = 4\mathbf{i}$ (b) $\nabla f \left(\frac{1}{4}\pi, \frac{1}{6}\pi \right) = \left(-1 - \dfrac{-1 + \sqrt{3}}{2\sqrt{2}} \right) \mathbf{i} + \left(-\dfrac{1}{2} - \dfrac{-1 + \sqrt{3}}{\sqrt{2}} \right) \mathbf{j}$

     (c) $\nabla f(1, e) = (1 - 2e)\mathbf{i} - 2\mathbf{j}$

28. (a) $\nabla f(1, 2, -3) = \dfrac{1}{8\sqrt{2}}\mathbf{i} + \dfrac{1}{2\sqrt{2}}\mathbf{j} - \dfrac{27}{8\sqrt{2}}\mathbf{k}$ (b) $\nabla f(1, -2, 3) = -\dfrac{5}{18}\mathbf{i} + \dfrac{1}{9}\mathbf{j} + \dfrac{1}{18}\mathbf{k}$

     (c) $\nabla f(1, e^2, \pi/6) = \dfrac{\sqrt{3}}{2}\mathbf{i} + \dfrac{\pi}{12e^2}\mathbf{j} + \mathbf{k}$

29. For the function $f(x, y) = 3x^2 - xy + y$, we have

$$f(\mathbf{x} + \mathbf{h}) - f(\mathbf{x}) = f(x + h_1, y + h_2) - f(x, y)$$
$$= 3(x + h_1)^2 - (x + h_1)(y + h_2) + (y + h_2) - \left[3x^2 - xy + y \right]$$
$$= [(6x - y)\mathbf{i} + (1 - x)\mathbf{j}] \cdot (h_1 \mathbf{i} + h_2 \mathbf{j}) + 3h_1^2 - h_1 h_2$$
$$= [(6x - y)\mathbf{i} + (1 - x)\mathbf{j}] \cdot \mathbf{h} + 3h_1^2 - h_1 h_2$$

The remainder $g(\mathbf{h}) = 3h_1^2 - h_1 h_2 = (3h_1 \mathbf{i} - h_1 \mathbf{j}) \cdot (h_1 \mathbf{i} + h_2 \mathbf{j}),$ and

$$\dfrac{|g(\mathbf{h})|}{\|\mathbf{h}\|} = \dfrac{\|3h_1 \mathbf{i} - h_1 \mathbf{j}\| \cdot \|\mathbf{h}\| \cdot \cos\theta}{\|\mathbf{h}\|} \leq \|3h_1 \mathbf{i} - h_1 \mathbf{j}\|$$

Since $\|3h_1 \mathbf{i} - h_1 \mathbf{j}\| \to 0$ as $\mathbf{h} \to \mathbf{0}$ it follows that

$$\nabla f = (6x - y)\mathbf{i} + (1 - x)\mathbf{j}$$

30. $f(\mathbf{x} + \mathbf{h}) - f(\mathbf{x}) = [(x + 2y)\mathbf{i} + (2x + 2y)\mathbf{j}] \cdot [h_1 \mathbf{i} + h_2 \mathbf{j}] + \frac{1}{2}h_1^2 + 2h_1 h_2 + h_2^2;$

$g(\mathbf{h}) = \frac{1}{2}h_1^2 + 2h_1 h_2 + h_2^2$ is $o(\mathbf{h})$.

31. For the function $f(x, y, z) = x^2y + y^2z + z^2x$, we have

$f(\mathbf{x} + \mathbf{h}) - f(\mathbf{x}) = f(x + h_1, y + h_2, z + h_3) - f(x, y, z)$

$= (x + h_1)^2 (y + h_2) + (y + h_2)^2 (z + h_3) + (z + h_3)^2 (x + h_1) - (x^2y + y^2z + z^2x)$

$= (2xy + z^2) h_1 + (2yz + x^2) h_2 + (2xz + y^2) h_3 + (2xh_2 + yh_1 + h_1h_2) h_1 +$

$\quad (2yh_3 + zh_2 + h_2h_3) h_2 + (2zh_1 + xh_3 + h_1h_3) h_3$

$= \left[(2xy + z^2) \mathbf{i} + (2yz + x^2) \mathbf{j} + (2xz + y^2) \mathbf{k} \right] \cdot \mathbf{h} + g(\mathbf{h}) \cdot \mathbf{h},$

where $\quad g(\mathbf{h}) = (2xh_2 + yh_1 + h_1h_2) \mathbf{i} + (2yh_3 + zh_2 + h_2h_3) \mathbf{j} + (2zh_1 + xh_3 + h_1h_3) \mathbf{k}$

Since $\quad \dfrac{|g(\mathbf{h})|}{\|\mathbf{h}\|} \to 0 \quad$ as $\quad \mathbf{h} \to \mathbf{0} \quad$ it follows that

$$\nabla f = (2xy + z^2) \mathbf{i} + (2yz + x^2) \mathbf{j} + (2xz + y^2) \mathbf{k}$$

32. $f(\mathbf{x} + \mathbf{h}) - f(\mathbf{x}) = \left[(2xy + 2h_2x + h_1y)\mathbf{i} + 2x^2\mathbf{j} + \dfrac{1}{z(z + h_3)}\mathbf{k} \right] \cdot (h_1\mathbf{i} + h_2\mathbf{j} + h_3\mathbf{k}) + h_1^2;$

$g(\mathbf{h}) = h_1^2 h_2$ is $o(h)$ and $\nabla f = 4xy\,\mathbf{i} = 2x^2\,\mathbf{j} + \dfrac{1}{z^2}\mathbf{k}$.

33. $\nabla f = \mathbf{F}(x, y) = 2xy\,\mathbf{i} + (1 + x^2)\,\mathbf{j} \quad \Rightarrow \quad \dfrac{\partial f}{\partial x} = 2xy \quad \Rightarrow \quad f(x, y) = x^2y + g(y)$ for some function g.

Now, $\dfrac{\partial f}{\partial y} = x^2 + g'(y) = 1 + x^2 \quad \Rightarrow \quad g'(y) = 1 \quad \Rightarrow \quad g(y) = y + C, \;\; C$ a constant.

Thus, $f(x, y) = x^2y + y + C$

34. $\nabla f = (2xy + x)\mathbf{i} + (x^2 + y)\mathbf{j} \quad \Longrightarrow \quad f_x = 2xy + x \quad \Longrightarrow \quad f(x, y) = x^2y + \dfrac{1}{2}x^2 + g(y)$

Now, $\quad f_y = x^2 + g'(y) = x^2 + y \quad \Longrightarrow \quad g'(y) = y \quad \Longrightarrow \quad g(y) = \frac{1}{2}y^2 + C$

Thus, $\quad f(x, y) = x^2y + \frac{1}{2}x^2 + \frac{1}{2}y^2 + C$

35. $\nabla f = \mathbf{F}(x, y) = (x + \sin y)\,\mathbf{i} + (x \cos y - 2y)\,\mathbf{j} \quad \Rightarrow \quad \dfrac{\partial f}{\partial x} = x + \sin y \quad \Rightarrow \quad f(x, y) = \frac{1}{2}x^2 + x \sin y + g(y)$

for some function g.

Now, $\dfrac{\partial f}{\partial y} = x \cos y + g'(y) = x \cos y - 2y \quad \Rightarrow \quad g'(y) = -2y \quad \Rightarrow \quad g(y) = -y^2 + C, \;\; C$ a constant.

Thus, $f(x, y) = \frac{1}{2}x^2 + x \sin y - y^2 + C$.

36. $\nabla f = yz\mathbf{i} + (xz + 2yz)\mathbf{j} + (xy + y^2)\mathbf{k} \quad \Longrightarrow \quad f_x = yz \quad \Longrightarrow \quad f(x, y, z) = xyz + g(y, z)$.

$f_y = xz + g_y = xz + 2yz \quad \Longrightarrow \quad g_y = 2yz \quad \Longrightarrow \quad g(y, z) = y^2z + h(z) \quad \Longrightarrow \quad f(x, y, z) = xyz + y^2z + h(z)$.

$f_x = xy + y^2 + h'(z) = xy + y^2 \quad \Longrightarrow \quad h'(z) = 0 \quad \Longrightarrow \quad h(z) = C$.

Thus, $\quad f(x, y, z) = xyz + y^2z + C$.

37. With $r = (x^2 + y^2 + z^2)^{1/2}$ we have

$$\frac{\partial r}{\partial x} = \frac{x}{r}, \quad \frac{\partial r}{\partial y} = \frac{y}{r}, \quad \frac{\partial r}{\partial z} = \frac{z}{r}.$$

(a)

$$\nabla(\ln r) = \frac{\partial}{\partial x}(\ln r)\,\mathbf{i} + \frac{\partial}{\partial y}(\ln r)\,\mathbf{j} + \frac{\partial}{\partial z}(\ln r)\mathbf{k}$$

$$= \frac{1}{r}\frac{\partial r}{\partial x}\,\mathbf{i} + \frac{1}{r}\frac{\partial r}{\partial y}\,\mathbf{j} + \frac{1}{r}\frac{\partial r}{\partial z}\,\mathbf{k}$$

$$= \frac{x}{r^2}\mathbf{i} + \frac{y}{r^2}\mathbf{j} + \frac{z}{r^2}\mathbf{k} = \frac{\mathbf{r}}{r^2}$$

(b)
$$\nabla(\sin r) = \frac{\partial}{\partial x}(\sin r)\,\mathbf{i} + \frac{\partial}{\partial y}(\sin r)\,\mathbf{j} + \frac{\partial}{\partial z}(\sin r)\mathbf{k}$$

$$= \cos r\,\frac{\partial r}{\partial x}\,\mathbf{i} + \cos r\,\frac{\partial r}{\partial y}\,\mathbf{j} + \cos r\,\frac{\partial r}{\partial z}\,\mathbf{k}$$

$$= (\cos r)\frac{x}{r}\,\mathbf{i} + (\cos r)\frac{y}{r}\,\mathbf{j} + (\cos r)\frac{z}{r}\,\mathbf{k}$$

$$= \left(\frac{\cos r}{r}\right)\mathbf{r}$$

(c) $\nabla e^r = \left(\dfrac{e^r}{r}\right)\mathbf{r}$ [same method as in (a) and (b)]

38. With $r^n = (x^2 + y^2 + z^2)^{n/2}$ we have
$$\frac{\partial r^n}{\partial x} = \frac{n}{2}(x^2 + y^2 + z^2)^{(n/2)-1}(2x) = n(x^2 + y^2 + z^2)^{(n-2)/2}x = nr^{n-2}x.$$

Similarly
$$\frac{\partial r^n}{\partial y} = nr^{n-2}y \quad \text{and} \quad \frac{\partial r^n}{\partial z} = nr^{n-2}z.$$

Therefore
$$\nabla r^n = nr^{n-2}x\mathbf{i} + nr^{n-2}y\mathbf{j} + nr^{n-2}z\mathbf{k} = nr^{n-2}(x\mathbf{i} + y\mathbf{j} + z\mathbf{k}) = nr^{n-2}\mathbf{r}$$

39. (a) $\nabla f = 2x\,\mathbf{i} + 2y\,\mathbf{j} = \mathbf{0} \implies x = y = 0; \quad \nabla f = \mathbf{0}$ at $(0,0)$.

(b)

(c) f has an absolute minimum at $(0,0)$

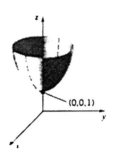

(0,0,1)

40. (a) $\nabla f = \dfrac{-1}{\sqrt{4 - x^2 - y^2}}(x\mathbf{i} + y\mathbf{j}) = \mathbf{0}$ at $(0,0)$ (b)

(c) f has a maximum at $(0,0)$

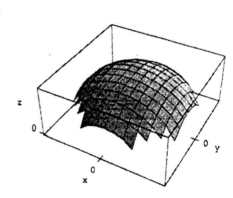

41. (a) Let $\mathbf{c} = c_1\mathbf{i} + c_2\mathbf{j} + c_3\mathbf{k}$. First, we take $\mathbf{h} = h\mathbf{i}$. Since $\mathbf{c} \cdot \mathbf{h}$ is $o(\mathbf{h})$,

$$0 = \lim_{\mathbf{h} \to \mathbf{0}} \frac{\mathbf{c} \cdot \mathbf{h}}{\|\mathbf{h}\|} = \lim_{h \to 0} \frac{c_1 h}{h} = c_1.$$

Similarly, $c_2 = 0$ and $c_3 = 0$.

(b) $(\mathbf{y} - \mathbf{z}) \cdot \mathbf{h} = [f(\mathbf{x} + \mathbf{h}) - f(\mathbf{x}) - \mathbf{z} \cdot \mathbf{h}] + [\mathbf{y} \cdot \mathbf{h} - f(\mathbf{x} + \mathbf{h}) + f(\mathbf{x})] = o(\mathbf{h}) + o(\mathbf{h}) = o(\mathbf{h})$,

so that, by part (a), $\mathbf{y} - \mathbf{z} = \mathbf{0}$.

42. $\displaystyle \lim_{\mathbf{h} \to \mathbf{0}} g(\mathbf{h}) = \lim_{\mathbf{h} \to \mathbf{0}} \left(\|\mathbf{h}\| \frac{\mathbf{g}(\mathbf{h})}{\|\mathbf{h}\|} \right) = \left(\lim_{\mathbf{h} \to \mathbf{0}} \|\mathbf{h}\| \right) \left(\lim_{\mathbf{h} \to \mathbf{0}} \frac{\mathbf{g}(\mathbf{h})}{\|\mathbf{h}\|} \right) = (\mathbf{0})(\mathbf{0}) = (\mathbf{0}).$

43. (a) In Section 15.6 we showed that f was not continuous at $(0,0)$. It is therefore not differentiable at $(0,0)$.

(b) For $(x, y) \neq (0,0)$, $\dfrac{\partial f}{\partial x} = \dfrac{2y(y^2 - x^2)}{(x^2 + y^2)^2}$. As (x, y) tends to $(0,0)$ along the positive y-axis,

$\dfrac{\partial f}{\partial x} = \dfrac{2y^3}{y^4} = \dfrac{2}{y}$ tends to ∞.

SECTION 16.2

1. $\nabla f = 2x\mathbf{i} + 6y\mathbf{j}$, $\nabla f(1, 1) = 2\mathbf{i} + 6\mathbf{j}$, $\mathbf{u} = \frac{1}{2}\sqrt{2}\,(\mathbf{i} - \mathbf{j})$, $f'_{\mathbf{u}}(1, 1) = \nabla f(1, 1) \cdot \mathbf{u} = -2\sqrt{2}$

2. $\nabla f = [1 + \cos(x + y)]\mathbf{i} + \cos(x + y)\mathbf{j}$, $\nabla f(0, 0) = 2\mathbf{i} + \mathbf{j}$, $\mathbf{u} = \dfrac{1}{\sqrt{5}}(2\mathbf{i} + \mathbf{j})$,

$f'_{\mathbf{u}}(0, 0) = \nabla f(0, 0) \cdot \mathbf{u} = \sqrt{5}$

3. $\nabla f = (e^y - ye^x)\,\mathbf{i} + (xe^y - e^x)\,\mathbf{j}$, $\nabla f(1, 0) = \mathbf{i} + (1 - e)\mathbf{j}$, $\mathbf{u} = \dfrac{1}{5}(3\mathbf{i} + 4\mathbf{j})$,

$f'_{\mathbf{u}}(1, 0) = \nabla f(1, 0) \cdot \mathbf{u} = \dfrac{1}{5}(7 - 4e)$

4. $\nabla f = \dfrac{1}{(x - y)^2}(-2y\mathbf{i} + 2x\mathbf{j})$, $\nabla f(1, 0) = 2\mathbf{j}$, $\mathbf{u} = \dfrac{1}{2}(\mathbf{i} - \sqrt{3}\mathbf{j})$,

$f'_{\mathbf{u}}(1, 0) = \nabla f(1, 0) \cdot \mathbf{u} = -\sqrt{3}$

5. $\nabla f = \dfrac{(a - b)y}{(x + y)^2}\,\mathbf{i} + \dfrac{(b - a)x}{(x + y)^2}\,\mathbf{j}$, $\nabla f(1, 1) = \dfrac{a - b}{4}(\mathbf{i} - \mathbf{j})$, $\mathbf{u} = \dfrac{1}{2}\sqrt{2}\,(\mathbf{i} - \mathbf{j})$,

$f'_{\mathbf{u}}(1, 1) = \nabla f(1, 1) \cdot \mathbf{u} = \dfrac{1}{4}\sqrt{2}\,(a - b)$

6. $\nabla f = \dfrac{1}{(cx + dy)^2}\,[(d - c)y\mathbf{i} + (c - d)x\mathbf{j}]$, $\nabla f(1, 1) = \dfrac{d - c}{(c + d)^2}(\mathbf{i} - \mathbf{j})$, $\mathbf{u} = \dfrac{1}{\sqrt{c^2 + d^2}}(c\mathbf{i} - d\mathbf{j})$,

$f'_{\mathbf{u}}(1, 1) = \nabla f(1, 1) \cdot \mathbf{u} = \dfrac{d - c}{(c + d)\sqrt{c^2 + d^2}}$

7. $\nabla f = \dfrac{2x}{x^2 + y^2}\,\mathbf{i} + \dfrac{2y}{x^2 + y^2}\,\mathbf{j}$, $\nabla f(0, 1) = 2\mathbf{j}$, $\mathbf{u} = \dfrac{1}{\sqrt{65}}(8\mathbf{i} + \mathbf{j})$,

$f'_{\mathbf{u}}(0, 1) = \nabla f(0, 1) \cdot \mathbf{u} = \dfrac{2}{\sqrt{65}}$

8. $\nabla f = 2xy\mathbf{i} + (x^2 + \sec^2 y)\mathbf{j}$, $\nabla f(-1, \frac{\pi}{4}) = -\frac{\pi}{2}\mathbf{i} + 3\mathbf{j}$, $\mathbf{u} = \frac{1}{\sqrt{5}}(\mathbf{i} - 2\mathbf{j})$

$f'_{\mathbf{u}}\left(-1, \frac{\pi}{4}\right) = \nabla f\left(-1, \frac{\pi}{4}\right) \cdot \mathbf{u} = -\frac{1}{\sqrt{5}}\left(\frac{\pi}{2} + 6\right)$

9. $\nabla f = (y + z)\mathbf{i} + (x + z)\mathbf{j} + (y + x)\mathbf{k}$, $\nabla f(1, -1, 1) = 2\mathbf{j}$, $\mathbf{u} = \frac{1}{6}\sqrt{6}(\mathbf{i} + 2\mathbf{j} + \mathbf{k})$,

$f'_{\mathbf{u}}(1, -1, 1) = \nabla f(1, -1, 1) \cdot \mathbf{u} = \frac{2}{3}\sqrt{6}$

10. $\nabla f = (z^2 + 2xy)\mathbf{i} + (x^2 + 2yz)\mathbf{j} + (y^2 + 2zx)\mathbf{k}$, $\nabla f(1, 0, 1) = \mathbf{i} + \mathbf{j} + 2\mathbf{k}$, $\mathbf{u} = \frac{1}{\sqrt{10}}(3\mathbf{j} - \mathbf{k})$

$f'_{\mathbf{u}}(1, 0, 1) = \nabla f(1, 0, 1) \cdot \mathbf{u} = \frac{\sqrt{10}}{10}$

11. $\nabla f = 2\left(x + y^2 + z^3\right)\left(\mathbf{i} + 2y\mathbf{j} + 3z^2\mathbf{k}\right)$, $\nabla f(1, -1, 1) = 6(\mathbf{i} - 2\mathbf{j} + 3\mathbf{k})$, $\mathbf{u} = \frac{1}{2}\sqrt{2}(\mathbf{i} + \mathbf{j})$,

$f'_{\mathbf{u}}(1, -1, 1) = \nabla f(1, -1, 1) \cdot \mathbf{u} = -3\sqrt{2}$

12. $\nabla f = (2Ax + Byz)\mathbf{i} + (Bxz + 2Cy)\mathbf{j} + Bxy\mathbf{k}$, $\nabla f(1, 2, 1) = 2(A + B)\mathbf{i} + (B + 4C)\mathbf{j} + 2B\mathbf{k}$

$\mathbf{u} = \frac{1}{\sqrt{A^2 + B^2 + C^2}}(A\mathbf{i} + B\mathbf{j} + C\mathbf{k})$; $f'_{\mathbf{u}}(1, 2, 1) = \nabla f(1, 2, 1) \cdot \mathbf{u} = \frac{2A^2 + B^2 + 2AB + 6BC}{\sqrt{A^2 + B^2 + C^2}}$

13. $\nabla f = \tan^{-1}(y + z)\mathbf{i} + \frac{x}{1 + (y + z)^2}\mathbf{j} + \frac{x}{1 + (y + z)^2}\mathbf{k}$, $\nabla f(1, 0, 1) = \frac{\pi}{4}\mathbf{i} + \frac{1}{2}\mathbf{j} + \frac{1}{2}\mathbf{k}$,

$\mathbf{u} = \frac{1}{\sqrt{3}}(\mathbf{i} + \mathbf{j} - \mathbf{k})$, $f'_{\mathbf{u}}(1, 0, 1) = \nabla f(1, 0, 1) \cdot \mathbf{u} = \frac{\pi}{4\sqrt{3}} = \frac{\sqrt{3}}{12}\pi$

14. $\nabla f = (y^2 \cos z - 2\pi yz^2 \cos \pi x + 6zx)\mathbf{i} + (2xy \cos z - 2z^2 \sin \pi x)\mathbf{j} + (-xy^2 \sin z - 4yz \sin \pi x + 3x^2)\mathbf{k}$

$\nabla f(0, -1, \pi) = (2\pi^3 - 1)\mathbf{i}$; $\mathbf{u} = \frac{1}{3}(2\mathbf{i} - \mathbf{j} + 2\mathbf{k})$, $f'_{\mathbf{u}}(0, -1, \pi) = \nabla f(0, -1, \pi) \cdot \mathbf{u} = \frac{2}{3}(2\pi^3 - 1)$.

15. $\nabla f = \frac{x}{x^2 + y^2}\mathbf{i} + \frac{y}{x^2 + y^2}\mathbf{j}$, $\mathbf{u} = \frac{1}{\sqrt{x^2 + y^2}}(-x\mathbf{i} - y\mathbf{j})$, $f'_{\mathbf{u}}(x, y) = \nabla f \cdot \mathbf{u} = -\frac{1}{\sqrt{x^2 + y^2}}$

16. $\nabla f = e^{xy}\left[(y^2 + xy^3 - y^3)\mathbf{i} + (x - 1)(2y + xy^2)\mathbf{j}\right]$, $\nabla f(0, 1) = -2\mathbf{j}$

$\mathbf{u} = \frac{1}{\sqrt{5}}(-\mathbf{i} + 2\mathbf{j})$, $f'_{\mathbf{u}}(0, 1) = \nabla f(0, 1) \cdot \mathbf{u} = -\frac{4}{5}\sqrt{5}$

17. $\nabla f = (2Ax + 2By)\mathbf{i} + (2Bx + 2Cy)\mathbf{j}$, $\nabla f(a, b) = (2aA + 2bB)\mathbf{i} + (2aB + 2bC)\mathbf{j}$

   (a) $\mathbf{u} = \frac{1}{2}\sqrt{2}(-\mathbf{i} + \mathbf{j})$, $f'_{\mathbf{u}}(a, b) = \nabla f(a, b) \cdot \mathbf{u} = \sqrt{2}\left[a(B - A) + b(C - B)\right]$

   (b) $\mathbf{u} = \frac{1}{2}\sqrt{2}(\mathbf{i} - \mathbf{j})$, $f'_{\mathbf{u}}(a, b) = \nabla f(a, b) \cdot \mathbf{u} = \sqrt{2}\left[a(A - B) + b(B - C)\right]$

18. $\nabla f = \frac{z}{x}\mathbf{i} - \frac{z}{y}\mathbf{j} + \ln\left(\frac{x}{y}\right)\mathbf{k}$, $\nabla f(1, 1, 2) = 2\mathbf{i} - 2\mathbf{j}$

$\mathbf{u} = \frac{1}{\sqrt{3}}(\mathbf{i} + \mathbf{j} - \mathbf{k})$; $f'_{\mathbf{u}}(1, 1, 2) = \nabla f(1, 1, 2) \cdot \mathbf{u} = 0$

19. $\nabla f = e^{y^2 - z^2}(\mathbf{i} + 2xy\mathbf{j} - 2xz\mathbf{k})$, $\nabla f(1, 2, -2) = \mathbf{i} + 4\mathbf{j} + 4\mathbf{k}$, $\mathbf{r}'(t) = \mathbf{i} - 2\sin(t-1)\mathbf{j} - 2e^{t-1}\mathbf{k}$,

at $(1, 2, -2)$ $t = 1$, $\mathbf{r}'(1) = \mathbf{i} - 2\mathbf{k}$, $\mathbf{u} = \frac{1}{5}\sqrt{5}\,(\mathbf{i} - 2\mathbf{k})$, $f_\mathbf{u}'(1, 2, -2) = \nabla f(1, 2, -2) \cdot \mathbf{u} = -\dfrac{7}{5}\sqrt{5}$

20. $\nabla f = 2x\mathbf{i} + z\mathbf{j} + y\mathbf{k}$, $\nabla f(1, -3, 2) = 2\mathbf{i} + 2\mathbf{j} - 3\mathbf{k}$

Direction: $\mathbf{r}'(-1) = -2\mathbf{i} + 3\mathbf{j} - 3\mathbf{k}$, $\mathbf{u} = \dfrac{1}{\sqrt{22}}(-2\mathbf{i} + 3\mathbf{j} - 3\mathbf{k})$, $f_\mathbf{u}'(1, -3, 2) = \nabla f(1, -3, 2) \cdot \mathbf{u} = \dfrac{1}{2}\sqrt{22}$

21. $\nabla f = (2x + 2yz)\mathbf{i} + (2xz - z^2)\mathbf{j} + (2xy - 2yz)\mathbf{k}$, $\nabla f(1, 1, 2) = 6\mathbf{i} - 2\mathbf{k}$

The vectors $\mathbf{v} = \pm(2\mathbf{i} + \mathbf{j} - 3\mathbf{k})$ are direction vectors for the given line; $\mathbf{u} = \pm\left(\dfrac{1}{\sqrt{14}}[2\mathbf{i} + \mathbf{j} - 3\mathbf{k}]\right)$

are corresponding unit vectors; $f_\mathbf{u}'(1, 1, 2) = \nabla f(1, 1, 2) \cdot (\pm\mathbf{u}) = \pm\dfrac{18}{\sqrt{14}}$

22. $\nabla f = e^x(\cos \pi yz\,\mathbf{i} - \pi z \sin \pi yz\,\mathbf{j} - \pi y \sin \pi yz\,\mathbf{k})$, $\nabla f(0, 1, \frac{1}{2}) = -\dfrac{\pi}{2}\mathbf{j} - \pi\mathbf{k}$

The vectors $\mathbf{v} = \pm(2\mathbf{i} + 3\mathbf{j} + 5\mathbf{k})$ are direction vectors for the line; $\mathbf{u} = \pm\left(\dfrac{1}{\sqrt{38}}[2\mathbf{i} + 3\mathbf{j} + 5\mathbf{k}]\right)$

are corresponding unit vectors; $f_\mathbf{u}'(0, 1, \frac{1}{2}) = \nabla f(0, 1, \frac{1}{2}) \cdot (\pm\mathbf{u}) = \mp\dfrac{13\pi}{2\sqrt{38}}$

23. $\nabla f = 2y^2 e^{2x}\,\mathbf{i} + 2y e^{2x}\,\mathbf{j}$, $\nabla f(0, 1) = 2\mathbf{i} + 2\mathbf{j}$, $\|\nabla f\| = 2\sqrt{2}$, $\dfrac{\nabla f}{\|\nabla f\|} = \dfrac{1}{\sqrt{2}}(\mathbf{i} + \mathbf{j})$

f increases most rapidly in the direction $\mathbf{u} = \dfrac{1}{\sqrt{2}}(\mathbf{i} + \mathbf{j})$; the rate of change is $2\sqrt{2}$.

f decreases most rapidly in the direction $\mathbf{v} = -\dfrac{1}{\sqrt{2}}(\mathbf{i} + \mathbf{j})$; the rate of change is $-2\sqrt{2}$.

24. $\nabla f = [1 + \cos(x + 2y)]\mathbf{i} + 2\cos(x + 2y)\mathbf{j}$, $\nabla f(0, 0) = 2\mathbf{i} + 2\mathbf{j}$

Fastest increase in direction $\mathbf{u} = \dfrac{1}{\sqrt{2}}(\mathbf{i} + \mathbf{j})$,   rate of change $\|\nabla f(0, 0)\| = 2\sqrt{2}$

Fastest decrease in direction $\mathbf{v} = -\dfrac{1}{\sqrt{2}}(\mathbf{i} + \mathbf{j})$,   rate of change $-2\sqrt{2}$

25. $\nabla f = \dfrac{x}{\sqrt{x^2 + y^2 + z^2}}\mathbf{i} + \dfrac{y}{\sqrt{x^2 + y^2 + z^2}}\mathbf{j} + \dfrac{z}{\sqrt{x^2 + y^2 + z^2}}\mathbf{k}$,

$\nabla f(1, -2, 1) = \dfrac{1}{\sqrt{6}}(\mathbf{i} - 2\mathbf{j} + \mathbf{k})$, $\|\nabla f\| = 1$

f increases most rapidly in the direction $\mathbf{u} = \dfrac{1}{\sqrt{6}}(\mathbf{i} - 2\mathbf{j} + \mathbf{k})$; the rate of change is 1.

f decreases most rapidly in the direction $\mathbf{v} = -\dfrac{1}{\sqrt{6}}(\mathbf{i} - 2\mathbf{j} + \mathbf{k})$; the rate of change is -1.

26. $\nabla f = (2xze^y + z^2)\mathbf{i} + x^2 ze^y\mathbf{j} + (x^2 e^y + 2xz)\mathbf{k}$, $\nabla f(1, \ln 2, 2) = 12\mathbf{i} + 4\mathbf{j} + 6\mathbf{k}$

Fastest increase in direction $\mathbf{u} = \dfrac{1}{7}(6\mathbf{i} + 2\mathbf{j} + 3\mathbf{k})$,   rate of change $\|\nabla f(1, \ln 2, 2)\| = 14$

Fastest decrease in direction $\mathbf{v} = -\dfrac{1}{7}(6\mathbf{i} + 2\mathbf{j} + 3\mathbf{k})$,   rate of change -14

27. $\nabla f = f'(x_0)\mathbf{i}$. If $f'(x_0) \neq 0$, the gradient points in the direction in which f increases: to the right if $f'(x_0) > 0$, to the left if $f'(x_0) < 0$.

28. 0;   the vector $c = \dfrac{\partial f}{\partial y}(x_0, y_0)\mathbf{i} - \dfrac{\partial f}{\partial x}(x_0, y_0)\mathbf{j}$ is perpendicular to the gradient $\nabla f(x_0, y_0)$ and

points along the level curve of f at (x_0, y_0).

29. (a) $\lim\limits_{h \to 0} \dfrac{f(h, 0) - f(0, 0)}{h} = \lim\limits_{h \to 0} \dfrac{\sqrt{h^2}}{h} = \lim\limits_{h \to 0} \dfrac{|h|}{h}$ does not exist

   (b)   no; by Theorem 16.2.5 f cannot be differentiable at $(0, 0)$

30. (a) $\dfrac{g(\mathbf{x} + \mathbf{h})o(\mathbf{h})}{\|\mathbf{h}\|} = g(\mathbf{x} + \mathbf{h}) \dfrac{o(\mathbf{h})}{\|\mathbf{h}\|} \to g(\mathbf{x})(0) = 0$

   (b) $\dfrac{|[g(\mathbf{x} + \mathbf{h}) - g(\mathbf{x})]\nabla f(\mathbf{x}) \cdot \mathbf{h}|}{\|\mathbf{h}\|} \leq \dfrac{\|[g(\mathbf{x} + \mathbf{h}) - g(\mathbf{x})]\nabla f(\mathbf{x})\|\|\mathbf{h}\|}{\|\mathbf{h}\|}$

   by Schwarz's inequality

$$= |g(\mathbf{x} + \mathbf{h}) - g(\mathbf{x})| \cdot \|\nabla f(\mathbf{x})\| \to 0$$

31. $\nabla \lambda(x, y) = -\frac{8}{3}x\mathbf{i} - 6y\mathbf{j}$

   (a) $\nabla \lambda(1, -1) = -\dfrac{8}{3}\mathbf{i} = 6\mathbf{j}$, $\mathbf{u} = \dfrac{-\nabla \lambda(1, -1)}{\|\nabla \lambda(1, -1)\|} = \dfrac{\frac{8}{3}\mathbf{i} - 6\mathbf{j}}{\frac{2}{3}\sqrt{97}}$, $\lambda'_{\mathbf{u}}(1, -1) = \nabla \lambda(1, -1) \cdot \mathbf{u} = -\dfrac{2}{3}\sqrt{97}$

   (b) $\mathbf{u} = \mathbf{i}$, $\lambda'_{\mathbf{u}}(1, 2) = \nabla \lambda(1, 2) \cdot \mathbf{u} = \left(-\frac{8}{3}\mathbf{i} - 12\mathbf{j}\right) \cdot \mathbf{i} = -\frac{8}{3}$

   (c) $\mathbf{u} = \frac{1}{2}\sqrt{2}\,(\mathbf{i} + \mathbf{j})$, $\lambda'_{\mathbf{u}}(2, 2) = \nabla \lambda(2, 2) \cdot \mathbf{u} = \left(-\frac{16}{3}\mathbf{i} - 12\mathbf{j}\right) \cdot \left[\frac{1}{2}\sqrt{2}\,(\mathbf{i} + \mathbf{j})\right] = -\frac{26}{3}\sqrt{2}$

32. $\nabla I = -4x\mathbf{i} - 2y\mathbf{j}$. We want the curve $\mathbf{r}(t) = x(t)\mathbf{i} + y(t)\mathbf{j}$ which begins at $(-2, 1)$ and has tangent vector $r'(t)$ in the direction ∇I. We can satisfy these conditions by setting

$$x'(t) = -4x(t),\ x(0) = -2;\quad y'(t) = -2y(t),\ y(0) = 1.$$

These equations imply that

$$x(t) = -2e^{-4t},\quad y(t) \doteq e^{-2t}.$$

Eliminating the parameter, we get $x = -2y^2$; the particle will follow the parabolic path $x = -2y^2$ toward the origin.

33. (a) The projection of the path onto the xy-plane is the curve

$$C:\ \mathbf{r}(t) = x(t)\mathbf{i} + y(t)\mathbf{j}$$

which begins at $(1, 1)$ and at each point has its tangent vector in the direction of $-\nabla f$. Since

$$\nabla f = 2x\mathbf{i} + 6y\mathbf{j},$$

we have the initial-value problems

$$x'(t) = -2x(t),\quad x(0) = 1\quad \text{and}\quad y'(t) = -6y(t),\quad y(0) = 1.$$

From Theorem 7.6.1 we find that

$$x(t) = e^{-2t}\quad \text{and}\quad y(t) = e^{-6t}.$$

Eliminating the parameter t, we find that C is the curve $y = x^3$ from $(1, 1)$ to $(0, 0)$.

(b) Here

$$x'(t) = -2x(t), \quad x(0) = 1 \qquad \text{and} \qquad y'(t) = -6y(t), \quad y(0) = -2$$

so that

$$x(t) = e^{-2t} \qquad \text{and} \qquad y(t) = -2e^{-6t}.$$

Eliminating the parameter t, we find that the projection of the path onto the xy-plane is the curve $y = -2x^3$ from $(1, -2)$ to $(0, 0)$.

34. $z = f(x, y) = \dfrac{1}{2}x^2 - y^2$; $\nabla f = x\mathbf{i} - 2y\mathbf{j}$, so we choose the projection $r(t) = x(t)\mathbf{i} + y(t)\mathbf{j}$ of the path onto the xy-plane such that $x'(t) = x(t)$, $y'(t) = -2y(t)$

(a) With initial point $(-1, 1, -\frac{1}{2})$, we get $x(t) = -e^t$, $y(t) = e^{-2t}$, or $y = \dfrac{1}{x^2}$ from $(-1, 1)$, in the direction of decreasing x.

(b) With initial point $(1, 0, \frac{1}{2})$, we get $x(t) = e^t$, $y(t) = 0$, or the x-axis from $(1, 0)$, in the direction of increasing x.

35. The projection of the path onto the xy-plane is the curve

$$C: \ \mathbf{r}(t) = x(t)\mathbf{i} + y(t)\mathbf{j}$$

which begins at (a, b) and at each point has its tangent vector in the direction of $-\nabla f = -\left(2a^2 x\mathbf{i} + 2b^2 y\mathbf{j}\right)$. We can satisfy these conditions by setting

$$x'(t) = -2a^2 x(t), \quad x(0) = a^2 \qquad \text{and} \qquad y'(t) = -2b^2 y(t), \quad y(0) = b$$

so that

$$x(t) = ae^{-2a^2 t} \qquad \text{and} \qquad y(t) = be^{-2b^2 t}.$$

Since

$$\left[\frac{x}{a}\right]^{b^2} = \left(e^{-2a^2 t}\right)^{b^2} = \left[\frac{y}{b}\right]^{a^2},$$

C is the curve $(b)^{a^2} x^{b^2} = (a)^{b^2} y^{a^2}$ from (a, b) to $(0, 0)$.

36. The particle must go in he direction $-\nabla T = -e^y \cos x\mathbf{i} - e^y \sin x\mathbf{j}$, so we set $x'(t) = -e^{y(t)} \cos x(t)$, $y'(t) = -e^{y(t)} \sin x(t)$. Dividing, we have $\dfrac{y'(t)}{x'(t)} = \dfrac{\sin x(t)}{\cos x(t)}$ or $\dfrac{dy}{dx} = \tan x$. With initial point $(0, 0)$, we get $y = \ln|\sec x|$, in the direction of decreasing x (since $x'(0) < 0$).

37. We want the curve

$$C: \ \mathbf{r}(t) = x(t)\mathbf{i} + y(t)\mathbf{j}$$

which begins at $(\pi/4, 0)$ and at each point has its tangent vector in the direction of

$$\nabla T = -\sqrt{2}\, e^{-y} \sin x\,\mathbf{i} - \sqrt{2}\, e^{-y} \cos x\,\mathbf{j}.$$

From

$$x'(t) = -\sqrt{2}\, e^{-y} \sin x \qquad \text{and} \qquad y'(t) = -\sqrt{2}\, e^{-y} \cos x$$

we obtain

$$\frac{dy}{dx} = \frac{y'(t)}{x'(t)} = \cot x$$

so that

$$y = \ln|\sin x| + C.$$

Since $y = 0$ when $x = \pi/4$, we get $C = \ln\sqrt{2}$ and $y = \ln|\sqrt{2}\sin x|$. As $\nabla T(\pi/4, 0) = -\mathbf{i} - \mathbf{j}$, the curve $y = \ln|\sqrt{2}\sin x|$ is followed in the direction of decreasing x.

38. $\nabla z = (1 - 2x)\mathbf{i} + (2 - 6y)\mathbf{j}$,   so the projection of the path onto the xy-plane satisfies $x'(t) = 1 - 2x(t)$, $y'(t) = 2 - 6y(t)$,   or $\dfrac{dy}{dx} = \dfrac{2 - 6y}{1 - 2x}$.   With initial point $(0, 0)$,   this gives the curve $3y = (2x - 1)^3 + 1$, in the direction of increasing x.

39.   (a)

$$\lim_{h \to 0} \frac{f\left(2 + h,\, (2 + h)^2\right) - f(2, 4)}{h} = \lim_{h \to 0} \frac{3(2 + h)^2 + (2 + h)^2 - 16}{h}$$

$$= \lim_{h \to 0} 4\left[\frac{4h + h^2}{h}\right] = \lim_{h \to 0} 4(4 + h) = 16$$

(b)

$$\lim_{h \to 0} \frac{f\left(\dfrac{h + 8}{4},\, 4 + h\right) - f(2, 4)}{h} = \lim_{h \to 0} \frac{3\left(\dfrac{h + 8}{4}\right)^2 + (4 + h) - 16}{h}$$

$$= \lim_{h \to 0} \frac{\frac{3}{16}h^2 + 3h + 12 + 4 + h - 16}{h}$$

$$= \lim_{h \to 0} \left(\tfrac{3}{16}h + 4\right) = 4$$

   (c) $\mathbf{u} = \frac{1}{17}\sqrt{17}\,(\mathbf{i} + 4\mathbf{j})$, $\nabla f(2, 4) = 12\mathbf{i} + \mathbf{j}$; $f'_{\mathbf{u}}(2, 4) = \nabla f(2, 4) \cdot \mathbf{u} = \frac{16}{17}\sqrt{17}$

   (d)   The limits computed in (a) and (b) are not directional derivatives. In (a) and (b) we have, in essence, computed $\nabla f(2, 4) \cdot \mathbf{r}_0$ taking $\mathbf{r}_0 = \mathbf{i} + 4\mathbf{j}$ in (a) and $\mathbf{r}_0 = \frac{1}{4}\mathbf{i} + \mathbf{j}$ in (b). In neither case is $\mathbf{r}_0$ a unit vector.

40. $\nabla f = \dfrac{-GMm}{(x^2 + y^2 + z^2)^{3/2}}(x\mathbf{i} + y\mathbf{j} + z\mathbf{k}) = \dfrac{-GMm}{\|\mathbf{r}\|^3}\mathbf{r}$

41.   (a) $\mathbf{u} = \cos\theta\,\mathbf{i} + \sin\theta\,\mathbf{j}$, $\nabla f(x, y) = \dfrac{\partial f}{\partial x}\mathbf{i} + \dfrac{\partial f}{\partial y}\mathbf{j}$;

$$f'_{\mathbf{u}}(x, y) = \nabla f \cdot \mathbf{u} = \left(\frac{\partial f}{\partial x}\mathbf{i} + \frac{\partial f}{\partial y}\mathbf{j}\right) \cdot (\cos\theta\,\mathbf{i} + \sin\theta\,\mathbf{j}) = \frac{\partial f}{\partial x}\cos\theta + \frac{\partial f}{\partial y}\sin\theta$$

   (b) $\nabla f = \left(3x^2 + 2y - y^2\right)\mathbf{i} + (2x - 2xy)\mathbf{j}$, $\nabla f(-1, 2) = 3\mathbf{i} + 2\mathbf{j}$

$$f'_{\mathbf{u}}(-1, 2) = 3\cos(2\pi/3) + 2\sin(2\pi/3) = \frac{2\sqrt{3} - 3}{2}$$

42. $f'_{\mathbf{u}}(x, y) = \dfrac{\partial f}{\partial x} \cos \dfrac{5\pi}{4} + \dfrac{\partial f}{\partial y} \sin \dfrac{5\pi}{4} = 2xe^{2y}\left(-\dfrac{\sqrt{2}}{2}\right) + 2x^2 e^{2y}\left(-\dfrac{\sqrt{2}}{2}\right) = -\sqrt{2}\,xe^{2y}(1+x)$

$f'_{\mathbf{u}}(2, \ln 2) = -\sqrt{2} \cdot 2 \cdot e^{2\ln 2}(1+2) = -24\sqrt{2}$

43. $\nabla(fg) = \dfrac{\partial(fg)}{\partial x}\mathbf{i} + \dfrac{\partial(fg)}{\partial y}\mathbf{j} + \dfrac{\partial(fg)}{\partial z}\mathbf{k} = \left(f\dfrac{\partial g}{\partial x} + g\dfrac{\partial f}{\partial x}\right)\mathbf{i} + \left(f\dfrac{\partial g}{\partial y} + g\dfrac{\partial f}{\partial y}\right)\mathbf{j} + \left(f\dfrac{\partial g}{\partial z} + g\dfrac{\partial f}{\partial z}\right)\mathbf{k}$

$$= f\left(\dfrac{\partial g}{\partial x}\mathbf{i} + \dfrac{\partial g}{\partial y}\mathbf{j} + \dfrac{\partial g}{\partial z}\mathbf{k}\right) + g\left(\dfrac{\partial f}{\partial x}\mathbf{i} + \dfrac{\partial f}{\partial y}\mathbf{j} + \dfrac{\partial f}{\partial z}\mathbf{k}\right)$$

$$= f\,\nabla g + g\,\nabla f$$

44.
$$\nabla\left(\dfrac{f}{g}\right) = \dfrac{\partial}{\partial x}\left(\dfrac{f}{g}\right)\mathbf{i} + \dfrac{\partial}{\partial y}\left(\dfrac{f}{g}\right)\mathbf{j} + \dfrac{\partial}{\partial z}\left(\dfrac{f}{g}\right)\mathbf{k}$$

$$= \dfrac{\dfrac{\partial f}{\partial x}g - f\dfrac{\partial g}{\partial x}}{g^2}\mathbf{i} + \dfrac{\dfrac{\partial f}{\partial y}g - f\dfrac{\partial g}{\partial y}}{g^2}\mathbf{j} + \dfrac{\dfrac{\partial f}{\partial z}g - f\dfrac{\partial g}{\partial z}}{g^2}\mathbf{k}$$

$$= \dfrac{g(\mathbf{x})\nabla f(\mathbf{x}) - f(\mathbf{x})\nabla g(\mathbf{x})}{g^2(\mathbf{x})}$$

45. $\nabla f^n = \dfrac{\partial f^n}{\partial x}\mathbf{i} + \dfrac{\partial f^n}{\partial y}\mathbf{j} + \dfrac{\partial f^n}{\partial z}\mathbf{k} = nf^{n-1}\dfrac{\partial f}{\partial x}\mathbf{i} + nf^{n-1}\dfrac{\partial f}{\partial y}\mathbf{j} + nf^{n-1}\dfrac{\partial f}{\partial z}\mathbf{k} = nf^{n-1}\nabla f$

SECTION 16.3

1. $f(\mathbf{b}) = f(1, 3) = -2$; $\ f(\mathbf{a}) = f(0, 1) = 0$; $\ \ f(\mathbf{b}) - f(\mathbf{a}) = -2$

$\nabla f = (3x^2 - y)\mathbf{i} - x\mathbf{j}$; $\ \ \mathbf{b} - \mathbf{a} = \mathbf{i} + 2\mathbf{j}$ $\ $ and $\ \ \nabla f \cdot (\mathbf{b} - \mathbf{a}) = 3x^2 - y - 2x$

The line segment joining $\mathbf{a}$ and $\mathbf{b}$ is parametrized by

$$x = t, \quad y = 1 + 2t, \quad 0 \le t \le 1$$

Thus, we need to solve the equation

$$3t^2 - (1 + 2t) - 2t = -2, \quad \text{which is the same as} \quad 3t^2 - 4t + 1 = 0, \ \ 0 \le t \le 1$$

The solutions are: $t = \frac{1}{3}, t = 1$. Thus, $\mathbf{c} = (\frac{1}{3}, \frac{5}{3})$ $\ $ satisfies the equation.
Note that the endpoint $\mathbf{b}$ also satisfies the equation.

2. $\nabla f = 4z\mathbf{i} - 2y\mathbf{j} + (4x + 2z)\mathbf{k}$, $\ \ f(\mathbf{a}) = f(0, 1, 1) = 0$, $\ \ f(\mathbf{b}) = f(1, 3, 2) = 3$

$\mathbf{b} - \mathbf{a} = \mathbf{i} + 2\mathbf{j} + \mathbf{k}$, $\ $ so we want $\ (x, y, z)$ $\ $ such that

$\nabla f \cdot (\mathbf{b} - \mathbf{a}) = 4z - 4y + 4x + 2z = 6z - 4y + 4x = f(\mathbf{b}) - f(\mathbf{a}) = 3$

Parametrizing the line segment from $\mathbf{a}$ to $\mathbf{b}$ by $x(t) = t$, $\ y(t) = 1 + 2t$, $\ z(t) = 1 + t$,
we get $t = \frac{1}{2}$, $\ $ or $\ \mathbf{c} = (\frac{1}{2}, 2, \frac{3}{2})$

3. (a) $\ f(x, y, z) = a_1 x + a_2 y + a_3 z + C$ $\qquad$ (b) $\quad f(x, y, z) = g(x, y, z) + a_1 x + a_2 y + a_3 z + C$

4. Using the mean-value theorem 16.3.1, there exists $\mathbf{c}$ such that $\quad \nabla f(\mathbf{c}) \cdot (\mathbf{b} - \mathbf{a}) = 0$

5. (a) U is not connected

(b) (i) $g(\mathbf{x}) = f(\mathbf{x}) - 1$ (ii) $g(\mathbf{x}) = -f(\mathbf{x})$

6. By the mean-value theorem

$$f(\mathbf{x}_1) - f(\mathbf{x}_2) = \nabla f(c) \cdot (\mathbf{x}_1 - \mathbf{x}_2)$$

for some point $\mathbf{c}$ on the line segment $\mathbf{x}_1\mathbf{x}_2$.   Since Ω is convex, $\mathbf{c}$ is in Ω.   Thus

$$|f(\mathbf{x}_1) - f(\mathbf{x}_2)| = |\nabla f(c) \cdot (\mathbf{x}_1 - \mathbf{x}_2)| \le \|\nabla f(\mathbf{c})\| \|\mathbf{x}_1 - \mathbf{x}_2\| \le M \|\mathbf{x}_1 - \mathbf{x}_2\|.$$

by Schwarz's inequality

7. $\nabla f = 2xy\mathbf{i} + x^2\mathbf{j}$;

$\nabla f(\mathbf{r}(t)) \cdot \mathbf{r}'(t) = (2\mathbf{i} + e^{2t}\mathbf{j}) \cdot (e^t\mathbf{i} - e^{-t}\mathbf{j}) = e^t$

8. $\nabla f = \mathbf{i} - \mathbf{j}$; $\nabla f(\mathbf{r}(t)) \cdot \mathbf{r}'(t) = (\mathbf{i} - \mathbf{j}) \cdot (a\mathbf{i} - ab\sin at\mathbf{j}) = a(1 + b\sin at)$

9. $\nabla f = \dfrac{-2x}{1 + (y^2 - x^2)^2}\mathbf{i} + \dfrac{2y}{1 + (y^2 - x^2)^2}\mathbf{j}$, $\nabla f(\mathbf{r}(t)) = \dfrac{-2\sin t}{1 + \cos^2 2t}\mathbf{i} + \dfrac{2\cos t}{1 + \cos^2 2t}\mathbf{j}$

$\nabla f(\mathbf{r}(t)) \cdot \mathbf{r}'(t) = \left(\dfrac{-2\sin t}{1 + \cos^2 2t}\mathbf{i} + \dfrac{2\cos t}{1 + \cos^2 2t}\mathbf{j}\right) \cdot (\cos t\,\mathbf{i} - \sin t\,\mathbf{j}) = \dfrac{-4\sin t\,\cos t}{1 + \cos^2 2t} = \dfrac{-2\sin 2t}{1 + \cos^2 2t}$

10. $\nabla f = \dfrac{1}{2x^2 + y^3}(4x\mathbf{i} + 3y^2\mathbf{j})$

$\nabla f(\mathbf{r}(t)) \cdot \mathbf{r}'(t) = \dfrac{1}{2e^{4t} + t}(4e^{2t}\mathbf{i} + 3t^{2/3}\mathbf{j}) \cdot (2e^{2t}\mathbf{i} + \dfrac{1}{3}t^{-2/3}\mathbf{j}) = \dfrac{8e^{4t} + 1}{2e^{4t} + t}$

11. $\nabla f = (e^y - ye^{-x})\,\mathbf{i} + (xe^y + e^{-x})\,\mathbf{j}$; $\nabla f(\mathbf{r}(t)) = (t^t - \ln t)\,\mathbf{i} + \left(t^t \ln t + \dfrac{1}{t}\right)\mathbf{j}$

$\nabla f(\mathbf{r}(t)) \cdot \mathbf{r}'(t) = \left((t^t - \ln t)\,\mathbf{i} + \left(t^t \ln t + \dfrac{1}{t}\right)\mathbf{j}\right) \cdot \left(\dfrac{1}{t}\mathbf{i} + [1 + \ln t]\mathbf{j}\right) = t^t\left(\dfrac{1}{t} + \ln t + [\ln t]^2\right) + \dfrac{1}{t}$

12. $\nabla f = \dfrac{2}{x^2 + y^2 + z^2}(x\mathbf{i} + y\mathbf{j} + z\mathbf{k})$

$\nabla f(\mathbf{r}(t)) \cdot \mathbf{r}'(t) = \dfrac{2}{1 + e^{4t}}(\sin t\mathbf{i} + \cos t\mathbf{j} + e^{2t}\mathbf{k}) \cdot (\cos t\mathbf{i} - \sin t\mathbf{j} + 2e^{2t}\mathbf{k}) = \dfrac{4e^{4t}}{1 + e^{4t}}$

13. $\nabla f = y\mathbf{i} + (x - z)\mathbf{j} - y\mathbf{k}$;

$\nabla f(\mathbf{r}(t)) \cdot \mathbf{r}'(t) = (t^2\mathbf{i} + (t - t^3)\,\mathbf{j} - t^2\mathbf{k}) \cdot (\mathbf{i} + 2t\mathbf{j} + 3t^2\mathbf{k}) = 3t^2 - 5t^4$

14. $\nabla f = 2x\,\mathbf{i} + 2y\,\mathbf{j}$

$\nabla f(\mathbf{r}(t)) \cdot \mathbf{r}'(t) = (2a\cos \omega t\mathbf{i} + 2b\sin \omega t\mathbf{j}) \cdot (-\omega a\sin \omega t\mathbf{i} + \omega b\cos \omega t\mathbf{j} + b\omega \mathbf{k}) = 2\omega(b^2 - a^2)\sin \omega t\,\cos \omega t$

15. $\nabla f = 2x\mathbf{i} + 2y\mathbf{j} + \mathbf{k}$;

$\nabla f(\mathbf{r}(t)) \cdot \mathbf{r}'(t) = (2a\cos \omega t\,\mathbf{i} + 2b\sin \omega t\,\mathbf{j} + \mathbf{k}) \cdot (-a\omega \sin \omega t\,\mathbf{i} + b\omega \cos \omega t\,\mathbf{j} + b\omega \mathbf{k})$

$= 2\omega\left(b^2 - a^2\right)\sin \omega t\,\cos \omega t + b\omega$

16. $\nabla f = y^2 \cos(x+z)\mathbf{i} + 2y\sin(x+z)\mathbf{j} + y^2\cos(x+z)\mathbf{k}$

$\nabla f(\mathbf{r}(t)) \cdot \mathbf{r}'(t)$

$= [\cos^2 t \cos(2t+t^3)\mathbf{i} + 2\cos t \sin(2t+t^3)\mathbf{j} + \cos^2 t \cos(2t+t^3)\mathbf{k}] \cdot (2\mathbf{i} - \sin t\mathbf{j} + 3t^2\mathbf{k})$

$= \cos t[(2+3t^2)\cos t \cos(2t+t^3) - 2\sin t \sin(2t+t^3)]$

17. $\dfrac{du}{dt} = \dfrac{\partial u}{\partial x}\dfrac{dx}{dt} + \dfrac{\partial u}{\partial y}\dfrac{dy}{dt} = (2x-3y)(-\sin t) + (4y-3x)(\cos t)$

$= 2\cos t \sin t + 3\sin^2 t - 3\cos^2 t = \sin 2t - 3\cos 2t$

18.
$$\dfrac{du}{dt} = \dfrac{\partial u}{\partial x}\cdot\dfrac{dx}{dt} + \dfrac{\partial u}{\partial y}\cdot\dfrac{dy}{dt} = \left(1 + 2\sqrt{\dfrac{y}{x}}\right)3t^2 + \left(2\sqrt{\dfrac{x}{y}} - 3\right)\left(-\dfrac{1}{t^2}\right)$$

$$= \left(1 + \dfrac{2}{t^2}\right)3t^2 + (2t^2 - 3)\left(-\dfrac{1}{t^2}\right) = 3t^2 + 4 + \dfrac{3}{t^2}$$

19. $\dfrac{du}{dt} = \dfrac{\partial u}{\partial x}\dfrac{dx}{dt} + \dfrac{\partial u}{\partial y}\dfrac{dy}{dt}$

$= (e^x \sin y + e^y \cos x)\left(\tfrac{1}{2}\right) + (e^x \cos y + e^y \sin x)(2)$

$= e^{t/2}\left(\tfrac{1}{2}\sin 2t + 2\cos 2t\right) + e^{2t}\left(\tfrac{1}{2}\cos \tfrac{1}{2}t + 2\sin \tfrac{1}{2}t\right)$

20. $\dfrac{du}{dt} = \dfrac{\partial u}{\partial x}\cdot\dfrac{dx}{dt} + \dfrac{\partial u}{\partial y}\cdot\dfrac{dy}{dt} = (4x-y)(-2\sin 2t) + (2y-x)\cos t$

$= 2\sin 2t(\sin t - 4\cos 2t) + \cos t(2\sin t - \cos 2t)$

21. $\dfrac{du}{dt} = \dfrac{\partial u}{\partial x}\dfrac{dx}{dt} + \dfrac{\partial u}{\partial y}\dfrac{dy}{dt} = (e^x \sin y)(2t) + (e^x \cos y)(\pi)$

$= e^{t^2}[2t\,\sin(\pi t) + \pi\,\cos(\pi t)]$

22. $\dfrac{du}{dt} = \dfrac{\partial u}{\partial x}\cdot\dfrac{dx}{dt} + \dfrac{\partial u}{\partial y}\cdot\dfrac{dy}{dt} + \dfrac{\partial u}{\partial z}\cdot\dfrac{dz}{dt} = -\dfrac{z}{x}2t + \dfrac{z}{y}\dfrac{1}{2\sqrt{t}} + \ln\left(\dfrac{y}{x}\right)e^t(1+t)$

$= -\dfrac{2t^2 e^t}{t^2+1} + \dfrac{e^t}{2} + \ln\left(\dfrac{\sqrt{t}}{t^2+1}\right)e^t(1+t)$

23. $\dfrac{du}{dt} = \dfrac{\partial u}{\partial x}\dfrac{dx}{dt} + \dfrac{\partial u}{\partial y}\dfrac{dy}{dt} + \dfrac{\partial u}{\partial z}\dfrac{dz}{dt}$

$= (y+z)(2t) + (x+z)(1-2t) + (y+x)(2t-2)$

$= (1-t)(2t) + (2t^2 - 2t + 1)(1-2t) + t(2t-2)$

$= 1 - 4t + 6t^2 - 4t^3$

24. $\dfrac{du}{dt} = \dfrac{\partial u}{\partial x}\cdot\dfrac{dx}{dt} + \dfrac{\partial u}{\partial y}\cdot\dfrac{dy}{dt} + \dfrac{\partial u}{\partial z}\cdot\dfrac{dz}{dt} = (\sin \pi y + \pi z \sin \pi x)2t + \pi x \cos \pi y(-1) - \cos \pi x(-2t)$

$= 2t\left[\sin[\pi(1-t)] + \pi(1-t^2)\sin(\pi t^2)\right] - \pi t^2 \cos[\pi(1-t)] + 2t\cos(\pi t^2)$

25. $V = \dfrac{1}{3}\pi r^2 h$, $\dfrac{dV}{dt} = \dfrac{\partial V}{\partial r}\dfrac{dr}{dt} + \dfrac{\partial V}{\partial h}\dfrac{dh}{dt} = \left(\dfrac{2}{3}\pi rh\right)\dfrac{dr}{dt} + \left(\dfrac{1}{3}\pi r^2\right)\dfrac{dh}{dt}.$

At the given instant,

$$\frac{dV}{dt} = \frac{2}{3}\pi(280)(3) + \frac{1}{3}\pi(196)(-2) = \frac{1288}{3}\pi.$$

The volume is increasing at the rate of $\dfrac{1288}{3}\pi$ in.^{3}/sec.

26. $v = \pi r^2 h$, $\dfrac{dv}{dt} = \dfrac{\partial v}{\partial r}\cdot\dfrac{dr}{dt} + \dfrac{\partial v}{\partial h}\cdot\dfrac{dh}{dt} = 2\pi rh\dfrac{dr}{dt} + \pi r^2\dfrac{dh}{dt}$

$\dfrac{dr}{dt} = -2$, $\dfrac{dh}{dt} = 3$, $r = 13$, $h = 18 \implies \dfrac{dv}{dt} = -429\pi$:   decreasing at the rate of

429π cm^3/sec.

27. $A = \frac{1}{2}xy\sin\theta$; $\dfrac{dA}{dt} = \dfrac{\partial A}{\partial x}\dfrac{dx}{dt} + \dfrac{\partial A}{\partial y}\dfrac{dy}{dt} + \dfrac{\partial A}{\partial\theta}\dfrac{d\theta}{dt} = \frac{1}{2}\left[(y\sin\theta)\dfrac{dx}{dt} + (x\sin\theta)\dfrac{dy}{dt} + (xy\cos\theta)\dfrac{d\theta}{dt}\right]$.

At the given instant

$$\frac{dA}{dt} = \frac{1}{2}\left[(2\sin 1)(0.25) + (1.5\sin 1)(0.25) - (2(1.5)\cos 1)(0.1)\right] \cong 0.2871\,\text{ft}^2/s \cong 41.34\,\text{in}^2/s$$

28. $\dfrac{dz}{dt} = 2x\dfrac{dx}{dt} + \dfrac{y}{2}\dfrac{dy}{dt}$. But $x^2 + y^2 = 13 \implies 2x\dfrac{dx}{dt} + 2y\dfrac{dy}{dt} = 0 \implies \dfrac{dy}{dt} = -\dfrac{x}{y}\dfrac{dx}{dt}$

$\implies \dfrac{dz}{dt} = 2x\dfrac{dx}{dt} + \dfrac{y}{2}\left(-\dfrac{x}{y}\dfrac{dx}{dt}\right) = \dfrac{3x}{2}\dfrac{dx}{dt} = 15.$ z is increasing 15 centimeters per second

29. $\dfrac{\partial u}{\partial s} = \dfrac{\partial u}{\partial x}\dfrac{\partial x}{\partial s} + \dfrac{\partial u}{\partial y}\dfrac{\partial y}{\partial s} = (2x - y)(\cos t) + (-x)(t\cos s)$

$= 2s\cos^2 t - t\sin s\cos t - st\cos s\cos t$

$\dfrac{\partial u}{\partial t} = \dfrac{\partial u}{\partial x}\dfrac{\partial x}{\partial t} + \dfrac{\partial u}{\partial y}\dfrac{\partial y}{\partial t} = (2x - y)(-s\sin t) + (-x)(\sin s)$

$= -2s^2\cos t\sin t + st\sin s\sin t - s\cos t\sin s$

30. $\dfrac{\partial u}{\partial s} = \dfrac{\partial u}{\partial x}\cdot\dfrac{\partial x}{\partial s} + \dfrac{\partial u}{\partial y}\cdot\dfrac{\partial y}{\partial s}$

$= [\cos(x - y) - \sin(x + y)]t + [-\cos(x - y) - \sin(x + y)]2s$

$= (t - 2s)\cos(st - s^2 + t^2) - (t + 2s)\sin(st + s^2 - t^2)$

$\dfrac{\partial u}{\partial t} = \dfrac{\partial u}{\partial x}\cdot\dfrac{\partial x}{\partial t} + \dfrac{\partial u}{\partial y}\cdot\dfrac{\partial y}{\partial t}$

$= [\cos(x - y) - \sin(x + y)]s + [-\cos(x - y) - \sin(x + y)](-2t)$

$= (s + 2t)\cos(st - s^2 + t^2) - (s - 2t)\sin(st + s^2 - t^2)$

31. $\dfrac{\partial u}{\partial s} = \dfrac{\partial u}{\partial x}\dfrac{\partial x}{\partial s} + \dfrac{\partial u}{\partial y}\dfrac{\partial y}{\partial s} = (2x\tan y)(2st) + (x^2\sec^2 y)(1)$

$= 4s^3t^2\tan\left(s + t^2\right) + s^4t^2\sec^2\left(s + t^2\right)$

$\dfrac{\partial u}{\partial t} = \dfrac{\partial u}{\partial x}\dfrac{\partial x}{\partial t} + \dfrac{\partial u}{\partial y}\dfrac{\partial y}{\partial t} = (2x\tan y)\left(s^2\right) + (x^2\sec^2 y)(2t)$

$= 2s^4t\tan\left(s + t^2\right) + 2s^4t^3\sec^2\left(s + t^2\right)$

32. $\dfrac{\partial u}{\partial s} = \dfrac{\partial u}{\partial x} \cdot \dfrac{\partial x}{\partial s} + \dfrac{\partial u}{\partial y} \cdot \dfrac{\partial y}{\partial s} + \dfrac{\partial u}{\partial z} \cdot \dfrac{\partial z}{\partial s}$

$\quad = z^2 y \sec xy \tan xy (2t) + z^2 x \sec xy \tan xy + 2z \sec xy (2st)$

$\quad = \sec[2st(s - t^2)] \left(2s^4 t^3 (s - t^2) \tan[2st(s - t^2)] + 2s^3 t^2 \tan[2st(s - t^2)] + 4s^3 t^2 \right)$

$\dfrac{\partial u}{\partial t} = z^2 y \sec xy \tan xy (2s) + z^2 x \sec xy \tan xy(-2t) + 2z \sec xy (s^2)$

$\quad = \sec[2st(s - t^2)] \left(2s^5 t^2 (s - t^2) \tan[2st(s - t^2)] - 4s^5 t^4 \tan[2st(s - t^2)] + 2s^4 t \right)$

33. $\dfrac{\partial u}{\partial s} = \dfrac{\partial u}{\partial x} \dfrac{\partial x}{\partial s} + \dfrac{\partial u}{\partial y} \dfrac{\partial y}{\partial s} + \dfrac{\partial u}{\partial z} \dfrac{\partial z}{\partial s}$

$\quad = (2x - y)(\cos t) + (-x)(-\cos(t - s)) + 2z(t \cos s)$

$\quad = 2s \cos^2 t - \sin(t - s) \cos t + s \cos t \cos(t - s) + 2t^2 \sin s \cos s$

$\dfrac{\partial u}{\partial t} = \dfrac{\partial u}{\partial x} \dfrac{\partial x}{\partial t} + \dfrac{\partial u}{\partial y} \dfrac{\partial y}{\partial t} + \dfrac{\partial u}{\partial z} \dfrac{\partial z}{\partial t}$

$\quad = (2x - y)(-s \sin t) + (-x)(\cos(t - s)) + 2z(\sin s)$

$\quad = -2s^2 \cos t \sin t + s \sin(t - s) \sin t - s \cos t \cos(t - s) + 2t \sin^2 s$

34. $\dfrac{\partial u}{\partial s} = \dfrac{\partial u}{\partial x} \cdot \dfrac{\partial x}{\partial s} + \dfrac{\partial u}{\partial y} \cdot \dfrac{\partial y}{\partial s} + \dfrac{\partial u}{\partial z} \cdot \dfrac{\partial z}{\partial s}$

$\quad = e^{yz^2} \dfrac{1}{s} + xz^2 e^{yz^2} \cdot 0 + 2xyz e^{yz^2} 2s$

$\quad = \dfrac{1}{s} e^{t^3(s^2 + t^2)^2} + 4st^3(s^2 + t^2) \ln(st) e^{t^3(s^2 + t^2)^2}$

$\dfrac{\partial u}{\partial t} = \dfrac{\partial u}{\partial x} \cdot \dfrac{\partial x}{\partial t} + \dfrac{\partial u}{\partial y} \cdot \dfrac{\partial y}{\partial t} + \dfrac{\partial u}{\partial z} \cdot \dfrac{\partial z}{\partial t}$

$\quad = e^{yz^2} \dfrac{1}{t} + xz^2 e^{yz^2} 3t^2 + 2xyz e^{yz^2} 2t$

$\quad = \dfrac{1}{t} e^{t^3(s^2 + t^2)^2} + t^2(s^2 + t^2)(3s^2 + 7t^2) \ln(st) e^{t^3(s^2 + t^2)^2}$

35. $\dfrac{d}{dt} [f(\mathbf{r}(t))] = \left[\boldsymbol{\nabla} f(\mathbf{r}(t)) \cdot \dfrac{\mathbf{r}'(t)}{\| \mathbf{r}'(t) \|} \right] \| \mathbf{r}'(t) \|$

$\quad\quad\quad = f'_{\mathbf{u}(t)}(\mathbf{r}(t)) \| \mathbf{r}'(t) \| \quad \text{where} \quad \mathbf{u}(t) = \dfrac{\mathbf{r}'(t)}{\| \mathbf{r}'(t) \|}$

36. $\dfrac{\partial}{\partial x}[f(r)] = \dfrac{d}{dr}[f(r)] \dfrac{\partial r}{\partial x} = f'(r) \dfrac{\partial r}{\partial x} = f'(r) \dfrac{x}{r}; \quad\quad \text{similarly}$

$\dfrac{\partial}{\partial y}[f(r)] = f'(r) \dfrac{y}{r} \quad \text{and} \quad \dfrac{\partial}{\partial z}[f(r)] = f'(r) \dfrac{z}{r}.$

Therefore $\quad \boldsymbol{\nabla} f(r) = f'(r) \dfrac{x}{r} \mathbf{i} + f'(r) \dfrac{y}{r} \mathbf{j} + f'(r) \dfrac{z}{r} \mathbf{k} = f'(r) \dfrac{\mathbf{r}}{r}.$

37. (a) $(\cos r) \dfrac{\mathbf{r}}{r}$ (b) $(r \cos r + \sin r) \dfrac{\mathbf{r}}{r}$

38. (a) $\boldsymbol{\nabla}(r \ln r) = (1 + \ln r) \dfrac{\mathbf{r}}{r}$ (b) $\boldsymbol{\nabla}(e^{1 - r^2}) = -2re^{1 - r^2} \dfrac{\mathbf{r}}{r} = -2e^{1 - r^2} \mathbf{r}$

39. (a) $(r\cos r - \sin r)\dfrac{\mathbf{r}}{r^3}$ (b) $\left(\dfrac{\sin r - r\cos r}{\sin^2 r}\right)\dfrac{\mathbf{r}}{r}$

40. (a)   (b) $\dfrac{du}{dt} = \dfrac{\partial u}{\partial x}\dfrac{dx}{ds}\dfrac{ds}{dt} + \dfrac{\partial u}{\partial y}\dfrac{dy}{ds}\dfrac{ds}{dt}$

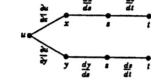

41. (a)

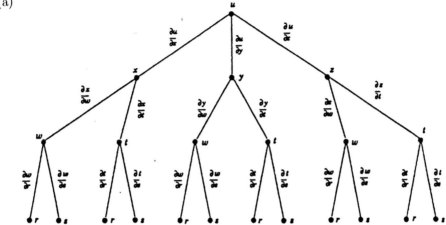

(b) $\dfrac{\partial u}{\partial r} = \dfrac{\partial u}{\partial x}\left(\dfrac{\partial x}{\partial w}\dfrac{\partial w}{\partial r} + \dfrac{\partial x}{\partial t}\dfrac{\partial t}{\partial r}\right) + \dfrac{\partial u}{\partial y}\left(\dfrac{\partial y}{\partial w}\dfrac{\partial w}{\partial r} + \dfrac{\partial y}{\partial t}\dfrac{\partial t}{\partial r}\right) + \dfrac{\partial u}{\partial z}\left(\dfrac{\partial z}{\partial w}\dfrac{\partial w}{\partial r} + \dfrac{\partial z}{\partial t}\dfrac{\partial t}{\partial r}\right).$

To obtain $\partial u/\partial s$, replace each r by s.

42. (a)

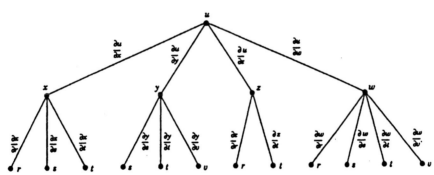

(b) $\dfrac{\partial u}{\partial r} = \dfrac{\partial u}{\partial x}\dfrac{\partial x}{\partial r} + \dfrac{\partial u}{\partial z}\dfrac{\partial z}{\partial r} + \dfrac{\partial u}{\partial w}\dfrac{\partial w}{\partial r}, \quad \dfrac{\partial u}{\partial v} = \dfrac{\partial u}{\partial y}\dfrac{\partial y}{\partial v} + \dfrac{\partial u}{\partial w}\dfrac{\partial w}{\partial v}$

43. $\dfrac{du}{dt} = \dfrac{\partial u}{\partial x}\dfrac{dx}{dt} + \dfrac{\partial u}{\partial y}\dfrac{dy}{dt}$

$\dfrac{d^2u}{dt^2} = \dfrac{\partial u}{\partial x}\dfrac{d^2x}{dt^2} + \dfrac{dx}{dt}\left[\dfrac{\partial^2 u}{\partial x^2}\dfrac{dx}{dt} + \dfrac{\partial^2 u}{\partial y \partial x}\dfrac{dy}{dt}\right] + \dfrac{\partial u}{\partial y}\dfrac{d^2y}{dt^2} + \dfrac{dy}{dt}\left[\dfrac{\partial^2 u}{\partial x\, \partial y}\dfrac{dx}{dt} + \dfrac{\partial^2 u}{\partial y^2}\dfrac{dy}{dt}\right]$

and the result follows.

44. $\dfrac{\partial u}{\partial s} = \dfrac{\partial u}{\partial x}\dfrac{\partial x}{\partial s} + \dfrac{\partial u}{\partial y}\dfrac{\partial y}{\partial s}$

$\dfrac{\partial^2 u}{\partial s^2} = \dfrac{\partial u}{\partial x}\dfrac{\partial^2 x}{\partial s^2} + \dfrac{\partial x}{\partial s}\left(\dfrac{\partial^2 u}{\partial x^2}\dfrac{\partial x}{\partial s} + \dfrac{\partial^2 u}{\partial y \partial x}\dfrac{\partial y}{\partial s}\right) + \dfrac{\partial u}{\partial y}\dfrac{\partial^2 y}{\partial s^2} + \dfrac{\partial y}{\partial s}\left(\dfrac{\partial^2 u}{\partial x \partial y}\dfrac{\partial x}{\partial s} + \dfrac{\partial^2 u}{\partial y^2}\dfrac{\partial y}{\partial s}\right)$

$= \dfrac{\partial^2 u}{\partial x^2}\left(\dfrac{\partial x}{\partial s}\right)^2 + 2\dfrac{\partial^2 u}{\partial x \partial y}\dfrac{\partial x}{\partial s}\dfrac{\partial y}{\partial s} + \dfrac{\partial^2 u}{\partial y^2}\left(\dfrac{\partial y}{\partial s}\right)^2 + \dfrac{\partial u}{\partial x}\dfrac{\partial^2 x}{\partial s^2} + \dfrac{\partial u}{\partial y}\dfrac{\partial^2 y}{\partial s^2}$

45. (a) $\dfrac{\partial u}{\partial r} = \dfrac{\partial u}{\partial x}\dfrac{\partial x}{\partial r} + \dfrac{\partial u}{\partial y}\dfrac{\partial y}{\partial r} = \dfrac{\partial u}{\partial x}\cos\theta + \dfrac{\partial u}{\partial y}\sin\theta$

$\dfrac{\partial u}{\partial \theta} = \dfrac{\partial u}{\partial x}\dfrac{\partial x}{\partial \theta} + \dfrac{\partial u}{\partial y}\dfrac{\partial y}{\partial \theta} = \dfrac{\partial u}{\partial x}(-r\sin\theta) + \dfrac{\partial u}{\partial y}(r\cos\theta)$

(b) $\left(\dfrac{\partial u}{\partial r}\right)^2 = \left(\dfrac{\partial u}{\partial x}\right)^2\cos^2\theta + 2\dfrac{\partial u}{\partial x}\dfrac{\partial u}{\partial y}\cos\theta\sin\theta + \left(\dfrac{\partial u}{\partial y}\right)^2\sin^2\theta,$

$\dfrac{1}{r^2}\left(\dfrac{\partial u}{\partial \theta}\right)^2 = \left(\dfrac{\partial u}{\partial x}\right)^2\sin^2\theta - 2\dfrac{\partial u}{\partial x}\dfrac{\partial u}{\partial y}\cos\theta\sin\theta + \left(\dfrac{\partial u}{\partial y}\right)^2\cos^2\theta,$

$\left(\dfrac{\partial u}{\partial r}\right)^2 + \dfrac{1}{r^2}\left(\dfrac{\partial u}{\partial \theta}\right)^2 = \left(\dfrac{\partial u}{\partial x}\right)^2(\cos^2\theta + \sin^2\theta) + \left(\dfrac{\partial u}{\partial y}\right)^2(\sin^2\theta + \cos^2\theta) = \left(\dfrac{\partial u}{\partial x}\right)^2 + \left(\dfrac{\partial u}{\partial y}\right)^2$

46. (a) By Exercise 45 (a)

$\dfrac{\partial w}{\partial r} = \dfrac{\partial w}{\partial x}\cos\theta + \dfrac{\partial w}{\partial y}\sin\theta, \quad \dfrac{\partial w}{\partial \theta} = -\dfrac{\partial w}{\partial x}r\sin\theta + \dfrac{\partial w}{\partial y}r\cos\theta.$

Solve these equations simultaneously for $\dfrac{\partial w}{\partial x}$ and $\dfrac{\partial w}{\partial y}$.

(b) To obtain the first pair of equations set $w = r$;

to obtain the second pair of equations set $w = \theta$.

(c) θ is not independent of x; $r = \sqrt{x^2 + y^2}$ gives

$\dfrac{\partial r}{\partial x} = \dfrac{x}{\sqrt{x^2 + y^2}} = \dfrac{r\cos\theta}{r} = \cos\theta$

47. Solve the equations in Exercise 45 (a) for $\dfrac{\partial u}{\partial x}$ and $\dfrac{\partial u}{\partial y}$:

$\dfrac{\partial u}{\partial x} = \dfrac{\partial u}{\partial r}\cos\theta - \dfrac{1}{r}\dfrac{\partial u}{\partial \theta}\sin\theta, \quad \dfrac{\partial u}{\partial y} = \dfrac{\partial u}{\partial r}\sin\theta + \dfrac{1}{r}\dfrac{\partial u}{\partial \theta}\cos\theta$

Then $\nabla u = \dfrac{\partial u}{\partial x}\mathbf{i} + \dfrac{\partial u}{\partial y}\mathbf{j} = \dfrac{\partial u}{\partial r}(\cos\theta\,\mathbf{i} + \sin\theta\,\mathbf{j}) + \dfrac{1}{r}\dfrac{\partial u}{\partial \theta}(-\sin\theta\,\mathbf{i} + \cos\theta\,\mathbf{j})$

48. $u(r, \theta) = r^2 \implies \nabla u = 2r\mathbf{e}_r$

49. $u(x,y) = x^2 - xy + y^2 = r^2 - r^2 \cos\theta \sin\theta = r^2\left(1 - \tfrac{1}{2}\sin 2\theta\right)$

$$\frac{\partial u}{\partial r} = r(2 - \sin 2\theta), \quad \frac{\partial u}{\partial \theta} = -r^2 \cos 2\theta$$

$$\nabla u = \frac{\partial u}{\partial r}\,\mathbf{e_r} + \frac{1}{r}\frac{\partial u}{\partial \theta}\,\mathbf{e_\theta} = r(2 - \sin 2\theta)\mathbf{e_r} - r\,\cos 2\theta\,\mathbf{e_\theta}$$

50. $$\frac{\partial u}{\partial \theta} = \frac{\partial u}{\partial x}\frac{\partial x}{\partial \theta} + \frac{\partial u}{\partial y}\frac{\partial y}{\partial \theta} = -r\sin\theta\frac{\partial u}{\partial x} + r\cos\theta\frac{\partial u}{\partial y}$$

$$\frac{\partial^2 u}{\partial r \partial \theta} = -\sin\theta\frac{\partial u}{\partial x} - r\sin\theta\left(\frac{\partial^2 u}{\partial x^2}\frac{\partial x}{\partial r} + \frac{\partial^2 u}{\partial y \partial x}\frac{\partial y}{\partial r}\right) + \cos\theta\frac{\partial u}{\partial y} + r\cos\theta\left(\frac{\partial^2 u}{\partial x \partial y}\frac{\partial x}{\partial r} + \frac{\partial^2 u}{\partial y^2}\frac{\partial y}{\partial r}\right)$$

$$= -\sin\theta\frac{\partial u}{\partial x} + \cos\theta\frac{\partial u}{\partial y} + r\sin\theta\cos\theta\left(\frac{\partial^2 u}{\partial y^2} - \frac{\partial^2 u}{\partial x^2}\right) + r(\cos^2\theta - \sin^2\theta)\frac{\partial^2 u}{\partial x \partial y}$$

51. From Exercise 45 (a),

$$\frac{\partial^2 u}{\partial r^2} = \frac{\partial^2 u}{\partial x^2}\cos^2\theta + 2\frac{\partial^2 u}{\partial y\,\partial x}\sin\theta\,\cos\theta + \frac{\partial^2 u}{\partial y^2}\sin^2\theta$$

$$\frac{\partial^2 u}{\partial \theta^2} = \frac{\partial^2 u}{\partial x^2}r^2\sin^2\theta - 2\frac{\partial^2 u}{\partial y\,\partial x}r^2\sin\theta\,\cos\theta + \frac{\partial^2 u}{\partial y^2}r^2\cos^2\theta - r\left(\frac{\partial u}{\partial x}\cos\theta + \frac{\partial u}{\partial y}\sin\theta\right).$$

The term in parentheses is $\dfrac{\partial u}{\partial r}$. Now divide the second equation by r^2 and add the two equations. The result follows.

52. $u(x,y) = x^2 - 2xy + y^4 - 4, \quad \dfrac{\partial u}{\partial x} = 2x - 2y, \quad \dfrac{\partial u}{\partial y} = -2x + 4y^3$

$$\frac{dy}{dx} = -\frac{\dfrac{\partial u}{\partial x}}{\dfrac{\partial u}{\partial y}} = \frac{2y - 2x}{4y^3 - 2x} = \frac{y - x}{2y^3 - x}$$

53. Set $u = xe^y + ye^x - 2x^2 y$. Then

$$\frac{\partial u}{\partial x} = e^y + ye^x - 4xy, \quad \frac{\partial u}{\partial y} = xe^y + e^x - 2x^2$$

$$\frac{dy}{dx} = -\frac{\partial u/\partial x}{\partial u/\partial y} = -\frac{e^y + ye^x - 4xy}{xe^y + e^x - 2x^2}.$$

54. $u(x,y) = x^{2/3} + y^{2/3}, \quad \dfrac{\partial u}{\partial x} = \dfrac{2}{3}x^{-1/3}, \quad \dfrac{\partial u}{\partial y} = \dfrac{2}{3}y^{-1/3}$

$$\implies \frac{dy}{dx} = -\frac{\dfrac{\partial u}{\partial x}}{\dfrac{\partial u}{\partial y}} = -\left(\frac{y}{x}\right)^{1/3}$$

55. Set $u = x\cos xy + y\cos x - 2$. Then

$$\frac{\partial u}{\partial x} = \cos xy - xy\sin xy - y\sin x, \quad \frac{\partial u}{\partial y} = -x^2\sin xy + \cos x$$

$$\frac{dy}{dx} = -\frac{\partial u/\partial x}{\partial u/\partial y} = \frac{\cos xy - xy\sin xy - y\sin x}{x^2\sin xy - \cos x}.$$

56. Set $u(x, y, z) = z^4 + x^2 z^3 + y^2 + xy - 2$. Then $\dfrac{\partial u}{\partial x} = 2xz^3 + y$, $\dfrac{\partial u}{\partial y} = 2y + x$, $\dfrac{\partial u}{\partial z} = 4z^3 + 3x^2 z^2$

$$\frac{\partial z}{\partial x} = -\frac{\dfrac{\partial u}{\partial x}}{\dfrac{\partial u}{\partial z}} = -\frac{2xz^3 + y}{4z^3 + 3x^2 z^2}, \qquad \frac{\partial z}{\partial y} = -\frac{\dfrac{\partial u}{\partial y}}{\dfrac{\partial u}{\partial z}} = -\frac{2y + x}{4z^3 + 3x^2 z^2}$$

57. Set $u = \cos xyz + \ln\left(x^2 + y^2 + z^2\right)$.   Then

$$\frac{\partial u}{\partial x} = -yz\,\sin xyz + \frac{2x}{x^2 + y^2 + z^2}, \qquad \frac{\partial u}{\partial y} = -xz\,\sin xyz + \frac{2y}{x^2 + y^2 + z^2}, \qquad \text{and}$$

$$\frac{\partial u}{\partial z} = -xy\,\sin xyz + \frac{2z}{x^2 + y^2 + z^2}.$$

$$\frac{\partial z}{\partial x} = -\frac{\partial u/\partial x}{\partial u/\partial z} = -\frac{2x - yz\left(x^2 + y^2 + z^2\right)\sin xyz}{2z - xy\left(x^2 + y^2 + z^2\right)\sin xyz},$$

$$\frac{\partial z}{\partial y} = -\frac{\partial u/\partial y}{\partial u/\partial z} = -\frac{2y - xz\left(x^2 + y^2 + z^2\right)\sin xyz}{2z - xy\left(x^2 + y^2 + z^2\right)\sin xyz}.$$

58. (a)   Use $\dfrac{d\mathbf{u}}{dt} = \dfrac{du_1}{dt}\mathbf{i} + \dfrac{du_2}{dt}\mathbf{j}$ and apply the chain rule to u_1, u_2.

    (b)   (i) $\dfrac{d\mathbf{u}}{dt} = t(e^x \cos y\,\mathbf{i} + e^x \sin y\,\mathbf{j}) + \pi(-e^x \sin y\,\mathbf{i} + e^x \cos y\,\mathbf{j})$

$$= te^{t^2/2}(\cos \pi t\,\mathbf{i} + \sin \pi t\,\mathbf{j}) + \pi e^{t^2/2}(-\sin \pi t\,\mathbf{i} + \cos \pi t\,\mathbf{j})$$

       (ii) $\mathbf{u}(t) = e^{t^2/2} \cos \pi t\,\mathbf{i} + e^{t^2/2} \sin \pi t\,\mathbf{j}$

$$\frac{d\mathbf{u}}{dt} = (-\pi e^{t^2/2} \sin \pi t + te^{t^2/2} \cos \pi t)\mathbf{i} + (\pi e^{t^2/2} \cos \pi t + te^{t^2/2} \sin \pi t)\mathbf{j}$$

59. $\dfrac{\partial u}{\partial s} = \dfrac{\partial u}{\partial x}\dfrac{\partial x}{\partial s} + \dfrac{\partial u}{\partial y}\dfrac{\partial y}{\partial s}, \qquad \dfrac{\partial u}{\partial t} = \dfrac{\partial u}{\partial x}\dfrac{\partial x}{\partial t} + \dfrac{\partial u}{\partial y}\dfrac{\partial y}{\partial t}$

60. $\dfrac{d u}{dt} = \dfrac{\partial u}{\partial x}\dfrac{dx}{dt} + \dfrac{\partial u}{\partial y}\dfrac{dy}{dt} + \dfrac{\partial u}{\partial z}\dfrac{dz}{dt}$ where $\dfrac{\partial \mathbf{u}}{\partial x} = \dfrac{\partial u_1}{\partial x}\mathbf{i} + \dfrac{\partial u_2}{\partial x}\mathbf{j} + \dfrac{\partial u_3}{\partial x}\mathbf{k}$,

$$\frac{\partial \mathbf{u}}{\partial y} = \frac{\partial u_1}{\partial y}\mathbf{i} + \frac{\partial u_2}{\partial y}\mathbf{j} + \frac{\partial u_3}{\partial y}\mathbf{k}, \qquad \frac{\partial \mathbf{u}}{\partial z} = \frac{\partial u_1}{\partial z}\mathbf{i} + \frac{\partial u_2}{\partial z}\mathbf{j} + \frac{\partial u_3}{\partial z}\mathbf{k}.$$

SECTION 16.4

1. Set $f(x, y) = x^2 + xy + y^2$.   Then,

$$\nabla f = (2x + y)\mathbf{i} + (x + 2y)\mathbf{j}, \qquad \nabla f(-1, -1) = -3\mathbf{i} - 3\mathbf{j}.$$

normal vector $\mathbf{i} + \mathbf{j}$;   tangent vector $\mathbf{i} - \mathbf{j}$

tangent line $x + y + 2 = 0$;   normal line $x - y = 0$

2. Set $f(x,y) = (y-x)^2 - 2x$, $\nabla f = -2(y-x+1)\mathbf{i} + 2(y-x)\mathbf{j}$, $\nabla f(2,4) = -6\mathbf{i} + 4\mathbf{j}$

normal vector $-3\mathbf{i} + 2\mathbf{j}$; tangent vector $2\mathbf{i} + 3\mathbf{j}$

tangent line $3x - 2y + 2 = 0$; normal line $2x + 3y - 16 = 0$

3. Set $f(x,y) = (x^2 + y^2)^2 - 9(x^2 - y^2)$.   Then,

$$\nabla f = [4x(x^2 + y^2) - 18x]\mathbf{i} + [4y(x^2 + y^2) + 18y]\mathbf{j}, \quad \nabla f\left(\sqrt{2}, 1\right) = -6\sqrt{2}\,\mathbf{i} + 30\mathbf{j}.$$

normal vector $\sqrt{2}\,\mathbf{i} - 5\mathbf{j}$; tangent vector $5\mathbf{i} + \sqrt{2}\,\mathbf{j}$

tangent line $\sqrt{2}x - 5y + 3 = 0$; normal line $5x + \sqrt{2}\,y - 6\sqrt{2} = 0$

4. Set $f(x,y) = x^3 + y^3$, $\nabla f = 3x^2\mathbf{i} + 3y^2\mathbf{j}$, $\nabla f(1,2) = 3\mathbf{i} + 12\mathbf{j}$

normal vector $\mathbf{i} + 4\mathbf{j}$;   tangent vector $4\mathbf{i} - \mathbf{j}$

tangent line $x + 4y - 9 = 0$;   normal line $4x - y - 2 = 0$

5. Set $f(x,y) = xy^2 - 2x^2 + y + 5x$.   Then,

$$\nabla f = (y^2 - 4x + 5)\,\mathbf{i} + (2xy + 1)\,\mathbf{j}, \quad \nabla f(4,2) = -7\mathbf{i} + 17\mathbf{j}.$$

normal vector $7\mathbf{i} - 17\mathbf{j}$; tangent vector $17\mathbf{i} + 7\mathbf{j}$

tangent line $7x - 17y + 6 = 0$; normal line $17x + 7y - 82 = 0$

6. Set $f(x,y) = x^5 + y^5 - 2x^3$. $\nabla f = (5x^4 - 6x^2)\mathbf{i} + 5y^4\mathbf{j}$, $\nabla f(1,1) = -\mathbf{i} + 5\mathbf{j}$

normal vector $\mathbf{i} - 5\mathbf{j}$;   tangent vector $5\mathbf{i} + \mathbf{j}$

tangent line $x - 5y + 4 = 0$;   normal line $5x + y - 6 = 0$

7. Set $f(x,y) = 2x^3 - x^2y^2 - 3x + y$.   Then,

$$\nabla f = (6x^2 - 2xy^2 - 3)\,\mathbf{i} + (-2x^2y + 1)\,\mathbf{j}, \quad \nabla f(1,-2) = -5\mathbf{i} + 5\mathbf{j}.$$

normal vector $\mathbf{i} - \mathbf{j}$; tangent vector $\mathbf{i} + \mathbf{j}$

tangent line $x - y - 3 = 0$; normal line $x + y + 1 = 0$

8. Set $f(x,y) = x^3 + y^2 + 2x$. $\nabla f = (3x^2 + 2)\mathbf{i} + 2y\mathbf{j}$, $\nabla f(-1,3) = 5\mathbf{i} + 6\mathbf{j}$

normal vector $5\mathbf{i} + 6\mathbf{j}$;   tangent vector $6\mathbf{i} - 5\mathbf{j}$

tangent line $5x + 6y - 13 = 0$;   normal line $6x - 5y + 21 = 0$

9. Set $f(x,y) = x^2y + a^2y$.   By (15.4.4)

$$m = -\frac{\partial f/\partial x}{\partial f/\partial y} = -\frac{2xy}{x^2 + a^2}.$$

At $(0,a)$ the slope is 0.

10. Set $f(x,y,z) = (x^2 + y^2)^2 - z$. $\nabla f = 4x(x^2 + y^2)\mathbf{i} + 4y(x^2 + y^2)\mathbf{j} - \mathbf{k}$, $\nabla f(1,1,4) = 8\mathbf{i} + 8\mathbf{j} - \mathbf{k}$

Tangent plane: $8x + 8y - z - 12 = 0$

Normal: $x = 1 + 8t$, $y = 1 + 8t$, $z = 4 - t$

11. Set $f(x, y, z) = x^3 + y^3 - 3xyz$.   Then,

$$\nabla f = (3x^2 - 3yz)\,\mathbf{i} + (3y^2 - 3xz)\,\mathbf{j} - 3xy\mathbf{k}, \quad \nabla f\left(1, 2, \tfrac{3}{2}\right) = -6\mathbf{i} + \tfrac{15}{2}\mathbf{j} - 6\mathbf{k};$$

tangent plane at $\left(1, 2, \tfrac{3}{2}\right)$: $-6(x - 1) + \tfrac{15}{2}(y - 2) - 6\left(z - \tfrac{3}{2}\right) = 0$, which reduces to $4x - 5y + 4z = 0$.
Normal: $x = 1 + 4t, \quad y = 2 - 5t, \quad z = \tfrac{3}{2} + 4t$

12. Set $f(x, y, z) = xy^2 + 2z^2$. $\nabla f = y^2\mathbf{i} + 2xy\mathbf{j} + 4z\mathbf{k}, \quad \nabla f(1, 2, 2) = 4\mathbf{i} + 4\mathbf{j} + 8\mathbf{k}$
Tangent plane: $x + y + 2z - 7 = 0$
Normal: $x = 1 + t, \quad y = 2 + t, \quad z = 2 + 2t$

13. Set $z = g(x, y) = axy$.   Then, $\nabla g = ay\mathbf{i} + ax\mathbf{j}, \quad \nabla g\left(1, \dfrac{1}{a}\right) = \mathbf{i} + a\mathbf{j}$.

tangent plane at $\left(1, \dfrac{1}{a}, 1\right)$: $z - 1 = 1(x - 1) + a\left(y - \dfrac{1}{a}\right)$, which reduces to $x + ay - z - 1 = 0$
Normal: $x = 1 + t, \quad y = \tfrac{1}{a} + at, \quad z = 1 - t$

14. Set $f(x, y, z) = \sqrt{x} + \sqrt{y} + \sqrt{z}$. $\nabla f = \dfrac{1}{2\sqrt{x}}\mathbf{i} + \dfrac{1}{2\sqrt{y}}\mathbf{j} + \dfrac{1}{2\sqrt{z}}\mathbf{k}, \quad \nabla f(1, 4, 1) = \dfrac{1}{2}\mathbf{i} + \dfrac{1}{4}\mathbf{j} + \dfrac{1}{2}\mathbf{k}$
Tangent plane: $2x + y + 2z - 8 = 0$
Normal: $x = 1 + 2t, \quad y = 4 + t, \quad z = 1 + 2t$

15. Set $z = g(x, y) = \sin x + \sin y + \sin(x + y)$.   Then,

$$\nabla g = [\cos x + \cos(x + y)]\,\mathbf{i} + [\cos y + \cos(x + y)]\,\mathbf{j}, \quad \nabla g(0, 0) = 2\mathbf{i} + 2\mathbf{j};$$

tangent plane at $(0, 0, 0)$: $z - 0 = 2(x - 0) + 2(y - 0), \quad 2x + 2y - z = 0$.
Normal: $x = 2t, \quad y = 2t, \quad z = -t$

16. Set $f(x, y, z) = x^2 + xy + y^2 - 6x + 2 - z$. $\nabla f = (2x + y - 6)\mathbf{i} + (x + 2y)\mathbf{j} - \mathbf{k}, \quad \nabla f(4, -2, -10) = -\mathbf{k}$

Tangent plane: $z = -10$
Normal: $x = 4, \quad y = -2, \quad z = -10 + t$

17. Set $f(x, y, z) = b^2c^2x^2 - a^2c^2y^2 - a^2b^2z^2$.   Then,

$$\nabla f(x_0, y_0, z_0) = 2b^2c^2x_0\mathbf{i} - 2a^2c^2y_0\mathbf{j} - 2a^2b^2z_0\mathbf{k};$$

tangent plane at (x_0, y_0, z_0):

$$2b^2c^2x_0(x - x_0) - 2a^2c^2y_0(y - y_0) - 2a^2b^2z_0(z - z_0) = 0,$$

which can be rewritten as follows:

$$b^2c^2x_0x - a^2c^2y_0y - a^2b^2z_0z = b^2c^2x_0{}^2 - a^2c^2y_0{}^2 - a^2b^2z_0{}^2$$

$$= f(x_0, y_0, z_0) = a^2b^2c^2.$$

Normal: $x = x_0 + 2b^2c^2x_0t, \quad y = y_0 - 2a^2c^2y_0t, \quad z = z_0 - 2a^2b^2z_0t$

18. Set $f(x, y, z) = \sin(x \cos y) - z$. $\nabla f = \cos y \cos(x \cos y)\mathbf{i} - x \sin y \cos(x \cos y)\mathbf{j} - \mathbf{k}$,

$\nabla f(1, \frac{\pi}{2}, 0) = \mathbf{j} - \mathbf{k}$

Tangent plane: $y + z = \dfrac{\pi}{2}$

Normal: $x = 1, \quad y = \dfrac{\pi}{2} + t, \quad z = t$

19. Set $z = g(x, y) = xy + a^3 x^{-1} + b^3 y^{-1}$.

$$\nabla g = \left(y - a^3 x^{-2}\right)\mathbf{i} + \left(x - b^3 y^{-2}\right)\mathbf{j}, \quad \nabla g = 0 \implies y = a^3 x^{-2} \text{ and } x = b^3 y^{-2}.$$

Thus,

$$y = a^3 b^{-6} y^4, \quad y^3 = b^6 a^{-3}, \quad y = b^2/a, \quad x = b^3 y^{-2} = a^2/b \quad \text{and} \quad g\left(a^2/b,\ b^2/a\right) = 3ab.$$

The tangent plane is horizontal at $\left(a^2/b,\ b^2/a,\ 3ab\right)$.

20. $z = g(x, y) = 4x + 2y - x^2 + xy - y^2$. $\nabla g = (4 - 2x + y)\mathbf{i} + (2 + x - 2y)\mathbf{j}$

$\nabla g = 0 \implies 4 - 2x + y = 0, \quad 2 + x - 2y = 0 \implies x = \dfrac{10}{3}, \; y = \dfrac{8}{3}$

The tangent plane is horizontal at $\left(\frac{10}{3}, \frac{8}{3}, \frac{28}{3}\right)$.

21. Set $z = g(x, y) = xy$. Then, $\nabla g = y\mathbf{i} + x\mathbf{j}$.

$$\nabla g = 0 \implies x = y = 0.$$

The tangent plane is horizontal at $(0, 0, 0)$.

22. $z = g(x, y) = x^2 + y^2 - x - y - xy$. $\nabla g = (2x - 1 - y)\mathbf{i} + (2y - 1 - x)\mathbf{j}$

$\nabla g = 0 \implies 2x - 1 - y = 0 = 2y - 1 - x = 0 \implies x = 1, \; y = 1$

The tangent plane is horizontal at $(1, 1, -1)$.

23. Set $z = g(x, y) = 2x^2 + 2xy - y^2 - 5x + 3y - 2$. Then,

$$\nabla g = (4x + 2y - 5)\,\mathbf{i} + (2x - 2y + 3)\,\mathbf{j}.$$

$$\nabla g = 0 \implies 4x + 2y - 5 = 0 = 2x - 2y + 3 \implies x = \tfrac{1}{3}, \quad y = \tfrac{11}{6}.$$

The tangent plane is horizontal at $\left(\frac{1}{3},\ \frac{11}{6},\ -\frac{1}{12}\right)$.

24. (a) Set $f(x, y, z) = xy - z$. $\nabla f = y\mathbf{i} + x\mathbf{j} - \mathbf{k}$, $\nabla f(1, 1, 1) = \mathbf{i} + \mathbf{j} - \mathbf{k}$

upper unit normal $= \dfrac{\sqrt{3}}{3}(-\mathbf{i} - \mathbf{j} + \mathbf{k})$

(b) Set $f(x, y, z) = \dfrac{1}{x} - \dfrac{1}{y} - z$. $\nabla f = -\dfrac{1}{x^2}\mathbf{i} + \dfrac{1}{y^2}\mathbf{j} - \mathbf{k}$, $\nabla f(1, 1, 0) = -\mathbf{i} + \mathbf{j} - \mathbf{k}$

lower unit normal: $= \dfrac{\sqrt{3}}{3}(-\mathbf{i} + \mathbf{j} - \mathbf{k})$

25. $\dfrac{x - x_0}{(\partial f/\partial x)(x_0, y_0, z_0)} = \dfrac{y - y_0}{(\partial f/\partial y)(x_0, y_0, z_0)} = \dfrac{z - z_0}{(\partial f/\partial z)(x_0, y_0, z_0)}$

26. All the tangent planes pass through the origin. To see this, write the equation of the surface as $xf(x/y) - z = 0$. The tangent plane at (x_0, y_0, z_0) has equation

$$(x - x_0)\left[\frac{x_0}{y_0}f'\left(\frac{x_0}{y_0}\right) + f\left(\frac{x_0}{y_0}\right)\right] - (y - y_0)\left[\frac{x_0^2}{y_0^2}f'\left(\frac{x_0}{y_0}\right)\right] - (z - z_0) = 0.$$

The plane passes through the origin:

$$-\frac{x_0^2}{y_0}f'\left(\frac{x_0}{y_0}\right) - x_0f\left(\frac{x_0}{y_0}\right) + \frac{x_0^2}{y_0}f'\left(\frac{x_0}{y_0}\right) + z_0 = z_0 - x_0f\left(\frac{x_0}{y_0}\right) = 0.$$

27. Since the tangent planes meet at right angles, the normals ∇F and ∇G meet at right angles:

$$\frac{\partial F}{\partial x}\frac{\partial G}{\partial x} + \frac{\partial F}{\partial y}\frac{\partial G}{\partial y} + \frac{\partial F}{\partial z}\frac{\partial G}{\partial z} = 0.$$

28. The sum of the intercepts is a. To see this, note that the equation of the tangent plane at (x_0, y_0, z_0) can be written

$$\frac{x - x_0}{\sqrt{x_0}} + \frac{y - y_0}{\sqrt{y_0}} + \frac{z - z_0}{\sqrt{z_0}} = 0.$$

Setting $y = z = 0$ we have

$$\frac{x - x_0}{\sqrt{x_0}} = \sqrt{y_0} + \sqrt{z_0}.$$

Therefore the x-intercept is given by

$$x = x_0 + \sqrt{x_0}(\sqrt{y_0} + \sqrt{z_0}) = x_0 + \sqrt{x_0}(\sqrt{a} - \sqrt{x_0}) = \sqrt{x_0}\sqrt{a}.$$

Similarly the y-intercept is $\sqrt{y_0}\sqrt{a}$ and the z-intercept is $\sqrt{z_0}\sqrt{a}$. The sum of the intercepts is

$$(\sqrt{x_0} + \sqrt{y_0} + \sqrt{z_0})\sqrt{a} = \sqrt{a}\sqrt{a} = a.$$

29. The tangent plane at an arbitrary point (x_0, y_0, z_0) has equation

$$y_0z_0(x - x_0) + x_0z_0(y - y_0) + x_0y_0(z - z_0) = 0,$$

which simplifies to

$$y_0z_0x + x_0z_0y + x_0y_0z = 3x_0y_0z_0 \quad \text{and thus to} \quad \frac{x}{3x_0} + \frac{y}{3y_0} + \frac{z}{3z_0} = 1.$$

The volume of the pyramid is

$$V = \frac{1}{3}Bh = \frac{1}{3}\left[\frac{(3x_0)(3y_0)}{2}\right](3z_0) = \frac{9}{2}x_0y_0z_0 = \frac{9}{2}a^3.$$

30. The equation of the tangent plane at (x_0, y_0, z_0) can be written

$$x_0^{-1/3}(x - x_0) + y_0^{-1/3}(y - y_0) + z_0^{-1/3}(z - z_0) = 0$$

Setting $y = z = 0$, we get the x-intercept $x = x_0 + x_0^{1/3}(y_0^{2/3} + z_0^{2/3}) = x_0 + x_0^{1/3}(a^{2/3} - x_0^{2/3})$
$\implies x = x_0^{1/3}a^{2/3}$

Similarly, the y-intercept is $y_0^{1/3}a^{2/3}$ and the z-intercept is $z_0^{1/3}a^{2/3}$.

The sum of the squares of the intercepts is

$$(x_0^{2/3} + y_0^{2/3} + z_0^{2/3})a^{4/3} = a^{2/3}a^{4/3} = a^2.$$

31. The point $(2, 3, -2)$ is the tip of $\mathbf{r}(1)$.

Since $\quad \mathbf{r}'(t) = 2\mathbf{i} - \dfrac{3}{t^2}\mathbf{j} - 4t\mathbf{k}, \quad$ we have $\quad \mathbf{r}'(1) = 2\mathbf{i} - 3\mathbf{j} - 4\mathbf{k}.$

Now set $\quad f(x, y, z) = x^2 + y^2 + 3z^2 - 25.$ The function has gradient $2x\mathbf{i} + 2y\mathbf{j} + 6z\mathbf{k}.$

At the point $(2, 3, -2)$,

$$\nabla f = 2(2\mathbf{i} + 3\mathbf{j} - 6\mathbf{k}).$$

The angle θ between $\mathbf{r}'(1)$ and the gradient gives

$$\cos \theta = \frac{(2\mathbf{i} - 3\mathbf{j} - 4\mathbf{k})}{\sqrt{29}} \cdot \frac{(2\mathbf{i} + 3\mathbf{j} - 6\mathbf{k})}{7} = \frac{19}{7\sqrt{29}} \cong 0.504.$$

Therefore $\theta \cong 1.043$ radians. The angle between the curve and the plane is

$$\frac{\pi}{2} - \theta \cong 1.571 - 1.043 \cong 0.528 \text{ radians.}$$

32. The curve passes through the point $(3, 2, 1)$ at $t = 1$, and its tangent vector is $\mathbf{r}'(1) = 3\mathbf{i} + 4\mathbf{j} + 3\mathbf{k}.$
For the ellipsoid, set $\quad f(x, y, z) = x^2 + 2y^2 + 3z^2. \quad \nabla f = 2x\mathbf{i} + 4y\mathbf{j} + 6z\mathbf{k},$
$\nabla f(3, 2, 1) = 6\mathbf{i} + 8\mathbf{j} + 6\mathbf{k}, \quad$ which is parallel to $\mathbf{r}'(1)$.

33. Set $\quad f(x, y, z) = x^2 y^2 + 2x + z^3. \quad$ Then,

$$\nabla f = (2xy^2 + 2)\,\mathbf{i} + 2x^2 y\mathbf{j} + 3z^2\mathbf{k}, \quad \nabla f(2, 1, 2) = 6\mathbf{i} + 8\mathbf{j} + 12\mathbf{k}.$$

The plane tangent to $f(x, y, z) = 16$ at $(2, 1, 2)$ has equation

$$6(x - 2) + 8(y - 1) + 12(z - 2) = 0, \quad \text{or} \quad 3x + 4y + 6z = 22.$$

Next, set $\quad g(x, y, z) = 3x^2 + y^2 - 2z. \quad$ Then,

$$\nabla g = 6x\mathbf{i} + 2y\mathbf{j} - 2\mathbf{k}, \quad \nabla g(2, 1, 2) = 12\mathbf{i} + 2\mathbf{j} - 2\mathbf{k}.$$

The plane tangent to $g(x, y, z) = 9$ at $(2, 1, 2)$ is

$$12(x - 2) + 2(y - 1) - 2(z - 2) = 0, \quad \text{or} \quad 6x + y - z = 11.$$

34. Sphere: $f(x, y, z) = x^2 + y^2 + z^2 - 8x - 8y - 6z + 24, \quad \nabla f = (2x - 8)\mathbf{i} + (2y - 8)\mathbf{j} + (2z - 6)\mathbf{k}$

$\nabla f(2, 1, 1) = -4\mathbf{i} - 6\mathbf{j} - 4\mathbf{k}$

Ellipsoid: $\quad g(x, y, z) = x^2 + 3y^2 + 2z^2, \quad \nabla g = 2x\mathbf{i} + 6y\mathbf{j} + 4z\mathbf{k}$

$\nabla g(2, 1, 1) = 4\mathbf{i} + 6\mathbf{j} + 4\mathbf{k}$

Since their normal vectors are parallel, the surfaces are tangent.

35. A normal vector to the sphere at $(1, 1, 2)$ is

$$2x\mathbf{i} + (2y - 4)\,\mathbf{j} + (2z - 2)\mathbf{k} = 2\mathbf{i} - 2\mathbf{j} + 2\mathbf{k}.$$

A normal vector to the paraboloid at $(1, 1, 2)$ is

$$6x\mathbf{i} + 4y\mathbf{j} - 2\mathbf{k} = 6\mathbf{i} + 4\mathbf{j} - 2\mathbf{k}.$$

Since

$$(2\mathbf{i} - 2\mathbf{j} + 2\mathbf{k}) \cdot (6\mathbf{i} + 4\mathbf{j} - 2\mathbf{k}) = 0,$$

the surfaces intersect at right angles.

36. Surface A: Set $f(x, y, z) = xy - az^2$, $\quad \nabla f = y\mathbf{i} + x\mathbf{j} - 2az\mathbf{k}$

Surface B: Set $g(x, y, z) = x^2 + y^2 + z^2$, $\quad \nabla g = 2x\mathbf{i} + 2y\mathbf{j} + 2z\mathbf{k}$

Surface C: Set $h(x, y, z) = z^2 + 2x^2 - c(z^2 + 2y^2)$. $\quad \nabla h = 4x\mathbf{i} - 4cy\mathbf{j} + 2(1 - c)z\mathbf{k}$

Where surface A and surface B intersect, $\quad \nabla f \cdot \nabla g = 4(xy - az^2) = 0$

Where surface A and surface C intersect, $\quad \nabla f \cdot \nabla h = 4(1 - c)(xy - az^2) = 0$

Where surface B and surface C intersect, $\quad \nabla g \cdot \nabla h = 4[2x^2 - 2cy^2 + (1 - c)z^2] = 0$

37. (a) $3x + 4y + 6 = 0$ since plane p is vertical.

(b) $y = -\frac{1}{4}(3x + 6) = -\frac{1}{4}[3(4t - 2) + 6] = -3t$

$z = x^2 + 3y^2 + 2 = (4t - 2)^2 + 3(-3t)^2 + 2 = 43t^2 - 16t + 6$

$\mathbf{r}(t) = (4t - 2)\mathbf{i} - 3t\mathbf{j} + (43t^2 - 16t + 6)\mathbf{k}$

(c) From part (b) the tip of $\mathbf{r}(1)$ is $(2, -3, 33)$.   We take

$\mathbf{r}'(1) = 4\mathbf{i} - 3\mathbf{j} + 70\mathbf{j}$ as $\mathbf{d}$ to write

$$\mathbf{R}(s) = (2\mathbf{i} - 3\mathbf{j} + 33\mathbf{k}) + s(4\mathbf{i} - 3\mathbf{j} + 70\mathbf{k}).$$

(d) Set $g(x, y) = x^2 + 3y^2 + 2$.   Then,

$$\nabla g = 2x\mathbf{i} + 6y\mathbf{j} \quad \text{and} \quad \nabla g(2, -3) = 4\mathbf{i} - 18\mathbf{j}.$$

An equation for the plane tangent to $z = g(x, y)$ at $(2, -3, 33)$ is

$$z - 33 = 4(x - 2) - 18(y + 3) \quad \text{which reduces to} \quad 4x - 18y - z = 29.$$

(e) Substituting t for x in the equations for p and p_1, we obtain

$$3t + 4y + 6 = 0 \quad \text{and} \quad 4t - 18y - z = 29.$$

From the first equation

$$y = -\tfrac{3}{4}(t + 2)$$

and then from the second equation

$$z = 4t - 18\left[-\tfrac{3}{4}(t + 2)\right] - 29 = \tfrac{35}{2}t - 2.$$

Thus,

(∗) $\mathbf{r}(t) = t\mathbf{i} - \left(\tfrac{3}{4}t + \tfrac{3}{2}\right)\mathbf{j} + \left(\tfrac{35}{2}t - 2\right)\mathbf{k}.$

Lines l and l' are the same. To see this, consider how l and l' are formed; to assure yourself, replace t in (∗) by $4s + 2$ to obtain $\mathbf{R}(s)$ found in part (c).

38. (a) normal vector: $\frac{12}{25}\mathbf{i} - \frac{14}{25}\mathbf{j}$;   normal line: $x = 2 + \frac{12}{25}t,\ y = 1 - \frac{14}{25}t$

(b) tangent line: $x = 2 + \frac{14}{25}t,\ y = 1 + \frac{12}{25}t$

(c)

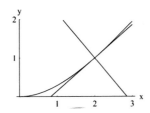

39. (a) normal vector: $2\mathbf{i} + 2\mathbf{j} + 4\mathbf{k}$;   normal line: $x = 1 + 2t,\ y = 2 + 2t,\ z = 2 + 4t$

(b) tangent plane: $2(x - 1) + 2(y - 2) + 4(z - 2) = 0$ or $x + y + 2z - 7 = 0$

(c)

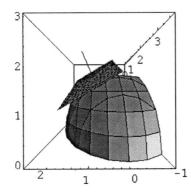

40. (a)

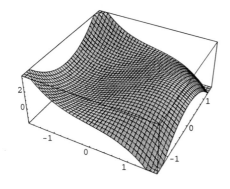

(b)

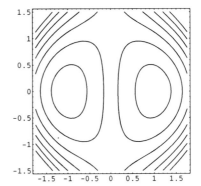

(c) $\nabla f = 0$ at $(\pm 1, 0),\ \left(0, \pm\sqrt{3/2}\right)$

41. (a)

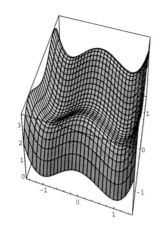

(b)

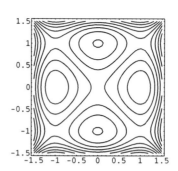

(c) $\nabla f = \left(4x^3 - 4x\right)\mathbf{i} - \left(4y^3 - 4y\right)\mathbf{j}$;

$\nabla f = 0$: $4x^3 - 4x = 0 \implies x = 0, \pm 1$; $4y^3 - 4y = 0 \implies y = 0, \pm 1$

$\nabla f = 0$ at $(0,0)$, $(\pm 1, 0)$, $(0, \pm 1)$, $(\pm 1, \pm 1)$

42. (a)

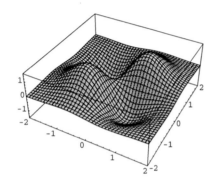

(b)

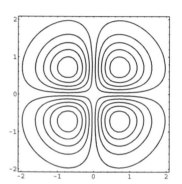

(c) $\nabla f = 0$ at $(0,0)$, $\left(\pm \sqrt{2}/2, \pm \sqrt{2}/2\right)$

SECTION 16.5

1. $\nabla f = (2 - 2x)\mathbf{i} - 2y\mathbf{j} = \mathbf{0}$ only at $(1, 0)$.

The difference

$$f(1 + h, k) - f(1, 0) = \left[2(1 + h) - (1 + h)^2 - k^2\right] - 1 = -h^2 - k^2 \leq 0$$

for all small h and k; there is a local maximum of 1 at $(1, 0)$.

2. $\nabla f = (2 - 2x)\mathbf{i} + (2 + 2y)\mathbf{j} = \mathbf{0}$ only at $(1, -1)$.

The difference

$f(1 + h, -1 + k) - f(1, -1)$

$\quad = [2(1 + h) + 2(-1 + k) - (1 + h)^2 + (-1 + k)^2 + 5] - 5 = -h^2 + k^2$

does not keep a constant sign for small h and k; $(1, -1)$ is a saddle point.

3. $\nabla f = (2x + y + 3)\mathbf{i} + (x + 2y)\mathbf{j} = \mathbf{0}$ only at $(-2, 1)$.

The difference

$f(-2 + h,\ 1 + k) - f(-2, 1)$
$$= [(-2 + h)^2 + (-2 + h)(1 + k) + (1 + k)^2 + 3(-2 + h) + 1] - (-2) = h^2 + hk + k^2$$

is nonnegative for all small h and k. To see this, note that

$$h^2 + hk + k^2 \geq h^2 - 2|h||k| + k^2 = (|h| - |k|)^2 \geq 0;$$

there is a local minimum of -2 at $(-2, 1)$.

4. $\nabla f = (3x^2 - 3)\mathbf{i} + \mathbf{j}$ is never $\mathbf{0}$;   there are no stationary points and no local extreme values.

NOTE: In Exercises 5–32, $A = f_{xx}(x_0, y_0), B = f_{xy}(x_0, y_0), C = f_{yy}(x_0, y_0), D = AC - B^2; (x_0, y_0)$ a stationary point of f.

5. $\nabla f = (2x + y - 6)\mathbf{i} + (x + 2y)\mathbf{j} = \mathbf{0}$ only at $(4, -2)$.

$f_{xx} = 2,\quad f_{xy} = 1,\quad f_{yy} = 2.$

At $(4, -2)$, $D = 3 > 0$ and $A = 2 > 0$ so we have a local min;   the value is -10.

6. $\nabla f = (2x + 2y + 2)\mathbf{i} + (2x + 6y + 10)\mathbf{j} = \mathbf{0}$ only at $(1, -2)$.

$\dfrac{\partial^2 f}{\partial x^2} = 2,\quad \dfrac{\partial^2 f}{\partial y \partial x} = 2,\quad \dfrac{\partial^2 f}{\partial y^2} = 6;\quad D = 2 \cdot 6 - 2^2 > 0,\ A = 2 \implies$ local min; the value is -8.

7. $\nabla f = (3x^2 - 6y)\mathbf{i} + (3y^2 - 6x)\mathbf{j} = \mathbf{0}$ at $(2, 2)$ and $(0, 0)$.

$f_{xx} = 6x,\quad f_{xy} = -6,\quad f_{yy} = 6y,\quad D = 36xy - 36.$

At $(2, 2)$, $D = 108 > 0$ and $A = 12 > 0$ so we have a local min;   the value is -8.

At $(0, 0)$, $D = -36 < 0$ so we have a saddle point.

8. $\nabla f = (6x + y + 5)\mathbf{i} + (x - 2y - 5)\mathbf{j} = \mathbf{0}$ at $\left(-\dfrac{5}{13}, -\dfrac{35}{13}\right)$.

$\dfrac{\partial^2 f}{\partial x^2} = 6,\quad \dfrac{\partial^2 f}{\partial y \partial x} = 1,\quad \dfrac{\partial^2 f}{\partial y^2} = -2;\quad D = 6 \cdot (-2) - 1^2 < 0;\quad (-5/13,\ -35/13)$ is a saddle point.

9. $\nabla f = (3x^2 - 6y + 6)\mathbf{i} + (2y - 6x + 3)\mathbf{j} = \mathbf{0}$ at $(5, \frac{27}{2})$ and $(1, \frac{3}{2})$.

$f_{xx} = 6x,\quad f_{xy} = -6,\quad f_{yy} = 2,\quad D = 12x - 36.$

At $(5, \frac{27}{2})$, $D = 24 > 0$ and $A = 30 > 0$ so we have a local min;   the value is $-\frac{117}{4}$.

At $(1, \frac{3}{2})$, $D = -24 < 0$ so we have a saddle point.

10. $\nabla f = (2x - 2y - 3)\mathbf{i} + (-2x + 4y + 5)\mathbf{j} = \mathbf{0}$ at $(\frac{1}{2}, -1)$.

$\dfrac{\partial^2 f}{\partial x^2} = 2,\quad \dfrac{\partial^2 f}{\partial y \partial x} = -2,\quad \dfrac{\partial^2 f}{\partial y^2} = 4;\quad D = 2 \cdot 4 - (-2)^2 > 0,\ A = 2 \implies$ local minimum;

the value is $-\frac{13}{4}$.

11. $\nabla f = \sin y\,\mathbf{i} + x \cos y\,\mathbf{j} = \mathbf{0}$ at $(0, n\pi)$ for all integral n.

$f_{xx} = 0,\quad f_{xy} = \cos y,\quad f_{yy} = -x \sin y.$

Since $D = -\cos^2 n\pi = -1 < 0$,   each stationary point is a saddle point.

12. $\nabla f = \sin y\,\mathbf{i} + (1 + x\cos y)\,\mathbf{j} = \mathbf{0}$ at $(-1, 2n\pi)$ and $(1, (2n+1)\pi)$ for all integral n.

$\dfrac{\partial^2 f}{\partial x^2} = 0, \quad \dfrac{\partial^2 f}{\partial y\partial x} = \cos y, \quad \dfrac{\partial^2 f}{\partial y^2} = -x\sin y; \quad D = 0\cdot(-x\sin y) - \cos^2 y < 0$ at the above points

so they are all saddle points

13. $\nabla f = (2xy + 1 + y^2)\,\mathbf{i} + (x^2 + 2xy + 1)\,\mathbf{j} = \mathbf{0}$ at $(1, -1)$ and $(-1, 1)$.

$f_{xx} = 2y, \quad f_{xy} = 2x + 2y, \quad f_{yy} = 2x, \quad D = 4xy - 4(x + y)^2.$

At both $(1, -1)$ and $(-1, 1)$ we have saddle points since $D = -4 < 0$.

14. $\nabla f = \left(\dfrac{1}{y} + \dfrac{y}{x^2}\right)\mathbf{i} + \left(-\dfrac{x}{y^2} - \dfrac{1}{x}\right)\mathbf{j} = \dfrac{x^2 + y^2}{x^2 y}\,\mathbf{i} - \dfrac{x^2 + y^2}{xy^2}\,\mathbf{j}$ is never $\mathbf{0}$;

no stationary points, no local extreme values.

15. $\nabla f = (y - x^{-2})\,\mathbf{i} + (x - 8y^{-2})\,\mathbf{j} = \mathbf{0}$ only at $\left(\frac{1}{2}, 4\right)$.

$f_{xx} = 2x^{-3}, \quad f_{xy} = 1, \quad f_{yy} = 16y^{-3}, \quad D = 32x^{-3}y^{-3} - 1.$

At $\left(\frac{1}{2}, 4\right)$, $D = 3 > 0$ and $A = 16 > 0$ so we have a local min;   the value is 6.

16. $\nabla f = (2x - 2y)\,\mathbf{i} + (-2x - 2y)\,\mathbf{j} = \mathbf{0}$ only at $(0, 0)$

$\dfrac{\partial^2 f}{\partial x^2} = 2, \quad \dfrac{\partial^2 f}{\partial y\partial x} = -2, \quad \dfrac{\partial^2 f}{\partial y^2} = -2; \quad D = 2\cdot(-2) - (-2)^2 < 0; \quad (0, 0)$ is a saddle point.

17. $\nabla f = (y - x^{-2})\,\mathbf{i} + (x - y^{-2})\,\mathbf{j} = \mathbf{0}$ only at $(1, 1)$.

$f_{xx} = 2x^{-3}, \quad f_{xy} = 1, \quad f_{yy} = 2y^{-3}, \quad D = 4x^{-3}y^{-3} - 1.$

At $(1, 1)$, $D = 3 > 0$ and $A = 2 > 0$ so we have a local min;   the value is 3.

18. $\nabla f = (2xy - y^2 - 1)\,\mathbf{i} + (x^2 - 2xy + 1)\,\mathbf{j} = \mathbf{0}$ at $(1, 1)$, $(-1, -1)$

$\dfrac{\partial^2 f}{\partial x^2} = 2y, \quad \dfrac{\partial^2 f}{\partial y\partial x} = 2(x - y), \quad \dfrac{\partial^2 f}{\partial y^2} = -2x; \quad D = -4xy - 4(x - y)^2 < 0$ at the above points;

$(1, 1)$ and $(-1, -1)$ are saddle points.

19. $\nabla f = \dfrac{2\left(x^2 - y^2 - 1\right)}{\left(x^2 + y^2 + 1\right)^2}\,\mathbf{i} + \dfrac{4xy}{\left(x^2 + y^2 + 1\right)^2}\,\mathbf{j} = \mathbf{0}$ at $(1, 0)$ and $(-1, 0)$.

$f_{xx} = \dfrac{-4x^3 + 12xy^2 + 12x}{\left(x^2 + y^2 + 1\right)^3}, \quad f_{xy} = \dfrac{4y^3 + 4y - 12x^2 y}{\left(x^2 + y^2 + 1\right)^3}, \quad f_{yy} = \dfrac{4x^3 + 4x - 12xy^2}{\left(x^2 + y^2 + 1\right)^3}.$

point	A	B	C	D	result
$(1, 0)$	1	0	1	1	loc. min.
$(-1, 0)$	-1	0	-1	1	loc. max.

$f(1, 0) = -1; \quad f(-1, 0) = 1$

20. $\nabla f = \left(\ln xy + 1 - \dfrac{3}{x} \right) \mathbf{i} + \dfrac{x-3}{y} \mathbf{j} = \mathbf{0}$ at $(3, 1/3)$

$\dfrac{\partial^2 f}{\partial x^2} = \dfrac{1}{x} + \dfrac{3}{x^2}, \quad \dfrac{\partial^2 f}{\partial y \partial x} = \dfrac{1}{y}, \quad \dfrac{\partial^2 f}{\partial y^2} = \dfrac{3-x}{y^2}$

At $(3, 1/3)$, $\dfrac{\partial^2 f}{\partial x^2} = \dfrac{2}{3}, \quad \dfrac{\partial^2 f}{\partial y \partial x} = 3, \quad \dfrac{\partial^2 f}{\partial y^2} = 0$ and $D = -9 < 0 \implies$ saddle point.

21. $\nabla f = \left(4x^3 - 4x \right) \mathbf{i} + 2y \mathbf{j} = \mathbf{0}$ at $(0,0), \ (1,0),$ and $(-1,0)$.

$f_{xx} = 12x^2 - 4, \quad f_{xy} = 0, \quad f_{yy} = 2.$

point	A	B	C	D	result
$(0,0)$	-4	0	2	-8	saddle
$(1,0)$	8	0	2	16	loc. min.
$(-1,0)$	8	0	2	16	loc. min.

$f(\pm 1, 0) = -3.$

22. $\nabla f = 2xe^{x^2-y^2}(1 + x^2 + y^2)\mathbf{i} + 2ye^{x^2-y^2}(1 - x^2 - y^2)\mathbf{j} = \mathbf{0}$ at $(0,0), \ (0,1), \ (0,-1)$

$A = \dfrac{\partial^2 f}{\partial x^2} = 2xe^{x^2-y^2}(2x + 2x^3 + 2xy^2) + e^{x^2-y^2}(2 + 6x^2 + 2y^2)$

$B = \dfrac{\partial^2 f}{\partial y \partial x} = -2ye^{x^2-y^2}(2x + 2x^3 + 2xy^2) + e^{x^2-y^2}(4xy)$

$C = \dfrac{\partial^2 f}{\partial y^2} = -2ye^{x^2-y^2}(2y - 2yx^2 - 2y^3) + e^{x^2-y^2}(2 - 2x^2 - 6y^2)$

At $(0,0)$, $AC - B^2 = (2)(2) = 4 > 0, A > 0$ local minimum; the value is 0.

At $(0, \pm 1)$, $AC - B^2 = (4e^{-1})(-4e^{-1}) = -8e^{-2} > 0$, saddle points

23. $\nabla f = \cos x \sin y \, \mathbf{i} + \sin x \cos y \, \mathbf{j} = \mathbf{0}$ at $\left(\frac{1}{2}\pi, \frac{1}{2}\pi \right), \ \left(\frac{1}{2}\pi, \frac{3}{2}\pi \right), \ (\pi, \pi), \ \left(\frac{3}{2}\pi, \frac{1}{2}\pi \right), \ \left(\frac{3}{2}\pi, \frac{3}{2}\pi \right).$

$f_{xx} = -\sin x \sin y, \quad f_{xy} = \cos x \cos y, \quad f_{yy} = -\sin x \sin y$

point	A	B	C	D	result
$\left(\frac{1}{2}\pi, \frac{1}{2}\pi \right)$	-1	0	-1	1	loc. max.
$\left(\frac{1}{2}\pi, \frac{3}{2}\pi \right)$	1	0	1	1	loc. min.
(π, π)	0	1	0	-1	saddle
$\left(\frac{3}{2}\pi, \frac{1}{2}\pi \right)$	1	0	1	1	loc. min.
$\left(\frac{3}{2}\pi, \frac{3}{2}\pi \right)$	-1	0	-1	1	loc. max.

$f\left(\frac{1}{2}\pi, \frac{1}{2}\pi \right) = f\left(\frac{3}{2}\pi, \frac{3}{2}\pi \right) = 1; \quad f\left(\frac{1}{2}\pi, \frac{3}{2}\pi \right) = f\left(\frac{3}{2}\pi, \frac{1}{2}\pi \right) = -1$

24. $\nabla f = -\sin x \cosh y \, \mathbf{i} + \cos x \sinh y \, \mathbf{j} = \mathbf{0}$ at $(-\pi, 0), \ (0,0), \ (\pi, 0)$.

$f_{xx} = -\cos x \cosh y, \quad f_{xy} = -\sin x \sinh y, \quad f_{yy} = \cos x \cosh y$

$D = -\cos^2 x \cosh^2 y - \sin^2 x \sinh^2 y < 0; \quad (-\pi, 0), \ (0,0), \ (\pi, 0)$ are saddle points.

25. (a) $\nabla f = (2x + ky)\mathbf{i} + (2y + kx)\mathbf{j}$ and $\nabla f(0,0) = \mathbf{0}$ independent of the value of k.

(b) $f_{xx} = 2,\quad f_{xy} = k,\quad f_{yy} = 2,\quad D = 4 - k^2$. Thus, $D < 0$ for $|k| > 2$ and $(0,0)$ is a saddle point

(c) $D = 4 - k^2 > 0$ for $|k| < 2$. Since $A = f_{xx} = 2 > 0$, $(0,0)$ is a local minimum.

(d) The test is inconclusive when $D = 4 - k^2 = 0$ i.e., for $k = \pm 2$. (If $k = \pm 2$, $f(x,y) = (x \pm y)^2$ and $(0,0)$ is a minimum.)

26. (a) $\nabla f = (2x + ky)\mathbf{i} + (kx + 8y)\mathbf{j} = \mathbf{0}$ at $(0,0)$.

(b) $\dfrac{\partial^2 f}{\partial x^2} = 2,\quad \dfrac{\partial^2 f}{\partial y \partial x} = k,\quad \dfrac{\partial^2 f}{\partial y^2} = 8$;   we want $16 - k^2 < 0$, or $|k| > 4$

(c) We want $16 - k^2 > 0$, or $|k| < 4$

(d) $k = \pm 4$. (If $k = \pm 4$, $f(x,y) = (x \pm 2y)^2$ and $(0,0)$ is a minimum.)

27. Let $P(x,y,z)$ be a point in the plane. We want to find the minimum of $f(x,y,z) = \sqrt{x^2 + y^2 + z^2}$. However, it is sufficient to minimize the square of the distance: $F(x,y,z) = x^2 + y^2 + z^2$. It is clear that F has a minimum value, but no maximum value. Since P lies in the plane, $2x - y + 2z = 16$ which implies $y = 2x + 2z - 16 = 2(x + z - 8)$. Thus, we want to find the minimum value of

$$F(x,z) = x^2 + 4(x + z - 8)^2 + z^2$$

Now,

$$\nabla F = [2x + 8(x + z - 8)]\mathbf{i} + [8(x + z - 8) + 2z]\mathbf{k}$$

The gradient is $\mathbf{0}$ when

$$2x + 8(x + z - 8) = 0 \quad \text{and} \quad 8(x + z - 8) + 2z = 0$$

The only solution to this pair of equations is: $x = z = \dfrac{32}{9}$, from which it follows that $y = -\dfrac{16}{9}$.

The point in the plane that is closest to the origin is $P\left(\frac{32}{9}, -\frac{16}{9}, \frac{32}{9}\right)$.

The distance from the origin to the plane is: $F(P) = \frac{16}{3}$.

Check using (13.6.5): $d(P,0) = \dfrac{|2 \cdot 0 - 0 + 2 \cdot 0 - 16|}{\sqrt{2^2 + (-1)^2 + 2^2}} = \dfrac{16}{3}$.

28. We want to minimize $(x + 1)^2 + (y - 2)^2 + (z - 4)^2$ on the plane. Since $z = -16 - \frac{3}{2}x + 2y$, we need to minimize $f(x,y) = (x + 1)^2 + (y - 2)^2 + (-20 - \frac{3}{2}x + 2y)^2$;

$$\nabla f = \left(\tfrac{13}{2}x - 6y + 62\right)\mathbf{i} + (-84 - 6x + 10y)\mathbf{j} = \mathbf{0} \text{ at } (-4,6)$$

Closest point $(-4, 6, 2)$,   distance$= \sqrt{(-1 - (-4))^2 + (2 - 6)^2 + (4 - 2)^2} = \sqrt{29}$

29. $f(x,y) = (x - 1)^2 + (y - 2)^2 + z^2 = (x - 1)^2 + (y - 2)^2 + x^2 + 2y^2$ $\left[\text{since } z = \sqrt{x^2 + 2y^2}\right]$

$\nabla f = [2(x - 1) + 2x]\mathbf{i} + [2(y - 2) + 4y]\mathbf{j} = \mathbf{0} \implies x = \dfrac{1}{2},\ y = \dfrac{2}{3}$.

$f_{xx} = 4 > 0,\quad f_{xy} = 0,\quad f_{yy} = 6,\quad D = 24 > 0$. Thus, f has a local minimum at $(1/2, 2/3)$.

The shortest distance from $(1,2,0)$ to the cone is $\sqrt{f\left(\frac{1}{2}, \frac{2}{3}\right)} = \frac{1}{6}\sqrt{114}$

30. $V = 8xyz, \quad x^2 + y^2 + z^2 = a^2 \implies V(x,y) = 8xy\sqrt{a^2 - x^2 - y^2}, \quad x > 0, \ y > 0, \ x^2 + y^2 < a^2$

$\nabla V = \dfrac{8y(a^2 - x^2 - y^2) - 8x^2 y}{\sqrt{a^2 - x^2 - y^2}}\,\mathbf{i} + \dfrac{8x(a^2 - x^2 - y^2) - 8xy^2}{\sqrt{a^2 - x^2 - y^2}}\,\mathbf{j} = \mathbf{0} \quad \text{at} \quad \left(a/\sqrt{3},\, a/\sqrt{3}\right)$

dimensions: $\dfrac{2a}{\sqrt{3}} \times \dfrac{2a}{\sqrt{3}} \times \dfrac{2a}{\sqrt{3}}, \quad$ maximum volume: $\dfrac{8}{9}a^3\sqrt{3}$

31.   (a)   (b)

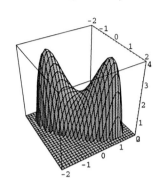   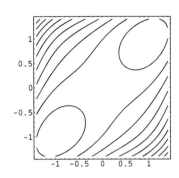

(c) $\nabla f = (4y - 4x^3)\,\mathbf{i} + (4x - 4y^3)\,\mathbf{j} = \mathbf{0}$ at $(0,0)$, $(1,1)$, $(-1,-1)$.

$f_{xx} = -12x^2, \quad f_{xy} = 4, \quad f_{yy} = -12y^2, \quad D = 144x^2 y^2 - 16$

point	A	B	C	D	result
$(0,0)$	0	4	0	-16	saddle
$(1,1)$	-12	4	-12	128	loc. max.
$(-1,-1)$	-12	4	-12	128	loc. max.

$f(1,1) = f(-1,-1) = 3$

32.   (a)   (b)

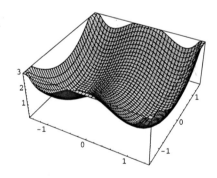   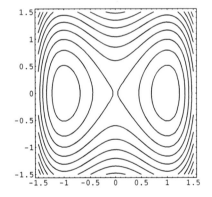

(c) $\nabla f = (4x^3 - 4x)\,\mathbf{i} + 2y\,\mathbf{j} = \mathbf{0}$ at $(0,0)$, $(1,0)$, $(-1,0)$.

$f_{xx} = 12x^2 - 4, \quad f_{x,y} = 0, \quad f_{yy} = 2, \quad D = 24x^2 - 8$

point	A	B	C	D	result
$(0,0)$	-8	0	2	-8	saddle
$(1,0)$	8	0	2	16	loc. min.
$(-1,0)$	8	0	2	16	loc. min.

$f(1,0) = f(-1,0) = 0$

33. (a) 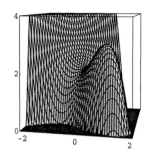   (b)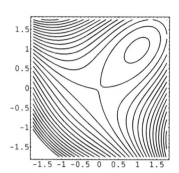

$f(1,1) = 3$ is a local max.; f has a saddle at $(0,0)$.

34. (a) 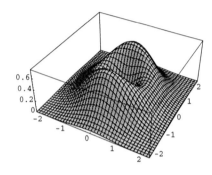   (b)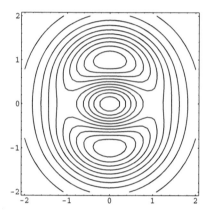

$f(0,0) = 0$ is a local min.; $f(0,1) = f(0,-1) = 2e^{-1}$ are local maxima; f has a saddle at $(\pm 1, 0)$.

35. (a) 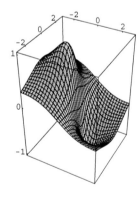   (b)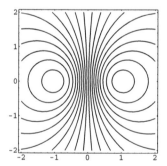

$f(1,0) = -1$ is a local min.; $f(-1,0) = 1$ is a loc. max.

36. (a)

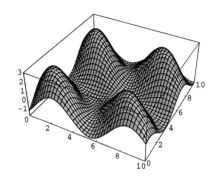

(b)

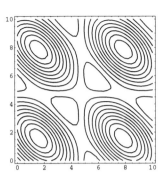

$f(\pi/2, \pi/2) = f(\pi/2, 5\pi/2) = f(5\pi/2, \pi/2) = f(5\pi/2, 5\pi/2) = 3$ are local maxima;

$(\pi/2, 3\pi/2), (3\pi/2, \pi/2), (3\pi/2, 3\pi/2), (3\pi/2, 5\pi/2), (5\pi/2, 3\pi/2)$ are saddle points of f;

$f(7\pi/6, 7\pi/6) = f(11\pi/6, 11\pi/6) = -\frac{3}{2}$ are local minima.

SECTION 16.6

1. $\nabla f = (4x - 4)\mathbf{i} + (2y - 2)\mathbf{j} = \mathbf{0}$ at $(1, 1)$ in D;

$f(1, 1) = -1$

Next we consider the boundary of D. We

parametrize each side of the triangle:

$$C_1: \mathbf{r}_1(t) = t\mathbf{i}, \quad t \in [0, 2],$$
$$C_2: \mathbf{r}_2(t) = 2\mathbf{i} + t\mathbf{j}, \quad t \in [0, 4],$$
$$C_3: \mathbf{r}_3(t) = t\mathbf{i} + 2t\mathbf{j}, \quad t \in [0, 2],$$

Now,

$$f_1(t) = f(\mathbf{r}_1(t)) = 2(t - 1)^2, \quad t \in [0, 2]; \quad \text{critical number: } t = 1,$$
$$f_2(t) = f(\mathbf{r}_2(t)) = (t - 1)^2 + 1, \quad t \in [0, 4]; \quad \text{critical number: } t = 1,$$
$$f_3(t) = f(\mathbf{r}_3(t)) = 6t^2 - 8t + 2, \quad t \in [0, 2]; \quad \text{critical number: } t = \frac{2}{3}.$$

Evaluating these functions at the endpoints of their domains and at the critical numbers, we find that:

$$f_1(0) = f_3(0) = f(0, 0) = 2; \quad f_1(1) = f(1, 0) = 0; \quad f_1(2) = f_2(0) = f(2, 0) = 2;$$
$$f_2(1) = f(2, 1) = 1; \quad f_2(4) = f_3(2) = f(2, 4) = 10; \quad f_3(2/3) = f(2/3, 4/3) = -\frac{2}{3}.$$

f takes on its absolute maximum of 10 at $(2, 4)$ and its absolute minimum of -1 at $(1, 1)$.

2. $\nabla f = -3\mathbf{i} + 2\mathbf{j} \neq \mathbf{0}$; no stationary points in D;

Next we consider the boundary of D. We parametrize each side of the triangle:

$$C_1: \mathbf{r}_1(t) = t\mathbf{i}, \quad t \in [0, 4],$$
$$C_2: \mathbf{r}_2(t) = t\mathbf{i} + \left(-\tfrac{3}{2}t + 6\right)\mathbf{j}, \quad t \in [0, 4],$$
$$C_3: \mathbf{r}_3(t) = t\mathbf{j}, \quad t \in [0, 6],$$

and evaluate f:

$$f_1(t) = f(\mathbf{r}_1(t)) = 2 - 3t, \quad t \in [0, 4]; \quad \text{no critical numbers,}$$
$$f_2(t) = f(\mathbf{r}_2(t)) = -6t + 14, \quad t \in [0, 4]; \quad \text{no critical numbers,}$$
$$f_3(t) = f(\mathbf{r}_3(t)) = 2 + 2t, \quad t \in [0, 6]; \quad \text{no critical numbers.}$$

Evaluating these functions at the endpoints of their domains, we find that:

$$f_1(0) = f_3(0) = f(0, 0) = 2; \qquad f_1(4) = f_2(4) = f(4, 0) = -10; \qquad f_2(0) = f_3(6) = f(0, 6) = 14;$$

f takes on its absolute maximum of 14 at $(0, 6)$ and its absolute minimum of -10 at $(4, 0)$.

3. $\nabla f = (2x + y - 6)\,\mathbf{i} + (x + 2y)\,\mathbf{j} = \mathbf{0}$ at $(4, -2)$ in

D; $\quad f(4, -2) = -13$

Next we consider the boundary of D. We parametrize each side of the rectangle:

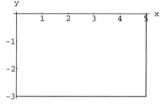

$$C_1 : \quad \mathbf{r}_1(t) = -t\,\mathbf{j}, \quad t \in [0, 3]$$
$$C_2 : \quad \mathbf{r}_2(t) = t\,\mathbf{i} - 3\,\mathbf{j}, \quad t \in [0, 5]$$
$$C_3 : \quad \mathbf{r}_3(t) = 5\,\mathbf{i} - t\,\mathbf{j}, \quad t \in [0, 3]$$
$$C_4 : \quad \mathbf{r}_4(t) = t\,\mathbf{i}, \quad t \in [0, 5]$$

Now,

$$f_1(t) = f(\mathbf{r}_1(t)) = t^2 - 1, \quad t \in [0, 3]; \quad \text{no critical numbers}$$
$$f_2(t) = f(\mathbf{r}_2(t)) = t^2 - 9t + 8, \quad t \in [0, 5]; \quad \text{critical number: } t = \tfrac{9}{2}$$
$$f_3(t) = f(\mathbf{r}_3(t)) = t^2 - 5t - 6, \quad t \in [0, 3]; \quad \text{critical number: } t = \tfrac{5}{2}$$
$$f_4(t) = f(\mathbf{r}_4(t)) = t^2 - 6t - 1, \quad t \in [0, 5]; \quad \text{critical number: } t = 3$$

Evaluating these functions at the endpoints of their domains and at the critical numbers, we find that:

$$f_1(0) = f_4(0) = f(0, 0) = -1; \qquad f_1(-3) = f_2(0) = f(0, -3) = 8; \qquad f_2(9/2) = f(9/2, -3) = -\tfrac{49}{4};$$
$$f_2(5) = f_3(3) = f(5, -3) = -12; \qquad f_3(5/2) = f(5, -5/2) = -\tfrac{49}{4}; \qquad f_3(0) = f_4(5) = f(5, 0) = -6.$$
$$f_4(3) = f(3, 0) = -10$$

f takes on its absolute maximum of 8 at $(0, -3)$ and its absolute minimum of -13 at $(4, -2)$.

4. $\nabla f = (2x + 2y)\,\mathbf{i} + (2x + 6y)\,\mathbf{j} = \mathbf{0}$ at $(0, 0)$ in D; $\quad f(0, 0) = 0$

Next we consider the boundary of D. We parametrize each side of the square:

$$C_1 : \quad \mathbf{r}_1(t) = t\,\mathbf{i} - 2\,\mathbf{j}, \quad t \in [-2, 2],$$
$$C_2 : \quad \mathbf{r}_2(t) = 2\,\mathbf{i} + t\,\mathbf{j}, \quad t \in [-2, 2],$$
$$C_3 : \quad \mathbf{r}_3(t) = t\,\mathbf{i} + 2\,\mathbf{j}, \quad t \in [-2, 2],$$
$$C_4 : \quad \mathbf{r}_4(t) = -2\,\mathbf{i} + t\,\mathbf{j}, \quad t \in [-2, 2],$$

and evaluate f:

$$f_1(t) = f(\mathbf{r}_1(t)) = t^2 - 4t + 12, \quad t \in [-2, 2]; \quad \text{no critical numbers,}$$
$$f_2(t) = f(\mathbf{r}_2(t)) = 4 + 4t + 3t^2, \quad t \in [-2, 2]; \quad \text{critical number: } t = -\tfrac{2}{3},$$
$$f_3(t) = f(\mathbf{r}_3(t)) = t^2 + 4t + 12, \quad t \in [-2, 2]; \quad \text{no critical numbers,}$$
$$f_4(t) = f(\mathbf{r}_4(t)) = 4 - 4t + 3t^2, \quad t \in [-2, 2]; \quad \text{critical number: } t = \tfrac{2}{3}.$$

Evaluating these functions at the endpoints of their domains and at the critical numbers, we find that:

$$f_1(-2) = f_4(-2) = f(-2, -2) = 24; \quad f_1(2) = f_2(-2) = f(2, -2) = 8; \quad f_2(-2/3) = f(2, -2/3) = \tfrac{8}{3};$$
$$f_2(2) = f_3(2) = f(2, 2) = 24; \quad f_3(-2) = f_4(2) = f(-2, 2) = 8; \quad f_4(2/3) = f(-2, 2/3) = \tfrac{8}{3}.$$

f takes on its absolute maximum of 24 at $(-2, -2)$ and $(2, 2)$ and its absolute minimum of 0 at $(0, 0)$.

Note that $x^2 + 2xy + 3y^2 = (x + y)^2 + 2y^2$. The results follow immediately from this observation.

5. $\nabla f = (2x + 3y)\mathbf{i} + (2y + 3x)\mathbf{j} = 0$ at $(0, 0)$ in D; $f(0, 0) = 2$

Next we consider the boundary of D. We parametrize the circle by:

$$C: \ \mathbf{r}(t) = 2\cos t\,\mathbf{i} + 2\sin t\,\mathbf{j}, \quad t \in [0, 2\pi]$$

The values of f on the boundary are given by the function

$$F(t) = f(\mathbf{r}(t)) = 6 + 12\sin t\,\cos t, \quad t \in [0, 2\pi]$$

$$F'(t) = 12\cos^2 t - 12\sin^2 t: \quad F'(t) = 0 \implies \cos t = \pm \sin t \implies t = \tfrac{1}{4}\pi, \tfrac{3}{4}\pi, \tfrac{5}{4}\pi, \tfrac{7}{4}\pi$$

Evaluating F at the endpoints and critical numbers, we have:

$$F(0) = F(2\pi) = f(2, 0) = 6; \quad F\left(\tfrac{1}{4}\pi\right) = F\left(\tfrac{5}{4}\pi\right) = f\left(\sqrt{2}, \sqrt{2}\right) = f\left(\left(-\sqrt{2}, -\sqrt{2}\right)\right) = 12;$$

$$F\left(\tfrac{3}{4}\pi\right) = f\left(-\sqrt{2}, \sqrt{2}\right) = F\left(\tfrac{7}{4}\pi\right) = f\left(\sqrt{2}, -\sqrt{2}\right) = 0$$

f takes on its absolute maximum of 12 at $\left(\sqrt{2}, \sqrt{2}\right)$ and at $\left(-\sqrt{2}, -\sqrt{2}\right)$; f takes on its absolute minimum of 0 at $\left(-\sqrt{2}, \sqrt{2}\right)$ and at $\left(\sqrt{2}, -\sqrt{2}\right)$.

6. $\nabla f = y\mathbf{i} + (x - 3)\mathbf{j} = 0$ at $(3, 0)$, which is not in the interior of D. The boundary is

$$\mathbf{r}(t) = 3\cos t\,\mathbf{i} + 3\sin t\,\mathbf{j}. \quad f(\mathbf{r}(t)) = 3\sin t(3\cos t - 3) = 9\sin t(\cos t - 1), \quad t \in [0, 2\pi].$$

$$\frac{df}{dt} = 9(2\cos^2 t - \cos t - 1); \quad \frac{df}{dt} = 0 \implies \cos t = 1, \ -\tfrac{1}{2} \quad \text{which yields the points } A\,(3, 0),$$

$B\left(-\tfrac{3}{2}, \tfrac{3\sqrt{3}}{2}\right), \ C\left(-\tfrac{3}{2}, -\tfrac{3\sqrt{3}}{2}\right): \quad f(A) = 0, \ f(B) = -\tfrac{27\sqrt{3}}{4} \text{ min}, \ f(C) = \tfrac{27\sqrt{3}}{4} \text{ max}$

7. $\nabla f = 2(x - 1)\mathbf{i} + 2(y - 1)\mathbf{j} = 0$ only at $(1, 1)$ in D. As the sum of two squares, $f(x, y) \geq 0$. Thus, $f(1, 1) = 0$ is a minimum. To examine the behavior of f on the boundary of D, we note that f represents the square of the distance between (x, y) and $(1, 1)$. Thus, f is maximal at the point of the boundary furthest from $(1, 1)$. This is the point $\left(-\sqrt{2}, -\sqrt{2}\right)$; the maximum value of f is $f\left(-\sqrt{2}, -\sqrt{2}\right) = 6 + 4\sqrt{2}$.

8. $\nabla f = (y+1)\mathbf{i} + (x-1)\mathbf{j} = \mathbf{0}$ at $(1,-1)$ which is not in the interior of D.

Next we consider the boundary of D.   We parametrize the boundary by:

$\quad\quad C_1: \ \mathbf{r}_1(t) = t\mathbf{j} + t^2\mathbf{j}, \quad t \in [-2,2],$

$\quad\quad C_2: \ \mathbf{r}_2(t) = t\mathbf{i} + 4\mathbf{j}, \quad t \in [-2,2],$

and evaluate f:

$\quad\quad f_1(t) = f(\mathbf{r}_1(t)) = t^3 - t^2 + t + 3, \quad t \in [-2,2]; \quad \text{no critical numbers,}$

$\quad\quad f_2(t) = f(\mathbf{r}_2(t)) = 5t - 1, \quad t \in [-2,2]; \quad \text{no critical numbers.}$

Evaluating these functions at the endpoints of their domains, we find that:

$\quad f_1(-2) = f_2(-2) = f(-2,4) = -11; \quad\quad f_1(2) = f_2(2) = f(2,4) = 9.$

f takes on its absolute maximum of 9 at $(2,4)$ and its absolute minimum of -11 at $(-2,4)$.

9. $\nabla f = \dfrac{2x^2 - 2y^2 - 2}{(x^2+y^2+1)^2}\mathbf{i} + \dfrac{4xy}{(x^2+y^2+1)^2}\mathbf{j} = \mathbf{0}$ at $(1,0)$ and $(-1,0)$ in D; $f(1,0) = -1, \quad f(-1,0) = 1$.

Next we consider the boundary of D. We parametrize each side of the square:

$\quad\quad C_1: \ \mathbf{r}_1(t) = -2\mathbf{i} + t\mathbf{j}, \quad t \in [-2,2]$

$\quad\quad C_2: \ \mathbf{r}_2(t) = t\mathbf{i} + 2\mathbf{j}, \quad t \in [-2,2]$

$\quad\quad C_3: \ \mathbf{r}_3(t) = 2\mathbf{i} + t\mathbf{j}, \quad t \in [-2,2]$

$\quad\quad C_4: \ \mathbf{r}_4(t) = t\mathbf{i}, \quad t \in [-2,2]$

Now,

$\quad\quad f_1(t) = f(\mathbf{r}_1(t)) = \dfrac{4}{t^2+5}, \quad t \in [-2,2]; \quad \text{critical number: } t = 0$

$\quad\quad f_2(t) = f(\mathbf{r}_2(t)) = \dfrac{-2t}{t^2+5}, \quad t \in [-2,2]; \quad \text{no critical numbers}$

$\quad\quad f_3(t) = f(\mathbf{r}_3(t)) = \dfrac{-4}{t^2+5}, \quad t \in [-2,2]; \quad \text{critical number: } t = 0$

$\quad\quad f_4(t) = f(\mathbf{r}_4(t)) = \dfrac{-2t}{t^2+5}, \quad t \in [-2,2]; \quad \text{no critical numbers}$

Evaluating these functions at the endpoints of their domains and at the critical numbers, we find that:

$\quad f_1(-2) = f_4(-2) = f(-2,-2) = \tfrac{4}{9}; \quad f_1(0) = f(-2,0) = \tfrac{4}{5}; \quad f_1(2) = f_2(-2) = f(-2,2) = \tfrac{4}{9};$

$\quad f_4(2) = f_3(-2) = f(2,-2) = -\tfrac{4}{9}; \quad f_3(0) = f(2,0) = -\tfrac{4}{5}; \quad f_2(2) = f_3(2) = f(2,2) = -\tfrac{4}{9}.$

f takes on its absolute maximum of 1 at $(-1,0)$ and its absolute minimum of -1 at $(1,0)$.

10. $\nabla f = \dfrac{2x^2 - 2y^2 - 2}{(x^2+y^2+1)^2}\mathbf{i} + \dfrac{4xy}{(x^2+y^2+1)^2}\mathbf{j} = \mathbf{0}$ at $(1,0)$ in D; $f(1,0) = -1$.

Next we consider the boundary of D. We parametrize each side of the triangle:

$\quad\quad C_1: \ \mathbf{r}_1(t) = t\mathbf{i} - t\mathbf{j}, \quad t \in [0,2]$

$\quad\quad C_2: \ \mathbf{r}_2(t) = 2\mathbf{i} + t\mathbf{j}, \quad t \in [-2,2]$

$\quad\quad C_3: \ \mathbf{r}_3(t) = t\mathbf{i} + t\mathbf{j}, \quad t \in [0,2],$

and evaluate f:

$$f_1(t) = f(\mathbf{r}_1(t)) = \frac{-2t}{2t^2 + 1}, \quad t \in [0, 2]; \quad \text{critical number: } t = 1/\sqrt{2},$$

$$f_2(t) = f(\mathbf{r}_2(t)) = \frac{-4}{t^2 + 5}, \quad t \in [-2, 2]; \quad \text{critical number: } t = 0$$

$$f_3(t) = f(\mathbf{r}_3(t)) = \frac{-2t}{2t^2 + 1}, \quad t \in [0, 2]; \quad \text{critical number: } t = 1/\sqrt{2}.$$

Evaluating these functions at the endpoints of their domains and at the critical numbers, we find that:

$$f_1(0) = f_3(0) = f(0,0) = 0; \quad f_1(1/\sqrt{2}) = f(1/\sqrt{2}, -1/\sqrt{2}) = -1/\sqrt{2};$$
$$f_1(2) = f_2(-2) = f(2, -2) = -\tfrac{4}{9}; \; f_2(0) = f(2,0) = -\tfrac{4}{5}; \quad f_2(2) = f_3(2) = f(2,2) = -\tfrac{4}{9};$$
$$f_3(1/\sqrt{2}) = f(1/\sqrt{2}, 1/\sqrt{2}) = -1/\sqrt{2}.$$

f takes on its absolute maximum of 0 at $(0,0)$ and its absolute minimum of -1 at $(1,0)$.

11. $\nabla f = (4 - 4x)\cos y\, \mathbf{i} - (4x - 2x^2)\sin y\, \mathbf{j} = \mathbf{0}$ at $(1,0)$ in D: $f(1,0) = 2$

Next we consider the boundary of D. We parametrize each side of the rectangle:

$$C_1 : \; \mathbf{r}_1(t) = t\,\mathbf{j}, \quad t \in \left[-\tfrac{1}{4}\pi, \tfrac{1}{4}\pi\right]$$
$$C_2 : \; \mathbf{r}_2(t) = t\,\mathbf{i} - \tfrac{1}{4}\pi\,\mathbf{j}, \quad t \in [0, 2]$$
$$C_3 : \; \mathbf{r}_3(t) = 2\,\mathbf{i} + t\,\mathbf{j}, \quad t \in \left[-\tfrac{1}{4}\pi, \tfrac{1}{4}\pi\right]$$
$$C_4 : \; \mathbf{r}_4(t) = t\,\mathbf{i} + \tfrac{1}{4}\pi\,\mathbf{j}, \quad t \in [0, 2]$$

Now,

$$f_1(t) = f(\mathbf{r}_1(t)) = 0;$$
$$f_2(t) = f(\mathbf{r}_2(t)) = \frac{\sqrt{2}}{2}(4t - 2t^2), \quad t \in [0, 2]; \quad \text{critical number: } t = 1;$$
$$f_3(t) = f(\mathbf{r}_3(t)) = 0;$$
$$f_4(t) = f(\mathbf{r}_4(t)) = \frac{\sqrt{2}}{2}(4t - 2t^2), \quad t \in [0, 2]; \quad \text{critical number: } t = 1;$$

f at the vertices of the rectangle has the value 0; $f_2(1) = f_4(1) = f\left(1, -\tfrac{1}{4}\pi\right) = f\left(1, \tfrac{1}{4}\pi\right) = \sqrt{2}.$

f takes on its absolute maximum of 2 at $(1,0)$ and its absolute minimum of 0 along the lines $x = 0$ and $x = 2$.

12. $\nabla f = 2(x - 3)\mathbf{i} + 2y\mathbf{j} = \mathbf{0}$ at $(3,0)$ which is not in the interior of D.

Boundary: On $y = x^2$, $f = (x-3)^2 + x^4$, $\dfrac{df}{dx} = 2(x-3) + 4x^3 = 0$ at $x = 1$ $\Longrightarrow$ $(1,1)$

On $y = 4x$, $f = (x-3)^2 + 16x^2$, $\dfrac{df}{dx} = 2(x-3) + 32x = 0$ at $x = \dfrac{3}{17}$ $\Longrightarrow$ $\left(\dfrac{3}{17}, \dfrac{12}{17}\right)$.

So the maximum and minimum occur at one or more of the following points:

$$(0,0), \; (4, 16), \; (1, 1), \; \left(\frac{3}{17}, \frac{12}{17}\right).$$

Evaluating f at these points, we find that $f(1,1) = 5$ is the absolute minimum of f; $f(4, 16) = 257$ is the absolute maximum of f.

13. $\nabla f = (3x^2 - 3y)\,\mathbf{i} + (-3x - 3y^2)\,\mathbf{j} = \mathbf{0}$ at $(-1, 1)$ in D;

$f(-1, 1) = 1$

Next we consider the boundary of D. We

parametrize each side of the triangle:

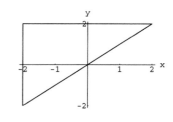

$$C_1:\ \mathbf{r}_1(t) = -2\,\mathbf{i} + t\,\mathbf{j}, \quad t \in [-2, 2],$$
$$C_2:\ \mathbf{r}_2(t) = t\,\mathbf{i} + t\,\mathbf{j}, \quad t \in [-2, 2],$$
$$C_3:\ \mathbf{r}_3(t) = t\,\mathbf{i} + 2\,\mathbf{j}, \quad t \in [-2, 2],$$

and evaluate f:

$$f_1(t) = f(\mathbf{r}_1(t)) = -8 + 6t - t^3, \quad t \in [-2, 2]; \quad \text{critical numbers: } t = \pm\sqrt{2},$$
$$f_2(t) = f(\mathbf{r}_2(t)) = -3t^2, \quad t \in [-2, 2]; \quad \text{critical number: } t = 0,$$
$$f_3(t) = f(\mathbf{r}_3(t)) = t^3 - 6t - 8, \quad t \in [-2, 2]; \quad \text{critical numbers: } t = \pm\sqrt{2}.$$

Evaluating these functions at the endpoints of their domains and at the critical numbers, we find that:

$$f_1(-2) = f_2(-2) = f(-2, -2) = -12; \qquad f_1(-\sqrt{2}) = f(-2, -\sqrt{2}) = -8 - 4\sqrt{2} \cong -13.66;$$
$$f_1(\sqrt{2}) = f(-2, \sqrt{2}) = -8 + 4\sqrt{2} \cong -2.34; \qquad f_1(2) = f_3(-2) = f(-2, 2) = -4;$$
$$f_2(0) = f(0, 0) = 0; \qquad f_2(2) = f_3(2) = f(2, 2) = -12;$$
$$f_3(-\sqrt{2}) = f(-\sqrt{2}, 2) = -8 + 4\sqrt{2}; \qquad f_3(\sqrt{2}) = f(\sqrt{2}, 2) = -8 - 4\sqrt{2}$$

f takes on its absolute maximum of 1 at $(-1, 1)$ and its absolute minimum of $-8 - 4\sqrt{2}$ at $(\sqrt{2}, 2)$ and $(-2, -\sqrt{2})$.

14. $\nabla f = 2(x - 4)\mathbf{i} + 2y\mathbf{j} = \mathbf{0}$ at $(4, 0)$ which is not in the interior of D. Next we examine f on the boundary of D:

$$C_1:\ \mathbf{r}_1(t) = t\mathbf{i} + 4t\mathbf{j}, \quad t \in [0, 2,],$$
$$C_2:\ \mathbf{r}_2(t) = t\mathbf{i} + t^3\mathbf{j}, \quad t \in [0, 2].$$

Note that

$$f_1(t) = f(\mathbf{r}_1(t)) = 17t^2 - 8t + 16,$$
$$f_2(t) = f(\mathbf{r}_2(t)) = (t - 4)^2 + t^6.$$

Next

$$f_1'(t) = 34t - 8 = 0 \quad \Longrightarrow \quad t = 4/17 \quad \text{and gives} \quad x = 4/17, \ y = 16/17$$

and

$$f_2'(t) = 6t^5 + 2t - 8 = 0 \quad \Longrightarrow \quad t = 1 \quad \text{and gives} \quad x = 1, \ y = 1.$$

The extreme values of f can be culled from the following list:

$$f(0, 0) = 16, \quad f(2, 8) = 68, \quad f\left(\tfrac{4}{17}, \tfrac{16}{17}\right) = \tfrac{256}{17}, \quad f(1, 1) = 10.$$

We see that $f(1, 1) = 10$ is the absolute minimum and $f(2, 8) = 68$ is the absolute maximum.

15. $\nabla f = \dfrac{4xy}{(x^2+y^2+1)^2}\,\mathbf{i} + \dfrac{2y^2 - 2x^2 - 2}{(x^2+y^2+1)^2}\,\mathbf{j} = \mathbf{0}$ at $(0,1)$ and $(0,-1)$ in D;

$f(0,1) = -1, \quad f(0,-1) = 1$

Next we consider the boundary of D.   We parametrize the circle by:

$\quad C: \; \mathbf{r}(t) = 2\cos t\,\mathbf{i} + 2\sin t\,\mathbf{j}, \quad t \in [0, 2\pi]$

The values of f on the boundary are given by the function

$$F(t) = f(\mathbf{r}(t)) = -\tfrac{4}{5}\sin t, \quad t \in [0, 2\pi]$$

$$F'(t) = -\tfrac{4}{5}\cos t: \quad F'(t) = 0 \implies \cos t = 0 \implies t = \tfrac{1}{2}\pi,\ \tfrac{3}{2}\pi.$$

Evaluating F at the endpoints and critical numbers, we have:

$$F(0) = F(2\pi) = f(2,0) = 0; \quad F\!\left(\tfrac{1}{2}\pi\right) = f(0,2) = -\tfrac{4}{5}; \quad F\!\left(\tfrac{3}{2}\pi\right) = f(0,-2) = \tfrac{4}{5}$$

f takes on its absolute maximum of 1 at $(0,-1)$ and its absolute minimum of -1 at $(0,1)$.

16. $\nabla f = (2x+1)\,\mathbf{i} + (8y-2)\,\mathbf{j} = \mathbf{0}$ at $\left(-\tfrac{1}{2}, \tfrac{1}{4}\right)$ in D; $f\!\left(-\tfrac{1}{2}, \tfrac{1}{4}\right) = -\tfrac{1}{2}$

Next we consider the boundary of D.   We parametrize the ellipse by:

$\quad C: \; \mathbf{r}(t) = 2\cos t\,\mathbf{i} + \sin t\,\mathbf{j}, \quad t \in [0, 2\pi]$

The values of f on the boundary are given by the function

$$F(t) = f(\mathbf{r}(t)) = 4\cos^2 t + 4\sin^2 t + 2\cos t - 2\sin t = 4 + 2\cos t - 2\sin t, \quad t \in [0, 2\pi]$$

$$F'(t) = -2\sin t - 2\cos t: \quad F'(t) = 0 \implies \cos t = -\sin t \implies t = \tfrac{3}{4}\pi,\ \text{or } \tfrac{7}{4}\pi$$

Evaluating F at the endpoints and critical numbers, we have:

$$F(0) = F(2\pi) = f(2,0) = 6;$$

$$F\!\left(\tfrac{3}{4}\pi\right) = f\!\left(-\sqrt{2}, \sqrt{2}/2\right) = 4 - 2\sqrt{2}; \quad F\!\left(\tfrac{7}{4}\pi\right) = f\!\left(\sqrt{2}, -\sqrt{2}/\right) = 4 + 2\sqrt{2}$$

f takes on its absolute maximum of $4 + 2\sqrt{2}$ at $\left(\sqrt{2}, -\sqrt{2}/2\right)$; f takes on its absolute minimum of $-\tfrac{1}{2}$ at $\left(-\tfrac{1}{2}, \tfrac{1}{4}\right)$.

17. $\nabla f = 2(x-y)\,\mathbf{i} - 2(x-y)\,\mathbf{j} = \mathbf{0}$ at each point of the line segment $y = x$ from $(0,0)$ to $(4,4)$. Since $f(x,x) = 0$ and $f(x,y) \geq 0$, f takes on its minimum of 0 at each of these points.

Next we consider the boundary of D. We parametrize each side of the triangle:

$\quad C_1: \; \mathbf{r}_1(t) = t\mathbf{j}, \quad t \in [0, 12]$

$\quad C_2: \; \mathbf{r}_2(t) = t\mathbf{i}, \quad t \in [0, 6]$

$\quad C_3: \; \mathbf{r}_3(t) = t\mathbf{i} + (12 - 2t)\,\mathbf{j}, \quad t \in [0, 6]$

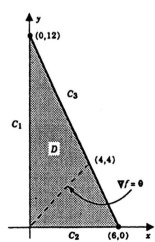

and observe from

$$f(\mathbf{r}_1(t)) = t^2, \quad t \in [0, 12]$$
$$f(\mathbf{r}_2(t)) = t^2, \quad t \in [0, 6]$$
$$f(\mathbf{r}_3(t)) = (3t - 12)^2, \quad t \in [0, 6]$$

that f takes on its maximum of 144 at the point $(0, 12)$.

18. $\nabla f = \dfrac{1}{(x^2 + y^2)^{3/2}} (-x\mathbf{i} - y\mathbf{j}) \neq \mathbf{0}$ in D. Note that $f(x, y)$ is the reciprocal of the distance of (x, y)

from the origin. The point of D closest to the origin (draw a figure) is $(1, 1)$. Therefore $f(1, 1) = 1/\sqrt{2}$
is the maximum value of f. The point of D furthest from the origin is $(3, 4)$. Therefore $f(3, 4) = 1/5$
is the least value taken on by f.

19. Using the hint, we want to find the maximum value of $f(x, y) = 18xy - x^2y - xy^2$ in the triangular
region. The gradient of f is:

$$\nabla D = \left(18y - 2xy - y^2\right) \mathbf{i} + \left(18x - x^2 - 2xy\right) \mathbf{j}$$

The gradient is $\mathbf{0}$ when

$$18y - 2xy - y^2 = 0 \quad \text{and} \quad 18x - x^2 - 2xy = 0$$

The solution set of this pair of equations is: $(0, 0)$, $(18, 0)$, $(0, 18)$, $(6, 6)$.

It is easy to verify that f is a maximum when $x = y = 6$. The three numbers that satisfy $x + y + z = 18$
and maximize the product xyz are: $x = 6$, $y = 6$, $z = 6$.

20. $f(y, z) = 30yz^2 - y^2z^2 - yz^3$, $\quad \nabla f = (30z^2 - 2yz^2 - z^3)\mathbf{j} + (60yz - 2y^2z - 3yz^2)\mathbf{k} = \mathbf{0}$ at $\left(\dfrac{15}{2}, 15\right)$

(other points are not in the interior); $\quad f\left(\dfrac{15}{2}, 15\right) = \dfrac{15^4}{4}$.

On the line $y + z = 30$, $f(y, z) = 0$ so the maximum of xyz^2 occurs at $x = y = \dfrac{15}{2}$, $z = 15$

21. $f(x, y) = xy(1 - x - y)$, $\quad 0 \leq x \leq 1$, $\quad 0 \leq y \leq 1 - x$.

[dom (f) is the triangle with vertices $(0, 0)$, $(1, 0)$, $(0, 1)$.]

$\nabla f = (y - 2xy - y^2)\mathbf{i} + (x - 2xy - x^2)\mathbf{j} = \mathbf{0} \implies x = y = 0$, $x = 1$, $y = 0$, $x = 0$, $y = 1$, $x = y = \frac{1}{3}$.

(Note that $[0, 0]$ is not an interior point of the domain of f.)

$f_{xx} = -2y$, $\quad f_{xy} = 1 - 2x - 2y$, $\quad f_{yy} = -2x$.

At $\left(\frac{1}{3}, \frac{1}{3}\right)$, $D = \frac{1}{3} > 0$ and $A < 0$ so we have a local max; the value is $1/27$.

Since $f(x, y) = 0$ at each point on the boundary of the domain, the local max of $1/27$ is also the
absolute max.

22. $V = xyz, \quad \dfrac{x}{a} + \dfrac{y}{b} + \dfrac{z}{c} = 1 \quad \Longrightarrow \quad V(x,y) = xyc\left(1 - \dfrac{x}{a} - \dfrac{y}{b}\right), \quad x > 0, \ y > 0, \ \dfrac{x}{a} + \dfrac{y}{b} < 1$

$\nabla V = yc\left(1 - \dfrac{2x}{a} - \dfrac{y}{b}\right)\mathbf{i} + xc\left(1 - \dfrac{x}{a} - \dfrac{2y}{b}\right)\mathbf{j} = \mathbf{0} \quad$ at $\quad \left(\dfrac{a}{3}, \dfrac{b}{3}\right)$

Maximum volume $\quad V = \dfrac{a}{3} \cdot \dfrac{b}{3} \cdot \dfrac{c}{3} = \dfrac{abc}{27}$

23. (a) $\nabla f = \frac{1}{2}x\,\mathbf{i} - \frac{2}{9}y\,\mathbf{j} = \mathbf{0}$ only at $(0,0)$.

  (b) The difference
$$f(h,k) - f(0,0) = \tfrac{1}{4}h^2 - \tfrac{1}{9}k^2$$

  does not keep a constant sign for all small h and k; $(0,0)$ is a saddle point. The function has no local extreme values.

  (c) Being the difference of two squares, f can be maximized by maximizing $\frac{1}{4}x^2$ and minimizing $\frac{1}{9}y^2$; $(1,0)$ and $(-1,0)$ give absolute maximum value $\frac{1}{4}$. Similarly, $(0,1)$ and $(0,-1)$ give absolute minimum value $-\frac{1}{9}$.

24. (a) $\nabla f = anx^{n-1}\mathbf{i} + cny^{n-1}\mathbf{j} = \mathbf{0} \quad$ at $\quad (0,0)$

  (b) $\dfrac{\partial^2 f}{\partial x^2} = an(n-1)x^{n-2}, \quad \dfrac{\partial^2 f}{\partial y \partial x} = 0, \quad \dfrac{\partial^2 f}{\partial y^2} = cn(n-1)y^{n-2}; \quad$ at $(0,0)$, $D = 0$.

  (c) (i) $(0,0)$ gives absolute min of 0 if n is even; no extreme value if n is odd.

   (ii) $(0,0)$ gives absolute max of 0 if n is even; no extreme value if n is odd.

   (iii) no extreme values.

25. Let x, y and z be the length, width and height of the box. The surface area is given by
$$S = 2xy + 2xz + 2yz, \quad \text{so} \quad z = \frac{S - 2xy}{2(x+y)}, \quad \text{where } S \text{ is a constant, and } x, y, z > 0.$$

Now, the volume $V = xyz$ is given by:
$$V(x,y) = xy\left[\frac{S - 2xy}{2(x+y)}\right]$$

and
$$\nabla V = \left\{ y\left[\frac{S - 2xy}{2(x+y)}\right] + xy\,\frac{2(x+y)(-2y) - (S - 2xy)(2)}{4(x+y)^2}\right\}\mathbf{i}$$
$$+ \left\{ x\left[\frac{S - 2xy}{2(x+y)}\right] + xy\,\frac{2(x+y)(-2x) - (S - 2xy)(2)}{4(x+y)^2}\right\}\mathbf{j}$$

Setting $\dfrac{\partial V}{\partial x} = \dfrac{\partial V}{\partial y} = 0$ and simplifying, we get the pair of equations
$$2S - 4x^2 - 8xy = 0$$
$$2S - 4y^2 - 8xy = 0$$

from which it follows that $x = y = \sqrt{S/6}$. From practical considerations, we conclude that V has a maximum value at $(\sqrt{S/6}, \sqrt{S/6})$. Substituting these values into the equation for z, we get $z = \sqrt{S/6}$ and so the box of maximum volume is a cube.

26. $V = xyz,\quad S = xy + 2xz + 2yz \quad\Longrightarrow\quad V(x,y) = xy\dfrac{(S - xy)}{2(x + y)},\quad x > 0,\ y > 0,\ xy < S.$

$$\nabla V = \frac{y^2(S - x^2 - 2xy)}{2(x + y)^2}\mathbf{i} + \frac{x^2(S - y^2 - 2xy)}{2(x + y)^2}\mathbf{j}$$

$$\nabla V = 0 \quad\Longrightarrow\quad x = \sqrt{\frac{S}{3}},\ y = \sqrt{\frac{S}{3}}\,;\qquad \text{dimensions for maximum volume:}\quad \sqrt{\frac{S}{3}} \times \sqrt{\frac{S}{3}} \times \frac{1}{2}\sqrt{\frac{S}{3}}$$

27.
$$f(x,y) = \sum_{i=1}^{3}\left[(x - x_i)^2 + (y - y_i)^2\right]$$

$$\nabla f(x,y) = 2\left[(3x - x_1 - x_2 - x_3)\,\mathbf{i} + (3y - y_1 - y_2 - y_3)\,\mathbf{j}\right]$$

$$\nabla f = 0 \quad\text{only at}\quad \left(\frac{x_1 + x_2 + x_3}{3},\ \frac{y_1 + y_2 + y_3}{3}\right) = (x_0, y_0)\,.$$

The difference $\quad f(x_0 + h,\ y_0 + k) - f(x_0, y_0)$

$$= \sum_{i=1}^{3}\left[(x_0 + h - x_i)^2 + (y_0 + k - y_i)^2 - (x_0 - x_i)^2 - (y_0 - y_i)^2\right]$$

$$= \sum_{i=1}^{3}\left[2h(x_0 - x_i) + h^2 + 2k(y_0 - y_i) + k^2\right]$$

$$= 2h(3x_0 - x_1 - x_2 - x_3) + 2k(3y_0 - y_1 - y_2 - y_3) + 3h^2 + 3k^2$$

$$= 3h^2 + 3k^2$$

is nonnegative for all h and k. Thus, f has its absolute minimum at (x_0, y_0).

28. Profit $\ P(x,y) = N_1(x - 50) + N_2(y - 60) = 250(y - x)(x - 50) + [32{,}000 + 250(x - 2y)](y - 60)$

$\nabla P = 250(2y - 2x - 10)\mathbf{i} + [32{,}000 + 250(2x + 70 - 4y)]\mathbf{j} = \mathbf{0}$

$\Longrightarrow x = 89,\quad y = 94$

29.

$$A = xy + \frac{1}{2}x\left(\frac{x}{2}\tan\theta\right),$$

$$P = x + 2y + 2\left(\frac{x}{2}\sec\theta\right),\ y = \tfrac{1}{2}(P - x - x\sec\theta)$$

$$0 < \theta < \frac{1}{2}\pi,\quad 0 < x < \frac{P}{1 + \sec\theta}.$$

$$A(x,\theta) = \tfrac{1}{2}x(P - x - x\sec\theta) + \tfrac{1}{4}x^2\tan\theta,$$

$$\nabla A = \left(\frac{P}{2} - x - x\sec\theta + \frac{x}{2}\tan\theta\right)\mathbf{i} + \left(\frac{x^2}{4}\sec^2\theta - \frac{x^2}{2}\sec\theta\tan\theta\right)\mathbf{j},$$

(Here $\mathbf{j}$ is the unit vector in the direction of increasing θ.)

$$\nabla A = \frac{1}{2}[P + x(\tan\theta - 2\sec\theta - 2)]\,\mathbf{i} + \frac{x^2}{4}\sec\theta\,(\sec\theta - 2\tan\theta)\,\mathbf{j}.$$

From $\dfrac{\partial A}{\partial \theta} = 0$ we get $\theta = \frac{1}{6}\pi$ and then from $\dfrac{\partial A}{\partial x} = 0$ we get

$$P + x\left(\tfrac{1}{3}\sqrt{3} - \tfrac{4}{3}\sqrt{3} - 2\right) = 0 \quad \text{so that} \quad x = (2 - \sqrt{3})P.$$

Next,

$$A_{xx} = \tfrac{1}{2}(\tan\theta - 2\sec\theta - 2),$$

$$A_{x\theta} = \frac{x}{2}\sec\theta\,(\sec\theta - 2\tan\theta),$$

$$A_{\theta\theta} = \frac{x^2}{2}\sec\theta\,\left(\sec\theta\tan\theta - \sec^2\theta - \tan^2\theta\right).$$

Apply the second-partials test:

$$A = -\tfrac{1}{2}(2 + \sqrt{3}), \quad B = 0, \quad C = -\tfrac{1}{3}P^2\sqrt{3}\,(2 - \sqrt{3})^2, \quad D < 0.$$

Since, $D > 0$ and $A < 0$, the area is a maximum when $\theta = \frac{1}{6}\pi$, $x = (2 - \sqrt{3})\,P$ and $y = \frac{1}{6}(3 - \sqrt{3})P$.

30. (a) $\nabla f = (2ax + by)\mathbf{i} + (bx + 2cy)\mathbf{j}$

$$\frac{\partial^2 f}{\partial x^2} = 2a, \quad \frac{\partial^2 f}{\partial y \partial x} = b, \quad \frac{\partial^2 f}{\partial y^2} = 2c; \quad D = 4ac - b^2.$$

(b) The point $(0,0)$ is the only stationary point. If $D < 0$, $(0,0)$ is a saddle point; if $D > 0$, $(0,0)$ is a local minimum if $a > 0$ and a local maximum if $a < 0$.

(c) (i) if $b > 0$, $f(x,y) = (\sqrt{a}x + \sqrt{c}y)^2$;   every point on the line $\sqrt{a}x + \sqrt{c}y = 0$

is a stationary point and at each such point f takes on a local and absolute min of 0

if $b < 0$, $f(x,y) = (\sqrt{a}x - \sqrt{c}y)^2$;   every point on the line $\sqrt{a}x - \sqrt{c}y = 0$

is a stationary point and at each such point f takes on a local and absolute min of 0

(ii) if $b > 0$, $f(x,y) = -(\sqrt{|a|}x - \sqrt{|c|}y)^2$;   every point on the line $\sqrt{|a|}x - \sqrt{|c|}y = 0$

is a stationary point and at each such point f takes on a local and absolute max of 0

if $b < 0$, $f(x,y) = -(\sqrt{|a|}x + \sqrt{|c|}y)^2$;   every point on the line $\sqrt{|a|}x + \sqrt{|c|}y = 0$

is a stationary point and at each such point f takes on a local and absolute max of 0

31. From

$$x = \tfrac{1}{2}y = \tfrac{1}{3}z = t \quad \text{and} \quad x = y - 2 = z = s$$

we take

$$(t, 2t, 3t) \quad \text{and} \quad (s, 2 + s, s)$$

as arbitrary points on the lines. It suffices to minimize the square of the distance between these points:

$$f(t, s) = (t - s)^2 + (2t - 2 - s)^2 + (3t - s)^2$$

$$= 14t^2 - 12ts + 3s^2 - 8t + 4s + 4, \qquad t, s \text{ real.}$$

Let $\mathbf{i}$ and $\mathbf{j}$ be the unit vectors in the direction of increasing t and s, respectively.

$$\nabla f = (28t - 12s - 8)\mathbf{i} + (-12t + 6s + 4)\mathbf{j}; \qquad \nabla f = 0 \quad \Longrightarrow \quad t = 0, \ s = -2/3.$$

$$f_{tt} = 28, \quad f_{ts} = -12, \quad f_{ss} = 6, \quad D = 6(28) - (-12)^2 = -24 < 0.$$

By the second-partials test, the distance is a minimum when $t = 0$, $s = -2/3$; the nature of the problem tells us the minimum is absolute. The distance is $\sqrt{f(0, -2/3)} = \frac{2}{3}\sqrt{6}$.

32. We want to minimize $S = 4\pi r^2 + 2\pi rh$ given that $V = \frac{4}{3}\pi r^3 + \pi r^2 h = 10,000$.

$$S(r) = 4\pi r^2 + 2\pi r \left(\frac{V}{\pi r^2} - \frac{4}{3}r \right) = \frac{4}{3}\pi r^2 + \frac{2V}{r}$$

$$S'(r) = \frac{8\pi r^3 - 6V}{3r^2} = 0 \implies r = \sqrt[3]{\frac{6V}{8\pi}}, \quad h = \frac{V}{\pi r^2} - \frac{4}{3}r = 0$$

The optimal container is a sphere of radius $r = \sqrt[3]{7500/\pi}$ meters.

33. (a) Let x and y be the cross-sectional measurements of the box, and let l be its length. Then

$$V = xyl, \quad \text{where} \quad 2x + 2y + l \le 108, \quad x, y > 0$$

To maximize V we will obviously take $2x+2y+l=108$. Therefore, $V(x,y)=xy(108-2x-2y)$ and

$$\nabla V = [y(108 - 2x - 2y) - 2xy]\,\mathbf{i} + [x(108 - 2x - 2y) - 2xy]\,\mathbf{j}$$

Setting $\dfrac{\partial V}{\partial x} = \dfrac{\partial V}{\partial y} = 0$, we get the pair of equations

$$\frac{\partial V}{\partial x} = 108y - 4xy - 2y^2 = 0$$

$$\frac{\partial V}{\partial y} = 108x - 4xy - 2x^2 = 0$$

from which it follows that $x = y = 18 \implies l = 36$.

Now, at $(18, 18)$, we have

$$A = V_{xx} = -4y = -72 < 0, \quad B = V_{xy} = 108 - 4x - 4y = -36,$$

$$C = V_{yy} = -4x = -72, \quad \text{and} \quad D = (36)^2 - (72)^2 < 0.$$

Thus, V is a maximum when $x = y = 18$ inches and $l = 36$ inches.

(b) Let r be the radius of the tube and let l be its length. Then

$$V = \pi r^2 l, \quad \text{where} \quad 2\pi r + l \le 108, \quad r > 0$$

To maximize V we take $2\pi r + l = 108$. Then $V(r) = \pi r^2 (108 - 2\pi r) = 108\pi r^2 - 2\pi^2 r^3$. Now

$$\frac{dV}{dr} = 216\pi r - 6\pi^2 r^2$$

Setting $\dfrac{dV}{dr} = 0$, we get

$$216\pi r - 6\pi^2 r^2 = 0 \implies r = \frac{36}{\pi} \implies l = 36$$

Now, at $r = 36/\pi$, we have

$$\frac{d^2 V}{dr^2} = 216\pi - 12\pi^2 \frac{36}{\pi} = -216\pi < 0$$

Thus, V is a maximum when $r = 36/\pi$ inches and $l = 36$ inches.

34. Let (x, y, z) be on the ellipsoid, $x > 0$, $y > 0$, $z > 0$.   Then
$$V = 2x \cdot 2y \cdot 2z = 8xyz.$$
Note that V achieves its maximum $\iff$ $x^2 y^2 z^2$ achieves its maximum.
Let $s = x^2 y^2 z^2$,   then
$$s = x^2 y^2 \left(1 - \frac{x^2}{9} - \frac{y^2}{4} \right),$$
$$\frac{\partial s}{\partial x} = 2xy^2 \left(1 - \frac{2x^2}{9} - \frac{y^2}{4} \right) = 0$$
$$\frac{\partial s}{\partial y} = 2x^2 y \left(1 - \frac{x^2}{9} - \frac{2y^2}{4} \right) = 0$$
$$\implies \quad \frac{2x^2}{9} + \frac{y^2}{4} = 1, \quad \frac{x^2}{9} + \frac{2y^2}{4} = 1 \quad \implies \quad x = \frac{3}{\sqrt{3}}, \quad y = \frac{2}{\sqrt{3}}, \quad z = \frac{1}{\sqrt{3}}$$
Thus,
$$V_{\max} = 8xyz = 8 \cdot \frac{3}{\sqrt{3}} \cdot \frac{2}{\sqrt{3}} \cdot \frac{1}{\sqrt{3}} = \frac{16\sqrt{3}}{3}.$$

35. Let S denote the cross-sectional area. Then
$$S = \frac{1}{2} \left(12 - 2x + 12 - 2x + 2x \cos\theta \right) x \sin\theta = 12x \sin\theta - 2x^2 \sin\theta + \frac{1}{2} x^2 \sin 2\theta,$$
where $0 < x < 6$, $0 < \theta < \pi/2$
Now, with $\mathbf{j}$ in the direction of increasing θ,
$$\nabla S = (12 \sin\theta - 4x \sin\theta + x \sin 2\theta)\, \mathbf{i} + (12x \cos\theta - 2x^2 \cos\theta + x^2 \cos 2\theta)\, \mathbf{j}$$
Setting $\dfrac{\partial S}{\partial x} = \dfrac{\partial S}{\partial \theta} = 0$, we get the pair of equations
$$12 \sin\theta - 4x \sin\theta + x \sin 2\theta = 0$$
$$12x \cos\theta - 2x^2 \cos\theta + x^2 \cos 2\theta = 0$$
from which it follows that $x = 4, \theta = \pi/3$.
Now, at $(4, \pi/3)$, we have
$$A = S_{xx} = -4 \sin\theta + \sin 2\theta = -\frac{3}{2}\sqrt{3}, \quad B = S_{x\theta} = 12 \cos\theta - 4x \cos\theta + 2x \cos 2\theta = -6,$$
$$C = S_{\theta\theta} = -12x \sin\theta + 2x^2 \sin\theta - 2x^2 \sin 2\theta = -24\sqrt{3} \quad \text{and} \quad D = 108 - 36 > 0.$$
Thus, S is a maximum when $x = 4$ inches and $\theta = \pi/3$.

36.

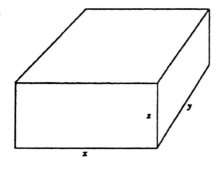

$$96 = xyz,$$
$$C = 30xy + 10(2xz + 2yz)$$
$$= 30xy + 20(x + y)\frac{96}{xy}.$$

$$C(x, y) = 30\left[xy + \frac{64}{x} + \frac{64}{y}\right],$$

$$\nabla C = 30(y - 64x^{-2})\mathbf{i} + 30(x - 64y^{-2})\mathbf{j} = 0 \quad \Longrightarrow \quad x = y = 4.$$

$$C_{xx} = 128x^{-3}, \quad C_{xy} = 1, \quad C_{yy} = 128y^{-3}.$$

When $x = y = 4$, we have $D = 3 > 0$ and $A = 2 > 0$ so the cost is minimized by making the dimensions of the crate $4 \times 4 \times 6$ meters.

37. (a) $f(m, b) = [2 - b]^2 + [-5 - (m + b)]^2 + [4 - (2m + b)]^2.$

$\quad f_m = 10m + 6b - 6, \quad f_b = 6m + 6b - 2; \quad f_m = f_b = 0 \quad \Longrightarrow \quad m = 1, \ b = -\frac{2}{3}.$

$\quad f_{mm} = 10, \quad f_{mb} = 6, \quad f_{bb} = 6, \quad D = 24 > 0 \quad \Longrightarrow \quad$ a minimum.

Answer: the line $y = x - \frac{2}{3}$.

(b) $f(\alpha, \beta) = [2 - \beta]^2 + [-5 - (\alpha + \beta)]^2 + [4 - (4\alpha + \beta)]^2.$

$\quad f_\alpha = 34\alpha + 10\beta - 22, \quad f_\beta = 10\alpha + 6\beta - 2; \quad f_\alpha = f_\beta = 0 \quad \Longrightarrow \quad \begin{cases} \alpha = \frac{14}{13} \\ \beta = -\frac{19}{13} \end{cases}.$

$\quad f_{\alpha\alpha} = 34, \quad f_{\alpha\beta} = 10, \quad f_{\beta\beta} = 6, \quad D = 104 > 0 \quad \Longrightarrow \quad$ a minimun.

Answer: the parabola $y = \frac{1}{13}\left(14x^2 - 19\right).$

38. (a) $f(m, b) = [2 - (-m + b)]^2 + [-1 - b]^2 + [1 - (m + b)]^2$

$\quad f_m = 4m + 2, \quad f_b = 6b - 4, \quad f_m = f_b = 0 \quad \Longrightarrow \quad m = -\frac{1}{2}, \ b = \frac{2}{3}$

$\quad f_{mm} = 4, \quad f_{mb} = 0, \quad f_{bb} = 6, \quad D = 24 > 0 \quad \Longrightarrow \quad$ minimum

Answer: the line $y = -\frac{1}{2}x + \frac{2}{3}$

(b) $f(\alpha, \beta) = [2 - (\alpha + \beta)]^2 + [-1 - \beta]^2 + [1 - (\alpha + \beta)]^2$

$\quad f_\alpha = 4\alpha + 4\beta - 6, \quad f_\beta = 4\alpha + 6\beta - 4; \quad f_\alpha = f_\beta = 0 \quad \Longrightarrow \quad \alpha = \frac{5}{2}, \ \beta = -1$

$\quad f_{\alpha\alpha} = 4, \quad f_{\alpha\beta} = 4, \quad f_{\beta\beta} = 6, \quad D = 8 > 0 \quad \Longrightarrow \quad$ minimum

Answer: the parabola $y = \frac{5}{2}x^2 - 1$

SECTION 16.7

1.

$$f(x, y) = x^2 + y^2, \qquad g(x, y) = xy - 1$$

$$\nabla f = 2x\mathbf{i} + 2y\mathbf{j}, \qquad \nabla g = y\mathbf{i} + x\mathbf{j}.$$

$$\nabla f = \lambda\nabla g \quad \Longrightarrow \quad 2x = \lambda y \ \text{ and } \ 2y = \lambda x.$$

Multiplying the first equation by x and the second equation by y, we get

$$2x^2 = \lambda xy = 2y^2.$$

Thus, $x = \pm y$. From $g(x, y) = 0$ we conclude that $x = y = \pm 1$. The points $(1, 1)$ and $(-1, -1)$ clearly give a minimum, since f represents the square of the distance of a point on the hyperbola from the origin. The minimum is 2.

2. $f(x,y) = xy, \quad g(x,y) = b^2x^2 + a^2y^2 - a^2b^2$

$$\nabla f = y\mathbf{i} + x\mathbf{j}, \quad \nabla g = 2b^2 x\mathbf{i} + 2a^2 y\mathbf{j}$$

$\nabla f = \lambda \nabla g \implies y = 2\lambda b^2 x, \quad x = 2\lambda a^2 y \implies a^2 y^2 = b^2 x^2$

From $g(x,y) = 0$ we get $2b^2 x^2 = a^2 b^2 \implies x = \pm\dfrac{a}{\sqrt{2}}, \quad y = \pm\dfrac{b}{\sqrt{2}}$

The maximum value of xy is $\frac{1}{2}ab$, achieved at $(a/\sqrt{2}, b\sqrt{2})$ and at $(-a/\sqrt{2}, -b\sqrt{2})$.

3.
$$f(x,y) = xy, \qquad\qquad g(x,y) = b^2 x^2 + a^2 y^2 - a^2 b^2$$
$$\nabla f = y\mathbf{i} + x\mathbf{j}, \qquad\qquad \nabla g = 2b^2 x\mathbf{i} + 2a^2 y\mathbf{j}.$$
$$\nabla f = \lambda \nabla g \implies y = 2\lambda b^2 x \text{ and } x = 2\lambda a^2 y.$$

Multiplying the first equation by $a^2 y$ and the second equation by $b^2 x$, we get

$$a^2 y^2 = 2\lambda a^2 b^2 xy = b^2 x^2.$$

Thus, $ay = \pm bx$. From $g(x,y) = 0$ we conclude that $x = \pm\frac{1}{2}a\sqrt{2}$ and $y = \pm\frac{1}{2}b\sqrt{2}$.

Since f is continuous and the ellipse is closed and bounded, the minimum exists. It occurs at $\left(\frac{1}{2}a\sqrt{2}, -\frac{1}{2}b\sqrt{2}\right)$ and $\left(-\frac{1}{2}a\sqrt{2}, \frac{1}{2}b\sqrt{2}\right)$; the minimum is $-\frac{1}{2}ab$.

4. $f(x,y) = xy^2, \quad g(x,y) = x^2 + y^2 - 1$

$\nabla f = y^2\mathbf{i} + 2xy\mathbf{j}, \quad \nabla g = 2x\mathbf{i} + 2y\mathbf{j}$

$\nabla f = \lambda \nabla g \implies y^2 = 2\lambda x, \quad 2xy = 2\lambda y \implies y = 0 \quad \text{or} \quad y^2 = 2x^2$

$y = 0$: From $g(x,y) = 0$ we get $x = \pm 1; \quad f(\pm 1, 0) = 0$

$y^2 = 2x^2$: from $g(x,y) = 0$ we get $3x^2 = 1, \implies x = \pm\dfrac{1}{\sqrt{3}}, \quad y = \pm\dfrac{\sqrt{2}}{\sqrt{3}}$

The minimum of xy^2 is: $\quad -\dfrac{2}{9}\sqrt{3}$ at $(-1/\sqrt{3}, \pm\sqrt{2}/\sqrt{3})$

5. Since f is continuous and the ellipse is closed and bounded, the maximum exists.

$$f(x,y) = xy^2, \qquad\qquad g(x,y) = b^2 x^2 + a^2 y^2 - a^2 b^2$$
$$\nabla f = y^2\mathbf{i} + 2xy\mathbf{j}, \qquad\qquad \nabla g = 2b^2 x\mathbf{i} + 2a^2 y\mathbf{j}.$$
$$\nabla f = \lambda \nabla g \implies y^2 = 2\lambda b^2 x \text{ and } 2xy = 2\lambda a^2 y.$$

Multiplying the first equation by $a^2 y$ and the second equation by $b^2 x$, we get

$$a^2 y^3 = 2\lambda a^2 b^2 xy = 2b^2 x^2 y.$$

We can exclude $y = 0$; it clearly cannot produce the maximum. Thus,

$$a^2 y^2 = 2b^2 x^2 \text{ and, from } g(x,y) = 0, \ 3b^2 x^2 = a^2 b^2.$$

This gives us $x = \pm\frac{1}{3}\sqrt{3}\,a$ and $y = \pm\frac{1}{3}\sqrt{6}\,b$. The maximum occurs at $x = \frac{1}{3}\sqrt{3}\,a, \ y = \pm\frac{1}{3}\sqrt{6}\,b$; the value there is $\frac{2}{9}\sqrt{3}\,ab^2$.

6. $f(x, y) = x + y, \quad g(x, y) = x^4 + y^4 - 1$

$\nabla f = \mathbf{i} + \mathbf{j}, \quad \nabla g = 4x^3 \mathbf{i} + 4y^3 \mathbf{j}$

$\nabla f = \lambda \nabla g \implies 1 = 4\lambda x^3, \quad 1 = 4\lambda y^3 \implies x = y$

From $g(x, y) = 0$ we get $2x^4 = 1 \implies x = y = \pm 2^{-1/4}$

The maximum of $x + y$ is: $2 \cdot 2^{-1/4} = 2^{3/4}$, achieved at $(2^{-1/4}, 2^{-1/4})$.

7. The given curve is closed and bounded. Since $x^2 + y^2$ represents the square of the distance from points on this curve to the origin, the maximum exists.

$$f(x, y) = x^2 + y^2, \qquad g(x, y) = x^4 + 7x^2 y^2 + y^4 - 1$$

$$\nabla f = 2x\mathbf{i} + 2y\mathbf{j}, \qquad \nabla g = \left(4x^3 + 14xy^2\right)\mathbf{i} + \left(4y^3 + 14x^2 y\right)\mathbf{j}.$$

We use the cross-product equation (16.7.4):

$$2x(4y^3 + 14x^2 y) - 2y(4x^3 + 14xy^2) = 0,$$

$$20x^3 y - 20xy^3 = 0,$$

$$xy(x^2 - y^2) = 0.$$

Thus, $x = 0$, $y = 0$, or $x = \pm y$. From $g(x, y) = 0$ we conclude that the points to examine are
$$(0, \pm 1), \quad (\pm 1, 0), \quad \left(\pm\tfrac{1}{3}\sqrt{3}, \pm\tfrac{1}{3}\sqrt{3}\right).$$
The value of f at each of the first four points is 1; the value at the last four points is 2/3. The maximum is 1.

8. $f(x, y, z) = xyz, \quad g(x, y, z) = x^2 + y^2 + z^2 - 1$

$\nabla f = yz\mathbf{i} + xz\mathbf{j} + xy\mathbf{k}, \quad \nabla g = 2x\mathbf{i} + 2y\mathbf{j} + 2z\mathbf{k}$

$\nabla f = \lambda \nabla g \implies yz = 2\lambda x, \quad xz = 2\lambda y, \quad xy = 2\lambda z \implies x^2 = y^2 = z^2$ or $\lambda = 0$.

$\lambda = 0$: In this case, at least two of x, y, z are 0 and $f = 0$.

$x^2 = y^2 = z^2$ From $g(x, y, z) = 0$ we get $3x^2 = 1 \implies x = \pm\frac{1}{\sqrt{3}}, \ y = \pm\frac{1}{\sqrt{3}}, \ z = \pm\frac{1}{\sqrt{3}}$

The minimum of xyz is: $-\dfrac{1}{9}\sqrt{3}$ at $(-1/\sqrt{3}, -1/\sqrt{3}, -1/\sqrt{3})$, $(-1/\sqrt{3}, 1/\sqrt{3}, 1/\sqrt{3})$,

$(1/\sqrt{3}, -1/\sqrt{3}, 1/\sqrt{3})$, $(1/\sqrt{3}, 1/\sqrt{3}, -1/\sqrt{3})$.

9. The maximum exists since xyz is continuous and the ellipsoid is closed and bounded.

$$f(x, y, z) = xyz, \qquad g(x, y, z) = \frac{x^2}{a^2} + \frac{y^2}{b^2} + \frac{z^2}{c^2} - 1$$

$$\nabla f = yz\mathbf{i} + xz\mathbf{j} + xy\mathbf{k}, \qquad \nabla g = \frac{2x}{a^2}\mathbf{i} + \frac{2y}{b^2}\mathbf{j} + \frac{2z}{c^2}\mathbf{k}.$$

$$\nabla f = \lambda \nabla g \implies yz = \frac{2x}{a^2}\lambda, \quad xz = \frac{2y}{b^2}\lambda, \quad xy = \frac{2z}{c^2}\lambda.$$

We can assume x, y, z are non-zero, for otherwise $f(x, y, z) = 0$, which is clearly not a maximum. Then from the first two equations

$$\frac{yza^2}{x} = 2\lambda = \frac{xzb^2}{y} \quad \text{so that} \quad a^2y^2 = b^2x^2 \quad \text{or} \quad \frac{x^2}{a^2} = \frac{y^2}{b^2}.$$

Similarly from the second and third equations we get

$$b^2z^2 = c^2y^2 \quad \text{or} \quad \frac{y^2}{b^2} = \frac{z^2}{c^2}.$$

From $g(x, y, z) = 0$, we get $\dfrac{3x^2}{a^2} = 1 \implies x \pm \dfrac{a}{\sqrt{3}}$, from which it follows that $y = \pm\dfrac{b}{\sqrt{3}}$, $z = \pm\dfrac{c}{\sqrt{3}}$. The maximum value is $\frac{1}{9}\sqrt{3}\, abc$.

10. $f(x, y, z) = x + 2y + 4z, \quad g(x, y, z) = x^2 + y^2 + z^2 - 7$

$\nabla f = \mathbf{i} + 2\mathbf{j} + 4\mathbf{k}, \quad \nabla g = 2x\mathbf{i} + 2y\mathbf{j} + 2z\mathbf{k}$

$\nabla f = \lambda\nabla g \implies 1 = 2\lambda x, \ 2 = 2\lambda y, \ 4 = 2\lambda z \implies y = 2x, \ z = 4x$

From $g(x, y, z) = 0$ we get $21x^2 = 7 \implies x = \pm\dfrac{1}{\sqrt{3}}$

Minimum of $x + 2y + 4z$ is: $-7\sqrt{3}$, achieved at $(-1/\sqrt{3}, -2/\sqrt{3}, -4/\sqrt{3})$.

11. Since the sphere is closed and bounded and $2x + 3y + 5z$ is continuous, the maximum exists.

$$f(x, y, z) = 2x + 3y + 5z, \qquad g(x, y, z) = x^2 + y^2 + z^2 - 19$$

$$\nabla f = 2\mathbf{i} + 3\mathbf{j} + 5\mathbf{k}, \qquad \nabla g = 2x\mathbf{i} + 2y\mathbf{j} + 2z\mathbf{k}.$$

$$\nabla f = \lambda\nabla g \implies 2 = 2\lambda x, \ 3 = 2\lambda y, \ 5 = 2\lambda z.$$

Since $\lambda \neq 0$ here, we solve the equations for x, y and z:

$$x = \frac{1}{\lambda}, \quad y = \frac{3}{2\lambda}, \quad z = \frac{5}{2\lambda},$$

and substitute these results in $g(x, y, z) = 0$ to obtain

$$\frac{1}{\lambda^2} + \frac{9}{4\lambda^2} + \frac{25}{4\lambda^2} - 19 = 0, \quad \frac{38}{4\lambda^2} - 19 = 0, \quad \lambda = \pm\frac{1}{2}\sqrt{2}.$$

The positive value of λ will produce positive values for x, y, z and thus the maximum for f. We get $x = \sqrt{2}, \ y = \frac{3}{2}\sqrt{2}, \ z = \frac{5}{2}\sqrt{2}$, and $2x + 3y + 5z = 19\sqrt{2}$.

12. $f(x, y, z) = x^4 + y^4 + z^4, \quad g(x, y, z) = x + y + z - 1$

$\nabla f = 4x^3\mathbf{i} + 4y^3\mathbf{j} + 4z^3\mathbf{k}, \quad \nabla g = \mathbf{i} + \mathbf{j} + \mathbf{k}$

$\nabla f = \lambda\nabla g \implies 4x^3 = \lambda, \ 4y^3 = \lambda, \ 4z^3 = \lambda \implies x = y = z$

From $g(x, y, z) = 0$ we get $3x = 1, \implies x = \frac{1}{3} = y = z$

Minimum is: $\dfrac{1}{27}$

13.

$$f(x, y, z) = xyz, \qquad g(x, y, z) = \frac{x}{a} + \frac{y}{b} + \frac{z}{c} - 1$$

$$\nabla f = yz\mathbf{i} + xz\mathbf{j} + xy\mathbf{k}, \qquad \nabla g = \frac{1}{a}\mathbf{i} + \frac{1}{b}\mathbf{j} + \frac{1}{c}\mathbf{k}.$$

$$\nabla f = \lambda\nabla g \implies yz = \frac{\lambda}{a}, \quad xz = \frac{\lambda}{b}, \quad xy = \frac{\lambda}{c}.$$

Multiplying these equations by x, y, z respectively, we obtain

$$xyz = \frac{\lambda x}{a}, \quad xyz = \frac{\lambda y}{b}, \quad xyz = \frac{\lambda z}{c}.$$

Adding these equations and using the fact that $g(x, y, z) = 0$, we have

$$3xyz = \lambda\left(\frac{x}{a} + \frac{y}{b} + \frac{z}{c}\right) = \lambda.$$

Since x, y, z are non-zero,

$$yz = \frac{\lambda}{a} = \frac{3xyz}{a}, \quad 1 = \frac{3x}{a}, \quad x = \frac{a}{3}.$$

Similarly, $y = \frac{b}{3}$ and $z = \frac{c}{3}$. The maximum is $\frac{1}{27}abc$.

14. Maximize area $A = xy$ given that the perimeter $P = 2x + 2y$

$f(x, y) = xy, \quad g(x, y) = 2x + 2y - P$

$\nabla f = y\mathbf{i} + x\mathbf{j}, \quad \nabla g = 2\mathbf{i} + 2\mathbf{j}; \qquad \nabla f = \lambda\nabla g \implies y = 2\lambda, \quad x = 2\lambda \implies x = y.$

The rectangle of maximum area is a square.

15. It suffices to minimize the square of the distance from $(0, 1)$ to a point on the parabola. Clearly, the minimum exists.

$$f(x, y) = x^2 + (y - 1)^2, \qquad g(x, y) = x^2 - 4y$$

$$\nabla f = 2x\mathbf{i} + 2(y - 1)\mathbf{j}, \qquad \nabla g = 2x\mathbf{i} - 4\mathbf{j}.$$

We use the cross-product equation (16.7.4):

$$2x(-4) - 2x(2y - 2) = 0, \quad 4x + 4xy = 0, \quad x(y + 1) = 0.$$

Since $y \geq 0$, we have $x = 0$ and thus $y = 0$. The minimum is 1.

16. Minimize $f(x, y) = (x - p)^2 + (y - 4p)^2$ subject to $g(x, y) = 2px - y^2 = 0$

$\nabla f = 2(x - p)\mathbf{i} + 2(y - 4p)\mathbf{j}, \quad \nabla g = 2p\mathbf{i} - 2y\mathbf{j}$

$\nabla f = \lambda\nabla g \implies 2(x - p) = 2\lambda p, \quad 2(y - 4p) = -2\lambda y \implies x = \frac{4p^2}{y}$

From $g(x, y) = 0$ we get $\frac{8p^3}{y} = y^2 \implies y = 2p, \; x = 2p$

Distance to parabola is: $\sqrt{f(x, y)} = \sqrt{5}p$

17. It suffices to maximize and minimize the square of the distance from $(2, 1, 2)$ to a point on the sphere. Clearly, these extreme values exist.

$$f(x, y, z) = (x - 2)^2 + (y - 1)^2 + (z - 2)^2, \qquad g(x, y, z) = x^2 + y^2 + z^2 - 1$$

$$\nabla f = 2(x - 2)\mathbf{i} + 2(y - 1)\mathbf{j} + 2(z - 2)\mathbf{k}, \qquad \nabla g = 2x\mathbf{i} + 2y\mathbf{j} + 2z\mathbf{k}.$$

$$\nabla f = \lambda\nabla g \implies 2(x - 2) = 2x\lambda, \; 2(y - 1) = 2y\lambda, \; 2(z - 2) = 2z\lambda$$

Thus,
$$x = \frac{2}{1-\lambda}, \quad y = \frac{1}{1-\lambda}, \quad z = \frac{2}{1-\lambda}.$$

Using the fact that $x^2 + y^2 + z^2 = 1$, we have
$$\left(\frac{2}{1-\lambda}\right)^2 + \left(\frac{1}{1-\lambda}\right)^2 + \left(\frac{2}{1-\lambda}\right)^2 = 1 \quad \Longrightarrow \quad \lambda = -2, 4$$

At $\lambda = -2$, $(x, y, z) = (2/3, 1/3, 2/3)$ and $f(2/3, 1/3, 2/3) = 4$

At $\lambda = 4$, $(x, y, z) = (-2/3, -1/3, -2/3)$ and $f(-2/3, -1/3, -2/3) = 16$

Thus, $(2/3, 1/3, 2/3)$ is the closest point and $(-2/3, -1/3, -2/3)$ is the furthest point.

18. $f(x, y, z) = \sin x \sin y \sin z, \quad g(x, y, z) = x + y + z - \pi$

$\nabla f = \cos x \sin y \sin z \mathbf{i} + \sin x \cos y \sin z \mathbf{j} + \sin x \sin y \cos z \mathbf{k}, \quad \nabla g = \mathbf{i} + \mathbf{j} + \mathbf{k}$

$\nabla f = \lambda \nabla g \implies \cos x \sin y \sin z = \lambda = \sin x \cos y \sin z = \sin z \sin y \cos z \implies \cos x = \cos y = \cos z$

$\implies x = y = z = \dfrac{\pi}{3}$

Maximum of $\sin x \sin y \sin z$ is: $\dfrac{3\sqrt{3}}{8}$

19.
$$f(x, y, z) = 3x - 2y + z, \qquad g(x, y, z) = x^2 + y^2 + z^2 - 14$$
$$\nabla f = 3\mathbf{i} - 2\mathbf{j} + \mathbf{k}, \qquad \nabla g = 2x\,\mathbf{i} + 2y\,\mathbf{j} + 2z\,\mathbf{k}.$$
$$\nabla f = \lambda \nabla g \implies 3 = 2x\lambda, \quad -2 = 2y\lambda, \quad 1 = 2z\lambda.$$

Thus,
$$x = \frac{3}{2\lambda}, \quad y = -\frac{1}{\lambda}, \quad z = \frac{1}{2\lambda}.$$

Using the fact that $x^2 + y^2 + z^2 = 14$, we have
$$\left(\frac{3}{2\lambda}\right)^2 + \left(-\frac{1}{\lambda}\right)^2 + \left(\frac{1}{2\lambda}\right)^2 = 14 \quad \Longrightarrow \quad \lambda = \pm\frac{1}{2}.$$

At $\lambda = \dfrac{1}{2}$, $(x, y, z) = (3, -2, 1)$ and $f(3, -2, 1) = 14$

At $\lambda = -\dfrac{1}{2}$, $(x, y, z) = (-3, 2, -1)$ and $f(-3, 2, -1) = -14$

Thus, the maximum value of f on the sphere is 14.

20. $f(x, y, z) = xyz, \quad g(x, y, z) = x^2 + y^2 + z - 4$

$\nabla f = yz\mathbf{i} + xz\mathbf{j} + xy\mathbf{k}, \quad \nabla g = 2x\mathbf{i} + 2y\mathbf{j} + \mathbf{k}$

$\nabla f = \lambda \nabla g \implies yz = 2\lambda x, \quad xz = 2\lambda y, \quad xy = \lambda \implies x^2 = y^2 = \dfrac{z}{2}$

From $g(x, y, z) = 0$ we get $4x^2 = 4 \implies x = 1, \; y = 1, \; z = 2$.

Maximum volume is 2

21. It's easier to work with the square of the distance;   the minimum certainly exists.

$$f(x, y, z) = x^2 + y^2 + z^2, \qquad g(x, y, z) = Ax + By + Cz + D$$

$$\nabla f = 2x\mathbf{i} + 2y\mathbf{j} + 2z\mathbf{k}, \qquad \nabla g = A\mathbf{i} + B\mathbf{j} + C\mathbf{k}.$$

$$\nabla f = \lambda \nabla g \implies 2x = A\lambda, \quad 2y = B\lambda, \quad 2z = C\lambda.$$

Substituting these equations in $g(x, y, z) = 0$, we have

$$\frac{1}{2}\lambda \left(A^2 + B^2 + C^2\right) + D = 0, \quad \lambda = \frac{-2D}{A^2 + B^2 + C^2}.$$

Thus, in turn,

$$x = \frac{-DA}{A^2 + B^2 + C^2}, \quad y = \frac{-DB}{A^2 + B^2 + C^2}, \quad z = \frac{-DC}{A^2 + B^2 + C^2}$$

so the minimum value of $\sqrt{x^2 + y^2 + z^2}$ is $|D| \left(A^2 + B^2 + C^2\right)^{-1/2}$.

22. $f(x, y, z) = xyz, \quad g(x, y, z) = 2xy + 2xz + 2yz - 6a^2$

$\nabla f = yz\mathbf{i} + xz\mathbf{j} + xy\mathbf{k}, \quad \nabla g = 2(y + z)\mathbf{i} + 2(x + z)\mathbf{j} + 2(x + y)\mathbf{k}$

$\nabla f = \lambda \nabla g \implies yz = 2\lambda(y + z), \quad xz = 2\lambda(x + z), \quad xy = 2\lambda(x + y) \implies x = y = z$

From $g(x, y, z) = 0$ we get $6x^2 = 6a^2 \implies x = y = z = a$

Maximum volume is a^3.

23.

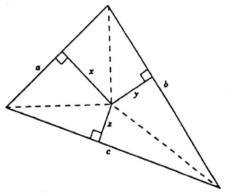

area $A = \frac{1}{2}ax + \frac{1}{2}by + \frac{1}{2}cz$.

The geometry suggests that

$$x^2 + y^2 + z^2$$

has a minimum.

$f(x, y, z) = x^2 + y^2 + z^2, \qquad g(x, y, z) = ax + by + cz - 2A$

$\nabla f = 2x\mathbf{i} + 2y\mathbf{j} + 2z\mathbf{k}, \qquad \nabla g = a\mathbf{i} + b\mathbf{j} + c\mathbf{k}.$

$$\nabla f = \lambda \nabla g \implies 2x = a\lambda, \quad 2y = b\lambda, \quad 2z = c\lambda.$$

Solving these equations for x, y, z and substituting the results in $g(x, y, z) = 0$, we have

$$\frac{a^2\lambda}{2} + \frac{b^2\lambda}{2} + \frac{c^2\lambda}{2} - 2A = 0, \quad \lambda = \frac{4A}{a^2 + b^2 + c^2}$$

and thus

$$x = \frac{2aA}{a^2 + b^2 + c^2}, \quad y = \frac{2bA}{a^2 + b^2 + c^2}, \quad z = \frac{2cA}{a^2 + b^2 + c^2}.$$

The minimum is $4A^2(a^2 + b^2 + c^2)^{-1}$.

24. Use figure 16.7.6 and write the side condition as $x + y + z = 2\pi$.

For (a) maximize $f(x, y, z) = 8R^3 \sin \frac{1}{2}x \sin \frac{1}{2}y \sin \frac{1}{2}z$.

For (b) maximize $f(x, y, z) = 4R^2(\sin^2 \frac{1}{2}x + \sin^2 \frac{1}{2}y + \sin^2 \frac{1}{2}z)$.

Each maximum occurs with $x = y = z = \dfrac{2\pi}{3}$.   This gives an equilateral triangle.

25. Since the curve is asymptotic to the line $y = x$ as $x \to -\infty$ and as $x \to \infty$, the maximum exists. The distance between the point (x, y) and the line $y - x = 0$ is given by

$$\frac{|y - x|}{\sqrt{1 + 1}} = \frac{1}{2}\sqrt{2}\,|y - x|. \qquad \text{(see Section 1.4)}$$

Since the points on the curve are below the line $y = x$, we can replace $|y - x|$ by $x - y$. To simplify the work we drop the constant factor $\frac{1}{2}\sqrt{2}$.

$$f(x, y) = x - y, \qquad\qquad g(x, y) = x^3 - y^3 - 1$$
$$\nabla f = \mathbf{i} - \mathbf{j}, \qquad\qquad \nabla g = 3x^2\mathbf{i} - 3y^2\mathbf{j}.$$

We use the cross-product equation (16.7.4):

$$1\left(-3y^2\right) - \left(3x^2\right)(-1) = 0, \quad 3x^2 - 3y^2 = 0, \quad x = -y \ (x \neq y).$$

Now $g(x, y) = 0$ gives us

$$x^3 - (-x)^3 - 1 = 0, \quad 2x^3 = 1, \quad x = 2^{-1/3}.$$

The point is $\left(2^{-1/3}, -2^{-1/3}\right)$.

26. Let r, s, t be the intercepts. We wish to minimize the volume

$$V = \frac{1}{6}rst \quad [\text{volume of pyramid} = \frac{1}{3}\text{base} \times \text{height}]$$

subject to the side condition $\dfrac{a}{r} + \dfrac{b}{s} + \dfrac{c}{t} = 1$. The minimum occurs when all the intercepts are:

$$r = 3a, \quad s = 3b, \quad t = 3c$$

27. It suffices to show that the square of the area is a maximum when $a = b = c$.

$$f(a, b, c) = s(s - a)(s - b)(s - c), \quad g(a, b, c) = a + b + c - 2s$$
$$\nabla f = -s(s - b)(s - c)\mathbf{i} - s(s - a)(s - c)\mathbf{j} - s(s - a)(s - b)\mathbf{k}, \quad \nabla g = \mathbf{i} + \mathbf{j} + \mathbf{k}.$$

     (Here $\mathbf{i}$, $\mathbf{j}$, $\mathbf{k}$ are the unit vectors in the directions of increasing a, b, c.)

$$\nabla f = \lambda \nabla g \implies -s(s - b)(s - c) = -s(s - a)(s - c) = -s(s - a)(s - b) = \lambda.$$

Thus, $s - b = s - a = s - c$ so that $a = b = c$. This gives us the maximum, as no minimum exists. [The area can be made arbitrarily small by taking a close to s.]

28. $f(x, y, z) = 8xyz, \quad g(x, y, z) = a^2 - x^2 - y^2 - z^2, \quad x > 0, \ y > 0, \ z > 0.$

$\nabla f = 8yz\mathbf{i} + 8xz\mathbf{j} + 8xy\mathbf{k}, \quad \nabla g = -2x\,\mathbf{i} - 2y\,\mathbf{j} - 2z\,\mathbf{k}$

$\nabla f = \lambda \nabla g \implies 8yz = -2\lambda x, \quad 8xz = -2\lambda y, \quad 8xy = -2\lambda z \implies x = y = z$

The rectangular box of maximum volume inscribed in the sphere is a cube.

29. (a) $f(x,y) = (xy)^{1/2}$, $g(x,y) = x + y - k$, $(x, y \geq 0$, k a nonnegative constant)

$$\nabla f = \frac{y^{1/2}}{2x^{1/2}}\mathbf{i} + \frac{x^{1/2}}{2y^{1/2}}\mathbf{j}, \qquad \nabla g = \mathbf{i} + \mathbf{j}.$$

$$\nabla f = \lambda \nabla g \quad \Longrightarrow \quad \frac{y^{1/2}}{2x^{1/2}} = \lambda = \frac{x^{1/2}}{2y^{1/2}} \quad \Longrightarrow \quad x = y = \frac{k}{2}.$$

Thus, the maximum value of f is: $f(k/2,\, k/2) = \dfrac{k}{2}$.

(b) For all x, y $(x, y \geq 0)$ we have

$$(xy)^{1/2} = f(x,y) \leq f(k/2,\, k/2) = \frac{k}{2} = \frac{x+y}{2}.$$

30. (a) The maximum occurs when $x = y = z = \dfrac{k}{3}$, where $(xyz)^{1/3} = \dfrac{k}{3}$.

(b) If $x + y + z = k$,   then, by (a), $(xyz)^{1/3} \leq \dfrac{k}{3} = \dfrac{x+y+z}{3}$.

31. Simply extend the arguments used in Exercises 29 and 30.

32. $T(x,y,z) = xy^2z$, $g(x,y,z) = x^2 + y^2 + z^2 - 1$

$\nabla T = y^2z\mathbf{i} + 2xyz\mathbf{j} + xy^2\mathbf{k}$, $\nabla g = 2x\mathbf{i} + 2y\mathbf{j} + 2z\mathbf{k}$

$\nabla T = \lambda \nabla g \implies y^2z = 2\lambda x$, $2xyz = 2\lambda y$, $xy^2 = 2\lambda z \implies x^2 = \dfrac{y^2}{2} = z^2$

From $g(x,y,z) = 0$ we get $4x^2 = 1 \implies x = \pm\frac{1}{2}$, $y = \pm\frac{1}{\sqrt{2}}, z = \pm\frac{1}{2}$

Maximum $\frac{1}{8}$ at $(\frac{1}{2}, \pm\frac{1}{\sqrt{2}}, \frac{1}{2})$, $(-\frac{1}{2}, \pm\frac{1}{\sqrt{2}}, -\frac{1}{2})$

Minimum $-\frac{1}{8}$ at $(\frac{1}{2}, \pm\frac{1}{\sqrt{2}}, -\frac{1}{2})$, $(-\frac{1}{2}, \pm\frac{1}{\sqrt{2}}, \frac{1}{2})$

33. $S(r,h) = 2\pi r^2 + 2\pi rh$, $g(r,h) = \pi r^2 h - V$,   (V constant)

$$\nabla S = (4\pi r + 2\pi h)\mathbf{i} + 2\pi r\mathbf{j}, \qquad \nabla g = 2\pi rh\mathbf{i} + \pi r^2\mathbf{j}.$$

$$\nabla S = \lambda \nabla g \quad \Longrightarrow \quad 4\pi r + 2\pi h = 2\pi rh\lambda, \quad 2\pi r = \pi r^2\lambda \quad \Longrightarrow \quad r = \frac{2}{\lambda}, \quad h = \frac{4}{\lambda}.$$

Now $\pi r^2 h = V, \quad \Longrightarrow \quad \lambda = \sqrt[3]{\dfrac{16\pi}{V}} \quad \Longrightarrow \quad r = \sqrt[3]{\dfrac{V}{2\pi}}, \quad h = \sqrt[3]{\dfrac{4V}{\pi}}.$

To minimize the surface area, take $r = \sqrt[3]{\dfrac{V}{2\pi}}$, and $h = \sqrt[3]{\dfrac{4V}{\pi}}$.

34. $f(x,y,z) = xyz$, $g(x,y,z) = x + y + z - 18$

$\nabla f = yz\mathbf{i} + xz\mathbf{j} + xy\mathbf{k}$, $\nabla g = \mathbf{i} + \mathbf{j} + \mathbf{k}$

$\nabla f = \lambda \nabla g \quad \Longrightarrow \quad yz = xz = xy \quad \Longrightarrow \quad x = y = z \quad \Longrightarrow \quad x = y = z = 6$

35. Same as Exercise 13.

36. $f(x, y, z) = xyz^2, \quad g(x, y, z) = x + y + z - 30$

$\nabla f = yz^2 \mathbf{i} + xz^2 \mathbf{j} + 2xyz \mathbf{k}, \quad \nabla g = \mathbf{i} + \mathbf{j} + \mathbf{k}$

$\nabla f = \lambda \nabla g \implies yz^2 = \lambda = xz^2 = 2xyz \implies x = y = \dfrac{z}{2} \implies x = y = \dfrac{15}{2}, \quad z = 15$

37. Let x, y, z denote the length, width and height of the box. We want to maximize the volume V of the box given that the surface area S is constant. That is:

maximize $V(x, y, z) = xyz$ subject to $S(x, y, z) = 2xy + 2xz + 2yz = S$ constant

Let $g(x, y, z) = 2xy + 2xz + 2yz - S$. Then

$$\nabla V = yz\,\mathbf{i} + xz\,\mathbf{j} + xy\,\mathbf{k}, \qquad \nabla g = (2y + 2z)\,\mathbf{i} + (2x + 2z)\,\mathbf{j} + (2x + 2y)\,\mathbf{k}$$

$\nabla V = \lambda \nabla g$ and the side condition yield the system of equations:

$$yz = \lambda\,(2y + 2z)$$

$$xz = \lambda\,(2x + 2z)$$

$$xy = \lambda\,(2x + 2y)$$

$$xy + 2xz + 2yz = S.$$

Multiply the first equation by x, the second by y and subtract. This gives

$$0 = 2\lambda\,z(x - y) \quad \implies \quad x = y \quad \text{since} \;\; z = 0 \quad \implies \quad V = 0.$$

Multiply the second equation by y, the third by z and subtract. This gives

$$0 = 2\lambda\,x(y - z) \quad \implies \quad y = z \quad \text{since} \;\; x = 0 \quad \implies \quad V = 0.$$

Thus the closed rectangular box of maximum volume is a cube. The cube has side length $x = \sqrt{S/6}$.

38. Let x, y, z denote the length, width and height of the box. We want to maximize the volume V of the box given that the surface area S is constant. That is:

maximize $V(x, y, z) = xyz$ subject to $S(x, y, z) = 2xy + 2xz + 2yz = S$ constant

Let $g(x, y, z) = xy + 2xz + 2yz - S$. Then

$$\nabla V = yz\,\mathbf{i} + xz\,\mathbf{j} + xy\,\mathbf{k}, \qquad \nabla g = (y + 2z)\,\mathbf{i} + (x + 2z)\,\mathbf{j} + (2x + 2y)\,\mathbf{k}$$

$\nabla V = \lambda \nabla g$ and the side condition yield the system of equations:

$$yz = \lambda\,(y + 2z)$$

$$xz = \lambda\,(x + 2z)$$

$$xy = \lambda\,(2x + 2y)$$

$$xy + 2xz + 2yz = S.$$

Multiplying the first equation by x, the second by y and subtracting, we get

$$0 = 2\lambda\,z(x - y) \quad \implies \quad x = y \quad \text{since} \;\; z = 0 \quad \implies \quad V = 0.$$

Now put $y = x$ in the third equation. This gives

$$x^2 = 4\lambda x \quad \Longrightarrow \quad x(x - 4\lambda) = 0 \quad \Longrightarrow \quad x = 4\lambda \quad \text{since} \ x = 0 \quad \Longrightarrow \quad V = 0.$$

Thus, $x = y = 4\lambda$. Substituting $x = 4\lambda$ in the second equation gives $z = 2\lambda$.

Finally, substituting these values for x, y, z in the fourth equation, we get

$$48\lambda^2 = S \quad \Longrightarrow \quad \lambda^2 = \frac{S}{48} \quad \Longrightarrow \quad \lambda = \frac{1}{4}\sqrt{\frac{S}{3}}$$

To maximize the volume, take $x = y = \sqrt{\frac{S}{3}}$ and $z = \frac{1}{2}\sqrt{\frac{S}{3}}$.

39. $S(r, h) = 4\pi r^2 + 2\pi rh, \quad g(r, h) = \frac{4}{3}\pi r^3 + \pi r^2 h - 10{,}000$

$\nabla S = (8\pi r + 2\pi h)\mathbf{i} + 2\pi r\mathbf{j}, \quad \nabla g = (4\pi r^2 + 2\pi rh)\mathbf{i} + \pi r^2 \mathbf{j}$

(Here $\mathbf{i}, \mathbf{j}$ are the unit vectors in the directions of increasing r and h.)

$\nabla S = \lambda \nabla g \quad \Longrightarrow \quad 2\pi(4r + h) = 2\pi r\lambda(2r + h), \quad 2\pi r = \lambda\pi r^2 \quad \Longrightarrow \quad h = 0$

Maximum volume for sphere of radius $r = \sqrt[3]{7500/\pi}$ meters.

40. (a) $f(x, y, l) = xyl, \qquad\qquad g(x, y, l) = 2x + 2y + l - 108,$

$\quad \nabla f = yl\,\mathbf{i} + xl\,\mathbf{j} + xy\,\mathbf{k}, \qquad \nabla g = 2\,\mathbf{i} + 2\,\mathbf{j} + \mathbf{k}.$

$\qquad \nabla f = \lambda \nabla g \quad \Longrightarrow \quad yl = 2\lambda, \quad xl = 2\lambda, \quad xy = \lambda \quad \Longrightarrow \quad y = x \text{ and } l = 2x.$

Now $2x + 2y + l = 108, \quad \Longrightarrow \quad x = 18 \text{ and } l = 36.$

To maximize the volume, take $x = y = 18$ in. and $l = 36$ in.

(b) $f(r, l) = \pi r^2 l, \qquad\qquad g(r, l) = 2\pi r + l - 108,$

$\quad \nabla f = 2\pi rl\,\mathbf{i} + \pi r^2\,\mathbf{j}, \qquad\qquad \nabla g = 2\pi\,\mathbf{i} + \mathbf{j}.$

$\qquad \nabla f = \lambda \nabla g \quad \Longrightarrow \quad 2\pi rl = 2\pi\lambda, \quad \pi r^2 = \lambda, \quad l = \pi r.$

Now $2\pi r + l = 108, \quad \Longrightarrow \quad r = \frac{36}{\pi} \text{ and } l = 36.$

To maximize the volume, take $r = 36/\pi$ in. and $l = 36$ in.

41. $f(x, y, z) = 8xyz, \qquad g(x, y, z) = 4x^2 + 9y^2 + 36z^2 - 36.$

$\nabla f(x, y, z) = 8yz\mathbf{i} + 8xz\mathbf{j} + 8xy\mathbf{k}, \qquad \nabla g(x, y, z) = 8x\mathbf{i} + 18y\mathbf{j} + 72z\mathbf{k}.$

$\nabla f = \lambda \nabla g \quad$ gives

$$yz = \lambda x, \quad 4xz = 9\lambda y, \quad xy = 9\lambda z.$$

$$4\frac{xyz}{\lambda} = 4x^2, \quad 4\frac{xyz}{\lambda} = 9y^2, \quad 4\frac{xyz}{\lambda} = 36z^2.$$

Also notice

$$4x^2 + 9y^2 + 36z^2 - 36 = 0$$

We have

$$12\frac{xyz}{\lambda} = 36 \quad \Longrightarrow \quad x = \sqrt{3}, \quad y = \frac{2}{\sqrt{3}}, \quad z = \frac{1}{\sqrt{3}}.$$

Thus,

$$V = 8xyz = 8 \cdot \sqrt{3} \cdot \frac{2}{\sqrt{3}} \cdot \frac{1}{\sqrt{3}} = \frac{16}{\sqrt{3}}.$$

42. Let x, y, z denote the length, width and height of the crate, and let C be the cost. Then

$$C(x, y, z) = 30xy + 20xz + 20yz \qquad \text{subject to} \qquad g(x, y, z) = xyz - 96 = 0$$

$$\nabla C = (30y + 20z)\,\mathbf{i} + (30x + 20z)\,\mathbf{j} + (20x + 20y)\,\mathbf{k}, \qquad \nabla g = yz\,\mathbf{i} + xz\,\mathbf{j} + xy\,\mathbf{k}$$

$\nabla C = \lambda \nabla g$ implies

$$30y + 20z = \lambda yz$$
$$30x + 20z = \lambda xz$$
$$20x + 20y = \lambda xy$$
$$xyz = 96$$

Multiplying the first equation by x, the second by y and subtracting, we get

$$20z(x - y) = 0 \quad \implies \quad x = y \quad \text{since } z \neq 0$$

Now put $y = x$ in the third equation. This gives

$$40x = \lambda x^2 \quad \implies \quad x(\lambda x - 40) = 0 \quad \implies \quad x = \frac{40}{\lambda} \quad \text{since } x \neq 0$$

Thus, $x = y = 40/\lambda$. Substituting $x = 40/\lambda$ in the second equation gives $z = 60/\lambda$. Finally, substituting these values for x, y, z in the fourth equation, we get

$$\frac{40}{\lambda} \frac{40}{\lambda} \frac{60}{\lambda} = 96 \quad \implies \quad 96\lambda^3 = 96,000 \quad \implies \quad \lambda^3 = 1000 \quad \implies \quad \lambda = 10$$

To minimize the cost, take $x = y = 4$ meters and $z = 6$ meters.

43. To simplify notation we set $x = Q_1, \quad y = Q_2, \quad z = Q_3$.

$$f(x, y, z) = 2x + 8y + 24z, \qquad g(x, y, z) = x^2 + 2y^2 + 4z^2 - 4,500,000,000$$

$$\nabla f = 2\mathbf{i} + 8\mathbf{j} + 24\mathbf{k}, \qquad \nabla g = 2x\mathbf{i} + 4y\mathbf{j} + 8z\mathbf{k}.$$

$$\nabla f = \lambda \nabla g \quad \implies \quad 2 = 2\lambda x, \quad 8 = 4\lambda y, \quad 24 = 8\lambda z.$$

Since $\lambda \neq 0$ here, we solve the equations for x, y, z:

$$x = \frac{1}{\lambda}, \quad y = \frac{2}{\lambda}, \quad z = \frac{3}{\lambda},$$

and substitute these results in $g(x, y, z) = 0$ to obtain

$$\frac{1}{\lambda^2} + 2\left(\frac{4}{\lambda^2}\right) + 4\left(\frac{9}{\lambda^2}\right) - 45 \times 10^8 = 0, \quad \frac{45}{\lambda^2} = 45 \times 10^8, \quad \lambda = \pm 10^{-4}.$$

Since x, y, z are non-negative, $\lambda = 10^{-4}$ and

$$x = 10^4 = Q_1, \quad y = 2 \times 10^4 = Q_2, \quad z = 3 \times 10^4 = Q_3.$$

44. $f(x, y, z) = 8xyz,$ $g(x, y, z) = \dfrac{x^2}{a^2} + \dfrac{y^2}{b^2} + \dfrac{z^2}{c^2} - 1.$ We take $a, b, c, x, y, z > 0$

$\nabla f(x, y, z) = 8yz\mathbf{i} + 8xz\mathbf{j} + 8xy\mathbf{k};$ $\nabla g(x, y, z) = \dfrac{2x}{a^2}\mathbf{i} + \dfrac{2y}{b^2}\mathbf{j} + \dfrac{2z}{c^2}\mathbf{k}.$

$\nabla f = \lambda \nabla g$ and the side condition yield the system of equations:

$$8yz = \frac{2x\lambda}{a^2}$$

$$8xz = \frac{2y\lambda}{b^2}$$

$$8xy = \frac{2z\lambda}{c^2}$$

$$\frac{x^2}{a^2} + \frac{y^2}{b^2} + \frac{z^2}{c^2} = 1$$

Multiply the first equation by x, the second by y and subtract. This gives

$$0 = \frac{2\lambda x^2}{a^2} - \frac{2\lambda y^2}{b^2} \quad \Longrightarrow \quad y = \frac{b}{a}x \quad \text{since } \lambda = 0 \quad \Longrightarrow \quad V = 0.$$

Multiply the second equation by y, the third by z and subtract. This gives

$$0 = \frac{2\lambda y^2}{b^2} - \frac{2\lambda z^2}{c^2} \quad \Longrightarrow \quad z = \frac{c}{b}y = \frac{c}{a}x.$$

Substituting these results into the side condition, we get:

$$\frac{3x^2}{a^2} = 1 \quad \Longrightarrow \quad x = \frac{a}{\sqrt{3}} \quad \Longrightarrow \quad y = \frac{b}{\sqrt{3}} \quad \text{and} \quad z = \frac{c}{\sqrt{3}}.$$

The volume of the largest rectangular box is: $V = 8 \left(\dfrac{a}{\sqrt{3}}\right) \left(\dfrac{b}{\sqrt{3}}\right) \left(\dfrac{c}{\sqrt{3}}\right) = \dfrac{8\sqrt{3}}{9} abc.$

PROJECT 16.7

1. $f(x, y, z) = xy + z^2,$ $g(x, y, z) = x^2 + y^2 + z^2 - 4,$ $h(x, y, z) = y - x$

$\nabla f = y\mathbf{i} + x\mathbf{j} + 2z\mathbf{k},$ $\nabla g = 2x\mathbf{i} + 2y\mathbf{j} + 2z\mathbf{k},$ $\nabla h = -\mathbf{i} + \mathbf{j}.$

$\nabla f = \lambda\nabla g + \mu\nabla h$ $\Longrightarrow$ $y = 2\lambda x - \mu,$ $x = 2\lambda y - \mu,$ $2z = 2\lambda z$

$2z = 2\lambda z$ $\Longrightarrow$ $\lambda = 0$ or $z = 1.$

$\lambda = 0$ $\Longrightarrow$ $y = -x$ which contradicts $y = x.$

$z = 1$ $\Longrightarrow$ $x^2 + y^2 = 3,$ which, with $y = x$ implies $x = \pm\sqrt{3/2};$ $\left(\pm\sqrt{3/2}, \pm\sqrt{3/2}\right)$

Adding the first two equations gives

$$x + y = 2\lambda(x + y) \qquad \Longrightarrow \qquad (x+y)[2\lambda - 1] = 0 \qquad \Longrightarrow \qquad \lambda = \frac{1}{c} \quad \text{or} \quad x = y = 0.$$

$$x = y = 0 \qquad \Longrightarrow \qquad z = \pm 2; \quad (0, 0, \pm 2).$$

$$\lambda = \frac{1}{2} \qquad \Longrightarrow \qquad z = 0 \text{ and } y = x \qquad \Longrightarrow \qquad 2x^4 = 4; \quad x = \pm\sqrt{2}; \quad (\pm\sqrt{2}, \pm\sqrt{2}, 0).$$

$$f\left(\pm\sqrt{3/2}, \pm\sqrt{3/2}, 1\right) = \frac{5}{2}; \quad f(0, 0, \pm 2) = 4; \quad f(\pm\sqrt{2}, \pm\sqrt{2}, 0) = 2.$$

The maximum value of f is 4; the minimum value is 2.

2. $D(x, y, z) = x^2 + y^2 + z^2, \quad g(x, y, z) = x + 2y + 3z, \quad h(x, y, z) = 2x + 3y + z - 4$

$$\nabla D = 2x\mathbf{i} + 2y\mathbf{j} + 2z\mathbf{k}, \quad \nabla g = \mathbf{i} + 2\mathbf{j} + 3\mathbf{k}, \quad \nabla h = 2\mathbf{i} + 3\mathbf{j} + \mathbf{k}$$

$$\nabla D = \lambda \nabla g + \mu \nabla h \quad \Longrightarrow \quad 2x = \lambda + 2\mu, \quad 2y = 2\lambda + 3\mu, \quad 2z = 3\lambda + \mu \quad \Longrightarrow \quad z = 5y - 7x$$

Then $g(x, y, z) = 0$ and $h(x, y, z) = 0$ give $x = \dfrac{68}{75}, \quad y = \dfrac{16}{15}, \quad z = -\dfrac{76}{75}$

Closest point $\left(\dfrac{68}{75}, \dfrac{16}{15}, -\dfrac{76}{75}\right)$

3. $f(x, y, z) = x^2 + y^2 + z^2, \qquad g(x, y, z) = x + y - z + 1, \qquad h(x, y, z) = x^2 + y^2 - z^2$

$$\nabla f = 2x\,\mathbf{i} + 2y\,\mathbf{j} + 2z\,\mathbf{k}, \qquad \nabla g = \mathbf{i} + \mathbf{j} - \mathbf{k}, \qquad \nabla h = 2x\,\mathbf{i} + 2y\,\mathbf{j} - 2z\,\mathbf{k}.$$

$$\nabla f = \lambda \nabla g + \mu \nabla h \qquad \Longrightarrow \qquad 2x = \lambda + 2x\mu, \quad 2y = \lambda + 2y\mu, \quad 2z = -\lambda - 2z\mu$$

Multiplying the first equation by y, the second equation by x and subtracting, yields

$$\lambda(y - x) = 0.$$

Now $\lambda = 0 \quad \Longrightarrow \quad \mu = 1 \quad \Longrightarrow \quad x = y = z = 0.$ This is impossible since $x + y - z = -1$.

Therefore, we must have $y = x \quad \Longrightarrow \quad z = \pm\sqrt{2}\,x.$

Substituting $y = x, \ z = \sqrt{2}\,x$ into the equation $x + y - z + 1 = 0$, we get

$$x = -1 - \frac{\sqrt{2}}{2} \qquad \Longrightarrow \qquad y = -1 - \frac{\sqrt{2}}{2}, \ z = -1 - \sqrt{2}$$

Substituting $y = x, \ z = -\sqrt{2}\,x$ into the equation $x + y - z + 1 = 0$, we get

$$x = -1 + \frac{\sqrt{2}}{2} \qquad \Longrightarrow \qquad y = -1 + \frac{\sqrt{2}}{2}, \ z = -1 + \sqrt{2}$$

Since

$$f\left(-1-\frac{\sqrt{2}}{2}, -1-\frac{\sqrt{2}}{2}, -1-\sqrt{2}\right) = 6+4\sqrt{2} \ \text{ and}$$

$$f\left(-1+\frac{\sqrt{2}}{2}, -1+\frac{\sqrt{2}}{2}, -1+\sqrt{2}\right) = 6-4\sqrt{2},$$

it follows that $\left(-1+\dfrac{\sqrt{2}}{2}, -1+\dfrac{\sqrt{2}}{2}, -1+\sqrt{2}\right)$ is closest to the origin and

$\left(-1-\dfrac{\sqrt{2}}{2}, -1-\dfrac{\sqrt{2}}{2}, -1-\sqrt{2}\right)$ is furthest from the origin.

SECTION 16.8

1. $df = \left(3x^2y - 2xy^2\right)\Delta x + \left(x^3 - 2x^2y\right)\Delta y$

2. $df = \dfrac{\partial f}{\partial x}\Delta x + \dfrac{\partial f}{\partial y}\Delta y + \dfrac{\partial f}{\partial z}\Delta z = (y+z)\Delta x + (x+z)\Delta y + (x+y)\Delta z$

3. $df = (\cos y + y\sin x)\,\Delta x - (x\sin y + \cos x)\,\Delta y$

4. $df = 2xye^{2z}\Delta x + x^2e^{2z}\Delta y + 2x^2ye^{2z}\Delta z$

5. $df = \Delta x - (\tan z)\,\Delta y - \left(y\sec^2 z\right)\Delta z$

6. $df = \left[\dfrac{x-y}{x+y} + \ln(x+y)\right]\Delta x + \left[\dfrac{x-y}{x+y} - \ln(x+y)\right]\Delta y$

7. $df = \dfrac{y(y^2+z^2-x^2)}{(x^2+y^2+z^2)^2}\Delta x + \dfrac{x(x^2+z^2-y^2)}{(x^2+y^2+z^2)^2}\Delta y - \dfrac{2xyz}{(x^2+y^2+z^2)^2}\Delta z$

8. $df = \left[\dfrac{2x}{x^2+y^2} + e^{xy}(1+xy)\right]\Delta x + \left[\dfrac{2y}{x^2+y^2} + x^2e^{xy}\right]\Delta y$

9. $df = [\cos(x+y) + \cos(x-y)]\,\Delta x + [\cos(x+y) - \cos(x-y)]\,\Delta y$

10. $df = \ln\left(\dfrac{1+y}{1-y}\right)\Delta x + \dfrac{2x}{1-y^2}\Delta y$

11. $df = \left(y^2ze^{xz} + \ln z\right)\Delta x + 2ye^{xz}\,\Delta y + \left(xy^2e^{xz} + \dfrac{x}{z}\right)\Delta z$

12. $df = y(1 - 2x^2)e^{-(x^2+y^2)}\Delta x + x(1 - 2y^2)e^{-(x^2+y^2)}\Delta y$

13.
$$\Delta u = \left[(x + \Delta x)^2 - 3(x + \Delta x)(y + \Delta y) + 2(y + \Delta y)^2\right] - \left(x^2 - 3xy + 2y^2\right)$$
$$= \left[(1.7)^2 - 3(1.7)(-2.8) + 2(-2.8)^2\right] - \left(2^2 - 3(2)(-3) + 2(-3)^2\right)$$
$$= (2.89 + 14.28 + 15.68) - 40 = -7.15$$
$$du = (2x - 3y)\,\Delta x + (-3x + 4y)\,\Delta y$$
$$= (4 + 9)(-0.3) + (-6 - 12)(0.2) = -7.50$$

14. $du = \left(\sqrt{x - y} + \dfrac{x + y}{2\sqrt{x - y}}\right)\Delta x + \left(\sqrt{x - y} - \dfrac{x + y}{2\sqrt{x - y}}\right)\Delta y = 1$

15. $\Delta u = \left[(x + \Delta x)^2(z + \Delta z) - 2(y + \Delta y)(z + \Delta z)^2 + 3(x + \Delta x)(y + \Delta y)(z + \Delta z)\right]$
$$- \left(x^2 z - 2yz^2 + 3xyz\right)$$
$$= \left[(2.1)^2(2.8) - 2(1.3)(2.8)^2 + 3(2.1)(1.3)(2.8)\right] - \left[(2)^2 3 - 2(1)(3)^2 + 3(2)(1)(3)\right] = 2.896$$
$$du = (2xz + 3yz)\,\Delta x + \left(-2z^2 + 3xz\right)\,\Delta y + \left(x^2 - 4yz + 3xy\right)\,\Delta z$$
$$= [2(2)(3) + 3(1)(3)](0.1) + [-2(3)^2 + 3(2)(3)](0.3) + [2^2 - 4(1)(3) + 3(2)(1)](-0.2) = 2.5$$

16. $du = \dfrac{y^3 + yz^2}{(x^2 + y^2 + z^2)^{3/2}}\,\Delta x + \dfrac{x^3 + xz^2}{(x^2 + y^2 + z^2)^{3/2}}\,\Delta y - \dfrac{xyz}{(x^2 + y^2 + z^2)^{3/2}}\,\Delta z = \dfrac{77}{4(14)^{3/2}}$

17. $f(x, y) = x^{1/2}y^{1/4}; \quad x = 121, \quad y = 16, \quad \Delta x = 4, \quad \Delta y = 1$
$$f(x + \Delta x, y + \Delta y) \cong f(x, y) + df$$
$$= x^{1/2}\,y^{1/4} + \tfrac{1}{2}x^{-1/2}\,y^{1/4}\,\Delta x + \tfrac{1}{4}x^{1/2}\,y^{-3/4}\,\Delta y$$
$$\sqrt{125}\,\sqrt[4]{17} \cong \sqrt{121}\,\sqrt[4]{16} + \tfrac{1}{2}(121)^{-1/2}\,(16)^{1/4}\,(4) + \tfrac{1}{4}(121)^{1/2}\,(16)^{-3/4}\,(1)$$
$$= 11(2) + \tfrac{1}{2}\left(\tfrac{1}{11}\right)(2)(4) + \tfrac{1}{4}(11)\left(\tfrac{1}{8}\right)$$
$$= 22 + \tfrac{4}{11} + \tfrac{11}{32} = 22\tfrac{249}{352} \cong 22.71$$

18. $f(x, y) = (1 - \sqrt{x})(1 + \sqrt{y}), \quad x = 9, \quad y = 25, \quad \Delta x = 1, \quad \Delta y = -1$
$$df = -\dfrac{1 + \sqrt{y}}{2\sqrt{x}}\Delta x + \dfrac{1 - \sqrt{x}}{2\sqrt{y}}\Delta y = -\dfrac{4}{5}$$
$$f(10, 24) \cong f(9, 25) - \dfrac{4}{5} = -12\dfrac{4}{5}$$

19.
$$f(x, y) = \sin x \cos y; \quad x = \pi, \quad y = \dfrac{\pi}{4}, \quad \Delta x = -\dfrac{\pi}{7}, \quad \Delta y = -\dfrac{\pi}{20}$$
$$df = \cos x \cos y\,\Delta x - \sin x \sin y\,\Delta y$$
$$f(x + \Delta x, y + \Delta y) \cong f(x, y) + df$$
$$\sin \dfrac{6}{7}\pi \cos \dfrac{1}{5}\pi \cong \sin \pi \cos \dfrac{\pi}{4} + \left(\cos \pi \cos \dfrac{\pi}{4}\right)\left(-\dfrac{\pi}{7}\right) - \left(\sin \pi \sin \dfrac{\pi}{4}\right)\left(-\dfrac{\pi}{20}\right)$$
$$= 0 + \left(\dfrac{1}{2}\sqrt{2}\right)\left(\dfrac{\pi}{7}\right) + 0 = \dfrac{\pi\sqrt{2}}{14} \cong 0.32$$

20. $f(x,y) = \sqrt{x}\tan y,$ $x = 9,$ $y = \dfrac{\pi}{4},$ $\Delta x = -1,$ $\Delta y = \dfrac{1}{16}\pi$

$$df = \frac{1}{2\sqrt{x}}\tan y\,\Delta x + \sqrt{x}\sec^2 y\,\Delta y = -\frac{1}{6} + \frac{3\pi}{8}$$

$$f\left(8,\frac{5}{16}\pi\right) \cong f\left(9,\frac{\pi}{4}\right) - \frac{1}{6} + \frac{3\pi}{8} = \frac{17}{6} + \frac{3}{8}\pi \cong 4.01$$

21. $f(2.9, 0.01) \cong f(3,0) + df,$ where df is to be evaluated at $x = 3,$ $y = 0,$ $\Delta x = -0.1,$ $\Delta y = 0.01.$

$$df = \left(2xe^{xy} + x^2 y e^{xy}\right)\Delta x + x^3 e^{xy}\,\Delta y = \left[2(3)e^0 + (3)^2(0)e^0\right](-0.1) + 3^3 e^0(0.01) = -0.33$$

Thus, $f(2.9, .01) \cong 3^2 e^0 - 0.33 = 8.67.$

22. $x = 2,$ $y = 3,$ $z = 3,$ $\Delta x = 0.12,$ $\Delta y = -0.08,$ $\Delta z = 0.02$

$$df = 2xy\cos\pi z\,\Delta x + x^2\cos\pi z\,\Delta y - \pi x^2 y\sin\pi z\,\Delta z = -12(0.12) + 4(0.08) = -1.12$$

$$f(2.12, 2.92, 3.02) \cong f(2,3,3) - 1.12 = -13.12$$

23. $f(2.94, 1.1, 0.92) \cong f(3,1,1) + df,$ where df is to be evaluated at $x = 3,$ $y = 1,$ $z = 1,$

$\Delta x = -0.06,$ $\Delta y = 0.1,$ $\Delta z = -0.08$

$$df = \tan^{-1} yz\,\Delta x + \frac{xz}{1 + y^2 z^2}\,\Delta y + \frac{xy}{1 + y^2 z^2}\,\Delta z = \frac{\pi}{4}(-0.06) + (1.5)(0.1) + (1.5)(-0.08) \cong -0.0171$$

Thus, $f(2.94, 1.1, 0.92) \cong \frac{3}{4}\pi - 0.0171 \cong 2.3391$

24. $x = 3,$ $y = 4,$ $\Delta x = 0.06,$ $\Delta y = -0.12$

$$df = \frac{x}{\sqrt{x^2 + y^2}}\Delta x + \frac{y}{\sqrt{x^2 + y^2}}\Delta y = \frac{3}{5}(0.06) + \frac{4}{5}(-0.12) = -0.06$$

$$f(3.06, 3.88) \cong f(3,4) - 0.06 = 4.94$$

25. $df = \dfrac{\partial z}{\partial x}\Delta x + \dfrac{\partial z}{\partial y}\Delta y = \dfrac{2y}{(x+y)^2}\Delta x - \dfrac{2x}{(x+y)^2}\Delta y$

With $x = 4,$ $y = 2,$ $\Delta x = 0.1,$ $\Delta y = 0.1,$ we get

$$df = \tfrac{4}{36}(0.1) - \tfrac{8}{36}(0.1) = -\tfrac{1}{90}.$$

The exact change is $\dfrac{4.1 - 2.1}{4.1 + 2.1} - \dfrac{4 - 2}{4 + 2} = \dfrac{2}{6.2} - \dfrac{1}{3} = -\dfrac{1}{93}.$

26. $V(r,h) = \pi r^2 h,$ $r = 8,$ $h = 12,$ $\Delta r = -0.3,$ $\Delta h = 0.2$

$dV = 2\pi rh\,\Delta r + \pi r^2\Delta h = 192\pi(-0.3) + 64\pi(0.2) = -44.8\pi$

decreases by approximately 44.8π cubic inches.

27. $S = 2\pi r^2 + 2\pi rh;$ $r = 8,$ $h = 12,$ $\Delta r = -0.3,$ $\Delta h = 0.2$

$$dS = \frac{\partial S}{\partial r}\Delta r + \frac{\partial S}{\partial h}\Delta h = (4\pi r + 2\pi h)\,\Delta r + (2\pi r)\,\Delta h$$

$$= 56\pi(-0.3) + 16\pi(0.2) = -13.6\pi.$$

The area decreases about 13.6π in.^{2}.

28. $dT = 2x \cos \pi z \Delta x - 2y \sin \pi z \Delta y - (\pi x^2 \sin \pi z + \pi y^2 \cos \pi z)\Delta z$

$\quad = 4(0.1) - 4\pi(0.2) = \dfrac{2}{5} - \dfrac{4}{5}\pi$

T decreases by about $\dfrac{4}{5}\pi - \dfrac{2}{5} \cong 2.11$

29. $S(9.98, 5.88, 4.08) \cong S(10, 6, 4) + dS = 248 + dS,$ where

$dS = (2w + 2h)\,\Delta l + (2l + 2h)\,\Delta w + (2l + 2w)\,\Delta h = 20(-0.02) + 28(-0.12) + 32(0.08) = -1.20$

Thus, $S(9.98, 5.88, 4.08) \cong 248 - 1.20 = 246.80.$

30. $f(r, h) = \dfrac{\pi r^2 h}{3}, \quad r = 7, \ h = 10, \ \Delta r = 0.2, \ \Delta h = 0.15$

$df = \dfrac{2\pi rh}{3}\Delta r + \dfrac{\pi r^2}{3}\Delta h = \dfrac{140}{3}\pi(0.2) + \dfrac{49}{3}\pi(0.15) = \dfrac{\pi}{3}(35.35)$

$f(7.2, 10.15) \cong f(7, 10) + \dfrac{\pi}{3}(35.35) = 525.35\dfrac{\pi}{3} \cong 550.15$

31. (a) $dV = yz\,\Delta x + xz\,\Delta y + xy\,\Delta z = (8)(6)(0.02) + (12)(6)(-0.05) + (12)(8)(0.03) = 0.24$

(b) $\Delta V = (12.02)(7.95)(6.03) - (12)(8)(6) = 0.22077$

32. (a) $S(x, y, z) = 2(xy + xz + yz), \quad x = 12, \ y = 8, \ z = 6, \ \Delta x = 0.02, \ \Delta y = -0.05, \ \Delta z = 0.03$

$dS = 2(y + z)\Delta x + 2(x + z)\Delta y + 2(x + y)\Delta z = 28(0.02) + 36(-0.05) + 40(0.03) = -0.04$

(b) $\Delta S = S(12.02, 7.95, 6.03) - S(12, 8, 6) = -0.0438$

33. $T(P) - T(Q) \cong dT = (-2x + 2yz)\,\Delta x + (-2y + 2xz)\,\Delta y + (-2z + 2xy)\,\Delta z$

Letting $x = 1, \ y = 3, \ z = 4, \ \Delta x = 0.15, \ \Delta y = -0.10, \ \Delta z = 0.10,$ we have

$$dT = (22)(0.15) + (2)(-0.10) + (-2)(0.10) = 2.9$$

34. Amount of paint is increase in volume. $f(x, y, z) = xyz, \ x = 48$ in, $y = 24$ in, $z = 36$ in,

$\Delta x = \Delta y = \Delta z = \frac{2}{16}$ in. $\qquad \Delta f \cong df = yz\Delta x + xz\Delta y + xy\Delta z = 3774(\frac{2}{16}) = 468$

The amount of paint is approximately 468 cubic inches.

35. (a) $\pi r^2 h = \pi(r + \Delta r)^2(h + \Delta h) \quad \Longrightarrow \quad \Delta h = \dfrac{r^2 h}{(r + \Delta r)^2} - h = -\dfrac{(2r + \Delta r)h}{(r + \Delta r)^2}\Delta r.$

$\qquad df = (2\pi rh)\,\Delta r + \pi r^2\,\Delta h, \qquad df = 0 \quad \Longrightarrow \quad \Delta h = \dfrac{-2h}{r}\Delta r.$

(b) $2\pi r^2 + 2\pi rh = 2\pi(r + \Delta r)^2 + 2\pi(r + \Delta r)(h + \Delta h).$

Solving for $\Delta h,$

$$\Delta h = \dfrac{r^2 + rh - (r + \Delta r)^2}{r + \Delta r} - h = -\dfrac{2r + h + \Delta r}{r + \Delta r}\Delta r.$$

$\qquad df = (4\pi r + 2\pi h)\,\Delta r + 2\pi r\,\Delta h, \qquad df = 0 \quad \Longrightarrow \quad \Delta h = -\left(\dfrac{2r + h}{r}\right)\Delta r.$

36. The area is given by $A = \frac{1}{2} x^2 \tan \theta$.

(a) The change in area is approximated by:

$$dA = x \tan \theta \, \Delta x + \frac{1}{2} x^2 \sec^2 \theta \, \Delta \theta = 3 \Delta x + \frac{25}{2} \Delta \theta.$$

(b) The actual change in area is

$$\frac{1}{2}(x + \Delta x)^2 \tan(\theta + \Delta \theta) - \frac{1}{2} x^2 \tan \theta = \frac{1}{2}[4 + \Delta x]^2 \tan[\arctan(3/4) + \Delta \theta] - 6.$$

(c) The area is more sensitive to a change in θ.

37. (a) $A = \frac{1}{2} x^2 \sin \theta;$ $\Delta A \cong dA = x \sin \theta \Delta x + \frac{x^2}{2} \cos \theta \Delta \theta$

(b) The area is more sensitive to changes in θ if $x > 2 \, \tan \theta$, otherwise it is more sensitive to changes in x.

38. (a) $dV \cong yz\Delta x + xz\Delta y + xy\Delta z,$ $x = 60$ in, $y = 36$ in, $z = 42$ in

Maximum possible error $= 6192(\frac{1}{12}) = 516$ cubic inches.

(b) $dS \cong 2(y + z)\Delta x + 2(x + z)\Delta y + 2(x + y)\Delta z$

Maximum possible error $= 552(\frac{1}{12}) = 46$ square inches

39. $s = \dfrac{A}{A - W};$ $A = 9,$ $W = 5,$ $\Delta A = \pm 0.01,$ $\Delta W = \pm 0.02$

$$ds = \frac{\partial s}{\partial A} \Delta A + \frac{\partial s}{\partial W} \Delta W = \frac{-W}{(A - W)^2} \Delta A + \frac{A}{(A - W)^2} \Delta W$$

$$= -\frac{5}{16}(\pm 0.01) + \frac{9}{16}(\pm 0.02) \cong \pm 0.014$$

The maximum possible error in the value of s is 0.014 lbs; $2.23 \leq s + \Delta s \leq 2.27$

40. Assuming $A > W$, s is more sensitive to change in A.

SECTION 16.9

1. $\dfrac{\partial f}{\partial x} = xy^2,$ $f(x, y) = \frac{1}{2}x^2y^2 + \phi(y),$ $\dfrac{\partial f}{\partial y} = x^2 y + \phi'(y) = x^2 y.$

Thus, $\phi'(y) = 0,$ $\phi(y) = C,$ and $f(x, y) = \frac{1}{2}x^2y^2 + C.$

2. $\dfrac{\partial f}{\partial x} = x,$ $\dfrac{\partial f}{\partial y} = y \implies f(x, y) = \frac{1}{2}(x^2 + y^2) + C$

3. $\dfrac{\partial f}{\partial x} = y,$ $f(x, y) = xy + \phi(y),$ $\dfrac{\partial f}{\partial y} = x + \phi'(y) = x.$

Thus, $\phi'(y) = 0,$ $\phi(y) = C,$ and $f(x, y) = xy + C.$

4. $\dfrac{\partial f}{\partial x} = x^2 + y \implies f(x, y) = \dfrac{x^3}{3} + xy + \phi(y);$

$\dfrac{\partial f}{\partial y} = x + \phi'(y) = y^3 + x \implies f(x, y) = \frac{1}{3}x^3 + \frac{1}{4}y^4 + xy + C$

5. No; $\dfrac{\partial}{\partial y}(y^3 + x) = 3y^2$ whereas $\dfrac{\partial}{\partial x}(x^2 + y) = 2x.$

6. $\dfrac{\partial f}{\partial x} = y^2 e^x - y \implies f(x,y) = y^2 e^x - xy + \phi(y);$

$\dfrac{\partial f}{\partial y} = 2ye^x - x + \phi'(y) = 2ye^x - x \implies f(x,y) = y^2 e^x - xy + C$

7. $\dfrac{\partial f}{\partial x} = \cos x - y\sin x, \quad f(x,y) = \sin x + y\cos x + \phi(y), \quad \dfrac{\partial f}{\partial y} = \cos x + \phi'(y) = \cos x.$

Thus, $\phi'(y) = 0, \ \phi(y) = C,$ and $f(x,y) = \sin x + y\cos x + C.$

8. $\dfrac{\partial f}{\partial x} = 1 + e^y \implies f(x,y) = x + xe^y + \phi(y);$

$\dfrac{\partial f}{\partial y} = xe^y + \phi'(y) = xe^y + y^2 \implies f(x,y) = x + xe^y + \dfrac{y^3}{3} + C$

9. $\dfrac{\partial f}{\partial x} = e^x \cos y^2, \quad f(x,y) = e^x \cos y^2 + \phi(y), \quad \dfrac{\partial f}{\partial y} = -2ye^x \sin y^2 + \phi'(y) = -2ye^x \sin y^2.$

Thus, $\phi'(y) = 0, \ \phi(y) = C,$ and $f(x,y) = e^x \cos y^2 + C.$

10. $\dfrac{\partial^2 f}{\partial y \partial x} = -e^x \sin y, \quad \dfrac{\partial^2 f}{\partial x \partial y} = e^x \sin y \neq \dfrac{\partial^2 f}{\partial y \partial x};$ not a gradient.

11. $\dfrac{\partial f}{\partial y} = xe^x - e^{-y}, \quad f(x,y) = xye^x + e^{-y} + \phi(x), \quad \dfrac{\partial f}{\partial x} = ye^x + xye^x + \phi'(x) = ye^x(1 + x).$

Thus, $\phi'(x) = 0, \ \phi(x) = C,$ and $f(x,y) = xye^x + e^{-y} + C.$

12. $\dfrac{\partial f}{\partial x} = e^x + 2xy \implies f(x,y) = e^x + x^2 y + \phi(y); \quad \dfrac{\partial f}{\partial y} = x^2 + \phi'(y) = x^2 + \sin y$

$\implies f(x,y) = e^x + x^2 y - \cos y + C$

13. No; $\dfrac{\partial}{\partial y}\left(xe^{xy} + x^2\right) = x^2 e^{xy}$ whereas $\dfrac{\partial}{\partial x}\left(ye^{xy} - 2y\right) = y^2 e^{xy}$

14. $\dfrac{\partial f}{\partial y} = x\sin x + 2y + 1 \implies f(x,y) = xy\sin x + y^2 + y + \phi(x)$

$\dfrac{\partial f}{\partial x} = y\sin x + xy\cos x + \phi'(x) = y\sin x + xy\cos x \implies f(x,y) = xy\sin x + y^2 + y + C$

15. $\dfrac{\partial f}{\partial x} = 1 + y^2 + xy^2, \ f(x,y) = x + xy^2 + \frac{1}{2}x^2y^2 + \phi(y), \ \dfrac{\partial f}{\partial y} = 2xy + x^2y + \phi'(y) = x^2y + y + 2xy + 1.$

Thus, $\phi'(y) = y + 1, \quad \phi(y) = \frac{1}{2}y^2 + y + C$ and $f(x,y) = x + xy^2 + \frac{1}{2}x^2y^2 + \frac{1}{2}y^2 + y + C.$

16. $\dfrac{\partial f}{\partial x} = 2\ln 3y + \dfrac{1}{x} \implies f(x,y) = 2x\ln 3y + \ln|x| + \phi(y); \quad \dfrac{\partial f}{\partial y} = \dfrac{2x}{y} + \phi'(y) = \dfrac{2x}{y} + y^2$

$f(x,y) = 2x\ln 3y + \ln|x| + \dfrac{y^3}{3} + C$

17. $\dfrac{\partial f}{\partial x} = \dfrac{x}{\sqrt{x^2 + y^2}}, \quad f(x,y) = \sqrt{x^2 + y^2} + \phi(y), \quad \dfrac{\partial f}{\partial y} = \dfrac{y}{\sqrt{x^2 + y^2}} + \phi'(y) = \dfrac{y}{\sqrt{x^2 + y^2}}.$

Thus, $\phi'(y) = 0, \ \phi(y) = C,$ and $f(x,y) = \sqrt{x^2 + y^2} + C.$

18. $\dfrac{\partial f}{\partial x} = x\tan y + \sec^2 x \implies f(x,y) = \dfrac{x^2}{2}\tan y + \tan x + \phi(y);$

$\dfrac{\partial f}{\partial y} = \dfrac{x^2}{2}\sec^2 y + \phi'(y) = \dfrac{x^2}{2}\sec^2 y + \pi y; \implies f(x,y) = \dfrac{x^2}{2}\tan y + \tan x + \dfrac{\pi}{2}y^2 + C.$

19. $\dfrac{\partial f}{\partial x} = x^2\sin^{-1} y, \quad f(x,y) = \tfrac{1}{3}x^3\sin^{-1} y + \phi(y), \quad \dfrac{\partial f}{\partial y} = \dfrac{x^3}{3\sqrt{1-y^2}} + \phi'(y) = \dfrac{x^3}{3\sqrt{1-y^2}} - \ln y.$

Thus, $\phi'(y) = -\ln y, \implies \phi(y) = y - y\ln y + C,$ and

$$f(x,y) = \dfrac{1}{3}x^3\sin^{-1} y + y - y\ln y + C.$$

20. $\dfrac{\partial f}{\partial x} = \dfrac{\tan^{-1} y}{\sqrt{1-x^2}} + \dfrac{x}{y} \implies f = \sin^{-1} x\tan^{-1} y + \dfrac{x^2}{2y} + \phi(y);$

$\dfrac{\partial f}{\partial y} = \dfrac{\sin^{-1} x}{1+y^2} - \dfrac{x^2}{2y^2} + \phi'(y) = \dfrac{\sin^{-1} x}{1+y^2} - \dfrac{x^2}{2y^2} + 1 \implies f(x,y) = \sin^{-1} x\tan^{-1} y + \dfrac{x^2}{2y} + y + C.$

21. (a) Yes   (b) Yes   (c) No

22. (a) $f(x,y) = (x-y)e^{-x^2 y} + C$

(b) $f(x,y) = \sin(x+y) - \cos(x-y) + C; \quad f(\pi/3, \pi/4) = 6 \implies C = 6$

$f(x,y) = \sin(x+y) - \cos(x-y) + 6.$

23. $\dfrac{\partial f}{\partial x} = f(x,y), \quad \dfrac{\partial f/\partial x}{f(x,y)} = 1, \quad \ln f(x,y) = x + \phi(y), \quad \dfrac{\partial f/\partial y}{f(x,y)} = 0 + \phi'(y), \quad \dfrac{\partial f}{\partial y} = f(x,y).$

Thus, $\phi'(y) = 1, \ \phi(y) = y + K,$ and $f(x,y) = e^{x+y+K} = Ce^{x+y}.$

24. $\dfrac{\partial f}{\partial x} = e^{g(x,y)}g_x(x,y) \implies f(x,y) = e^{g(x,y)} + \phi(y);$

$\dfrac{\partial f}{\partial y} = e^{g(x,y)}g_y(x,y) + \phi'(y) = e^{g(x,y)}g_y(x,y) \implies f(x,y) = e^{g(x,y)} + C.$

25. (a) $P = 2x, \ Q = z, \ R = y; \quad \dfrac{\partial P}{\partial y} = 0 = \dfrac{\partial Q}{\partial x}, \quad \dfrac{\partial P}{\partial z} = 0 = \dfrac{\partial R}{\partial x}, \quad \dfrac{\partial Q}{\partial z} = 1 = \dfrac{\partial R}{\partial y}$

(b), (c), and (d)

$$\dfrac{\partial f}{\partial x} = 2x, \quad f(x,y,z) = x^2 + g(y,z).$$

$$\dfrac{\partial f}{\partial y} = 0 + \dfrac{\partial g}{\partial y} \quad \text{with} \quad \dfrac{\partial f}{\partial y} = z \implies \dfrac{\partial g}{\partial y} = z.$$

Then,

$$g(y,z) = yz + h(z) \implies f(x,y,z) = x^2 + yz + h(z),$$

$$\dfrac{\partial f}{\partial z} = 0 + y + h'(z) \quad \text{and} \quad \dfrac{\partial f}{\partial z} = y \implies h'(z) = 0.$$

Thus, $h(z) = C$ and $f(x,y,z) = x^2 + yz + C.$

26. $\dfrac{\partial f}{\partial x} = yz \implies f(x,y,z) = xyz + g(y,z); \quad \dfrac{\partial f}{\partial y} = xz + \dfrac{\partial g}{\partial y} = xz \implies f = xyz + h(z)$

$\dfrac{\partial f}{\partial z} = xy + h'(z) = xy \implies f(x,y,z) = xyz + C$

27. The function is a gradient by the test stated before Exercise 25.

Take $P = 2x + y$, $Q = 2y + x + z$, $R = y - 2z$.   Then

$$\frac{\partial P}{\partial y} = 1 = \frac{\partial Q}{\partial x}, \quad \frac{\partial P}{\partial z} = 0 = \frac{\partial R}{\partial x}, \quad \frac{\partial Q}{\partial z} = 1 = \frac{\partial R}{\partial y}.$$

Next, we find f where $\nabla f = P\mathbf{i} + Q\mathbf{j} + R\mathbf{k}$.

$$\frac{\partial f}{\partial x} = 2x + y \quad \Longrightarrow \quad f(x, y, z) = x^2 + xy + g(y, z).$$

$$\frac{\partial f}{\partial y} = x + \frac{\partial g}{\partial y} \quad \text{with} \quad \frac{\partial f}{\partial y} = 2y + x + z \quad \Longrightarrow \quad \frac{\partial g}{\partial y} = 2y + z.$$

Then,

$$g(y, z) = y^2 + yz + h(z),$$

$$f(x, y, z) = x^2 + xy + y^2 + yz + h(z).$$

$$\frac{\partial f}{\partial z} = y + h'(z) = y - 2z \quad \Longrightarrow \quad h'(z) = -2z.$$

Thus, $h(z) = -z^2 + C$ and $f(x, y, z) = x^2 + xy + y^2 + yz - z^2 + C$.

28. $\dfrac{\partial f}{\partial x} = 2x \sin 2y \cos z \implies f(x, y, z) = x^2 \sin 2y \cos z + g(y, z);$

$\dfrac{\partial f}{\partial y} = 2x^2 \cos 2y \cos z + \dfrac{\partial g}{\partial y} = 2x^2 \cos 2y \cos z \implies f(x, y, z) = x^2 \sin 2y \cos z + h(z)$

$\dfrac{\partial f}{\partial z} = -x^2 \sin 2y \sin z + h'(z) = -x^2 \sin 2y \sin z \implies f(x, y, z) = x^2 \sin 2y \cos z + C$

29. The function is a gradient by the test stated before Exercise 25.

Take $P = y^2 z^3 + 1$, $Q = 2xyz^3 + y$, $R = 3xy^2 z^2 + 1$.   Then

$$\frac{\partial P}{\partial y} = 2yz^3 = \frac{\partial Q}{\partial x}, \quad \frac{\partial P}{\partial z} = 3y^2 z^2 = \frac{\partial R}{\partial x}, \quad \frac{\partial Q}{\partial z} = 6xyz^2 = \frac{\partial R}{\partial y}.$$

Next, we find f where $\nabla f = P\mathbf{i} + Q\mathbf{j} + R\mathbf{k}$.

$$\frac{\partial f}{\partial x} = y^2 z^3 + 1,$$

$$f(x, y, z) = xy^2 z^3 + x + g(y, z).$$

$$\frac{\partial f}{\partial y} = 2xyz^3 + \frac{\partial g}{\partial y} \quad \text{with} \quad \frac{\partial f}{\partial y} = 2xyz^3 + y \quad \Longrightarrow \quad \frac{\partial g}{\partial y} = y.$$

Then,

$$g(y, z) = \tfrac{1}{2} y^2 + h(z),$$

$$f(x, y, z) = xy^2 z^3 + x + \tfrac{1}{2} y^2 + h(z).$$

$$\frac{\partial f}{\partial z} = 3xy^2 z^2 + h'(z) = 3xy^2 z^2 + 1 \quad \Longrightarrow \quad h'(z) = 1.$$

Thus, $h(z) = z + C$ and $f(x, y, z) = xy^2 z^3 + x + \tfrac{1}{2} y^2 + z + C$.

30. $\dfrac{\partial f}{\partial x} = \dfrac{y}{z} - e^z \implies f(x,y,z) = \dfrac{xy}{z} - xe^z + g(y,z)$

$\dfrac{\partial f}{\partial y} = \dfrac{x}{z} + \dfrac{\partial g}{\partial y} = \dfrac{x}{z} + 1 \implies f(x,y,z) = \dfrac{xy}{z} + y - xe^z + h(z)$

$\dfrac{\partial f}{\partial z} = -\dfrac{xy}{z^2} - xe^z + h'(z) = -xe^z - \dfrac{xy}{z^2} \implies f(x,y,z) = \dfrac{xy}{z} - xe^z + y + C$

31. $\mathbf{F}(\mathbf{r}) = \nabla \left(\dfrac{GmM}{r} \right)$

32.
$$h(\mathbf{r}) = \begin{cases} \nabla\left(\dfrac{k}{n+2} r^{n+2} \right), & n \neq 2 \\[4mm] \nabla\left(k \ln r \right), & n = -2. \end{cases}$$

REVIEW EXERCISES

1. $\nabla f(x,y) = (4x - 4y)\mathbf{i} + (3y^2 - 4x)\mathbf{j}$

2. $\nabla f(x,y) = \dfrac{y^3 - x^2 y}{(x^2 + y^2)^2}\mathbf{i} + \dfrac{x^3 - xy^2}{(x^2 + y^2)^2}\mathbf{j}$

3. $\nabla f(x,y) = (ye^{xy}\tan 2x + 2e^{xy}\sec^2 2x)\mathbf{i} + xe^{xy}\tan 2x\,\mathbf{j}$

4. $\nabla f = \dfrac{1}{x^2 + y^2 + z^2}(x\mathbf{i} + y\mathbf{j} + z\mathbf{k})$

5. $\nabla f(x,y) = 2xe^{-yz}\sec z\,\mathbf{i}, -zx^2 e^{-yz}\sec z\,\mathbf{j} - (x^2 y e^{-yz}\sec z - x^2 e^{-yz}\sec z \tan z)\mathbf{k}$

6. $\nabla f(x,y) = ye^{-3z}\cos xy\,\mathbf{i} + e^{-3z}(x\cos xy + \sin y)\mathbf{j}, -3e^{-3z}(\sin xy - \cos y)\,\mathbf{k}$

7. $\nabla f(x,y) = (2x - 2y)\mathbf{i} - 2x\mathbf{j}, \quad \nabla f(1,-2) = 6\mathbf{i} - 2\mathbf{j}; \quad \mathbf{u_a} = \dfrac{1}{\sqrt{5}}\mathbf{i} + \dfrac{2}{\sqrt{5}}\mathbf{j};$

$f'_{\mathbf{u_a}}(1,-2) = \nabla f(1,-2) \cdot \mathbf{u_a} = \dfrac{2}{\sqrt{5}}.$

8. $\nabla f(x,y) = (e^{xy} + xye^{xy})\mathbf{i} + x^2 e^{xy}\mathbf{j}, \quad \nabla f(2,0) = \mathbf{i} + 4\mathbf{j}; \quad \mathbf{u_a} = \dfrac{1}{2}\mathbf{i} + \dfrac{\sqrt{3}}{2}\mathbf{j}$

$f'_{\mathbf{u_a}}(2,0) = \nabla f(2,0) \cdot \mathbf{u_a} = \dfrac{1}{2} + 2\sqrt{3}.$

9. $\nabla f(x,y,z) = (y^2 + 6xz)\mathbf{i} + (2xy + 2z)\mathbf{j} + (2y + 3x^2)\mathbf{k}, \quad \nabla f(1,-2,3) = 22\mathbf{i} + 2\mathbf{j} - \mathbf{k};$

$\mathbf{u_a} = \dfrac{1}{3}\mathbf{i} - \dfrac{2}{3}\mathbf{j} + \dfrac{2}{3}\mathbf{k}; f'_{\mathbf{u_a}}(1,-2,3) = \nabla f(1,-2,3) \cdot \mathbf{u_a} = \dfrac{16}{3}.$

10. $\nabla f(x,y,z) = \dfrac{2x}{x^2 + y^2 + z^2}\mathbf{i} + \dfrac{2y}{x^2 + y^2 + z^2}\mathbf{j} + \dfrac{2z}{x^2 + y^2 + z^2}\mathbf{k}, \quad \nabla f(1,2,3) = \dfrac{1}{7}(\mathbf{i} + 2\mathbf{j} + 3\mathbf{k});$

$\mathbf{u_a} = \dfrac{1}{\sqrt{3}}\mathbf{i} - \dfrac{1}{\sqrt{3}}\mathbf{j} + \dfrac{1}{\sqrt{3}}\mathbf{k}; \quad f'_{\mathbf{u_a}}(1,2,3) = \nabla f(1,2,3) \cdot \mathbf{u_a} = \dfrac{2}{7\sqrt{3}}.$

11. $\nabla f(x,y) = (6x - 2y^2)\,\mathbf{i} - 4xy\,\mathbf{j}, \quad \nabla f(3,-2) = 10\,\mathbf{i} + 24\,\mathbf{j};$

$\mathbf{a} = (0,0) - (3,-2) = (-3,2) = -3\,\mathbf{i} + 2\,\mathbf{j}, \quad \mathbf{u_a} = \dfrac{-3}{\sqrt{13}}\,\mathbf{i} + \dfrac{2}{\sqrt{13}}\,\mathbf{j};$

$f'_{\mathbf{u_a}}(3,-2) = \nabla f(3,-2) \cdot \mathbf{u_a} = \dfrac{18}{\sqrt{13}}.$

12. $\nabla f(x,y,z) = (y^2 z - 3yz)\,\mathbf{i} + (2xyz - 3xz)\,\mathbf{j} + (xy^2 - 3xy)\,\mathbf{k}, \quad \nabla f(1,-1,2) = 8\,\mathbf{i} - 10\,\mathbf{j} + 4\,\mathbf{k};$

$\mathbf{r}'(t) = \mathbf{i} - \pi \sin \pi t\,\mathbf{j} + 2e^{t-1}\,\mathbf{k}, \quad \mathbf{a} = \mathbf{r}'(1) = \mathbf{i} + 2\mathbf{k}, \quad \mathbf{u_a} = \dfrac{1}{\sqrt{5}}\,\mathbf{i} + \dfrac{2}{\sqrt{5}}\,\mathbf{k};$

$f'_{\mathbf{u_a}}(1,-1,2) = \nabla f(1,-1,2) \cdot \mathbf{u_a} = \dfrac{16}{\sqrt{5}}.$

13. $\nabla f(x,y,z) = \dfrac{1}{\sqrt{x^2+y^2+z^2}}(x\,\mathbf{i} + y\,\mathbf{j} + z\,\mathbf{k}), \quad \nabla f(3,-1,4) = \dfrac{1}{\sqrt{26}}(3\,\mathbf{i} - \mathbf{j} + 4\,\mathbf{k});$

$\mathbf{a} = \pm(4\,\mathbf{i} - 3\,\mathbf{j} + \mathbf{k}), \quad \mathbf{u_a} = \pm\dfrac{1}{\sqrt{26}}(4\,\mathbf{i} - 3\,\mathbf{j} + \mathbf{k}); \quad f'_{\mathbf{u_a}}(3,-1,4) = \nabla f(3,-1,4) \cdot \mathbf{u_a} = \pm\dfrac{19}{26}.$

14. $\nabla f(x,y) = 2e^{2x}(\cos y - \sin y)\,\mathbf{i} - e^{2x}(\sin y + \cos y)\,\mathbf{j}, \quad \nabla f\left(\tfrac{1}{2}, -\tfrac{1}{2}\pi\right) = 2e\,\mathbf{i} + e\,\mathbf{j};$

maximum directional derivative: $\left\| \nabla f\left(\tfrac{1}{2}, -\tfrac{1}{2}\pi\right) \right\| = e\sqrt{5}.$

15. $\nabla f(x,y,z) = \cos xyz\,(yz\,\mathbf{i} + xz\,\mathbf{j} + xy\,\mathbf{k}), \quad \nabla f(\tfrac{1}{2}, \tfrac{1}{3}, \pi) = \tfrac{\pi\sqrt{3}}{6}\,\mathbf{i} + \tfrac{\pi\sqrt{3}}{4}\,\mathbf{j} + \tfrac{\sqrt{3}}{12}\,\mathbf{k};$

minimum directional derivative: $f'_{\mathbf{u}} = -\|\nabla f(\tfrac{1}{2}, \tfrac{1}{3}, \pi)\| = -\dfrac{\sqrt{39\pi^2+3}}{12}$

16. Let $\mathbf{r}(t) = x(t)\,\mathbf{i} + y(t)\,\mathbf{j}$ be the path of the particle. $\nabla I(x,y) = -2x\,\mathbf{i} - 6y\,\mathbf{j}.$ Then

$$x'(t) = -2x(t), \quad y'(t) = -6y(t) \quad \Longrightarrow \quad x(t) = C_1 e^{-2t}, \quad y(t) = C_2 e^{-6t}.$$

$$\mathbf{r}(0) = (4,3) \quad \Longrightarrow \quad C_1 = 4, \quad C_2 = 3.$$

Therefore the path of the particle is: $\mathbf{r}(t) = 4e^{-2t}\,\mathbf{i} + 3e^{-6t}\,\mathbf{j}, \; t \ge 0, \quad$ or, $\quad y = \tfrac{3}{64}x^3, \; 0 < x \le 4$

17. Let $\mathbf{r}(t) = x(t)\,\mathbf{i} + y(t)\,\mathbf{j}$ be the path of the particle. $\nabla T = -e^{-x}\cos y\,\mathbf{i} - e^{-x}\sin y\,\mathbf{j}.$ Then

$$x'(t) = -e^{-x(t)}\cos y(t), \quad y'(t) = -e^{-x(t)}\sin y(t) \quad \Longrightarrow \quad \dfrac{y'(t)}{x'(t)} = \tan y(t) \quad \Longrightarrow \quad \dfrac{dy}{dx} = \tan y$$

The solution is $\sin y = Ce^x.$ Since $\mathbf{r}(0) = 0, \; C = 0$ and $y = 0.$ The particle moves to the right the x-axis.

18. $\nabla z = 8x\,\mathbf{i} + 2y\,\mathbf{j}; \quad \mathbf{r}(t) = x(t)\,\mathbf{i} + y(t)\,\mathbf{j}.$

$$x'(t) = -8x(t), \quad y'(t) = -2y(t) \quad \Longrightarrow \quad x(t) = C_1 e^{-8t}, \quad y(t) = C_2 e^{-2t}.$$

(a) $\mathbf{r}(0) = (1,1) \quad \Longrightarrow \quad C_1 = 1, \; C_2 = 1; \quad x = e^{-8t}, \; y = e^{-2t} \quad$ or $\quad x = y^4.$

(b) $\mathbf{r}(0) = (1,-2) \quad \Longrightarrow \quad C_1 = 1, \; C_2 = -2; \quad x = e^{-8t}, \; y = -2e^{-2t} \quad$ or $\quad x = y^4/16.$

19. $\nabla f(x, y) = e^x \arctan y\, \mathbf{i} + e^x \dfrac{1}{1 + y^2}\, \mathbf{j}; \quad \nabla f(0, 1) = \dfrac{\pi}{4}\, \mathbf{i} + \dfrac{1}{2}\, \mathbf{j}.$

$\mathbf{u} = \dfrac{\nabla f(0, 1)}{\|\nabla f(0, 1)\|} = \dfrac{1}{\sqrt{4 + \pi^2}}(\pi\, \mathbf{i} + 2\, \mathbf{j}); \quad \text{rate: } \|\nabla f(0, 1)\| = \dfrac{\sqrt{\pi^2 + 4}}{4}$

20. $\nabla f(x, y, z) = \dfrac{1}{(y + z)^2}[(y + z)\, \mathbf{i} + (z - x)\, \mathbf{j} - (x + y)\, \mathbf{k}]; \quad \nabla f(-1, 1, 3) = \dfrac{1}{4}\, \mathbf{i} + \dfrac{1}{4}\, \mathbf{j}.$

$\mathbf{u} = \dfrac{\nabla f(-1, 1, 3)}{\|\nabla f(-1, 1, 3)\|} = \tfrac{1}{2}\sqrt{2}\, \mathbf{i} + \tfrac{1}{2}\sqrt{2}\, \mathbf{j}; \quad \text{rate: } \|\nabla f(-1, 1, 3)\| = \tfrac{1}{4}\sqrt{2}$

21. rate: $\dfrac{df}{dt} = \nabla f \cdot \mathbf{r}' = \left(4x\, \mathbf{i} - 9y^2\, \mathbf{j}\right) \cdot \left(\dfrac{1}{2}t^{-1/2}\, \mathbf{i} + 2e^{2t}\, \mathbf{j}\right) = 2 - 18e^{6t}$

22. $f(\mathbf{r}(t)) = \sin t^2 + \cos t^2, \quad \text{rate: } \quad f'(\mathbf{r}(t)) = 2t\cos t^2 - 2t\sin t^2$

23. rate: $\dfrac{df}{dt} = \nabla f \cdot \mathbf{r}' = \left[\left(\dfrac{1}{y} + \dfrac{z}{x^2}\right)\mathbf{i} - \dfrac{x}{y^2}\mathbf{j} - \dfrac{1}{x}\mathbf{k}\right] \cdot (\cos t\, \mathbf{i} - \sin t\, \mathbf{j} + \sec^2 t\, \mathbf{k}) = \dfrac{1 - \sin t}{\cos^2 t}$

24. $\dfrac{du}{dt} = \nabla u \cdot \mathbf{r}' = \dfrac{1}{1 + x^2 y^2}(y\, \mathbf{i} + x\, \mathbf{j}) \cdot (\sec^2 t\, \mathbf{i} + 2e^{2t}\, \mathbf{j}) = \dfrac{e^{2t}}{1 + e^{4t}\tan^2 t}(\sec^2 t + 2\tan t)$

25. $\dfrac{du}{dt} = \nabla u \cdot \mathbf{r}' = \left[(3y^2 - 2x)\, \mathbf{i} + 6xy\, \mathbf{j}\right] \cdot [(2t + 2)\, \mathbf{i} + 3\, \mathbf{j}] = 104t^3 + 150t^2 - 8t$

26. $u(\mathbf{r}(t)) = \dfrac{1}{\sqrt{1 + t^2}}, \quad \dfrac{du}{dt} = \dfrac{-t}{(1 + t^2)^{3/2}}$

27. area $A = \tfrac{1}{2}x(t)y(t)\sin\theta(t)$

$\dfrac{dA}{dt} = 0 = \tfrac{1}{2}y(t)x'(t)\sin\theta(t) + \tfrac{1}{2}x(t)y'(t)\sin\theta(t) + \tfrac{1}{2}\theta'(t)x(t)y(t)\cos\theta(t) = 0$

At $x = 4, \; y = 5, \; \theta = \pi/3, \; \dfrac{dx}{dt} = \dfrac{dy}{dt} = 2,$ we have

$$5\dfrac{d\theta}{dt} + 2\sqrt{3} + \dfrac{5\sqrt{3}}{2} = 0 \quad \Longrightarrow \quad \dfrac{d\theta}{dt} = -\dfrac{9\sqrt{3}}{10}.$$

28. $V = \pi r^2 h; \quad \dfrac{dV}{dt} = 2\pi r h\dfrac{dr}{dt} + \pi r^2\dfrac{dh}{dt}$

Measure in centimeters: at $r = 12, \; h = 1000, \; \dfrac{dr}{dt} = 4, \; \dfrac{dh}{dt} = 150,$

$$\dfrac{dV}{dt} = 2\pi(12)(1000)(4) + \pi(144)(150) = 117{,}600\pi \text{ cu.cm/yr} \cong 0.37 \text{ cu m/yr.}$$

29. $\dfrac{\partial u}{\partial s} = \dfrac{\partial u}{\partial x} + \dfrac{\partial u}{\partial y}; \qquad \dfrac{\partial u}{\partial t} = \dfrac{\partial u}{\partial x} - \dfrac{\partial u}{\partial y}$

$\dfrac{\partial u}{\partial s}\dfrac{\partial u}{\partial t} = \left(\dfrac{\partial u}{\partial x} + \dfrac{\partial u}{\partial y}\right)\left(\dfrac{\partial u}{\partial x} - \dfrac{\partial u}{\partial y}\right) = \left(\dfrac{\partial u}{\partial x}\right)^2 - \left(\dfrac{\partial u}{\partial y}\right)^2$

30. $\dfrac{\partial u}{\partial s} = u_x e^s \cos t + u_y e^s \sin t$

$\dfrac{\partial^2 u}{\partial s^2} = u_{xx} e^{2s} \cos^2 t + u_{xy} e^{2s} \sin t \cos t + u_x e^s \cos t + u_y e^s \sin t + u_{yx} e^{2s} \cos t \sin t + u_{yy} e^{2s} \sin^2 t$

$\dfrac{\partial u}{\partial t} = -u_x e^s \sin t + u_y e^s \cos t$

$\dfrac{\partial^2 u}{\partial t^2} = u_{xx} e^{2s} \sin^2 t - u_{xy} e^{2s} \sin t \cos t - u_x e^s \cos t - u_y e^s \sin t - u_{yx} e^{2s} \cos t \sin t + u_{yy} e^{2s} \cos^2 t$

$\dfrac{\partial^2 u}{\partial s^2} + \dfrac{\partial^2 u}{\partial t^2} = e^{2s}(u_{xx} + u_{yy}) \quad \Longrightarrow \quad \dfrac{\partial^2 u}{\partial x^2} + \dfrac{\partial^2 u}{\partial y^2} = e^{-2s}\left[\dfrac{\partial^2 u}{\partial s^2} + \dfrac{\partial^2 u}{\partial t^2}\right]$

31. $\nabla f(x, y) = (3x^2 - 6xy)\,\mathbf{i} + (-3x^2 + 2y)\,\mathbf{j}; \quad \nabla f(1, -1) = \mathbf{N} = 9\,\mathbf{i} - 5\,\mathbf{j}$

normal line: $x = 1 + 9t, \quad y = -1 - 5t; \qquad$ tangent line: $x = 1 + 5t, \quad y = -1 + 9t$

32. $\nabla f(x, y) = -\pi y \sin \pi xy\,\mathbf{i} - \pi x \sin \pi xy\,\mathbf{j}; \quad \nabla f(1/3, 2) = -\pi\sqrt{3}\,\mathbf{i} - \dfrac{\pi\sqrt{3}}{6}\,\mathbf{j}, \quad$ take $\mathbf{N} = 6\mathbf{i} + \mathbf{j};$

normal line: $x = 1/3 + 6t, \quad y = 2 + t; \qquad$ tangent line: $x = 1/3 + t, \quad y = 2 - 6t$

33. Set $f(x, y, z) = x^{1/2} + y^{1/2} - z$

$\nabla f(x, y, z) = \dfrac{1}{2\sqrt{x}}\,\mathbf{i} + \dfrac{1}{2\sqrt{y}}\,\mathbf{j} - \mathbf{k}; \quad \nabla f(1, 1, 2) = \tfrac{1}{2}\,\mathbf{i} + \tfrac{1}{2}\,\mathbf{j} - \mathbf{k}. \quad$ Take $\mathbf{N} = \mathbf{i} + \mathbf{j} - 2\,\mathbf{k}.$

tangent plane: $(x - 1) + (y - 1) - 2(z - 2) = 0; \qquad$ normal line: $x = 1 + t, \ y = 1 + t, \ z = 2 - 2t$

34. Set $f(x, y, z) = x^2 + y^2 + z^2.$

$\nabla f(x, y, z) = 2x\,\mathbf{i} + 2y\,\mathbf{j} + 2z\,\mathbf{k}; \quad \nabla f(1, 2, -2) = 2\,\mathbf{i} + 4\,\mathbf{j} - 4\,\mathbf{k}. \quad$ Take $\mathbf{N} = \mathbf{i} + 2\,\mathbf{j} - 2\,\mathbf{k}.$

tangent plane: $(x - 1) + 2(y - 2) - 2(z + 2) = 0; \qquad$ normal line: $x = 1 + t, \ y = 2 + 2t, \ z = -2 - 2t$

35. Set $f(x, y, z) = z^3 + xyz - 2.$

$\nabla f(x, y, z) = yz\,\mathbf{i} + xz\,\mathbf{j} + (3z^2 + xy)\,\mathbf{k}; \quad \nabla f(1, 1, 1) = \mathbf{i} + \mathbf{j} + 4\mathbf{k}.$

tangent plane: $(x - 1) + (y - 1) + 4(z - 1) = 0; \qquad$ normal line: $x = 1 + t; \ y = 1 + t; \ z = 1 + 4t$

36. Set $f(x, y, z) = e^{3x} \sin 3y - z.$

$\nabla f(x, y, z) = 3e^{3x} \sin 3y\,\mathbf{i} + 3e^{3x} \cos 3y\,\mathbf{j} - \mathbf{k}; \quad \nabla f(0, \pi/6, 1) = 3\mathbf{i} - \mathbf{k}.$

tangent plane: $3(x - 0) - (z - 1) = 0 \quad$ or $\quad 3x - z + 1 = 0; \quad$ normal line: $x = 3t, \ y = \pi/6, \ z = 1 - t$

37. The point $(2, 2, 1)$ is on each hyperboloid. Set $f(x, y, z) = x^2 + 2y^2 - 4z^2, \ g(x, y, z) = 4x^2 - y^2 + 2z^2.$

$\nabla f = 2x\,\mathbf{i} + 4y\,\mathbf{j} - 8z\,\mathbf{k}, \quad \nabla f(2, 2, 1) = (4, 8, -8); \quad \nabla g = 8x\,\mathbf{i} - 2y\,\mathbf{j} + 4z\,\mathbf{k}, \ \nabla g(2, 2, 1) = (16, -4, 4).$

Since $\nabla f(2, 2, 1) \cdot \nabla g(2, 2, 1) = 0,$ the hyperboloids are mutually perpendicular at $(2, 2, 1).$

38. Set $f(x, y, z) = x^2 + y^2 + z^2$.   At each point (x_0, y_0, z_0), $\nabla f(x_0, y_0, z_0) = 2x_0\,\mathbf{i} + 2y_0\,\mathbf{j} + 2z_0\,\mathbf{k}$.
normal line to the sphere: $x = x_0 + x_0 t$, $y = y_0 + y_0 t$, $z = z_0 + z_0 t$.   At $t = -1$, $x = y = z = 0$.

39. $\nabla f(x, y) = (2xy - 2y)\,\mathbf{i} + (x^2 - 2x + 4y - 15)\,\mathbf{j} = \mathbf{0}$ at $(5, 0)$, $(-3, 0)$, $(1, 4)$.
$f_{xx} = 2y$, $f_{xy} = 2x - 2$, $f_{yy} = 4$.

point	A	B	C	D	result
$(5, 0)$	0	8	4	-64	saddle
$(-3, 0)$	0	-8	4	-64	saddle
$(1, 4)$	8	0	4	32	loc. min.

$f(1, 4) = -34$

40. $\nabla f(x, y) = (6x - 3y^2)\,\mathbf{i} + (3y^2 + 6y - 6xy)\,\mathbf{j} = \mathbf{0}$ at $(0, 0)$, $(2, 2)$, $(\tfrac{1}{2}, -1)$.
$f_{xx} = 6$, $f_{xy} = -6y$, $f_{yy} = 6y - 6x + 6$.

point	A	B	C	D	result
$(0, 0)$	6	0	6	36	loc. min.
$(2, 2)$	6	-12	6	-108	saddle
$(\tfrac{1}{2}, -1)$	6	6	-3	-54	saddle

$f(0, 0) = 0$

41. $\nabla f(x, y) = (3x^2 - 18y)\,\mathbf{i} + (3y^2 - 18x)\,\mathbf{j} = \mathbf{0}$ at $(0, 0)$, $(6, 6)$.
$f_{xx} = 6x$, $f_{xy} = -18$, $f_{yy} = 6y$.

point	A	B	C	D	result
$(0, 0)$	0	-18	0	-18^2	saddle
$(6, 6)$	36	-18	36	> 0	loc. min.

$f(6, 6) = -216$

42. $\nabla f(x, y) = (3x^2 - 12x)\,\mathbf{i} + (2y + 1)\,\mathbf{j} = \mathbf{0}$ at $(0, -\tfrac{1}{2})$, $(4, -\tfrac{1}{2})$.
$f_{xx} = 6x - 12$, $f_{xy} = 0$, $f_{yy} = 2$.

point	A	B	C	D	result
$(0, -\tfrac{1}{2})$	-12	0	2	-24	saddle
$(4, -\tfrac{1}{2})$	12	0	2	24	loc. min.

$f(4, -\tfrac{1}{2}) = -\tfrac{145}{4}$

43. $\nabla f(x,y) = (1 - 2xy + y^2)\mathbf{i} + (-1 - x^2 + 2xy)\mathbf{j} = \mathbf{0}$ at $(1,1)$, $(-1,-1)$.

$f_{xx} = -2y$, $f_{xy} = -2x + 2y$, $f_{yy} = 2x$.

point	A	B	C	D	result
$(1,1)$	-2	0	2	-4	saddle
$(-1,-1)$	2	0	-2	-4	saddle

44. $\nabla f(x,y) = e^{-(x^2+y^2)/2}\left[(y^2 - x^2 y^2)\mathbf{i} + (2xy - xy^3)\mathbf{j}\right] = \mathbf{0}$ at $(\pm 1, \pm\sqrt{2})$, $(x,0)$, x any real number.

$$f_{xx} = e^{-(x^2+y^2)/2}(-3xy^2 + x^3 y^2), \quad f_{xy} = e^{-(x^2+y^2)/2}(2y - y^3 - 2x^2 y + x^2 y^3),$$

$$f_{yy} = e^{-(x^2+y^2)/2}(2x - 5xy^2 + xy^4).$$

point	A	B	C	D	result
$(1, \sqrt{2})$	$-4e^{-3/2}$	0	$-4e^{-3/2}$	$16e^{-3}$	loc. max
$(1, -\sqrt{2})$	$-4e^{-3/2}$	0	$-4e^{-3/2}$	$16e^{-3}$	loc. max
$(-1, \sqrt{2})$	$4e^{-3/2}$	0	$4e^{-3/2}$	$16e^{-3}$	loc. min
$(-1, -\sqrt{2})$	$4e^{-3/2}$	0	$4e^{-3/2}$	$16e^{-3}$	loc. min

local maxima: $f(1, \sqrt{2}) = f(1, -\sqrt{2}) = 2e^{-3/2}$; local minima: $f(-1, \sqrt{2}) = f(-1, -\sqrt{2}) = -2e^{-3/2}$.
At $(x,0)$, $D = 0$ and $f(x,0) \equiv 0$. For $x < 0$, $f(x,y) < f(x,0)$ for all $y \neq 0$; for $x > 0$,
$f(x,y) > f(x,0)$ for all $y > 0$; $(0,0)$ is a saddle point.

Here is a graph of the surface.

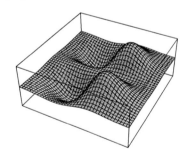

45. $\nabla f = (2x - 2)\mathbf{i} + (2y + 2)\mathbf{j} = \mathbf{0}$ at $(1, -1)$ in D; $f(1, -1) = 0$

Next we consider the boundary of D. We parametrize the circle by:

$C : \mathbf{r}(t) = 2\cos t\,\mathbf{i} + 2\sin t\,\mathbf{j}, \quad t \in [0, 2\pi]$

The values of f on the boundary are given by the function

$$F(t) = f(\mathbf{r}(t)) = 6 - 4\cos t + 4\sin t, \quad t \in [0, 2\pi]$$

$$F'(t) = 4\sin t + 4\cos t : \quad F'(t) = 0 \implies \sin t = -\cos t \implies t = \frac{3}{4}\pi, \frac{7}{4}\pi$$

Evaluating F at the endpoints and critical numbers, we have:

$$F(0) = F(2\pi) = f(2,0) = 2; \quad F\left(\tfrac{3}{4}\pi\right) = f\left(-\sqrt{2}, \sqrt{2}\right) = 6 + 4\sqrt{2};$$

$$F\left(\tfrac{7}{4}\pi\right) = f\left(\sqrt{2}, -\sqrt{2}\right) = 6 - 4\sqrt{2}.$$

f takes on its absolute maximum of $6 + 4\sqrt{2}$ at $\left(-\sqrt{2}, \sqrt{2}\right)$; f takes on its absolute minimum of 0 at $(1, -1)$.

46. $\nabla f(x,y) = (4x - 4)\mathbf{i} + (2y - 4)\mathbf{j} = \mathbf{0}$ at $(1,2)$ on the boundry of D; no critical points in D.

Next we consider the boundary of D. We parametrize each side of the triangle:

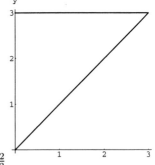

$$C_1 : \; \mathbf{r}_1(t) = t\,\mathbf{i} + 2t\,\mathbf{j}, \quad t \in [0,1]$$

$$C_2 : \; \mathbf{r}_2(t) = (1-t)\,\mathbf{i} + 2\,\mathbf{j}, \quad t \in [0,1]$$

$$C_3 : \; \mathbf{r}_3(t) = (2-t)\,\mathbf{j}, \quad t \in [0,2]$$

Now,

$$f_1(t) = f(\mathbf{r}_1(t)) = 6t^2 - 8t + 3, \quad t \in [0,1]; \quad \text{critical number: } t = \tfrac{2}{3}$$

$$f_2(t) = f(\mathbf{r}_2(t)) = 2t^2 - 3+, \quad t \in [0,1]; \quad \text{critical number}$$

$$f_3(t) = f(\mathbf{r}_3(t)) = t^2 - 1, \quad t \in [0,2]; \quad \text{critical number}$$

Evaluating these functions at the endpoints of their domains and at the critical numbers, we find that:

$$f_1(0) = f_3(2) = f(0,0) = 3; \quad f_1(2/3) = f(2/3, 4/3) = -\tfrac{7}{3}; \quad f_1(1) = f_2(0) = f(1,2) = -3;$$

$$f_2(1) = f_3(0) = f(0,2) = -1.$$

f takes on its absolute maximum of 3 at $(0,0)$ and its absolute minimum of -3 at $(1,2)$.

47. $\nabla f(x,y) = (8x - y)\mathbf{i} + (-x + 2y + 1)\mathbf{j} = \mathbf{0}$ at $(-1/15, -8/15)$ in D; $f(-1/15, -8/15) = -4/15$.

On the boundary of D: $x = \cos t$, $y = 2\sin t$. Set

$$F(t) = f(\cos t, 2\sin t) = 4 + 2\sin t - 2\sin t \cos t, \quad 0 \le t \le 2\pi.$$

Then

$$F'(t) = 2\cos t - 4\cos^2 t + 2 = -2(2\cos t + 1)(\cos t - 1); \quad F'(t) = 0 \implies t = \frac{2\pi}{3}, \frac{4\pi}{3}.$$

Evaluating F at the endpoints of the interval and at the critical points, we get

$$F(0) = F(2\pi) = f(1,0) = 4, \quad F(2\pi/3) = f(-1/2, \sqrt{3}) = 4 + \frac{3\sqrt{3}}{2},$$

$$F(4\pi/3) = f(-1/2, -\sqrt{3}) = 4 - \frac{3\sqrt{3}}{2} > -\frac{4}{15}$$

f takes on its absolute maximum of 2 at $(0,1)$; f takes on its absolute minimum of $-4/15$ at $(-1/15, -8/15)$.

48. $\nabla f(x, y) = 4x^3 \mathbf{i} + 6y^2 \mathbf{j} = \mathbf{0}$ at $(0, 0)$ in D; $f(0, 0) = 0$.

On the boundary of D: $x = \cos t$, $y = \sin t$. Set

$$F(t) = f(\cos t, \sin t) = \cos^4 t + 2\sin^3 t, \quad 0 \le t \le 2\pi.$$

Then

$F'(t) = 4\cos^3 t \sin t + 6\sin^2 t \cos t = 2\sin t \cos t(2\sin t - 1)(\sin t + 2);$

$F'(t) = 0 \implies t = \pi/6, \ \pi/2, \ 5\pi/6, \ \pi, \ 3\pi/2$

Evaluating F at the endpoints of the interval and at the critical points, we get

$$F(0) = F(2\pi) = f(1, 0) = 1, \ F(\pi/6) = f(\sqrt{3}/2, 1/2) = 13/16, \ F(\pi/2) = f(0, 1) = 2,$$

$$F(5\pi/6) = f(-\sqrt{3}/2, 1/2) = 13/16, \ F(\pi) = f(-1, 0) = 1, \ F(3\pi/2) = f(0, -1) = -2.$$

f takes on its absolute maximum of 2 at $(0, 1)$; f takes on its absolute minimum of -2 at $(0, -1)$.

49. Set $f(x, y, z) = D^2 = (x - 1)^2 + (y + 2)^2 + (z - 3)^2$, $g(x, y) = 3x + 2y - z - 5$.

$\nabla f = 2(x - 1)\mathbf{i} + 2(y + 2)\mathbf{j} + 2(z - 3)\mathbf{k}$, $\nabla g = 3\mathbf{i} + 2\mathbf{j} - \mathbf{k}$.

Set $\nabla f = \lambda \nabla g$:

$$2(x - 1) = 3\lambda \implies x = \tfrac{3}{2}\lambda + 1,$$

$$2(y + 2) = 2\lambda \implies y = \lambda - 2,$$

$$2(z - 3) = -\lambda \implies z = -\tfrac{1}{2}\lambda + 3.$$

Substituting these values in $3x + 2y - z = 5$ gives $\lambda = \dfrac{9}{7} \implies x = \dfrac{41}{14}$, $y = -\dfrac{5}{7}$, $z = \dfrac{33}{14}$.

The point on the plane that is closest to $(1, -2, 3)$ is $(41/14, -5/7, 33/14)$. The distance from the point to the plane is $\dfrac{9}{\sqrt{14}}$.

50. Set $f(x, y, z) = 3x - 2y + z$, $g(x, y, z) = x^2 + y^2 + z^2 - 14$,

$\nabla f = 3\mathbf{i} - 2\mathbf{j} + \mathbf{k}$, $\nabla g = 2x\mathbf{i} + 2y\mathbf{j} + 2z\mathbf{k}$.

Set $\nabla f = \lambda \nabla g$:

$$3 = 2\lambda x \implies x = 3/2\lambda, \quad -2 = 2\lambda y \implies y = -1/\lambda, \quad 1 = 2\lambda z \implies z = 1/2\lambda.$$

Substituting these values in $x^2 + y^2 + z^2 = 14$ gives $\lambda = \pm\tfrac{1}{2} \implies x = 3, \ y = -2, \ z = 1$ or

$x = -3, \ y = 2, \ z = -1$. Evaluating f: $f(3, -2, 1) = 14$, $f(-3, 2, -1) = -14$. The maximum value of f on the sphere is 14.

51. Set $f(x, y, z) = x + y - z$, $g(x, y, z) = x^2 + y^2 + 4z^2 - 4$,

$\nabla f = \mathbf{i} + \mathbf{j} - \mathbf{k}$, $\nabla g = 2x\mathbf{i} + 2y\mathbf{j} + 8z\mathbf{k}$.

Set $\nabla f = \lambda \nabla g$:

$$1 = 2\lambda x \implies x = 1/2\lambda, \quad 1 = 2\lambda y \implies y = 1/2\lambda, \quad -1 = 8\lambda z \implies z = -1/8\lambda.$$

Substituting these values in $x^2 + y^2 + 4z^2 = 4$ gives $\lambda = \pm\tfrac{3}{8} \implies x = 4/3, \ y = 4/3, \ z = -1/3$

or $x = -4/3, \ y = -4/3, \ z = 1/3$. Evaluating f: $f(\tfrac{4}{3}, \tfrac{4}{3}, -\tfrac{1}{3}) = 3$, $f(-\tfrac{4}{3}, -\tfrac{4}{3}, \tfrac{1}{3}) = -3$. The maximum

value of f is 3, the minimum value is -3.

52. Let the length, width and height be x, y, z respectively. Then the total cost is

$$f(x, y, z) = \tfrac{1}{2} xy + \tfrac{1}{2} xz + \tfrac{1}{2} yz + \tfrac{1}{10} xy = \tfrac{3}{5} xy + \tfrac{1}{2} xz + \tfrac{1}{2} yz$$

with the condition

$$g(x, y, z) = xyz - 16 = 0.$$

Note first that $\quad xyz = 16 \quad \Longrightarrow \quad x \neq 0, \quad y \neq 0 \quad z \neq 0.$

$$\nabla f = \left(\tfrac{3}{5} y + \tfrac{1}{2} z\right) \mathbf{i} + \left(\tfrac{3}{5} x + \tfrac{1}{2} z\right) \mathbf{j} + \left(\tfrac{1}{2} y + \tfrac{1}{2} x\right) \mathbf{k}, \qquad \nabla g = yz\,\mathbf{i} + xz\,\mathbf{j} + xy\,\mathbf{k}$$

$$\nabla f = \lambda \nabla g \quad \Longrightarrow \quad \tfrac{3}{5} y + \tfrac{1}{2} z = \lambda yz, \quad \tfrac{3}{5} x + \tfrac{1}{2} z = \lambda xz, \quad \tfrac{1}{2} x + \tfrac{1}{2} y = \lambda xy$$

Multiply the first equation by x, the second equation by y and subtract. This gives:

$$\tfrac{1}{2}(xz - yz) = 0 \quad \Longrightarrow \quad z(x - y) = 0 \quad \Longrightarrow \quad y = x$$

Substituting $y = x$ in the third equation yields $x = \lambda x^2 \quad \Longrightarrow \quad x = \dfrac{1}{\lambda}.$

Substituting $x = y = \dfrac{1}{\lambda}$ in the first equation yields $z = \dfrac{6}{5\lambda}.$

Finally, substituting these values for x, y and z into the equation $xyz = 16$, we get $\lambda = \dfrac{\sqrt[3]{3}}{2\sqrt[3]{5}}.$

Therefore, $\quad x = y = \dfrac{2\sqrt[3]{5}}{\sqrt[3]{3}} = \dfrac{10}{\sqrt[3]{75}} \cong 2.37 \quad$ and $\quad z = \dfrac{6}{5} x = \dfrac{12}{\sqrt[3]{75}} \cong 2.85.$

53. $df = (9x^2 - 10xy^2 + 2)\,dx + (-10x^2 y - 1)\,dy$

54. $df = (2xy \sec^2 x^2 - 2y^2)\,dx + (\tan x^2 - 4xy)\,dy$

55. $df = \dfrac{y^2 z + z^2 y}{(x + y + z)^2}\,dx + \dfrac{xz^2 + zx^2}{(x + y + z)^2}\,dy + \dfrac{x^2 y + y^2 x}{(x + y + z)^2}\,dz$

56. $df = -\dfrac{z}{y^2 + xz}\,dx + \left(ze^{yz} - \dfrac{2y}{y^2 + xz}\right)dy + \left(ye^{yz} - \dfrac{x}{y^2 + xz}\right)dz$

57. Set $f(x, y, z) = e^x \sqrt{y + z^3}$. Then

$$df = e^x \sqrt{y + z^3}\,\Delta x + \dfrac{e^x}{2} \dfrac{1}{\sqrt{y + z^3}}\,\Delta y + \dfrac{e^x}{2} \dfrac{3z^2}{\sqrt{y + z^3}}\,\Delta z.$$

With $x = 0$, $y = 15$, $z = 1$, $\Delta x = 0.02$, $\Delta y = 0.2$, $\Delta z = 0.01$, $df = 4\,\Delta x + \tfrac{1}{8}\,\Delta y + \tfrac{3}{8}\,\Delta z \cong 0.1088.$

Therefore, $e^{0.02} \sqrt{15.2 + (1.01)^3} \cong e^0 \sqrt{15 + 1} + 0.1088 = 4.1088.$

58. Set $f(x, y) = x^{1/3} \cos^2 y$. Then

$$df = \tfrac{1}{3} x^{-2/3} \cos^2 y\,\Delta x - 2x^{1/3} \cos y \sin y\,\Delta y.$$

With $x = 64$, $y = 30° = \pi/6$, $\Delta x = 0.5$, $\Delta y = -2° = -\dfrac{\pi}{90}$, $df = \dfrac{1}{64}\,\Delta x - 2\sqrt{3}\,\Delta y \cong 0.1287.$

Therefore, $(64.5)^{1/3} \cos^2(28°) \cong 64^{1/3} \cos^2(30°) + 0.1287 = 3.1287.$

59. $V = \pi r^2 h; \quad r = 5 \,\text{ft.}, \quad h = 22 \,\text{ft.}, \quad \Delta r = 0.01 \,\text{in.} = \frac{1}{1200} \,\text{ft.}, \quad \Delta h = 0.01 = \frac{1}{1200}$

$$dV = 2\pi r h \,\Delta r + \pi r^2 \,\Delta h$$

Using the values given above,

$$dV = 2\pi(5)(22)\,\frac{1}{1200} + \pi(25)\,\frac{1}{1200} \cong 0.6414 \,\text{cu. ft.} \cong 1108.35 \,\text{cu. in.}; \qquad \frac{1108.35}{231} \cong 4.80.$$

Approximately 4.80 gallons will be needed.

60. $\dfrac{\partial P}{\partial y} = 12x^2 y - 8x = \dfrac{\partial Q}{\partial x};$ the vector function is a gradient.

$\dfrac{\partial f}{\partial x} = 6x^2 y^2 - 8xy + 2x, \quad f(x,y) = 2x^3 y^2 - 4x^2 y + x^2 + \phi(y),$

$\dfrac{\partial f}{\partial y} = 4x^3 y - 4x^2 + \phi'(y) = 4x^3 y - 4x^2 - 8.$

Thus, $\phi'(y) = -8, \ \phi(y) = -8y + C,$ and $f(x,y) = 2x^3 y^2 - 4x^2 y + x^2 - 8y + C.$

61. $\dfrac{\partial P}{\partial y} = 2x - \sin x = \dfrac{\partial Q}{\partial x};$ the vector function is a gradient.

$\dfrac{\partial f}{\partial x} = 2xy + 3 - y \sin x, \quad f(x,y) = x^2 y + 3x + y \cos x + \phi(y),$

$\dfrac{\partial f}{\partial y} = x^2 + \cos x + \phi'(y) = x^2 + 2y + 1 + \cos x.$

Thus, $\phi'(y) = 2y + 1, \ \phi(y) = y^2 + y + C,$ and $f(x,y) = x^2 y + 3x + y \cos x + y^2 + y + C.$

62. $\dfrac{\partial P}{\partial y} = 2xy + 4y; \quad \dfrac{\partial Q}{\partial x} = -2xy + 2; \quad \dfrac{\partial P}{\partial y} \neq \dfrac{\partial Q}{\partial x};$ the vector function is not a gradient.

63. $\dfrac{\partial P}{\partial y} = e^y \sin z = \dfrac{\partial Q}{\partial x}, \quad \dfrac{\partial P}{\partial z} = e^y \cos z = \dfrac{\partial R}{\partial x}, \quad \dfrac{\partial Q}{\partial z} = xe^y \cos z = \dfrac{\partial R}{\partial y};$

the vector function is a gradient.

$f(x,y,z) = \displaystyle\int (e^y \sin z + 2x)\, dx = xe^y \sin z + x^2 + \phi(y,z),$

$f_y = xe^y \sin z + \dfrac{\partial \phi}{\partial y} = xe^y \sin z - y^2 \implies \dfrac{\partial \phi}{\partial y} = -y^2 \implies \phi = -\dfrac{1}{3}y^3 + \psi(z),$

$f(x,y,z) = xe^y \sin z + x^2 - \dfrac{1}{3}y^3 + \psi(z), \quad f_z = xe^y \cos z + \psi'(z) = xe^y \cos z \implies \psi'(x) = 0 \implies$
$\psi(x) = C$

Therefore $f(x,y,z) = xe^y \sin z + x^2 - \frac{1}{3}y^3 + C.$

SECTION 17.1

1. $\displaystyle\sum_{i=1}^{3}\sum_{j=1}^{3} 2^{i-1}3^{j+1} = \left(\sum_{i=1}^{3} 2^{i-1}\right)\left(\sum_{j=1}^{3} 3^{j+1}\right) = (1 + 2 + 4)(9 + 27 + 81) = 819$

2. $2 + 2^2 + 3 + 3^2 + 4 + 4^2 + 5 + 5^2 = 68$

3. $\displaystyle\sum_{i=1}^{4}\sum_{j=1}^{3}(i^2 + 3i)(j - 2) = \left[\sum_{i=1}^{4}(i^2 + 3i)\right]\left[\sum_{j=1}^{3}(j - 2)\right] = (4 + 10 + 18 + 28)(-1 + 0 + 1) = 0$

4. $\dfrac{2}{2} + \dfrac{2}{3} + \dfrac{2}{4} + \dfrac{2}{5} + \dfrac{2}{6} + \dfrac{2}{7} + \dfrac{4}{2} + \dfrac{4}{3} + \dfrac{4}{4} + \dfrac{4}{5} + \dfrac{4}{6} + \dfrac{4}{7} + \dfrac{6}{2} + \dfrac{6}{3} + \dfrac{6}{4} + \dfrac{6}{5} + \dfrac{6}{6} + \dfrac{6}{7} = 19\dfrac{4}{35}.$

5. $\displaystyle\sum_{i=1}^{m}\Delta x_i = \Delta x_1 + \Delta x_2 + \cdots + \Delta x_m = (x_1 - x_0) + (x_2 - x_1) + \cdots + (x_m - x_{m-1})$

$$= x_m - x_0 = a_2 - a_1$$

6. $(y_1 - y_0) + (y_2 - y_1) + \cdots + (y_n - y_{n-1}) = y_n - y_0 = b_2 - b_1$

7. $\displaystyle\sum_{i=1}^{m}\sum_{j=1}^{n}\Delta x_i\,\Delta y_j = \left(\sum_{i=1}^{m}\Delta x_i\right)\left(\sum_{j=1}^{n}\Delta y_j\right) = (a_2 - a_1)\,(b_2 - b_1)$

8. $\displaystyle\sum_{j=1}^{n}\sum_{k=1}^{q}\Delta y_j\,\Delta z_k = \left(\sum_{j=1}^{n}\Delta y_j\right)\left(\sum_{k=1}^{q}\Delta z_k\right) = (b_2 - b_1)\,(c_2 - c_1)$

9. $\displaystyle\sum_{i=1}^{m}(x_i + x_{i-1})\Delta x_i = \sum_{i=1}^{m}(x_i + x_{i-1})(x_i - x_{i-1}) = \sum_{i=1}^{m}(x_i^2 - x_{i-1}^2)$

$$= x_m{}^2 - x_0{}^2 = a_2{}^2 - a_1{}^2$$

10. $\displaystyle\sum_{j=1}^{n}\frac{1}{2}(y_j{}^2 + y_j y_{j-1} + y_{j-1}{}^2)\Delta y_j = \frac{1}{2}\sum_{j=1}^{n}(y_j{}^3 - y_{j-1}{}^3) = \frac{1}{2}(b_2{}^3 - b_1{}^3)$

11. $\displaystyle\sum_{i=1}^{m}\sum_{j=1}^{n}(x_i + x_{i-1})\Delta x_i \Delta y_j = \left(\sum_{i=1}^{m}(x_i + x_{i-1})\Delta x_i\right)\left(\sum_{j=1}^{n}\Delta y_j\right)$

(Exercise 9)⤴

$$= \left(a_2{}^2 - a_1{}^2\right)(b_2 - b_1)$$

12. $\displaystyle\sum_{i=1}^{m}\sum_{j=1}^{n}(y_i + y_{j-1})\Delta x_i \Delta y_j = \left(\sum_{i=1}^{m}\Delta x_i\right)\left[\sum_{j=1}^{n}(y_j{}^2 - y_{j-1}{}^2)\right] = (a_2 - a_1)(b_2{}^2 - b_1{}^2)$

13. $\displaystyle\sum_{i=1}^{m}\sum_{j=1}^{n}(2\Delta x_i - 3\Delta y_j) = 2\left(\sum_{i=1}^{m}\Delta x_i\right)\left(\sum_{j=1}^{n}1\right) - 3\left(\sum_{i=1}^{m}1\right)\left(\sum_{j=1}^{n}\Delta y_j\right)$

$$= 2n(a_2 - a_1) - 3m(b_2 - b_1)$$

14. $\displaystyle\sum_{i=1}^{m}\sum_{j=1}^{n}(3\Delta x_i - 2\Delta y_j) = 3\sum_{i=1}^{m}\sum_{j=1}^{n}\Delta x_i - 2\sum_{i=1}^{m}\sum_{j=1}^{n}\Delta y_j = 3n(a_2 - a_1) - 2m(b_2 - b_1).$

15. $\displaystyle\sum_{i=1}^{m}\sum_{j=1}^{n}\sum_{k=1}^{q}\Delta x_i\,\Delta y_j\,\Delta z_k = \left(\sum_{i=1}^{m}\Delta x_i\right)\left(\sum_{j=1}^{n}\Delta y_j\right)\left(\sum_{k=1}^{q}\Delta z_k\right)$

$$= (a_2 - a_1)(b_2 - b_1)(c_2 - c_1)$$

16. $\displaystyle\sum_{i=1}^{m}\sum_{j=1}^{n}\sum_{k=1}^{q}(x_i + x_{i-1})\Delta x_i\Delta y_j\Delta z_k = \left[\sum_{i=1}^{m}(x_i{}^2 - x_{i-1}{}^2)\right]\left(\sum_{j=1}^{n}\Delta y_j\right)\left(\sum_{k=1}^{q}\Delta z_k\right)$

$$= (a_2{}^2 - a_1{}^2)(b_2 - b_1)(c_2 - c_1)$$

17. $\displaystyle\sum_{i=1}^{n}\sum_{j=1}^{n}\sum_{k=1}^{n}\delta_{ijk}a_{ijk} = a_{111} + a_{222} + \cdots + a_{nnn} = \sum_{p=1}^{n}a_{ppp}$

18. Start with $\displaystyle\sum_{i=1}^{m}\sum_{j=1}^{n}a_{ij}$.   Take all the a_{ij} (there are only a finite number of them) and order them in any order you chose. Call the first one b_1, the second b_2, and so on. Then

$$\sum_{i=1}^{m}\sum_{j=1}^{n}a_{ij} = \sum_{p=1}^{r}b_p \quad\text{where}\quad r = m \times n.$$

SECTION 17.2

1. $L_f(P) = 2\frac{1}{4}, \quad U_f(P) = 5\frac{3}{4}$

2. $L_f(P) = 3, \quad U_f(P) = 5$

3. (a) $\displaystyle L_f(P) = \sum_{i=1}^{m}\sum_{j=1}^{n}(x_{i-1} + 2y_{j-1})\,\Delta x_i\,\Delta y_j, \quad U_f(P) = \sum_{i=1}^{m}\sum_{j=1}^{n}(x_i + 2y_j)\,\Delta x_i\,\Delta y_j$

(b) $\displaystyle L_f(P) \le \sum_{i=1}^{m}\sum_{j=1}^{n}\left[\frac{x_{i-1} + x_i}{2} + 2\left(\frac{y_{j-1} + y_j}{2}\right)\right]\Delta x_i\,\Delta y_j \le U_f(P).$

The middle expression can be written

$$\sum_{i=1}^{m}\sum_{j=1}^{n}\frac{1}{2}\left(x_i{}^2 - x_{i-1}^2\right)\Delta y_j + \sum_{i=1}^{m}\sum_{j=1}^{n}\left(y_j{}^2 - y_{j-1}^2\right)\Delta x_i.$$

The first double sum reduces to

$$\sum_{i=1}^{m}\sum_{j=1}^{n}\frac{1}{2}\left(x_i{}^2 - x_{i-1}^2\right)\Delta y_j = \frac{1}{2}\left(\sum_{i=1}^{m}\left(x_i{}^2 - x_{i-1}^2\right)\right)\left(\sum_{j=1}^{n}\Delta y_j\right) = \frac{1}{2}\,(4 - 0)\,(1 - 0) = 2.$$

In like manner the second double sum also reduces to 2.   Thus, $I = 4$; the volume of the prism bounded above by the plane $z = x + 2y$ and below by R.

4. $L_f(P) = -7/16, \quad U_f(P) = 7/16$

5. $L_f(P) = -7/24, \quad U_f(P) = 7/24$

6. (a) $L_f(p) = \sum_{i=1}^{m}\sum_{j=1}^{n}(x_{i-1} - y_j)\Delta x_i \Delta y_j, \quad U_f(P) = \sum_{i=1}^{m}\sum_{j=1}^{n}(x_i - y_{j-1})\Delta x_i \Delta y_j$

(b) $L_f(P) \le \sum_{i=1}^{m}\sum_{j=1}^{n}\left(\dfrac{x_i + x_{i-1}}{2} - \dfrac{y_j + y_{j-1}}{2}\right)\Delta x_i \Delta y_j \le U_f(P)$

The middle expression can be written

$$\sum_{i=1}^{m}\sum_{j=1}^{n}\frac{1}{2}(x_i{}^2 - x_{i-1}{}^2)\Delta y_j - \sum_{i=1}^{m}\sum_{j=1}^{n}\frac{1}{2}(y_j{}^2 - y_{j-1}{}^2)\Delta x_i.$$

The first sum reduces to

$$\frac{1}{2}\left(\sum_{i=1}^{m}(x_i{}^2 - x_{i-1}{}^2)\right)\left(\sum_{j=1}^{n}\Delta y_j\right) = \frac{1}{2}(1 - 0)(1 - 0) = \frac{1}{2}.$$

In like manner the second sum also reduces to $\frac{1}{2}$.   Thus $I = \frac{1}{2} - \frac{1}{2} = 0.$

7. (a) $L_f(P) = \sum_{i=1}^{m}\sum_{j=1}^{n}(4x_{i-1}\,y_{j-1})\,\Delta x_i\,\Delta y_j, \quad U_f(P) = \sum_{i=1}^{m}\sum_{j=1}^{n}(4x_i\,y_j)\,\Delta x_i\,\Delta y_j$

(b) $L_f(P) \le \sum_{i=1}^{m}\sum_{j=1}^{n}(x_i + x_{i-1})(y_j + y_{j-1})\,\Delta x_1\,\Delta y_j \le U_f(P).$

The middle expression can be written

$$\sum_{i=1}^{m}\sum_{j=1}^{n}(x_i{}^2 - x_{i-1}^2)(y_j{}^2 - y_{j-1}^2) = \left(\sum_{i=1}^{m}x_i{}^2 - x_{i-1}^2\right)\left(\sum_{j=1}^{n}y_j{}^2 - y_{j-1}^2\right)$$

by (17.1.5)

$$= (b^2 - 0^2)(d^2 - 0^2) = b^2 d^2.$$

It follows that $I = b^2 d^2.$

8. (a) $L_f(P) = \sum_{i=1}^{m}\sum_{j=1}^{n}3(x_{i-1}{}^2 + y_{j-1}{}^2)\Delta x_i \Delta y_j, \quad U_f(P) = \sum_{i=1}^{m}\sum_{j=1}^{n}3(x_i{}^2 + y_j{}^2)\Delta x_i \Delta y_j$

(b) $L_f(P) \le \sum_{i=1}^{m}\sum_{j=1}^{n}\left[(x_i{}^2 + x_i x_{i-1} + x_{i-1}{}^2) + (y_j{}^2 + y_j y_{j-1} + y_{j-1}{}^2)\right]\Delta x_i \Delta y_j \le U_f(P)$

Since in general $(A^2 + AB + B^2)(A - B) = A^3 - B^3,$ the middle expression can be written

$$\sum_{i=1}^{m}\sum_{j=1}^{n}(x_i{}^3 - x_{i-1}{}^3)\Delta y_j + \sum_{i=1}^{m}\sum_{j=1}^{n}(y_j{}^3 - y_{j-1}{}^3)\Delta x_i,$$

which reduces to

$$\left(\sum_{i=1}^{m}x_i{}^3 - x_{i-1}{}^3\right)\left(\sum_{j=1}^{n}\Delta y_j\right) + \left(\sum_{i=1}^{m}\Delta x_i\right)\left(\sum_{j=1}^{n}y_j{}^3 - y_{j-1}{}^3\right).$$

This can be evaluated as $b^3 d + bd^3 = bd(b^2 + d^2).$ It follows that $I = bd(b^2 + d^2).$

9. (a) $L_f(P) = \sum_{i=1}^{m} \sum_{j=1}^{n} 3 \left(x_{i-1}^2 - y_j^2 \right) \Delta x_i \, \Delta y_j, \quad U_f(P) = \sum_{i=1}^{m} \sum_{j=1}^{n} 3 \left(x_i^2 - y_{j-1}^2 \right) \Delta x_i \, \Delta y_j$

(b) $L_f(P) \le \sum_{i=1}^{m} \sum_{j=1}^{n} \left[\left(x_i^2 + x_i x_{i-1} + x_{i-1}^2 \right) - \left(y_j^2 + y_j y_{j-1} + y_{j-1}^2 \right) \right] \Delta x_i \, \Delta y_j \le U_f(P).$

Since in general $\left(A^2 + AB + B^2 \right) (A - B) = A^3 - B^3,$ the middle expression can be written

$$\sum_{i=1}^{m} \sum_{j=1}^{n} \left(x_i^3 - x_{i-1}^3 \right) \Delta y_j - \sum_{i=1}^{m} \sum_{j=1}^{n} \left(y_j^3 - y_{j-1}^3 \right) \Delta x_i,$$

which reduces to

$$\left(\sum_{i=1}^{m} x_i^3 - x_{i-1}^3 \right) \left(\sum_{j=1}^{n} \Delta y_j \right) - \left(\sum_{i=1}^{m} \Delta x_i \right) \left(\sum_{j=1}^{n} y_j^3 - y_{j-1}^3 \right).$$

This can be evaluated as $b^3 d - b d^3 = bd \left(b^2 - d^2 \right)$. It follows that $I = bd \left(b^2 - d^2 \right)$.

10. On each subrectangle, the minimum and the maximum of f are equal, so f is constant on each subrectangle and therefore (since f is continuous) on the entire rectangle R. Then

$$\iint_R f(x, y) \, dx dy = f(a, c)(b - a)(d - c).$$

11. $\iint_\Omega dx dy = \int_a^b \phi(x) \, dx$

12. Suppose that there is a point (x_0, y_0) on the boundary of Ω at which f is not zero. As (x, y) tends to (x_0, y_0) through that part of R which is outside Ω, $f(x, y)$, being zero, tends to zero. Since $f(x_0, y_0)$ is not zero, $f(x, y)$ does not tend to $f(x_0, y_0)$. Thus the extended function f can not be continuous at (x_0, y_0).

13. Suppose $f(x_0, y_0) \neq 0$. Assume $f(x_0, y_0) > 0$. Since f is continuous, there exists a disc Ω_ϵ with radius ϵ centered at (x_0, y_0) such that $f(x, y) > 0$ on Ω_ϵ. Let R be a rectangle contained in Ω_ϵ. Then $\iint_R f(x, y) \, dx dy > 0$, which contradicts the hypothesis.

14. $\iint_R (x + 2y) \, dx \, dy = 4; \quad \text{area}(R) = (2)(1) = 2, \quad \text{so average value} = \dfrac{4}{2} = 2$

15. By Exercise 7, Section 17.2, $\iint_R 4xy \, dx dy = 2^2 3^2 = 36$. Thus

$$f_{avg} = \frac{1}{\text{area}(R)} \iint_R 4xy \, dx dy = \frac{1}{6} (36) = 6$$

16. $\iint_R (x^2 + y^2) \, dx dy = \dfrac{bd(b^2 + d^2)}{3}; \quad \text{area}(R) = bd, \quad \text{so average value} = \dfrac{b^2 + d^2}{3}$

17. By Theorem 16.2.10, there exists a point $(x_1, y_1) \in D_r$ such that

$$\iint_{D_r} f(x,y)\,dxdy = f(x_1,y_1)\iint_R dxdy = f(x_1,y_1)\pi r^2 \quad \Longrightarrow \quad f(x_1,y_1) = \frac{1}{\pi r^2}\iint_{D_r} f(x,y)\,dxdy$$

As $r \to 0$, $(x_1, y_1) \to (x_0, y_0)$ and $f(x_1, y_1) \to f(x_0, y_0)$ since f is continuous.

The result follows.

18. $0 \le \sin(x+y) \le 1$ for all $(x,y) \in R$. Thus, $0 \le \iint_R \sin(x+y)\,dxdy \le \iint_R dxdy = 1$

19. $z = \sqrt{4 - x^2 - y^2}$ on $\Omega : x^2 + y^2 \le 4$, $x \ge 0$, $y \ge 0$; $\iint_\Omega \sqrt{4 - x^2 - y^2}\,dxdy$ is the volume V of one

quarter of a hemisphere; $V = \frac{4}{3}\pi$.

20. $8 - 4\sqrt{x^2 + y^2}$ on Ω is a cone with height $h = 8$ and radius $r = 2$; $V = \dfrac{32\pi}{3}$

21. $z = 6 - 2x - 3y \Rightarrow \dfrac{x}{3} + \dfrac{y}{2} + \dfrac{z}{6} = 1$;   the solid is the tetrahedron bounded by the coordinate planes

and the plane: $\dfrac{x}{3} + \dfrac{y}{2} + \dfrac{z}{6} = 1$; $V = \frac{1}{6}(3)(2)(6) = 6$

22. (a) $L_f(P) \cong 35.4603$; $U_f(P) \cong 36.5403$ (c) $\displaystyle\iint_R (3y^2 - 2x)\,dxdy = 36$

SECTION 17.3

1. $\displaystyle\int_0^1 \int_0^3 x^2\,dy\,dx = \int_0^1 3x^2\,dx = 1$

2. $\displaystyle\int_0^3 \int_0^1 e^{x+y}\,dx\,dy = \int_0^3 (e^{1+y} - e^y)\,dy = \left[e^{1+y} - e^y\right]_0^3 = e^4 - e^3 - e + 1$

3. $\displaystyle\int_0^1 \int_0^3 xy^2\,dy\,dx = \int_0^1 x\left[\frac{1}{3}y^3\right]_0^3 dx = \int_0^1 9x\,dx = \frac{9}{2}$

4. $\displaystyle\int_0^1 \int_0^x x^3 y\,dy\,dx = \int_0^1 x^3\frac{x^2}{2}\,dx = \frac{1}{12}$

5. $\displaystyle\int_0^1 \int_0^x xy^3\,dy\,dx = \int_0^1 x\left[\frac{1}{4}y^4\right]_0^x dx = \int_0^1 \frac{1}{4}x^5\,dx = \frac{1}{24}$

6. $\displaystyle\int_0^1 \int_0^x x^2 y^2\,dy\,dx = \int_0^1 x^2\frac{x^3}{3}\,dx = \frac{1}{18}$

7. $\displaystyle\int_0^{\pi/2} \int_0^{\pi/2} \sin(x+y)\,dy\,dx = \int_0^{\pi/2} [-\cos(x+y)]_0^{\pi/2}\,dx = \int_0^{\pi/2} \left[\cos x - \cos\left(x + \frac{\pi}{2}\right)\right]dx = 2$

8. $\displaystyle\int_0^{\pi/2}\int_0^{\pi/2}\cos(x+y)\,dx\,dy = \int_0^{\pi/2}\left[\sin\left(\frac{\pi}{2}+y\right)-\sin y\right]dy = \left[\cos y - \cos\left(\frac{\pi}{2}+y\right)\right]_0^{\pi/2} = 0$

9. $\displaystyle\int_0^{\pi/2}\int_0^{\pi/2}(1+xy)\,dy\,dx = \int_0^{\pi/2}\left[y+\frac{1}{2}xy^2\right]_0^{\pi/2}dx = \int_0^{\pi/2}\left(\frac{1}{2}\pi+\frac{1}{8}\pi^2 x\right)dx = \frac{1}{4}\pi^2+\frac{1}{64}\pi^4$

10. $\displaystyle\int_{-1}^1\int_{-\sqrt{1-y^2}}^{\sqrt{1-y^2}}(x+3y^3)\,dx\,dy = \int_{-1}^1 6y^3\sqrt{1-y^2}\,dy = 0$ (integrand is odd)

11. $\displaystyle\int_0^1\int_{y^2}^y\sqrt{xy}\,dx\,dy = \int_0^1\sqrt{y}\left[\frac{2}{3}x^{3/2}\right]_{y^2}^y dy = \int_0^1\frac{2}{3}\left(y^2-y^{7/2}\right)dy = \frac{2}{27}$

12. $\displaystyle\int_0^1\int_0^{y^2}ye^x\,dx\,dy = \int_0^1 y(e^{y^2}-1)\,dy = \left[\frac{e^{y^2}}{2}-\frac{y^2}{2}\right]_0^1 = \frac{1}{2}(e-2)$

13. $\displaystyle\int_{-2}^2\int_{\frac{1}{2}y^2}^{4-\frac{1}{2}y^2}(4-y^2)\,dx\,dy = \int_{-2}^2(4-y^2)\left[\left(4-\frac{1}{2}y^2\right)-\left(\frac{1}{2}y^2\right)\right]dy$

$$= 2\int_0^2(16-8y^2+y^4)\,dy = \frac{512}{15}$$

14. $\displaystyle I = \int_0^1\int_{x^3}^{x^2}(x^4+y^2)\,dy\,dx = \int_0^1\left[x^4 y+\frac{y^3}{3}\right]_{x^3}^{x^2}dx = \int_0^1\left(\frac{4x^6}{3}-x^7-\frac{x^9}{3}\right)dx$

$$= \left[\frac{4x^7}{21}-\frac{x^8}{8}-\frac{x^{10}}{30}\right]_0^1 = \frac{9}{280}$$

15.   0   by symmetry (integrand odd in y, Ω symmetric about x-axis)

16. $\displaystyle\int_0^1\int_0^{2y}e^{-y^2/2}\,dx\,dy = \int_0^1 2ye^{-y^2/2}\,dy = \left[-2e^{-y^2/2}\right]_0^1 = 2\left(1-\frac{1}{\sqrt{e}}\right)$

17. $\displaystyle\int_0^2\int_0^{x/2}e^{x^2}\,dy\,dx = \int_0^2\frac{1}{2}xe^{x^2}\,dx = \left[\frac{1}{4}e^{x^2}\right]_0^2 = \frac{1}{4}\left(e^4-1\right)$

18. $\displaystyle\int_{-1}^0\int_{x^3}^{x^4}(x+y)\,dy\,dx + \int_0^1\int_{x^4}^{x^3}(x+y)\,dy\,dx = \int_{-1}^0\left[xy+\frac{y^2}{2}\right]_{x^3}^{x^4}dx + \int_0^1\left[xy+\frac{y^2}{2}\right]_{x^4}^{x^3}dx$

$$= \int_{-1}^0\left(x^5+\frac{x^8}{2}-x^4-\frac{x^6}{2}\right)dx + \int_0^1\left(x^4+\frac{x^6}{2}-x^5-\frac{x^8}{2}\right)dx$$

$$= \left[\frac{x^6}{6}+\frac{x^9}{18}-\frac{x^5}{5}-\frac{x^7}{14}\right]_{-1}^0 + \left[\frac{x^5}{5}+\frac{x^7}{14}-\frac{x^6}{6}-\frac{x^9}{18}\right]_0^1 = -\frac{1}{3}$$

19.

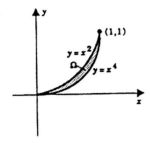

$$\int_0^1 \int_{y^{1/2}}^{y^{1/4}} f(x,y) \, dx \, dy$$

20.

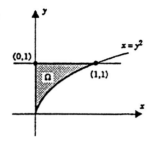

$$\int_0^1 \int_{\sqrt{x}}^1 f(x,y) \, dy \, dx$$

21.

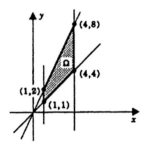

$$\int_{-1}^0 \int_{-x}^1 f(x,y) \, dy \, dx + \int_0^1 \int_x^1 f(x,y) \, dy \, dx$$

22.

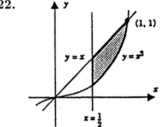

$$\int_{\sqrt[3]{1/2}}^{1/2} \int_{1/2}^{y^{1/3}} f(x,y) \, dx \, dy + \int_{1/2}^1 \int_y^{y^{1/3}} f(x,y) \, dx \, dy$$

23.

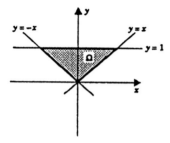

$$\int_1^2 \int_1^y f(x,y) \, dx \, dy + \int_2^4 \int_{y/2}^y f(x,y) \, dx \, dy$$

$$+ \int_4^8 \int_{y/2}^4 f(x,y) \, dx \, dy$$

24.

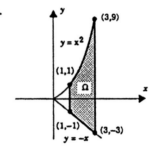

$$\int_{-3}^{-1} \int_{-y}^3 f(x,y) \, dx \, dy + \int_{-1}^1 \int_1^3 f(x,y) \, dx \, dy$$

$$+ \int_1^9 \int_{\sqrt{y}}^3 f(x,y) \, dx \, dy$$

25. $\displaystyle \int_{-2}^4 \int_{1/4x^2}^{\frac{1}{2}x+2} dy \, dx = \int_{-2}^4 \left[\frac{1}{2}x + 2 - \frac{1}{4}x^2\right] dx = 9$

26. $\displaystyle \int_0^3 \int_y^{4y-y^2} dx \, dy = \int_0^3 (3y - y^2) \, dy = \frac{9}{2}$

27. $\displaystyle \int_0^{1/4} \int_{2y^{3/2}}^y dx \, dy = \int_0^{1/4} \left[y - 2y^{3/2}\right] dy = \frac{1}{160}$

28. $\displaystyle\int_2^3 \int_{6/x}^{5-x} dy\, dx = \int_2^3 (5 - x - 6/x)\, dx = \frac{5}{2} + 6\ln\frac{2}{3}$

29.

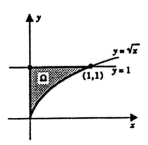

$\displaystyle\int_0^1 \int_0^{y^2} \sin\left(\frac{y^3 + 1}{2}\right) dx\, dy = \int_0^1 y^2 \sin\left(\frac{y^3 + 1}{2}\right) dy$

$\displaystyle\qquad\qquad = \left[-\frac{2}{3}\cos\left(\frac{y^3 + 1}{2}\right)\right]_0^1$

$\displaystyle\qquad\qquad = \frac{2}{3}\left(\cos\frac{1}{2} - \cos 1\right)$

30.

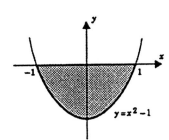

$\displaystyle\int_{-1}^1 \int_{x^2-1}^0 x^2\, dy\, dx = \int_{-1}^1 x^2(1 - x^2)\, dx$

$\displaystyle\qquad\qquad = \frac{4}{15}$

31.

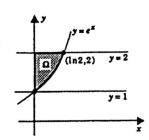

$\displaystyle\int_0^{\ln 2} \int_{e^x}^2 e^{-x}\, dy\, dx = \int_0^{\ln 2} e^{-x}\left(2 - e^x\right) dx$

$\displaystyle\qquad\qquad = \left[-2e^{-x} - x\right]_0^{\ln 2} = 1 - \ln 2$

32.

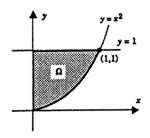

$\displaystyle\int_0^1 \int_0^{\sqrt{y}} \frac{x^3}{\sqrt{x^4 + y^2}}\, dx\, dy = \int_0^1 \frac{1}{2}(\sqrt{2} - 1)y\, dy$

$\displaystyle\qquad\qquad = \frac{1}{4}(\sqrt{2} - 1)$

33. $\displaystyle\int_1^2 \int_{y-1}^{2/y} dx\, dy = \int_1^2 \left[\frac{2}{y} - (y - 1)\right] dy = \ln 4 - \frac{1}{2}$

34. $\displaystyle\int_0^1 \int_0^{1-x} (x + y)\, dy\, dx = \int_0^1 \left[x(1 - x) + \frac{(1 - x)^2}{2}\right] dx = \frac{1}{3}$

35. $\displaystyle\int_0^2\int_0^{3-\frac{3}{2}x}\left(4-2x-\frac{4}{3}y\right)dy\,dx = \int_0^3\int_0^{2-\frac{2}{3}y}\left(4-2x-\frac{4}{3}y\right)dx\,dy = 4$

36. $\displaystyle\int_0^1\int_0^1(2x+3y)\,dy\,dx = \int_0^1\left(2x+\frac{3}{2}\right)dx = \frac{5}{2}$

37. $\displaystyle\int_0^2\int_0^{1-\frac{1}{2}x}x^3y\,dy\,dx = \int_0^2\int_0^{2-2y}x^3y\,dx\,dy = \frac{2}{15}$

38. $\displaystyle\int_{-1}^1\int_{-\sqrt{1-x^2}}^{\sqrt{1-x^2}}(x^2+y^2)\,dy\,dx = 2\int_{-1}^1\left[x^2\sqrt{1-x^2}+\frac{1}{3}(1-x^2)^{3/2}\right]dx = \frac{\pi}{2}$

39. $\displaystyle\int_0^2\int_{-\sqrt{2x-x^2}}^{\sqrt{2x-x^2}}(2x+1)\,dy\,dx = \int_{-1}^1\int_{1-\sqrt{1-y^2}}^{1+\sqrt{1-y^2}}(2x+1)\,dx\,dy$

$$= \int_{-1}^1\left[x^2+x\right]_{1-\sqrt{1-y^2}}^{1+\sqrt{1-y^2}}dy$$

$$= 6\int_{-1}^1\sqrt{1-y^2}\,dy = 6\left(\frac{\pi}{2}\right) = 3\pi$$

40. $\displaystyle\int_{-1}^1\int_{1-\sqrt{1-x^2}}^{1+\sqrt{1-x^2}}\left(4-y^2-\frac{1}{4}x^2\right)dy\,dx = \int_{-1}^1\left[6\sqrt{1-x^2}-\frac{1}{2}x^2\sqrt{1-x^2}-\frac{2}{3}(1-x^2)^{3/2}\right]dx = \frac{43}{16}\pi.$

41. $\displaystyle\int_0^1\int_0^{1-x}(x^2+y^2)\,dy\,dx = \int_0^1\left(2x^2-\frac{4}{3}x^3-x+\frac{1}{3}\right)dx = \frac{1}{6}$

42. $\displaystyle\int_{-1}^1\int_{-\sqrt{1-x^2}}^{\sqrt{1-x^2}}(1-x)\,dy\,dx = \int_{-1}^1\left(2\sqrt{1-x^2}-2x\sqrt{1-x^2}\right)dx = \pi$

43. $\displaystyle\int_0^1\int_{x^2}^x(x^2+3y^2)\,dy\,dx = \int_0^1\left(2x^3-x^4-x^6\right)dx = \frac{11}{70}$

44. $\displaystyle\int_1^2\int_{2y-1}^{5-y}(1+xy)\,dx\,dy = \int_1^2\left(6+9y-3y^2-\frac{3}{2}y^3\right)dy = \frac{55}{8}$

45. $\displaystyle\int_0^a\int_0^{\sqrt{a^2-x^2}}\sqrt{a^2-x^2}\,dy\,dx = \int_0^a(a^2-x^2)\,dx = \frac{2}{3}a^3$

46. $\displaystyle\int_0^a\int_0^{b(1-x/a)}c\left(1-\frac{x}{a}-\frac{y}{b}\right)dy\,dx = \int_0^a\frac{bc}{2}\left(1-\frac{x}{a}\right)^2dx = \frac{abc}{6}$

47. $\displaystyle\int_0^1\int_y^1 e^{y/x}\,dx\,dy = \int_0^1\int_0^x e^{y/x}\,dy\,dx = \int_0^1\left[xe^{y/x}\right]_0^x dx = \int_0^1 x(e-1)\,dx = \frac{1}{2}(e-1)$

48. $\displaystyle\int_0^1\int_0^{\cos^{-1}y}e^{\sin x}\,dx\,dy = \int_0^{\pi/2}\int_0^{\cos x}e^{\sin x}\,dy\,dx = \int_0^{\pi/2}\cos x\,e^{\sin x}\,dx = e-1$

49. $\displaystyle\int_0^1 \int_x^1 x^2 e^{y^4}\,dy\,dx = \int_0^1 \int_0^y x^2 e^{y^4}\,dx\,dy = \int_0^1 \left[\frac{1}{3}x^3 e^{y^4}\right]_0^y dy = \frac{1}{3}\int_0^1 y^3 e^{y^4}\,dy = \frac{1}{12}(e-1)$

50. $\displaystyle\int_0^1 \int_0^y e^{y^2}\,dx\,dy = \int_0^1 y e^{y^2}\,dy = \frac{1}{2}(e-1)$

51. $\displaystyle f_{avg} = \frac{1}{8}\int_{-1}^1 \int_0^4 x^2 y\,dy\,dx = \frac{1}{8}\int_{-1}^1 8x^2\,dx = \int_{-1}^1 x^2\,dx = \frac{2}{3}$

52. area of $\Omega = \frac{1}{4}\pi$ (quarter circle of radius 1)

Average value $= \dfrac{4}{\pi}\displaystyle\int_0^1 \int_0^{\sqrt{1-x^2}} xy\,dy\,dx = \frac{2}{\pi}\int_0^1 x(1-x^2)\,dx = \dfrac{1}{2\pi}$

53. $\displaystyle f_{avg} = \frac{1}{(\ln 2)^2}\int_{\ln 2}^{2\ln 2}\int_{\ln 2}^{2\ln 2} \frac{1}{xy}\,dy\,dx = \frac{1}{(\ln 2)^2}\int_{\ln 2}^{2\ln 2} \frac{1}{x}\ln 2\,dx = 1$

54. area of $\Omega = 1$ (parallelogram)

Average value $= \displaystyle\int_0^1 \int_{x-1}^{x+1} e^{x+y}\,dy\,dx = \int_0^1 (e^{2x+1}-e^{2x-1})\,dx = \frac{1}{2}(e^3 - 2e + e^{-1}).$

55. $\displaystyle\iint_R f(x)g(y)\,dx\,dy = \int_c^d \int_a^b f(x)g(y)\,dx\,dy = \int_c^d \left(\int_a^b f(x)g(y)\,dx\right) dy$

$\displaystyle = \int_c^d g(y)\left(\int_a^b f(x)\,dx\right) dy = \left(\int_a^b f(x)\,dx\right)\left(\int_c^d g(y)\,dy\right)$

56. Symmetry about the origin [we want Ω to contain $(-x,-y)$ whenever it contains (x,y)].

57. Note that $\Omega = \{(x,y): 0 \le x \le y,\ \ 0 \le y \le 1\}.$

Set $\Omega' = \{(x,y): 0 \le y \le x,\ \ 0 \le x \le 1\}.$

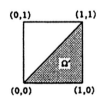

$\displaystyle\iint_\Omega f(x)f(y)\,dx\,dy = \int_0^1 \int_0^y f(x)f(y)\,dx\,dy$

$\displaystyle = \int_0^1 \int_0^x f(y)f(x)\,dy\,dx$

x and y are dummy variables

$\displaystyle = \int_0^1 \int_0^x f(x)f(y)\,dy\,dx = \iint_{\Omega'} f(x)f(y)\,dx\,dy.$

Note that Ω and Ω' don't overlap and their union is the unit square

$R = \{(x,y): 0 \le x \le 1,\ \ 0 \le y \le 1\}.$

If $\displaystyle\int_0^1 f(x)\,dx = 0,$ then

$$0 = \left(\int_0^1 f(x)\,dx\right)\left(\int_0^1 f(y)\,dy\right) = \iint_R f(x)f(y)\,dx dy$$

by Exercise 55

$$= \iint_\Omega f(x)f(y)\,dx dy + \iint_{\Omega'} f(x)f(y)\,dx dy$$

$$= 2\iint_\Omega f(x)f(y)\,dx dy$$

and therefore $\displaystyle\iint_\Omega f(x)f(y)\,dx dy = 0.$

58.
$$\int_0^x \int_a^b \frac{\partial f}{\partial x}(t,y)\,dy\,dt = \int_a^b \int_0^x \frac{\partial f}{\partial x}(t,y)\,dt\,dy$$

$$= \int_a^b [f(x,y) - f(0,y)]\,dy$$

$$= \int_a^b f(x,y)\,dy - \int_a^b f(0,y)\,dy.$$

Thus
$$\int_a^b f(x,y)\,dy = \int_0^x \int_a^b \frac{\partial f}{\partial x}(t,y)\,dy\,dt + \int_a^b f(0,y)\,dy$$

and

$$\frac{d}{dx}\left[\int_a^b f(x,y)\,dy\right] = \frac{d}{dx}\left[\int_0^x \int_a^b \frac{\partial f}{\partial x}(t,y)\,dy\,dt\right] + \frac{d}{dx}\left[\int_a^b f(0,y)\,dy\right]$$

$$= \frac{d}{dx}\left[\int_0^x H(t)\,dt\right] + 0 = H(x) = \int_a^b \frac{\partial f}{\partial x}(x,y)\,dy.$$

59. Let M be the maximum value of $|f(x,y)|$ on Ω.

$$\int_{\phi_1(x+h)}^{\phi_2(x+h)} = \int_{\phi_1(x+h)}^{\phi_1(x)} + \int_{\phi_1(x)}^{\phi_2(x)} + \int_{\phi_2(x)}^{\phi_2(x+h)}$$

$$|F(x+h) - F(x)| = \left| \int_{\phi_1(x+h)}^{\phi_2(x+h)} f(x,y)\,dy - \int_{\phi_1(x)}^{\phi_2(x)} f(x,y)\,dy \right|$$

$$= \left| \int_{\phi_1(x+h)}^{\phi_1(x)} f(x,y)\,dy + \int_{\phi_2(x)}^{\phi_2(x+h)} f(x,y)\,dy \right|$$

$$\leq \left| \int_{\phi_1(x+h)}^{\phi_1(x)} f(x,y)\,dy \right| + \left| \int_{\phi_2(x)}^{\phi_2(x+h)} f(x,y)\,dy \right|$$

$$\leq |\phi_1(x) - \phi_1(x+h)|\,M + |\phi_2(x+h) - \phi_2(x)|\,M.$$

The expression on the right tends to 0 as h tends to 0 since ϕ_1 and ϕ_2 are continuous.

60. (a) $\displaystyle\int_{-1}^{3}\int_{2}^{5} xe^{-xy}\, dy\, dx \cong -25.9893$ (b) $\displaystyle\int_{3}^{7}\int_{1}^{4} \frac{xy}{x^2+y^2}\, dy\, dx \cong 4.5720$

61. (a) $\displaystyle\int_{1}^{2}\int_{x^2-2x+2}^{1+\sqrt{x-1}} 1\, dy\, dx = \tfrac{1}{3}$ (b) $\displaystyle\int_{1}^{2}\int_{y^2-2y+2}^{1+\sqrt{y-1}} 1\, dx\, dy = \tfrac{1}{3}$

62. (a) $\displaystyle\int_{0}^{1}\int_{2y}^{3-y} \sqrt{2x+y}\, dx\, dy \cong 2.8133$ (b) $\displaystyle\int_{0}^{2}\int_{0}^{x/2} \sqrt{2x+y}\, dy\, dx + \int_{2}^{3}\int_{0}^{3-x} \sqrt{2x+y}\, dy\, dx$

$$\cong 2.8133$$

SECTION 17.4

1. $\displaystyle\int_{0}^{\pi/2}\int_{0}^{\sin\theta} r\cos\theta\, dr\, d\theta = \int_{0}^{\pi/2} \frac{1}{2}\sin^2\theta\cos\theta\, d\theta = \left[\frac{1}{6}\sin^3\theta\right]_{0}^{\pi/2} = \frac{1}{6}$

2. $\displaystyle\int_{0}^{\pi/4}\int_{0}^{\cos 2\theta} r\, dr\, d\theta = \int_{0}^{\pi/4}\frac{\cos^2 2\theta}{2}\, d\theta = \frac{\pi}{16}$

3. $\displaystyle\int_{0}^{\pi/2}\int_{0}^{3\sin\theta} r^2\, dr\, d\theta = \int_{0}^{\pi/2} 9\sin^3\theta\, d\theta = 9\int_{0}^{\pi/2}(1-\cos^2\theta)\sin\theta\, d\theta = 9\left[-\cos\theta + \frac{1}{3}\cos^3\theta\right]_{0}^{\pi/2} = 6$

4. $\displaystyle\int_{-\pi/3}^{2\pi/3}\int_{0}^{2\cos\theta} r\sin\theta\, dr\, d\theta = \int_{-\pi/3}^{2\pi/3} 2\cos^2\theta\sin\theta\, d\theta = \frac{1}{6}$

5. (a) $\Gamma: \ 0 \le \theta \le 2\pi, \quad 0 \le r \le 1$

$$\iint_{\Gamma} (\cos r^2)r\, dr d\theta = \int_{0}^{2\pi}\int_{0}^{1}(\cos r^2)r\, dr\, d\theta = 2\pi\int_{0}^{1} r\cos r^2\, dr = \pi\sin 1$$

(b) $\Gamma: \ 0 \le \theta \le 2\pi, \quad 1 \le r \le 2$

$$\iint_{\Gamma} (\cos r^2)r\, dr d\theta = \int_{0}^{2\pi}\int_{1}^{2}(\cos r^2)r\, dr\, d\theta = 2\pi\int_{1}^{2} r\cos r^2\, dr = \pi(\sin 4 - \sin 1)$$

6. (a) $\Gamma: \ 0 \le \theta \le 2\pi, \quad 0 \le r \le 1.$

$$\iint_{\Gamma} (\sin r)r\, dr\, d\theta = \int_{0}^{2\pi}\int_{0}^{1}(\sin r)r\, dr\, d\theta = 2\pi\int_{0}^{1} r\sin r\, dr = 2\pi(\sin 1 - \cos 1)$$

(b) $\Gamma: \ 0 \le \theta \le 2\pi, \quad 1 \le r \le 2.$

$$\iint_{\Gamma} (\sin r)r\, dr\, d\theta = \int_{0}^{2\pi}\int_{1}^{2}(\sin r)r\, dr\, d\theta = 2\pi\int_{1}^{2} r\sin r\, dr = 2\pi[\cos 1 - 2\cos 2 + \sin 2 - \sin 1]$$

7. (a) $\Gamma: \ 0 \le \theta \le \pi/2, \quad 0 \le r \le 1$

$$\iint_{\Gamma} (r\cos\theta + r\sin\theta)r\, dr d\theta = \int_{0}^{\pi/2}\int_{0}^{1} r^2(\cos\theta + \sin\theta)\, dr\, d\theta$$

$$= \left(\int_{0}^{\pi/2}(\cos\theta + \sin\theta)\, d\theta\right)\left(\int_{0}^{1} r^2\, dr\right) = 2\left(\frac{1}{3}\right) = \frac{2}{3}$$

(b) $\Gamma:\ 0 \le \theta \le \pi/2,\quad 1 \le r \le 2$

$$\iint_{\Gamma} (r\cos\theta + r\sin\theta) r\,dr d\theta = \int_0^{\pi/2} \int_1^2 r^2 (\cos\theta + \sin\theta)\,dr\,d\theta$$

$$= \left(\int_0^{\pi/2} (\cos\theta + \sin\theta)\,d\theta \right) \left(\int_1^2 r^2\,dr \right) = 2\left(\frac{7}{3}\right) = \frac{14}{3}$$

8. $\Gamma:\ 0 \le \theta \le \frac{\pi}{3},\quad 0 \le r \le \frac{1}{\cos\theta}$

$$\iint_{\Gamma} \sqrt{x^2 + y^2}\,dx\,dy = \int_0^{\pi/3} \int_0^{1/\cos\theta} r\cdot r\,dr\,d\theta = \int_0^{\pi/3} \frac{d\theta}{3\cos^3\theta} = \frac{1}{3}\int_0^{\pi/3} \sec^3\theta\,d\theta$$

$$= \frac{1}{3}\left[\frac{1}{2}\sec\theta\tan\theta + \frac{1}{2}\ln|\sec\theta + \tan\theta| \right]_0^{\pi/3} = \frac{1}{3}\sqrt{3} + \frac{1}{6}\ln(2 + \sqrt{3})$$

9. $\displaystyle \int_{-\pi/2}^{\pi/2} \int_0^1 r^2\,dr\,d\theta = \frac{1}{3}\pi$ 10. $\displaystyle \int_0^{\pi/2} \int_0^2 r^2\,dr\,d\theta = \frac{\pi}{2}\int_0^2 r^2\,dr = \frac{4}{3}\pi$

11. $\displaystyle \int_{1/2}^1 \int_0^{\sqrt{1-x^2}} dy\,dx = \int_0^{\pi/3} \int_{\frac{1}{2}\sec\theta}^1 r\,dr\,d\theta = \int_0^{\pi/3} \left(\frac{1}{2} - \frac{1}{8}\sec^2\theta \right) d\theta = \frac{1}{6}\pi - \frac{\sqrt{3}}{8}$

12. $\displaystyle \int_0^{\pi/3} \int_0^{1/2\sec\theta} r^4 \cos\theta\sin\theta\,dr\,d\theta + \int_{\pi/3}^{\pi/2} \int_0^1 r^4 \cos\theta\sin\theta\,dr\,d\theta$

$$= \int_0^{\pi/3} \frac{\sin\theta}{5(32)\cos^4\theta}\,d\theta + \int_{\pi/3}^{\pi/2} \frac{1}{5}\cos\theta\sin\theta\,d\theta = \frac{7}{480} + \frac{1}{40} = \frac{19}{480}$$

13. $\displaystyle \int_0^1 \int_0^{\sqrt{1-x^2}} \sin\sqrt{x^2+y^2}\,dy\,dx = \int_0^{\pi/2} \int_0^1 \sin(r)\,r\,dr\,d\theta = \int_0^{\pi/2} (\sin 1 - \cos 1)\,d\theta = \frac{\pi}{2}(\sin 1 - \cos 1)$

14. $\displaystyle \int_0^{2\pi} \int_0^1 e^{-r^2} r\,dr\,d\theta = 2\pi \int_0^1 re^{-r^2}\,dr = \pi\left(1 - \frac{1}{e}\right)$

15. $\displaystyle \int_0^2 \int_0^{\sqrt{2x-x^2}} x\,dy\,dx = \int_0^{\pi/2} \int_0^{2\cos\theta} r\cos\theta\,r\,dr\,d\theta = \frac{8}{3}\int_0^{\pi/2} \cos^4\theta\,d\theta = \frac{8}{3}\cdot\frac{3}{4}\cdot\frac{1}{2}\cdot\frac{\pi}{2} = \frac{\pi}{2}$

(See Exercise 62, Section 8.3)

16. The region is the inside of the circle $(x - 1/2)^2 + y^2 = 1/4$, which has polar equation $r = \cos\theta$.
So the integral becomes

$$\int_{-\pi/2}^{\pi/2} \int_0^{\cos\theta} r^3\,dr\,d\theta = \int_{-\pi/2}^{\pi/2} \frac{1}{4}\cos^4\theta\,d\theta = \frac{3\pi}{32}$$

17. $A = \displaystyle \int_0^{\pi/3} \int_0^{3\sin 3\theta} r\,dr\,d\theta = \frac{9}{2}\int_0^{\pi/3} \sin^2 3\theta\,d\theta = \frac{9}{4}\int_0^{\pi/3} (1 - 6\cos\theta)\,d\theta = \frac{3\pi}{4}$

18. $A = \displaystyle\int_0^{2\pi}\int_0^{2(1-\cos\theta)} r\,dr\,d\theta = \int_0^{2\pi} 2(1-\cos\theta)^2\,d\theta = 6\pi$

19. First we find the points of intersection:

$r = 4\cos\theta = 2 \implies \cos\theta = \dfrac{1}{2}$

$\implies \theta = \pm\dfrac{\pi}{3}.$

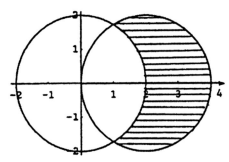

$A = \displaystyle\int_{-\pi/3}^{\pi/3}\int_2^{4\cos\theta} r\,dr\,d\theta = \int_{-\pi/3}^{\pi/3}(8\cos^2\theta - 2)\,d\theta = \int_{-\pi/3}^{\pi/3}(2 + 4\cos 2\theta)\,d\theta = \dfrac{4\pi}{3} + 2\sqrt{3}$

20. $A = \displaystyle\int_0^{2\pi}\int_0^{1+2\cos\theta} r\,dr\,d\theta = \int_0^{2\pi}\frac{1}{2}(1+2\cos\theta)^2\,d\theta = 3\pi$

21. $A = 4\displaystyle\int_0^{\pi/4}\int_0^{2\sqrt{\cos 2\theta}} r\,dr\,d\theta = 8\int_0^{\pi/4}\cos 2\theta\,d\theta = 4$

22. $A = \displaystyle\int_{-\pi/3}^{\pi/3}\int_{1+\cos\theta}^{3\cos\theta} r\,dr\,d\theta = \int_{-\pi/3}^{\pi/3}\frac{1}{2}\left[9\cos^2\theta - (1+\cos\theta)^2\right]d\theta = \left[\frac{3\theta}{2} + \sin 2\theta - \sin\theta\right]_{-\pi/3}^{\pi/3} = \pi$

23. $\displaystyle\int_0^{2\pi}\int_0^b (r^2\sin\theta + br)\,dr\,d\theta = \int_0^{2\pi}\left[\frac{1}{3}r^3\sin\theta + \frac{b}{2}r^2\right]_0^b d\theta$

$\qquad\qquad = b^3\displaystyle\int_0^{2\pi}\left(\frac{1}{3}\sin\theta + \frac{1}{2}\right)d\theta = b^3\pi$

24. $V = \displaystyle\int_0^{2\pi}\int_0^1 (1-r^2)r\,dr\,d\theta = 2\pi\int_0^1(r - r^3)\,dr = \dfrac{\pi}{2}$

25. $8\displaystyle\int_0^{\pi/2}\int_0^2 \frac{r}{2}\sqrt{12 - 3r^2}\,dr\,d\theta = 8\int_0^{\pi/2}\left[-\frac{1}{18}(12 - 3r^2)^{3/2}\right]_0^2 d\theta$

$\qquad\qquad = 8\displaystyle\int_0^{\pi/2}\frac{4}{3}\sqrt{3}\,d\theta = \dfrac{16}{3}\sqrt{3}\,\pi$

26. $V = \displaystyle\int_0^{2\pi}\int_0^{\sqrt{5}} \sqrt[6]{5 - r^2}\;r\,dr\,d\theta = 2\pi\int_0^{\sqrt{5}} r\sqrt[6]{5 - r^2}\,dr = \dfrac{30}{7}(5)^{1/6}\pi$

27. $\displaystyle\int_0^{2\pi}\int_0^1 r\sqrt{4 - r^2}\,dr\,d\theta = \int_0^{2\pi}\left[-\frac{1}{3}(4 - r^2)^{3/2}\right]_0^1 d\theta$

$\qquad\qquad = \displaystyle\int_0^{2\pi}\left(\frac{8}{3} - \sqrt{3}\right)d\theta = \dfrac{2}{3}(8 - 3\sqrt{3})\pi$

28. $V = \displaystyle\int_{-\pi/2}^{\pi/2}\int_0^{\cos\theta}(1 - r^2)r\,dr\,d\theta = \int_{-\pi/2}^{\pi/2}\left(\frac{\cos^2\theta}{2} - \frac{\cos^4\theta}{4}\right)d\theta = \dfrac{5\pi}{32}$

29. $\displaystyle\int_{-\pi/2}^{\pi/2}\int_0^{2\cos\theta} 2r^2\cos\theta\,dr\,d\theta = \int_{-\pi/2}^{\pi/2}\left[\frac{2}{3}r^3\cos\theta\right]_0^{2\cos\theta} d\theta$

$$= \int_{-\pi/2}^{\pi/2}\frac{16}{3}\cos^4\theta\,d\theta = \frac{32}{3}\int_0^{\pi/2}\cos^4\theta\,d\theta = \frac{32}{3}\left(\frac{3}{16}\pi\right) = 2\pi$$

Ex. 46, Sect. 8.3

30. $\displaystyle\int_{-\pi/2}^{\pi/2}\int_0^{2a\cos\theta} r^2\,dr\,d\theta = \int_{-\pi/2}^{\pi/2}\frac{8a^3}{3}\cos^3\theta\,d\theta = \frac{32}{9}a^3$

31. $\displaystyle\frac{b}{a}\int_0^\pi\int_0^{a\sin\theta} r\sqrt{a^2-r^2}\,dr\,d\theta = \frac{b}{a}\int_0^\pi\left[-\frac{1}{3}\left(a^2-r^2\right)^{3/2}\right]_0^{a\sin\theta} d\theta$

$$= \frac{1}{3}a^2b\int_0^\pi\left(1-\cos^3\theta\right)d\theta = \frac{1}{3}\pi a^2 b$$

32. (a) $V = 2\displaystyle\int_0^{2\pi}\int_0^r \sqrt{R^2-s^2}\,s\,ds\,d\theta = 4\pi\int_0^r S\sqrt{R^2-S^2}\,dS = \frac{4\pi}{3}\left[R^3-(R^2-r^2)^{3/2}\right].$

(b) $\displaystyle\frac{4}{3}\pi R^3 - V = \frac{4}{3}\pi(R^2-r^2)^{3/2}$

33. $A = 2\displaystyle\int_0^{\pi/4}\int_0^{2\cos 2\theta} r\,dr\,d\theta = 2\int_0^{\pi/4}\frac{1}{2}(2\cos 2\theta)^2\,d\theta = 2\int_0^{\pi/4}(1+\cos 4\theta)\,d\theta = \frac{\pi}{2}$

34. $\displaystyle\iint_\Omega e^{x^2+y^2}\,dx\,dy = \int_0^{2\pi}\int_2^4 re^{r^2}\,dr\,d\theta = \pi\left(e^{16}-e^4\right)$

SECTION 17.5

1. $M = \displaystyle\int_{-1}^1\int_0^1 x^2\,dy\,dx = \frac{2}{3}$

$x_M M = \displaystyle\int_{-1}^1\int_0^1 x^3\,dy\,dx = 0 \quad\Longrightarrow\quad x_M = 0$

$y_M M = \displaystyle\int_{-1}^1\int_0^1 x^2 y\,dy\,dx = \int_{-1}^1\frac{1}{2}x^2\,dx = \frac{1}{3} \quad\Longrightarrow\quad y_M = \frac{1/3}{2/3} = \frac{1}{2}$

2. $M = \displaystyle\int_0^1\int_0^{\sqrt{x}}(x+y)\,dy\,dx = \int_0^1\left(x^{3/2}+\frac{x}{2}\right)dx = \frac{13}{20}$

$x_M M = \displaystyle\int_0^1\int_0^{\sqrt{x}} x(x+y)\,dy\,dx = \int_0^1\left(x^{5/2}+\frac{x^2}{2}\right)dx = \frac{19}{42} \quad\Longrightarrow\quad x_M = \frac{190}{273}$

$y_M M = \displaystyle\int_0^1\int_0^{\sqrt{x}} y(x+y)\,dy\,dx = \int_0^1\left(\frac{x^2}{2}+\frac{x^{3/2}}{3}\right)dx = \frac{3}{10} \quad\Longrightarrow\quad y_M = \frac{6}{13}$

3. $M = \int_0^1 \int_{x^2}^1 xy \, dy \, dx = \frac{1}{2} \int_0^1 (x - x^5) \, dx = \frac{1}{6}$

$x_M M = \int_0^1 \int_{x^2}^1 x^2 y \, dy \, dx = \frac{1}{2} \int_0^1 (x^2 - x^6) \, dx = \frac{2}{21} \implies x_M = \frac{2/21}{1/6} = \frac{4}{7}$

$y_M M = \int_0^1 \int_{x^2}^1 xy^2 \, dy \, dx = \frac{1}{3} \int_0^1 (x - x^7) \, dx = \frac{1}{8} \implies y_M = \frac{1/8}{1/6} = \frac{3}{4}$

4. $M = \int_0^\pi \int_0^{\sin x} y \, dy \, dx = \int_0^\pi \frac{\sin^2 x}{2} \, dx = \frac{\pi}{4}$

$x_M M = \int_0^\pi \int_0^{\sin x} xy \, dy \, dx = \int_0^\pi x \frac{\sin^2 x}{2} \, dx = \frac{\pi^2}{8} \implies x_M = \frac{\pi}{2}$

$y_M M = \int_0^\pi \int_0^{\sin x} y^2 \, dy \, dx = \int_0^\pi \frac{\sin^3 x}{3} \, dx = \frac{4}{9} \implies y_M = \frac{16}{9\pi}$

5. $M = \int_0^8 \int_0^{x^{1/3}} y^2 \, dy \, dx = \frac{1}{3} \int_0^8 x \, dx = \frac{32}{3}$

$x_M M = \int_0^8 \int_0^{x^{1/3}} xy^2 \, dy \, dx = \frac{1}{3} \int_0^8 x^2 \, dx = \frac{512}{9} \implies x_M = \frac{512/9}{32/3} = \frac{16}{3}$

$y_M M = \int_0^8 \int_0^{x^{1/3}} y^3 \, dy \, dx = \frac{1}{4} \int_0^8 x^{4/3} \, dx = \frac{96}{7} \implies y_M = \frac{96/7}{32/3} = \frac{9}{7}$

6. $M = \int_0^a \int_0^{\sqrt{a^2 - x^2}} xy \, dy \, dx = \int_0^a \frac{x}{2}(a^2 - x^2) \, dx = \frac{a^4}{8}$

$x_M M = \int_0^a \int_0^{\sqrt{a^2 - x^2}} x^2 y \, dy \, dx = \int_0^a \frac{x^2}{2}(a^2 - x^2) \, dx = \frac{a^5}{15} \implies x_M = \frac{8}{15}a$

$y_M M = \int_0^a \int_0^{\sqrt{a^2 - x^2}} xy^2 \, dy \, dx = \int_0^a \frac{x}{3}(a^2 - x^2)^{3/2} \, dx = \frac{a^5}{15} \implies y_M = \frac{8}{15}a$

7. $M = \int_0^1 \int_{2x}^{3x} xy \, dy \, dx = \frac{5}{2} \int_0^1 x^3 \, dx = \frac{5}{8}$

$x_M M = \int_0^1 \int_{2x}^{3x} x^2 y \, dy \, dx = \frac{5}{2} \int_0^1 x^4 \, dx = \frac{1}{2} \implies x_M = \frac{1/2}{5/8} = \frac{4}{5}$

$y_M M = \int_0^1 \int_{2x}^{3x} xy^2 \, dy \, dx = \frac{19}{3} \int_0^1 x^4 \, dx = \frac{19}{15} \implies y_M = \frac{19/15}{5/8} = \frac{152}{75}$

8. $M = \int_0^2 \int_0^{3-\frac{3}{2}x} (x + y) \, dy \, dx = \int_0^2 \left[x(3 - \frac{3}{2}x) + \frac{1}{2}(3 - \frac{3}{2}x)^2 \right] dx = 5$

$x_M M = \int_0^2 \int_0^{3-\frac{3}{2}x} x(x + y) \, dy \, dx = \int_0^2 \left[x^2(3 - \frac{3}{2}x) + \frac{x}{2}(3 - \frac{3}{2}x)^2 \right] dx = \frac{7}{2} \implies x_M = \frac{7}{10}$

$y_M M = \int_0^2 \int_0^{3-\frac{3}{2}x} y(x + y) \, dy \, dx = \int_0^2 \left[\frac{x}{2}(3 - \frac{3}{2}x)^2 + \frac{(3 - \frac{3}{2}x)^3}{3} \right] dx = 6 \implies y_M = \frac{6}{5}$

9. $M = \displaystyle\int_0^{2\pi}\int_0^{1+\cos\theta} r^2\,dr\,d\theta = \frac{1}{3}\int_0^{2\pi}(1+3\cos\theta+3\cos^2\theta+\cos^3\theta)\,d\theta = \frac{5\pi}{3}$

$x_M\,M = \displaystyle\int_0^{2\pi}\int_0^{1+\cos\theta} r^3\cos\theta\,dr\,d\theta = \frac{1}{4}\int_0^{2\pi}(1+\cos\theta)^4\cos\theta\,d\theta$

$\qquad\qquad = \dfrac{1}{4}\displaystyle\int_0^{2\pi}\left[\cos\theta+4\cos^2\theta+6\cos^3\theta+4\cos^4\theta+\cos^5\theta\right]\,d\theta$

$\qquad\qquad = \dfrac{7\pi}{4}$

Therefore, $x_M = \dfrac{7\pi/4}{5\pi/3} = \dfrac{21}{20}$.

$y_M\,M = \displaystyle\int_0^{2\pi}\int_0^{1+\cos\theta} r^3\sin\theta\,dr\,d\theta = \frac{1}{4}\int_0^{2\pi}(1+\cos\theta)^4\sin\theta\,d\theta = \frac{1}{4}\left[\frac{1}{5}(1+\cos\theta)^5\right]_0^{2\pi} = 0$

Therefore, $y_M = 0$.

10. $M = \displaystyle\iint_\Omega y\,dx\,dy = \int_{\pi/6}^{5\pi/6}\int_1^{2\sin\theta} r\sin\theta\,r\,dr\,d\theta = \int_{\pi/6}^{5\pi/6}\left(\frac{8}{3}\sin^4\theta - \frac{1}{3}\sin\theta\right)d\theta = \frac{3\sqrt{3}+8\pi}{12}$

$x_M = 0 \quad$ by symmetry

$y_M\,M = \displaystyle\int_{\pi/6}^{5\pi/6}\int_1^{2\sin\theta} r^2\sin^2\theta\,r\,dr\,d\theta = \int_{\pi/6}^{5\pi/6}\left(4\sin^6\theta - \frac{1}{4}\sin^2\theta\right)d\theta = \frac{11\sqrt{3}+12\pi}{16}$

Therefore, $y_M = \dfrac{33\sqrt{3}+36\pi}{12\sqrt{3}+32\pi}$

11. $\Omega: \ -L/2 \le x \le L/2, \quad -W/2 \le y \le W/2$

$I_x = \displaystyle\iint_\Omega \frac{M}{LW}y^2\,dx\,dy = \frac{4M}{LW}\int_0^{W/2}\int_0^{L/2} y^2\,dx\,dy = \frac{1}{12}MW^2$

symmetry $\rightarrow\!\uparrow$

$I_y = \displaystyle\iint_\Omega \frac{M}{LW}x^2\,dx\,dy = \frac{1}{12}ML^2, \quad I_z = \iint_\Omega \frac{M}{LW}(x^2+y^2)\,dx\,dy = \frac{1}{12}M(L^2+W^2)$

$K_x = \sqrt{I_x/M} = \dfrac{W\sqrt{3}}{6}, \quad K_y = \sqrt{I_y/M} = \dfrac{L\sqrt{3}}{6}$

$K_z = \sqrt{I_z/M} = \dfrac{\sqrt{3}\,\sqrt{L^2+W^2}}{6}$

12. $\lambda(x,y) = k\left(x+\dfrac{L}{2}\right) \qquad M = \displaystyle\int_{-W/2}^{W/2}\int_{-L/2}^{L/2} k\left(x+\frac{L}{2}\right)dx\,dy = \frac{kWL^2}{2}$

$I_x = \displaystyle\int_{-W/2}^{W/2}\int_{-L/2}^{L/2} k\left(x+\frac{L}{2}\right)y^2\,dx\,dy = \left(\int_{-W/2}^{W/2} y^2\,dy\right)\left(\int_{-L/2}^{L/2} k\left(x+\frac{L}{2}\right)dx\right)$

$\qquad = \left(\dfrac{W^3}{12}\right)\left(\dfrac{M}{W}\right) = \dfrac{1}{12}MW^2$

$$I_y = \int_{-W/2}^{W/2} \int_{-L/2}^{L/2} k\left(x + \frac{L}{2}\right) x^2 \, dx \, dy = \frac{kL^4W}{24} = \frac{1}{12}ML^2$$

$$I_z = \int_{-W/2}^{W/2} \int_{-L/2}^{L/2} k\left(x + \frac{L}{2}\right)(x^2 + y^2) \, dx \, dy = I_x + I_y = \frac{1}{12}M(L^2 + W^2).$$

13. $\quad M = \iint_\Omega k\left(x + \frac{L}{2}\right) dx dy = \iint_\Omega \frac{1}{2}kL \, dx dy = \frac{1}{2}kL(\text{ area of } \Omega) = \frac{1}{2}kL^2W$

$$\text{symmetry}\underline{}\uparrow$$

$$x_M M = \iint_\Omega x\left[k\left(x + \frac{L}{2}\right)\right] dx dy = \iint_\Omega \left(kx^2 + \frac{1}{2}Lx\right) dx dy$$

$$= \iint_\Omega kx^2 \, dx dy = 4k \int_0^{W/2} \int_0^{L/2} x^2 \, dx \, dy = \frac{1}{12}kWL^3$$

$$\text{symmetry}\underline{}\uparrow \qquad \text{symmetry}\underline{}\uparrow$$

$$= \tfrac{1}{6}\left(\tfrac{1}{2}kL^2W\right) L = \tfrac{1}{6} ML; \quad x_M = \tfrac{1}{6}L$$

$$y_M M = \iint_\Omega y\left[k\left(x + \frac{L}{2}\right)\right] dx dy = 0; \quad y_M = 0$$

$$\text{by symmetry}\underline{}\uparrow$$

14. $\quad I_z = \iint_\Omega \lambda(x, y)[x^2 + y^2] \, dx \, dy = \iint_\Omega \lambda(x, y)x^2 \, dx \, dy + \iint_\Omega \lambda(x, y)y^2 \, dx \, dy = I_x + I_y.$

Since $\quad I_z = I_x + I_y, \quad$ we have $\quad MK_z^2 = MK_x^2 + MK_y^2 \quad$ therefore $\quad K_z^2 = K_x^2 + K_y^2.$

15. $\quad I_x = \iint_\Omega \frac{4M}{\pi R^2} y^2 \, dx dy = \frac{4M}{\pi R^2} \int_0^{\pi/2} \int_0^R r^3 \sin^2\theta \, dr \, d\theta$

$$= \frac{4M}{\pi R^2}\left(\int_0^{\pi/2} \sin^2\theta \, d\theta\right)\left(\int_0^R r^3 \, dr\right) = \frac{4M}{\pi R^2}\left(\frac{\pi}{4}\right)\left(\frac{1}{4}R^4\right) = \frac{1}{4}MR^2$$

$$I_y = \tfrac{1}{4}MR^2, \quad I_z = \tfrac{1}{2}MR^2$$

$$K_x = K_y = \tfrac{1}{2}R, \quad K_z = R/\sqrt{2}$$

16. $\quad I_z = I_M + d^2M.$ Rotation doesn't change d, doesn't change M, and doesn't change I_M.

17. I_M, the moment of inertia about the vertical line through the center of mass, is

$$\iint_\Omega \frac{M}{\pi R^2} \left(x^2 + y^2\right) dx\, dy$$

where Ω is the disc of radius R centered at the origin. Therefore

$$I_M = \frac{M}{\pi R^2} \int_0^{2\pi} \int_0^R r^3\, dr\, d\theta = \frac{1}{2} MR^2.$$

We need $I_0 = \frac{1}{2} MR^2 + d^2 M$ where d is the distance from the center of the disc to the origin. Solving this equation for d, we have $d = \sqrt{I_0 - \frac{1}{2} MR^2} \Big/ \sqrt{M}$.

18. $I_x = \displaystyle\int_a^b \int_0^{f(x)} \lambda y^2\, dy\, dx = \frac{\lambda}{3} \int_a^b [f(x)]^3\, dx$

$I_y = \displaystyle\int_a^b \int_0^{f(x)} \lambda x^2\, dy\, dx = \lambda \int_a^b x^2 f(x)\, dx.$

19. $\Omega : 0 \le x \le a, \quad 0 \le y \le b$

$$I_x = \iint_\Omega \frac{4M}{\pi ab} y^2\, dx\, dy = \frac{4M}{\pi ab} \int_0^a \int_0^{\frac{b}{a}\sqrt{a^2 - x^2}} y^2\, dy\, dx = \frac{1}{4} Mb^2$$

$$I_y = \iint_\Omega \frac{4M}{\pi ab} x^2\, dx\, dy = \frac{4M}{\pi ab} \int_0^a \int_0^{\frac{b}{a}\sqrt{a^2 - x^2}} x^2\, dy\, dx = \frac{1}{4} Ma^2$$

$I_z = \frac{1}{4} M \left(a^2 + b^2\right)$

20. $I_x = \displaystyle\int_0^1 \int_0^{\sqrt{x}} (x + y) y^2\, dy\, dx = \int_0^1 \left(\frac{x^{5/2}}{3} + \frac{x^2}{4}\right) dx = \frac{5}{28}$

$I_y = \displaystyle\int_0^1 \int_0^{\sqrt{x}} (x + y) x^2\, dy\, dx = \int_0^1 \left(x^{7/2} + \frac{x^3}{2}\right) dx = \frac{25}{72}; \qquad I_z = I_x + I_y.$

21. $I_x = \displaystyle\int_0^1 \int_{x^2}^1 xy^3\, dy\, dx = \frac{1}{4} \int_0^1 (x - x^9)\, dx = \frac{1}{10}$

$I_y = \displaystyle\int_0^1 \int_{x^2}^1 x^3 y\, dy\, dx = \frac{1}{2} \int_0^1 (x^3 - x^7)\, dx = \frac{1}{16}$

$I_z = \displaystyle\int_0^1 \int_{x^2}^1 xy(x^2 + y^2)\, dy\, dx = I_x + I_y = \frac{13}{80}$

22. $I_x = \displaystyle\int_0^8 \int_0^{\sqrt[3]{x}} y^2 \cdot y^2\, dy\, dx = \int_0^8 \frac{x^{5/3}}{5}\, dx = \frac{96}{5}$

$I_y = \displaystyle\int_0^8 \int_0^{\sqrt[3]{x}} y^2 \cdot x^2\, dy\, dx = \int_0^8 \frac{x^3}{3}\, dx = \frac{1024}{3}; \qquad I_z = I_x + I_y$

23. $I_x = \int_0^{2\pi} \int_0^{1+\cos\theta} r^4 \sin^2\theta \, dr \, d\theta = \frac{1}{5} \int_0^{2\pi} (1+\cos\theta)^5 \sin^2\theta \, d\theta = \frac{33\pi}{40}$

$I_y = \int_0^{2\pi} \int_0^{1+\cos\theta} r^4 \cos^2\theta \, dr \, d\theta = \frac{1}{5} \int_0^{2\pi} (1+\cos\theta)^5 \cos^2\theta \, d\theta = \frac{93\pi}{40}$

$I_z = \int_0^{2\pi} \int_0^{1+\cos\theta} r^4 \, dr \, d\theta = I_x + I_y = \frac{63\pi}{20}$

24. $x_M = \dfrac{x_1 M_1 + x_2 M_2}{M_1 + M_2}, \quad y_M = \dfrac{y_1 M_1 + y_2 M_2}{M_1 + M_2}$

25. $\Omega: \ r_1^2 \le x^2 + y^2 \le r_2^2, \quad A = \pi\left(r_2^2 - r_1^2\right)$

   (a) Place the diameter on the x-axis.

$$I_x = \iint_\Omega \frac{M}{A} y^2 \, dx\,dy = \frac{M}{A} \int_0^{2\pi} \int_{r_1}^{r_2} \left(r^2 \sin^2\theta\right) r \, dr \, d\theta = \frac{1}{4} M\left(r_2^2 + r_1^2\right)$$

   (b) $\frac{1}{4}M\left(r_2^2 + r_1^2\right) + Mr_1^2 = \frac{1}{4}M\left(r_2^2 + 5r_1^2\right)$ (parallel axis theorem)

   (c) $\frac{1}{4}M\left(r_2^2 + r_1^2\right) + Mr_2^2 = \frac{1}{4}M\left(5r_2^2 + r_1^2\right)$

26. Set $r_1 = r_2 = r$ in the proceeding problem. Then the required moments of inertia are

   (a) $\frac{1}{2}Mr^2$ (b) $\frac{3}{2}Mr^2.$

27. $\Omega: \ r_1^2 \le x^2 + y^2 \le r_2^2, \quad A = \pi\left(r_2^2 - r_1^2\right)$

$$I = \iint_\Omega \frac{M}{A}\left(x^2 + y^2\right) dx\,dy = \frac{M}{A} \int_0^{2\pi} \int_{r_1}^{r_2} r^3 \, dr \, d\theta = \frac{1}{2}M(r_2^2 + r_1^2)$$

28. Let l be the x-axis and let the plane of the plate be the xy-plane. Then

$$I - I_M = \iint_\Omega \lambda(x,y)y^2 \, dx\,dy - \iint_\Omega \lambda(x,y)(y - y_M)^2 \, dx\,dy$$

$$= \iint_\Omega \lambda(x,y)[2y_M y - y_M^2] \, dx\,dy$$

$$= 2y_M \iint_\Omega y\lambda(x,y)\,dx\,dy - y_M^2 \iint_\Omega \lambda(x,y)\,dx\,dy$$

$$= 2y_M^2 M - y_M^2 M = y_M^2 M = d^2 M.$$

29. $M = \iint_\Omega k\left(R - \sqrt{x^2 + y^2}\right) dx\,dy = k \int_0^\pi \int_0^R (Rr - r^2)\, dr\, d\theta = \frac{1}{6}k\pi R^3$

   $x_M = 0$ by symmetry

   $y_M M = \iint_\Omega y\left[k\left(R - \sqrt{x^2 + y^2}\right)\right] dx\,dy = k \int_0^\pi \int_0^R (Rr^2 - r^3)\sin\theta \, dr\, d\theta = \frac{1}{6}kR^4$

   $y_M = R/\pi$

30. $I_x = \displaystyle\iint\limits_{\Omega} k(R - \sqrt{x^2+y^2})y^2\,dx\,dy = k\int_0^\pi \int_0^R (R-r)\,r^2\sin^2\theta\,r\,dr\,d\theta = \dfrac{k\pi R^5}{40} = \dfrac{3MR^2}{20}.$

$I_y = k\displaystyle\int_0^\pi \int_0^R (R-r)r^2\cos^2\theta\,r\,dr\,d\theta = \dfrac{k\pi R^5}{40} = \dfrac{3MR^2}{20}$

$I_z = I_x + I_y = \dfrac{3MR^2}{10}.$

31. Place P at the origin.

$$M = \iint\limits_{\Omega} k\sqrt{x^2+y^2}\,dx\,dy$$

$$= k\int_0^\pi \int_0^{2R\sin\theta} r^2\,dr\,d\theta = \dfrac{32}{9}kR^3$$

$x_M = 0 \quad$ by symmetry

$$y_M M = \iint\limits_{\Omega} y\left(k\sqrt{x^2+y^2}\right)dx\,dy = k\int_0^\pi \int_0^{2R\sin\theta} r^3\sin\theta\,dr\,d\theta = \dfrac{64}{15}kR^4$$

$$y_M = 6R/5$$

Answer: the center of mass lies on the diameter through P at a distance $6R/5$ from P.

32. Putting the right angle at the origin, we have $\lambda(x,y) = k(x^2+y^2).$

$M = \displaystyle\int_0^b \int_0^{h-\frac{h}{b}x} k(x^2+y^2)\,dy\,dx = \dfrac{1}{12}kbh(b^2+h^2)$

$x_M M = \displaystyle\int_0^b \int_0^{h-\frac{h}{b}x} kx(x^2+y^2)\,dy\,dx = \dfrac{kb^2h(3b^2+h^2)}{60} \implies x_M = \dfrac{b(3b^2+h^2)}{5(b^2+h^2)}$

$y_M M = \displaystyle\int_0^b \int_0^{h-\frac{h}{b}x} ky(x^2+y^2)\,dy\,dx = \dfrac{kbh^2(b^2+3h^2)}{60} \implies y_M = \dfrac{h(b^2+3h^2)}{5(b^2+h^2)}$

33. Suppose Ω, a basic region of area A, is broken up into n basic regions $\Omega_1, \cdots, \Omega_n$ with areas $A_1, \cdots, A_n$. Then

$$\overline{x}A = \iint\limits_{\Omega} x\,dx\,dy = \sum_{i=1}^n \left(\iint\limits_{\Omega_i} x\,dx\,dy\right) = \sum_{i=1}^n \overline{x}_i\,A_i = \overline{x}_1\,A_1 + \cdots + \overline{x}_n\,A_n.$$

The second formula can be derived in a similar manner.

34. (a) $M = \displaystyle\int_0^2 \int_{x/2}^1 (x+y)\,dy\,dx = \dfrac{4}{3}$

$x_M M = \displaystyle\int_0^2 \int_{x/2}^1 x(x+y)\,dy\,dx = \dfrac{7}{6}; \qquad x_M = \dfrac{7}{8}$

$y_M M = \displaystyle\int_0^2 \int_{x/2}^1 y(x+y)\,dy\,dx = 1; \qquad y_M = \dfrac{3}{4}$

(b) $I_x = \displaystyle\int_0^2 \int_{x/2}^1 y^2(x+y)\,dy\,dx = \dfrac{4}{5} \qquad I_y = \displaystyle\int_0^2 \int_{x/2}^1 x^2(x+y)\,dy\,dx = \dfrac{4}{3}$

SECTION 17.6

1. They are equal; they both give the volume of T.

2. (a) $L_f(P) = \sum\limits_{i=1}^{m}\sum\limits_{j=1}^{n}\sum\limits_{k=1}^{q} x_{i-1}y_{j-1}z_{k-1}\Delta x_i\Delta y_j\Delta z_k, \quad U_f(P) = \sum\limits_{i=1}^{m}\sum\limits_{j=1}^{n}\sum\limits_{k=1}^{q} x_i y_j z_k\Delta x_i\Delta y_j\Delta z_k$

   (b) $x_{i-1}y_{j-1}z_{k-1} \le \left(\dfrac{x_i + x_{i-1}}{2}\right)\left(\dfrac{y_j + y_{j-1}}{2}\right)\left(\dfrac{z_k + z_{k-1}}{2}\right) \le x_i y_j z_k$

   $x_{i-1}y_{j-1}z_{k-1}\Delta x_i\Delta y_j\Delta z_k \le \dfrac{1}{8}\left({x_i}^2 - {x_{i-1}}^2\right)\left({y_j}^2 - {y_{j-1}}^2\right)\left({z_k}^2 - {z_{k-1}}^2\right) \le x_i y_j z_k\Delta x_i\Delta y_j\Delta z_k$

   $L_f(P) \le \dfrac{1}{8}\sum\limits_{i=1}^{m}\sum\limits_{j=1}^{n}\sum\limits_{k=1}^{q}\left({x_i}^2 - {x_{i-1}}^2\right)\left({y_j}^2 - {y_{j-1}}^2\right)\left({z_k}^2 - {z_{k-1}}^2\right) \le U_f(P).$

   The middle term can be written

   $$\dfrac{1}{8}\left(\sum\limits_{i=1}^{m}{x_i}^2 - {x_{i-1}}^2\right)\left(\sum\limits_{j=1}^{n}{y_j}^2 - {y_{j-1}}^2\right)\left(\sum\limits_{k=1}^{q}{z_k}^2 - {z_{k-1}}^2\right) = \dfrac{1}{8}(1)(1)(1) = \dfrac{1}{8}.$$

   Therefore $I = \dfrac{1}{8}.$

3. $\iiint\limits_{\Pi} \alpha\, dx\, dy\, dz = \alpha \iiint\limits_{\Pi} dx\, dy\, dz = \alpha\,(\text{volume of } \Pi) = \alpha(a_2 - a_1)(b_2 - b_1)(c_2 - c_1)$

4. Since the volume is 1, the average value is $\iiint\limits_{\Omega} xyz\, dx\, dy\, dz = \dfrac{1}{8}.$

5. Let $P_1 = \{x_0, \cdots, x_m\}$, $\quad P_2 = \{y_0, \cdots, y_n\}$, $\quad P_3 = \{z_0, \cdots, z_q\}$ be partitions of $[0, a]$, $[0, b]$, $[0, c]$ respectively and let $P = P_1 \times P_2 \times P_3$. Note that

   $$x_{i-1}y_{j-1} \le \left(\dfrac{x_i + x_{i-1}}{2}\right)\left(\dfrac{y_j + y_{j-1}}{2}\right) \le x_i y_j$$

   and therefore

   $$x_{i-1}y_{j-1}\,\Delta x_i\,\Delta y_j\,\Delta z_k \le \tfrac{1}{4}\left({x_i}^2 - x_{i-1}^2\right)\left({y_j}^2 - y_{j-1}^2\right)\Delta z_k \le x_i y_j\,\Delta x_i\,\Delta y_j\Delta z_k.$$

   It follows that

   $$L_f(P) \le \dfrac{1}{4}\sum\limits_{i=1}^{m}\sum\limits_{j=1}^{n}\sum\limits_{k=1}^{q}\left({x_i}^2 - x_{i-1}^2\right)\left({y_j}^2 - y_{j-1}^2\right)\Delta z_k \le U_f(P).$$

   The middle term can be written

   $$\dfrac{1}{4}\left(\sum\limits_{i=1}^{m}{x_i}^2 - x_{i-1}^2\right)\left(\sum\limits_{j=1}^{n}{y_j}^2 - y_{j-1}^2\right)\left(\sum\limits_{k=1}^{q}\Delta z_k\right) = \dfrac{1}{4}a^2 b^2 c.$$

6. $I_x = I_{xy} + I_{xz}, \quad I_y = I_{xy} + I_{yz}, \quad I_z = I_{xz} + I_{yz}$

7. $\bar{x}_1 = a, \quad \bar{y}_1 = b, \quad \bar{z}_1 = c; \qquad \bar{x}_0 = A, \quad \bar{y}_0 = B, \quad \bar{z}_0 = C$

$$\bar{x}_1 V_1 + \bar{x} V = \bar{x}_0 V_0 \quad \Longrightarrow \quad a^2 bc + (ABC - abc)\,\bar{x} = A^2 BC$$

$$\Longrightarrow \quad \bar{x} = \frac{A^2 BC - a^2 bc}{ABC - abc}$$

similarly

$$\bar{y} = \frac{AB^2 C - ab^2 c}{ABC - abc}, \quad \bar{z} = \frac{ABC^2 - abc^2}{ABC - abc}$$

8. Encase T in a box Π.   A partition P of Π breaks up Π into little boxes Π_{ijk}.   Since f is nonnegative on Π, all the m_{ijk} are nonnegative. Therefore

$$0 \leq L_f(P) \leq \iiint\limits_T f(x,y,z)\,dx\,dy\,dz.$$

9. $M = \displaystyle\iiint\limits_\Pi Kz\,dxdydz$

Let $P_1 = \{x_0, \cdots, x_m\}, \quad P_2 = \{y_0, \cdots, y_n\}, \quad P_3 = \{z_0, \cdots, z_q\}$ be partitions of $[0,a]$ and let $P = P_1 \times P_2 \times P_3$.   Note that

$$z_{k-1} \leq \tfrac{1}{2}(z_k + z_{k-1}) \leq z_k$$

and therefore

$$Kz_{k-1}\,\Delta x_i\,\Delta y_j\,\Delta z_k \leq \tfrac{1}{2}K\,\Delta x_i\,\Delta y_j\left(z_k^2 - z_{k-1}^2\right) \leq Kz_k\,\Delta x_i\,\Delta y_j\Delta z_k.$$

It follows that

$$L_f(P) \leq \frac{1}{2}K\sum_{i=1}^m\sum_{j=1}^n\sum_{k=1}^q \Delta x_i\,\Delta y_j\left(z_k^2 - z_{k-1}^2\right) \leq U_f(P).$$

The middle term can be written

$$\frac{1}{2}K\left(\sum_{i=1}^m \Delta x_i\right)\left(\sum_{j=1}^n \Delta y_j\right)\left(\sum_{k=1}^q z_k^2 - z_{k-1}^2\right) = \frac{1}{2}K(a)(a)(a^2) = \frac{1}{2}Ka^4.$$

$M = \tfrac{1}{2}Ka^4$ where K is the constant of proportionality for the density function.

10. $x_M M = \displaystyle\iiint\limits_\Pi Kzx\,dx\,dy\,dz = \frac{Ka^5}{4} \quad \Longrightarrow \quad x_M = \frac{1}{2}a$

$$y_M M = \iiint\limits_\Pi Kzy\,dx\,dy\,dz = \frac{Ka^5}{4} \quad \Longrightarrow \quad y_M = \frac{1}{2}a$$

$$z_M M = \iiint\limits_\Pi Kz^2\,dx\,dy\,dz = \frac{Ka^5}{3} \quad \Longrightarrow \quad z_M = \frac{2}{3}a.$$

11.
$$I_z = \iiint_\Pi Kz\left(x^2 + y^2\right) dx dy dz$$

$$= \underbrace{\iiint_\Pi Kx^2 z \, dx dy dz}_{I_1} + \underbrace{\iiint_\Pi Ky^2 z \, dx dy dz}_{I_2}.$$

We will calculate I_1 using the partitions we used in doing Exercise 9.   Note that

$$x_{i-1}^2 z_{k-1} \le \left(\frac{x_i^2 + x_i x_{i-1} + x_{i-1}^2}{3}\right)\left(\frac{z_k + z_{k-1}}{2}\right) \le x_i^2 z_k$$

and therefore

$$Kx_{i-1}^2 z_{k-1} \, \Delta x_i \, \Delta y_j \, \Delta z_k \le \tfrac{1}{6} K\left(x_i^3 - x_{i-1}^3\right) \Delta y_j \left(z_k^2 - z_{k-1}^2\right) \le Kx_i^2 z_k^2 \, \Delta x_i \, \Delta y_j \, \Delta z_k.$$

It follows that

$$L_f(P) \le \frac{1}{6} K \sum_{i=1}^m \sum_{j=1}^n \sum_{k=1}^q \left(x_i^3 - x_{i-1}^3\right) \Delta y_j \left(z_k^2 - z_{k-1}^2\right) \le U_f(P).$$

The middle term can be written

$$\frac{1}{6} K \left(\sum_{i=1}^m x_i^3 - x_{i-1}^3\right)\left(\sum_{j=1}^n \Delta y_j\right)\left(\sum_{k=1}^q z_k^2 - z_{k-1}^2\right) = \frac{1}{6} Ka^3 \, (a)(a^2) = \frac{1}{6} Ka^6.$$

Similarly $I_2 = \frac{1}{6} Ka^6$ and therefore $I_z = \frac{1}{3} Ka^6 = \frac{2}{3}\left(\frac{1}{2} Ka^4\right)a^2 = \frac{2}{3} Ma^2.$

by Exercise 9 $\overset{\uparrow}{\rule{0pt}{1em}}$

12. (a) $L_f(P) \cong 56.4803$ $U_f(P) \cong 57.5603$ (c) $\displaystyle\iint_R \left(3y^2 - 2x\right) dx dy = 57$

SECTION 17.7

1. $\displaystyle\int_0^a \int_0^b \int_0^c dx\, dy\, dz = \int_0^a \int_0^b c\, dy\, dz = \int_0^a bc\, dz = abc$

2. $\displaystyle\int_0^1 \int_0^x \int_0^y y\, dz\, dy\, dx = \int_0^1 \int_0^x y^2\, dy\, dx = \int_0^1 \frac{x^3}{3}\, dx = \frac{1}{12}.$

3.
$$\int_0^1 \int_1^{2y} \int_0^x (x + 2z)\, dz\, dx\, dy = \int_0^1 \int_1^{2y} \left[xz + z^2\right]_0^x dx\, dy = \int_0^1 \int_1^{2y} 2x^2\, dx\, dy$$

$$= \int_0^1 \left[\frac{2}{3} x^3\right]_1^{2y} dy = \int_0^1 \left(\frac{16}{3} y^3 - \frac{2}{3}\right) dy = \frac{2}{3}$$

4. $\displaystyle\int_0^1 \int_{1-x}^{1+x} \int_0^{xy} 4z\, dz\, dy\, dx = \int_0^1 \int_{1-x}^{1+x} 2x^2 y^2\, dy\, dx = \int_0^1 \frac{2x^2}{3}\left[(1+x)^3 - (1-x)^3\right] dx = \frac{11}{9}$

5. $\displaystyle\int_0^2 \int_{-1}^1 \int_0^3 (z - xy)\, dz\, dy\, dx = \int_0^2 \int_{-1}^1 \left[\frac{1}{2} z^2 - xyz\right]_1^3 dy\, dx$

$$= \int_0^2 \int_{-1}^1 (4 - 2xy)\, dy\, dx = \int_0^2 \left[2y - xy^2\right]_{-1}^1 dx = \int_0^2 8\, dy = 16$$

6. $\displaystyle\int_0^2\int_{-1}^1\int_1^3 (z-xy)\,dy\,dx\,dz = \int_0^2\int_{-1}^1 (2z-4x)\,dx\,dz = \int_0^2 4z\,dz = 8$

7. $\displaystyle\int_0^{\pi/2}\int_0^1\int_0^{\sqrt{1-x^2}} x\cos z\,dy\,dx\,dz = \int_0^{\pi/2}\int_0^1 [xy\cos z]_0^{\sqrt{1-x^2}}\,dx\,dz$

$\displaystyle = \int_0^{\pi/2}\int_0^1 x\sqrt{1-x^2}\cos z\,dx\,dz = \int_0^{\pi/2}\left[-\frac{1}{3}(1-x^2)^{3/2}\cos z\right]_0^1\,dz = \frac{1}{3}\int_0^{\pi/2}\cos z\,dz = \frac{1}{3}$

8. $\displaystyle\int_{-1}^2\int_1^{y-2}\int_e^{e^2} \frac{x+y}{z}\,dz\,dx\,dy = \int_{-1}^2\int_1^{y-2} (x+y)\,dx\,dy = \int_{-1}^2\left[\frac{(y-2)^2-1}{2}+y(y-3)\right]\,dy = \frac{3}{2}$

9. $\displaystyle\int_1^2\int_y^{y^2}\int_0^{\ln x} ye^z\,dz\,dx\,dy = \int_1^2\int_y^{y^2} [ye^z]_0^{\ln x}\,dx\,dy$

$\displaystyle = \int_1^2\int_y^{y^2} y(x-1)\,dx\,dy = \int_1^2\left[\frac{1}{2}x^2y-xy\right]_y^{y^2}\,dy = \int_1^2\left(\frac{1}{2}y^5-\frac{3}{2}y^3+y^2\right)\,dy = \frac{47}{24}$

10. $\displaystyle\int_0^{\pi/2}\int_0^{\pi/2}\int_0^1 e^z\cos x\sin y\,dz\,dy\,dx = \int_0^{\pi/2}\int_0^{\pi/2} (e-1)\cos x\sin y\,dy\,dx$

$\displaystyle = \int_0^{\pi/2} (e-1)\cos x\,dx = e-1$

11. $\displaystyle\iiint_\Pi f(x)g(y)h(z)\,dx\,dy\,dz = \int_{c_1}^{c_2}\left[\int_{b_1}^{b_2}\left(\int_{a_1}^{a_2} f(x)g(y)h(z)\,dx\right)dy\right]dz$

$\displaystyle = \int_{c_1}^{c_2}\left[\int_{b_1}^{b_2} g(y)h(z)\left(\int_{a_1}^{a_2} f(x)\,dx\right)dy\right]dz$

$\displaystyle = \int_{c_1}^{c_2}\left[h(z)\left(\int_{a_1}^{a_2} f(x)\,dx\right)\left(\int_{b_1}^{b_2} g(y)\,dy\right)dz\right]$

$\displaystyle = \left(\int_{a_1}^{a_2} f(x)\,dx\right)\left(\int_{b_1}^{b_2} g(y)\,dy\right)\left(\int_{c_1}^{c_2} h(z)\,dz\right)$

12. $\displaystyle\left(\int_0^1 x^3\,dx\right)\left(\int_0^2 y^2\,dy\right)\left(\int_0^3 z\,dz\right) = \left(\frac{1}{4}\right)\left(\frac{8}{3}\right)\left(\frac{9}{2}\right) = 3$

13. $\displaystyle\left(\int_0^1 x^2\,dx\right)\left(\int_0^2 y^2\,dy\right)\left(\int_0^3 z^2\,dz\right) = \left(\frac{1}{3}\right)\left(\frac{8}{3}\right)\left(\frac{27}{3}\right) = 8$

14. $\displaystyle M = \iiint_\Pi kxyz\,dx\,dy\,dz = k\left(\int_0^a x\,dx\right)\left(\int_0^b y\,dy\right)\left(\int_0^c z\,dz\right) = \frac{1}{8}ka^2b^2c^2$

15. $\displaystyle x_M M = \iiint_\Pi kx^2yz\,dx\,dy\,dz = k\left(\int_0^a x^2\,dx\right)\left(\int_0^b y\,dy\right)\left(\int_0^c z\,dz\right)$

$\displaystyle = k\left(\tfrac{1}{3}a^3\right)\left(\tfrac{1}{2}b^2\right)\left(\tfrac{1}{2}c^2\right) = \tfrac{1}{12}ka^3b^2c^2.$

By Exercise 14, $M = \frac{1}{8}ka^2b^2c^2$. Therefore $\bar{x} = \frac{2}{3}a$. Similarly, $\bar{y} = \frac{2}{3}b$ and $\bar{z} = \frac{2}{3}c$.

16. (a)

$$I = \iiint\limits_{\Pi} kxyz \left[(x-a)^2 + (y-b)^2\right] \, dx \, dy \, dz$$

$$= k \left(\int_0^a x(x-a)^2 \, dx\right) \left(\int_0^b y \, dy\right) \left(\int_0^c z \, dz\right) + k \left(\int_0^a x \, dx\right) \left(\int_0^b y(y-b)^2 \, dy\right) \left(\int_0^c z \, dz\right)$$

$$= \frac{1}{48} ka^2 b^2 c^2 (a^2 + b^2).$$

Since $M = \frac{1}{8} ka^2 b^2 c^2$, $I = \frac{1}{6} M(a^2 + b^2)$.

(b) $I_M - \dfrac{1}{18}(a^2 + b^2)$ by parallel axis theorem

17.

18. $V = \iiint\limits_{T} dx \, dy \, dz = \displaystyle\int_0^1 \int_0^1 \int_0^{1-y} dz \, dy \, dx = \dfrac{1}{2}$

19. center of mass is the centroid

$$\bar{x} = \tfrac{1}{2} \quad \text{by symmetry}$$

$$\bar{y} V = \iiint\limits_{T} y \, dx dy dz = \int_0^1 \int_0^1 \int_0^{1-y} y \, dz \, dy \, dx = \int_0^1 \int_0^1 (y - y^2) \, dy \, dx$$

$$= \int_0^1 \left[\frac{1}{2} y^2 - \frac{1}{3} y^3\right]_0^1 dx = \int_0^1 \frac{1}{6} \, dx = \frac{1}{6}$$

$$\bar{z} V = \iiint\limits_{T} z \, dx dy dz = \int_0^1 \int_0^1 \int_0^{1-y} z \, dz \, dy \, dx = \int_0^1 \int_0^1 \frac{1}{2}(1-y)^2 \, dy \, dx$$

$$= \frac{1}{2} \int_0^1 \int_0^1 (1 - 2y + y^2) \, dy \, dx = \frac{1}{2} \int_0^1 \left[y - y^2 \frac{1}{3} y^3\right]_0^1 dx = \frac{1}{2} \int_0^1 \frac{1}{3} \, dx = \frac{1}{6}$$

$V = \frac{1}{2}$ (by Exercise 18); $\quad \bar{y} = \frac{1}{3}, \quad \bar{z} = \frac{1}{3}$

20. $I_x = \iiint\limits_{T} \dfrac{M}{V}(y^2 + z^2) \, dx \, dy \, dz = \dfrac{1}{3} M$

$$I_y = \iiint\limits_{T} \frac{M}{V}(x^2 + z^2) \, dx \, dy \, dz = \frac{1}{2} M \qquad I_z = \iiint\limits_{T} \frac{M}{V}(x^2 + y^2) \, dx \, dy \, dz = \frac{1}{2} M$$

21. $\displaystyle\int_{-r}^{r}\int_{-\phi(x)}^{\phi(x)}\int_{-\psi(x,y)}^{\psi(x,y)} k\left(r-\sqrt{x^2+y^2+z^2}\right)dz\,dy\,dx$ with $\phi(x)=\sqrt{r^2-x^2}$,

$\psi(x,y)=\sqrt{r^2-(x^2+y^2)}$, k the constant of proportionality

22. $\displaystyle\int_{-1}^{1}\int_{-\sqrt{1-x^2}}^{\sqrt{1-x^2}}\int_{\sqrt{x^2+y^2}}^{1} k\sqrt{x^2+y^2+z^2}\,dz\,dy\,dx$

23. $\displaystyle\int_{0}^{1}\int_{-\sqrt{x-x^2}}^{\sqrt{x-x^2}}\int_{-2x-3y-10}^{1-y^2} dz\,dy\,dx$

24. $\displaystyle\int_{-\sqrt{2}}^{\sqrt{2}}\int_{-\sqrt{1-x^2/2}}^{\sqrt{1-x^2/2}}\int_{2+y^2}^{4-x^2-y^2} dz\,dy\,dx$

25. $\displaystyle\int_{-1}^{1}\int_{-2\sqrt{2-2x^2}}^{2\sqrt{2-2x^2}}\int_{3x^2+y^2/4}^{4-x^2-y^2/4} k\left(z-3x^2-\frac{1}{4}y^2\right)dz\,dy\,dx$

26. $\displaystyle\int_{-\sqrt{2}}^{\sqrt{2}}\int_{-\sqrt{2-y^2}}^{\sqrt{2-y^2}}\int_{z^2+2y^2}^{4-z^2} k\sqrt{x^2+y^2}\,dx\,dz\,dy$

27. $\displaystyle\iiint_{T}(x^2z+y)\,dx\,dy\,dz=\int_{0}^{2}\int_{1}^{3}\int_{0}^{1}(x^2z+y)\,dx\,dy\,dz=\int_{0}^{2}\int_{1}^{3}\left[\frac{1}{3}x^3z+xy\right]_{0}^{1}dy\,dz$

$\displaystyle =\int_{0}^{2}\int_{1}^{3}\left(\frac{1}{3}z+y\right)dy\,dz=\int_{0}^{2}\left[\frac{1}{3}zy+\frac{1}{2}y^2\right]_{1}^{3}dz=\int_{0}^{2}\left(\frac{2}{3}z+4\right)dz=\frac{28}{3}$

28. $\displaystyle\int_{0}^{1}\int_{0}^{y}\int_{0}^{x+y}2ye^x\,dz\,dx\,dy=\int_{0}^{1}\int_{0}^{y}2y(x+y)e^x\,dx\,dy=\int_{0}^{1}(4y^2e^y-2ye^y+2y-2y^2)\,dy=4e-\frac{29}{3}$

29. $\displaystyle\iiint_{T}x^2y^2z^2\,dx\,dy\,dz=\int_{-1}^{0}\int_{0}^{y+1}\int_{0}^{1}x^2y^2z^2\,dx\,dz\,dy+\int_{0}^{1}\int_{0}^{1-y}\int_{0}^{1}x^2y^2z^2\,dx\,dz\,dy$

$\displaystyle =\int_{-1}^{0}\int_{0}^{y+1}\left[\frac{1}{3}x^3y^2z^2\right]_{0}^{1}dz\,dy+\int_{0}^{1}\int_{0}^{1-y}\left[\frac{1}{3}x^3y^2z^2\right]_{0}^{1}dz\,dy$

$\displaystyle =\frac{1}{3}\int_{-1}^{0}\int_{0}^{y+1}y^2z^2\,dz\,dy+\frac{1}{3}\int_{0}^{1}\int_{0}^{1-y}[y^2z^2]_{0}^{1}\,dz\,dy$

$\displaystyle =\frac{1}{3}\int_{-1}^{0}\left[\frac{1}{3}y^2z^3\right]_{0}^{y+1}dy+\frac{1}{3}\int_{0}^{1}\left[\frac{1}{3}y^2z^3\right]_{0}^{1-y}dy$

$\displaystyle =\frac{1}{9}\int_{-1}^{0}(y^5+3y^4+3y^3+y^2)\,dy+\frac{1}{9}\int_{0}^{1}(y^2-3y^3+3y^4-y^5)\,dy=\frac{1}{270}$

30. $\displaystyle\int_0^2 \int_0^{\sqrt{4-x^2}} \int_0^{\sqrt{4-x^2-y^2}} xy \, dz \, dy \, dx = \int_0^2 \int_0^{\sqrt{4-x^2}} xy\sqrt{4-x^2-y^2} \, dy \, dx$

$$= \int_0^{\pi/2} \int_0^2 r^2 \cos\theta \sin\theta \sqrt{4-r^2} \, r \, dr \, d\theta$$

$$= \frac{1}{2}\int_0^2 r^3\sqrt{4-r^2} \, dr = \frac{1}{4}\int_0^4 (4\sqrt{u} - u^{3/2}) \, du = \frac{32}{15}$$

31. $\displaystyle\iiint_T y^2 \, dx \, dy \, dz = \int_0^3 \int_0^{2-2x/3} \int_0^{6-2x-3y} y^2 \, dz \, dy \, dx = \int_0^3 \int_0^{2-2x/3} [y^2 z]_0^{6-2x-3y} \, dy \, dx$

$$= \int_0^3 \int_0^{2-2x/3} (6y^2 - 2xy^2 - 3y^3) \, dy \, dx$$

$$= \int_0^3 \left[2y^3 - \frac{2}{3}xy^3 - \frac{3}{4}y^4\right]_0^{2-2x/3} dx$$

$$= \frac{1}{4}\int_0^3 \left(2 - \frac{2}{3}x\right) dx = \frac{12}{5}$$

32. $\displaystyle\int_0^1 \int_0^{1-x^2} \int_0^{\sqrt{1-y}} y^2 \, dz \, dy \, dx = \int_0^1 \int_0^{1-x^2} y^2\sqrt{1-y} \, dy \, dx$

$$= \int_0^1 \left[-\frac{2}{3}x^3 + \frac{4}{5}x^5 - \frac{2}{7}x^7 + \frac{16}{105}\right] dx = \frac{1}{12}$$

33. $\displaystyle V = \int_0^2 \int_{x^2}^{x+2} \int_0^x dz \, dy \, dx = \int_0^2 \int_{x^2}^{x+2} x \, dy \, dx = \int_0^2 (x^2 + 2x - x^3) \, dx = \frac{8}{3}$

$\displaystyle\overline{x}V = \int_0^2 \int_{x^2}^{x+2} \int_0^x x \, dz \, dy \, dx = \int_0^2 \int_{x^2}^{x+2} x^2 \, dy \, dx = \int_0^2 (x^3 + 2x^2 - x^4) \, dx = \frac{44}{15}$

$\displaystyle\overline{y}V = \int_0^2 \int_{x^2}^{x+2} \int_0^x y \, dz \, dy \, dx = \int_0^2 \int_{x^2}^{x+2} xy \, dy \, dx = \int_0^2 \frac{1}{2}(x^3 + 4x^2 + 4x - x^5) \, dx = 6$

$\displaystyle\overline{z}V = \int_0^2 \int_{x^2}^{x+2} \int_0^x z \, dz \, dy \, dx = \int_0^2 \int_{x^2}^{x+2} \frac{1}{2}x^2 \, dy \, dx = \int_0^2 \frac{1}{2}(x^3 + 2x^2 - x^4) \, dx = \frac{22}{15}$

$$\overline{x} = \frac{11}{10}, \quad \overline{y} = \frac{9}{4}, \quad \overline{z} = \frac{11}{20}$$

34. (a) $\displaystyle M = \int_0^1 \int_0^1 \int_0^1 kz \, dx \, dy \, dz = \frac{1}{2}k$

   (b) $\displaystyle M = \int_0^1 \int_0^1 \int_0^1 k(x^2 + y^2 + z^2) \, dz \, dy \, dx = k$

35. $\displaystyle V = \int_{-1}^2 \int_0^3 \int_{2-x}^{4-x^2} dz \, dy \, dx = \frac{27}{2}; \quad (\overline{x}, \overline{y}, \overline{z}) = \left(\frac{1}{2}, \frac{3}{2}, \frac{12}{5}\right)$

36. $\displaystyle\iiint_T (x - \bar{x})\, dx\, dy\, dz = \iiint_T x\, dx\, dy\, dz - \bar{x} \iiint_T dx\, dy\, dz = \bar{x} V - \bar{x} V = 0.$

Similarly, the other two integrals are zero.

37. $\displaystyle V = \int_0^a \int_0^{\phi(x)} \int_0^{\psi(x,y)} dz\, dy\, dx = \frac{1}{6}\, abc \;\; \text{with} \;\; \phi(x) = b\left(1 - \frac{x}{a}\right), \quad \psi(x,y) = c\left(1 - \frac{x}{a} - \frac{y}{b}\right)$

$(\bar{x},\, \bar{y},\, \bar{z}) = \left(\tfrac{1}{4}\, a,\; \tfrac{1}{4}\, b,\; \tfrac{1}{4}\, c\right)$

38. $\displaystyle I_z = \iiint_T \frac{M}{V}(x^2 + y^2)\, dx\, dy\, dz = \frac{1}{30}\left(\frac{M}{V}\right) = \frac{1}{5} M$

39. $\Pi: 0 \le x \le a, \quad 0 \le y \le b, \quad 0 \le z \le c$

   (a) $\displaystyle I_z = \int_0^a \int_0^b \int_0^c \frac{M}{abc}(x^2 + y^2)\, dz\, dy\, dx = \frac{1}{3} M(a^2 + b^2)$

   (b) $I_M = I_z - d^2 M = \tfrac{1}{3} M(a^2 + b^2) - \tfrac{1}{4}(a^2 + b^2)M = \tfrac{1}{12} M(a^2 + b^2)$

         ⌐ parallel axis theorem (17.5.7)

   (c) $I = I_M + d^2 M = \tfrac{1}{12} M(a^2 + b^2) + \tfrac{1}{4} a^2 M = \tfrac{1}{3} M a^2 + \tfrac{1}{12} M b^2$

         ⌐ parallel axis theorem (17.5.7)

40. $\displaystyle V = \int_1^2 \int_1^2 \int_{-2}^{1+x+y} dz\, dy\, dx = \int_1^2 \int_1^2 (3 + x + y)\, dy\, dx = 6$

$\displaystyle \bar{x} V = \int_1^2 \int_1^2 \int_{-2}^{1+x+y} x\, dz\, dy\, dx = \frac{109}{12} \implies \bar{x} = \frac{109}{72} = \bar{y} \;\; \text{by symmetry}$

$\displaystyle \bar{z} V = \int_1^2 \int_1^2 \int_{-2}^{1+x+y} z\, dz\, dy\, dx = \frac{73}{12} \implies \bar{z} = \frac{73}{72}.$

41. $\displaystyle M = \int_0^1 \int_0^1 \int_0^y k\,(x^2 + y^2 + z^2)\, dz\, dy\, dx = \int_0^1 \int_0^1 k\left(x^2 y + y^3 + \frac{1}{3} y^3\right) dy\, dx$

$\displaystyle = \int_0^1 k\left(\frac{1}{2} x^2 + \frac{1}{3}\right) dx = \frac{1}{2} k$

$(x_M,\, y_M,\, z_M) = \left(\tfrac{7}{12},\; \tfrac{34}{45},\; \tfrac{37}{90}\right)$

42. T is symmetric   (a) about the yz-plane,   (b) about the xz-plane,   (c) about the xy-plane,

   (d) about the origin.

43. (a)   0 by symmetry

(b) $\displaystyle \iiint_T (a_1 x + a_2 y + a_3 z + a_4)\, dx\,dy\,dz = \iiint_T a_4\, dx\,dy\,dz = a_4 \text{ (volume of ball) } = \tfrac{4}{3}\pi a_4$

by symmetry $\uparrow$

44. $\displaystyle \int_0^2 \int_0^2 \int_{2-y}^{4-y^2} x^2 y^2\, dz\,dy\,dx = \frac{352}{45}$

45. $\displaystyle V = 8 \int_0^a \int_0^{\sqrt{a^2-x^2}} \int_0^{\sqrt{a^2-x^2-y^2}} dz\,dy\,dx = 8 \int_0^a \int_0^{\sqrt{a^2-x^2}} \sqrt{a^2-x^2-y^2}\, dy\,dx$

polar coordinates $\uparrow$

$\displaystyle = 8 \int_0^{\pi/2} \int_0^a \sqrt{a^2-r^2}\, r\, dr\, d\theta$

$\displaystyle = -4 \int_0^{\pi/2} \left[\frac{2}{3}(a^2-r^2)^{3/2} \right]_0^a d\theta$

$\displaystyle = \frac{8}{3} \int_0^{\pi/2} d\theta = \frac{4}{3}\pi a^3$

46. $\displaystyle 8 \int_0^a \int_0^{b\sqrt{1-x^2/a^2}} \int_0^{c\sqrt{1-x^2/a^2-y^2/b^2}} dz\,dy\,dx = \frac{4}{3}\pi abc.$

47. $\displaystyle M = \int_{-2}^2 \int_{-\sqrt{4-x^2}/2}^{\sqrt{4-x^2}/2} \int_{x^2+3y^2}^{4-y^2} k|x|\, dz\,dy\,dx = 4 \int_0^2 \int_0^{\sqrt{4-x^2}/2} \int_{x^2+3y^2}^{4-y^2} kx\, dz\,dy\,dx$

$\displaystyle = 4k \int_0^2 \int_0^{\sqrt{4-x^2}/2} \left(4x - x^3 - 4xy^2 \right) dy\,dx = \frac{4}{3}k \int_0^2 x\left(4-x^2\right)^{3/2} dx = \frac{128}{15}k$

48. using polar coordinates $\displaystyle \quad V = 2 \int_0^{2\pi} \int_0^1 (r - r^3)\, dr\, d\theta = \pi$

49. $\displaystyle M = \int_{-1}^2 \int_0^3 \int_{2-x}^{4-x^2} k(1+y)\, dz\,dy\,dx = \frac{135}{4}k; \quad (x_M, y_M, z_M) = \left(\frac{1}{2}, \frac{9}{5}, \frac{12}{5}\right)$

50. (a) $\displaystyle V = \int_0^2 \int_0^{2-z} \int_0^{9-x^2} dy\,dx\,dz$

(b) $\displaystyle V = \int_0^2 \int_0^{2-x} \int_0^{9-x^2} dy\,dz\,dx$

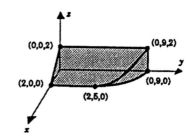

(c) $\displaystyle V = \int_0^5 \int_0^2 \int_0^{2-x} dz\,dx\,dy + \int_5^9 \int_0^{\sqrt{9-y}} \int_0^{2-x} dz\,dx\,dy$

51. (a) $V = \displaystyle\int_0^6 \int_{z/2}^3 \int_x^{6-x} dy\, dx\, dz$

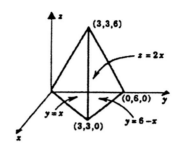

(b) $V = \displaystyle\int_0^3 \int_0^{2x} \int_x^{6-x} dy\, dz\, dx$

(c) $V = \displaystyle\int_0^6 \int_{z/2}^3 \int_{z/2}^y dx\, dy\, dz + \int_0^6 \int_3^{(12-z)/2} \int_{z/2}^{6-y} dx\, dy\, dz$

52. (a) $V = \displaystyle\iint_{\Omega_{xy}} 2\sqrt{4-y}\, dx\, dy$
    (b) $V = \displaystyle\iint_{\Omega_{xy}} \left(\int_{-\sqrt{4-y}}^{\sqrt{4-y}} dz \right) dx\, dy$

(c) $V = \displaystyle\int_{-4}^4 \int_{|x|}^4 \int_{-\sqrt{4-y}}^{\sqrt{4-y}} dz\, dy\, dx$
    (d) $V = \displaystyle\int_0^4 \int_{-y}^y \int_{-\sqrt{4-y}}^{\sqrt{4-y}} dz\, dx\, dy$

53. (a) $V = \displaystyle\iint_{\Omega_{xy}} 2y\, dy\, dz$
    (b) $V = \displaystyle\iint_{\Omega_{xy}} \left(\int_{-y}^y dx \right) dy\, dz$

(c) $V = \displaystyle\int_0^4 \int_{-\sqrt{4-y}}^{\sqrt{4-y}} \int_{-y}^y dx\, dz\, dy$
    (d) $V = \displaystyle\int_{-2}^2 \int_0^{4-z^2} \int_{-y}^y dx\, dy\, dz$

54. (a) $V = \displaystyle\iint_{\Omega_{xy}} (4 - z^2 - |x|)\, dx\, dz$
    (b) $V = \displaystyle\iint_{\Omega_{xy}} \left(\int_{|x|}^{4-z^2} dy \right) dx\, dz$

(c) $V = \displaystyle\int_{-2}^2 \int_{z^2-4}^{4-z^2} \int_{|x|}^{4-z^2} dy\, dx\, dz$

(d) $V = \displaystyle\int_{-2}^0 \int_{-\sqrt{4+x}}^{\sqrt{4+x}} \int_{|x|}^{4-z^2} dy\, dz\, dx + \int_0^2 \int_{-\sqrt{4-x}}^{\sqrt{4-x}} \int_{|x|}^{4-z^2} dy\, dz\, dx$

55. (a) $\displaystyle\int_2^4 \int_3^5 \int_1^2 \frac{\ln xy}{z}\, dz\, dy\, dx \cong 6.80703$
    (b) $\displaystyle\int_0^4 \int_1^2 \int_0^3 x\sqrt{yz}\, dz\, dy\, dx = \frac{16\sqrt{3}}{3}\left(4\sqrt{2} - 2 \right)$

56. Let $f(x,z) = 36 - 9x^2 - 4z^2$ and $g(x,z) = 1 - \frac{1}{2}x - \frac{1}{3}z$. Then

$$V = \int_0^2 \int_0^{3-(3/2)x} \int_{g(x,z)}^{f(x,z)} 1\, dy\, dz\, dx = 71$$

SECTION 17.8

1. $r^2 + z^2 = 9$
        2. $r = 2$
        3. $z = 2r$

4. $r\cos\theta = 4z$
        5. $4r^2 = z^2$
        6. $r^2 \sin^2\theta + z^2 = 8$

7. $\displaystyle\int_0^{\pi/2}\int_0^2\int_0^{4-r^2} r\,dz\,dr\,d\theta$

$= \displaystyle\int_0^{\pi/2}\int_0^2 \left(4r - r^2\right)\,dr\,d\theta$

$= \displaystyle\int_0^{\pi/2} 4\,d\theta = 2\pi$

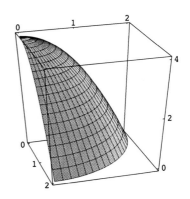

8. $\displaystyle\int_0^{\pi/4}\int_0^1\int_0^{\sqrt{1-r^2}} r\,dz\,dr\,d\theta = \frac{\pi}{12}$

9. $\displaystyle\int_0^{2\pi}\int_0^2\int_0^{r^2} r\,dz\,dr\,d\theta$

$= \displaystyle\int_0^{2\pi}\int_0^2 r^3\,dr\,d\theta$

$= \displaystyle\int_0^{2\pi} 4\,d\theta = 8\pi$

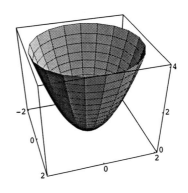

10. $\displaystyle\int_0^3\int_0^{2\pi}\int_r^3 r\,dz\,d\theta\,dr = 9\pi$

11. $\displaystyle\int_0^1\int_0^{\sqrt{1-x^2}}\int_0^{\sqrt{4-(x^2+y^2)}} dz\,dy\,dx = \int_0^{\pi/2}\int_0^1\int_0^{\sqrt{4-r^2}} r\,dz\,dr\,d\theta$

$= \displaystyle\int_0^{\pi/2}\int_0^1 r\sqrt{4-r^2}\,dr\,d\theta$

$= \displaystyle\int_0^{\pi/2}\left(\frac{8}{3} - \sqrt{3}\right)\,d\theta = \frac{1}{6}\left(8 - 3\sqrt{3}\right)\pi$

12. $\displaystyle\int_0^{\pi}\int_0^1\int_r^1 z^3 r\,dz\,dr\,d\theta = \int_0^{\pi}\int_0^1 \frac{1}{4}(1-r^4)\,r\,dr\,d\theta = \frac{\pi}{12}$

13. $\displaystyle\int_0^3 \int_0^{\sqrt{9-y^2}} \int_0^{\sqrt{9-x^2-y^2}} \frac{1}{\sqrt{x^2+y^2}}\, dz\, dx\, dy = \int_0^{\pi/2} \int_0^3 \int_0^{\sqrt{9-r^2}} \frac{1}{r}\cdot r\, dz\, dr\, d\theta$

$$= \int_0^{\pi/2} \int_0^3 \sqrt{9-r^2}\, dr\, d\theta$$

$$= \int_0^{\pi/2} \left[\frac{r}{2}\sqrt{9-r^2} + \frac{9}{2}\sin^{-1}\frac{r}{3}\right]_0^3 d\theta$$

$$= \frac{9\pi}{4}\int_0^{\pi/2} d\theta = \frac{9}{8}\pi^2$$

14. $\displaystyle\int_0^{\pi/2} \int_0^1 \int_0^{\sqrt{1-r^2}} zr\, dz\, dr\, d\theta = \int_0^{\pi/2} \int_0^1 \frac{r}{2}(1-r^2)\, dr\, d\theta = \frac{\pi}{16}$

15. $\displaystyle\int_0^1 \int_0^{\sqrt{1-x^2}} \int_0^2 \sin(x^2+y^2)\, dz\, dy\, dx = \int_0^{\pi/2} \int_0^1 \int_0^2 \sin(r^2)r\, dz\, dr\, d\theta$

$$= \int_0^{\pi/2} \int_0^1 2r\sin(r^2)\, dr\, d\theta = \tfrac{1}{2}\pi(1-\cos 1) \cong 0.7221$$

16. $\displaystyle\int_0^{2\pi} \int_0^1 \int_{r^2}^{2-r^2} r^2 dz\, dr\, d\theta = 4\pi\int_0^1 (1-r^2)r^2\, dr = \frac{8\pi}{15}$

17. $(0,1,2) \to (1,\frac{1}{2}\pi,2)$ **18.** $(0,1,-2) \to (1,\frac{1}{2}\pi,-2)$

19. $(0,-1,2) \to (1,\frac{3}{2}\pi,2)$ **20.** $(0,0,0) \to (0,\text{arbitrary},0)$

21. $\displaystyle V = \int_{-\pi/2}^{\pi/2} \int_0^{2a\cos\theta} \int_0^r r\, dz\, dr\, d\theta = \int_{-\pi/2}^{\pi/2} \int_0^{2a\cos\theta} r^2\, dr\, d\theta$

$$= \int_{-\pi/2}^{\pi/2} \frac{8}{3}a^3\cos^3\theta\, d\theta = \frac{32}{9}a^3$$

22. $\displaystyle V = \int_{-\pi/2}^{\pi/2} \int_0^{2a\cos\theta} \int_0^{r^2/a} r\, dz\, dr\, d\theta = \frac{3}{2}\pi a^3$

23. $\displaystyle V = \int_{-\pi/2}^{\pi/2} \int_0^{a\cos\theta} \int_0^{a-r} r\, dz\, dr\, d\theta = \int_{-\pi/2}^{\pi/2} \int_0^{a\cos\theta} r(a-r)\, dr\, d\theta$

$$= \int_{-\pi/2}^{\pi/2} a^3\left(\frac{1}{2}\cos^2\theta - \frac{1}{3}\cos^3\theta\right) d\theta = \frac{1}{36}a^3(9\pi-16)$$

24. $\displaystyle V = \int_{-\pi/2}^{\pi/2} \int_0^{2\cos\theta} \int_0^{2+\frac{1}{2}r\cos\theta} r\, dz\, dr\, d\theta = \frac{5}{2}\pi$

25. $V = \int_{-\pi/2}^{\pi/2} \int_0^{\cos\theta} \int_{r^2}^{r\cos\theta} r \, dz \, dr \, d\theta = \int_{-\pi/2}^{\pi/2} \int_0^{\cos\theta} \left(r^2 \cos\theta - r^3 \right) dr \, d\theta$

$\qquad = \int_{-\pi/2}^{\pi/2} \frac{1}{12} \cos^4\theta \, d\theta = \frac{1}{32}\pi$

26. $V = \int_0^{2\pi} \int_0^3 \int_{r+1}^{\sqrt{25-r^2}} r \, dz \, dr \, d\theta = \frac{41}{3}\pi$

27. $V = \int_0^{2\pi} \int_0^{1/2} \int_{r\sqrt{3}}^{\sqrt{1-r^2}} r \, dz \, dr \, d\theta = \int_0^{2\pi} \int_0^{1/2} \left(r\sqrt{1-r^2} - r^2\sqrt{3} \right) dr \, d\theta = \frac{1}{3}\pi\left(2 - \sqrt{3}\right)$

28. $V = \int_0^{2\pi} \int_0^a \int_{\sqrt{2}r}^{\sqrt{a^2+r^2}} r \, dz \, dr \, d\theta = \frac{2\pi}{3}a^3(\sqrt{2}-1)$

29. $V = \int_0^{2\pi} \int_1^3 \int_0^{\sqrt{9-r^2}} r \, dz \, dr \, d\theta = \int_0^{2\pi} \int_1^3 r\sqrt{9-r^2} \, dr \, d\theta = \frac{32}{3}\pi\sqrt{2}$

30. $V = \int_0^{2\pi} \int_1^2 \int_0^{\frac{1}{2}\sqrt{36-r^2}} r \, dz \, dr \, d\theta = \frac{1}{3}\pi(35\sqrt{35} - 128\sqrt{2}).$

31. Set the lower base of the cylinder on the xy-plane so that the axis of the cylinder coincides with the z-axis. Assume that the density varies directly as the distance from the lower base.

$$M = \int_0^{2\pi} \int_0^R \int_0^h kzr \, dz \, dr \, d\theta = \frac{1}{2} k\pi R^2 h^2$$

32. $x_M = y_M = 0$ by symmetry

$z_M M = \int_0^{2\pi} \int_0^R \int_0^h kz^2 r \, dz \, dr \, d\theta = \frac{1}{3} k\pi R^2 h^3$

$M = \frac{1}{2} k\pi R^2 h^2, \quad z_M = \frac{2}{3}h$

The center of mass lies on the axis of the cylinder at a distance $\frac{2}{3}h$ from the base of zero mass density.

33. $I = I_z = k \int_0^{2\pi} \int_0^R \int_0^h zr^3 \, dr \, d\theta \, dz$

$\qquad = \frac{1}{4} k\pi R^4 h^2 = \frac{1}{2}\left(\frac{1}{2} k\pi R^2 h^2\right) R^2 = \frac{1}{2} MR^2$

$\qquad\qquad\qquad$ from Exercise 31

34. (a) $I = \dfrac{M}{\pi R^2 h} \displaystyle\int_0^{2\pi} \int_0^R \int_0^h r^3 \, dz \, dr \, d\theta = \dfrac{1}{2}MR^2$

(b) $I = \dfrac{M}{\pi R^2 h} \displaystyle\int_0^{2\pi} \int_0^R \int_0^h (r^2 \sin^2\theta + z^2) r \, dz \, dr \, d\theta = \dfrac{1}{4}MR^2 + \dfrac{1}{3}Mh^2$

(c) $I = \dfrac{1}{4}MR^2 + \dfrac{1}{3}Mh^2 - M(\tfrac{1}{2}h)^2 = \dfrac{1}{4}MR^2 + \dfrac{1}{12}Mh^2$

35. Inverting the cone and placing the vertex at the origin, we have

$$V = \int_0^h \int_0^{2\pi} \int_0^{(R/h)z} r \, dr \, d\theta \, dz = \dfrac{1}{3}\pi R^2 h.$$

36. $x_M = y_M = 0$ by symmetry

$$z_M M = \int_0^h \int_0^{2\pi} \int_0^{(R/h)z} \left(\dfrac{M}{V}\right) zr \, dr \, d\theta \, dz = \left(\dfrac{M}{V}\right) \dfrac{\pi R^2 h^2}{4} \implies z_M = \dfrac{\pi R^2 h^2}{4V} = \dfrac{3}{4}h$$

On the axis of the cone at a distance $\frac{3}{4}h$ from the vertex.

37. $I = \dfrac{M}{V} \displaystyle\int_0^h \int_0^{2\pi} \int_0^{(R/h)z} r^3 \, dr \, d\theta \, dz = \dfrac{3}{10}MR^2$

38. $I = \dfrac{M}{V} \displaystyle\int_0^h \int_0^{2\pi} \int_0^{(R/h)z} z^2 r \, dr \, d\theta \, dz = \dfrac{3}{5}Mh^2.$

39. $V = \displaystyle\int_0^{2\pi} \int_0^1 \int_0^{1-r^2} r \, dz \, dr \, d\theta = \dfrac{1}{2}\pi$ **40.** $M = \displaystyle\int_0^{2\pi} \int_0^1 \int_0^{1-r^2} kzr \, dz \, dr \, d\theta = \dfrac{1}{6}\pi k$

41. $M = \displaystyle\int_0^{2\pi} \int_0^1 \int_0^{1-r^2} k(r^2 + z^2) r \, dz \, dr \, d\theta = \dfrac{1}{4}k\pi$

SECTION 17.9

1. $\left(\sqrt{3}, \; \tfrac{1}{4}\pi, \; \cos^{-1}\left[\tfrac{1}{3}\sqrt{3}\right]\right)$ **2.** $\left(\tfrac{1}{2}\sqrt{6}, \; \tfrac{1}{2}\sqrt{2}, \; \sqrt{2}\right)$

3. $\left(\tfrac{3}{4}, \; \tfrac{3}{4}\sqrt{3}, \; \tfrac{3}{2}\sqrt{3}\right)$ **4.** $\left(2\sqrt{10}, \; \tfrac{2}{3}\pi, \; \cos^{-1}\left[\tfrac{3}{10}\sqrt{10}\right]\right)$

5. $\rho = \sqrt{2^2 + 2^2 + (2\sqrt{6}/3)^2} = \dfrac{4\sqrt{6}}{3}$ **6.** $\left(8, \; -\dfrac{\pi}{4}, \; \dfrac{5\pi}{6}\right)$

$\phi = \cos^{-1}\left(\dfrac{2\sqrt{6}/3}{4\sqrt{6}/3}\right) = \cos^{-1}(1/2) = \dfrac{\pi}{3}$

$\theta = \tan^{-1}(1) = \dfrac{\pi}{4}$

$(\rho, \theta, \phi) = \left(\dfrac{4\sqrt{6}}{3}, \; \dfrac{\pi}{4}, \; \dfrac{\pi}{3}\right)$

7. $x = \rho \sin \phi \cos \theta = 3 \sin 0 \cos(\pi/2) = 0$

     $z = \rho \cos \phi = 3 \cos 0 = 3$

     $y = \rho \sin \phi \sin \theta = 3 \sin 0 \sin(\pi/2) = 0$

     $(x, y, z) = (0, 0, 3)$

8.   (a) $(5, \frac{1}{2}\pi, \arccos \frac{4}{5})$

     (b) $(5, \frac{3}{2}\pi, \arccos \frac{4}{5})$

9.   The circular cylinder $x^2 + y^2 = 1$;   the radius of the cylinder is 1 and the axis is the z-axis.

10.   The xy-plane.

11.   The lower nappe of the circular cone $z^2 = x^2 + y^2$.

12.   Vertical plane which bisects the first and third quadrants of the xy-plane.

13.   Horizontal plane one unit above the xy-plane.

14.   sphere $x^2 + y^2 + (z - \frac{1}{2})^2 = \frac{1}{4}$ of radius $\frac{1}{2}$ and center $(0, 0, \frac{1}{2})$

15.   Sphere of radius 2 centered at the origin:

$$\int_0^{2\pi} \int_0^{\pi} \int_0^2 \rho^2 \sin \phi \, d\rho \, d\phi \, d\theta = \frac{8}{3} \int_0^{2\pi} \int_0^{\pi} \sin \phi \, d\phi \, d\theta = \frac{16}{3} \int_0^{2\pi} d\theta = \frac{32\pi}{3}$$

16.   That part of the sphere of radius 1 that lies in the first quadrant between the x, z-plane and the plane $y = x$

$$\int_0^{\pi/4} \int_0^{\pi/2} \int_0^1 \rho^2 \sin \phi \, d\rho \, d\phi \, d\theta = \frac{\pi}{12}$$

17.   The first quadrant portion of the sphere that lies between the x, y-plane and the plane $z = \frac{3}{2}\sqrt{3}$.

$$\int_{\pi/6}^{\pi/2} \int_0^{\pi/2} \int_0^3 \rho^2 \sin \phi \, d\rho \, d\theta \, d\phi = 9 \int_{\pi/6}^{\pi/2} \int_0^{\pi/2} \sin \phi \, d\theta \, d\phi$$

$$= \frac{9}{2} \pi \int_{\pi/6}^{\pi/2} \sin \phi \, d\phi$$

$$= \frac{9}{2} \pi \, [- \cos \phi]_{\pi/6}^{\pi/2} = \frac{9}{4}\pi\sqrt{3}$$

18.   A cone of radius 1 and height 1; $\int_0^{\pi/4} \int_0^{2\pi} \int_0^{\sec \phi} \rho^2 \sin \phi \, d\rho \, d\theta \, d\phi = \frac{1}{3}\pi$

19. $\int_0^1 \int_0^{\sqrt{1-x^2}} \int_{\sqrt{x^2+y^2}}^{\sqrt{2-x^2-y^2}} dz \, dy \, dx = \int_0^{\pi/4} \int_0^{\pi/2} \int_0^{\sqrt{2}} \rho^2 \sin \phi \, d\rho \, d\theta \, d\phi$

$$= \frac{2}{3} \sqrt{2} \int_0^{\pi/4} \int_0^{\pi/2} \sin \phi \, d\theta \, d\phi$$

$$= \frac{\sqrt{2}}{3} \pi \int_0^{\pi/4} \sin \phi \, d\phi = \frac{\sqrt{2}}{6} \pi \, (2 - \sqrt{2})$$

20. $\left(\int_0^{\pi/4}\right) \int_0^{\pi/2} \int_0^2 \rho^4 \sin\phi \, d\rho \, d\theta \, d\phi = \dfrac{16\pi}{5} \int_0^{\pi/4} \sin\phi \, d\phi = \dfrac{8\pi}{5}(2 - \sqrt{2})$

21. $\displaystyle\int_0^3 \int_0^{\sqrt{9-y^2}} \int_0^{\sqrt{9-x^2-y^2}} z\sqrt{x^2 + y^2 + x^2} \, dz \, dx \, dy$

$$= \int_0^{\pi/2} \int_0^{\pi/2} \int_0^3 \rho\cos\phi \cdot \rho \cdot \rho^2 \sin\phi \, d\rho \, d\theta \, d\phi$$

$$= \int_0^{\pi/2} \frac{1}{2}\sin 2\phi \, d\phi \int_0^{\pi/2} d\theta \int_0^3 \rho^4 \, d\rho = \left[-\frac{1}{4}\cos 2\phi\right]_0^{\pi/2} \left(\frac{\pi}{2}\right) \left[\frac{1}{5}\rho^5\right]_0^3$$

$$= \frac{243\pi}{20}$$

22. $\displaystyle\int_0^{\pi/2} \int_0^{\pi/2} \int_0^1 \frac{1}{\rho^2}\rho^2 \sin\phi \, d\rho \, d\theta \, d\phi = \frac{\pi}{2}$

23. $V = \displaystyle\int_0^{2\pi} \int_0^{\pi} \int_0^R \rho^2 \sin\phi \, d\rho \, d\phi \, d\theta = \frac{4}{3}\pi R^3$

24. $r = \rho\sin\phi, \quad \theta = \theta, \quad z = \rho\cos\phi$

25. $V = \displaystyle\int_0^{\alpha} \int_0^{\pi} \int_0^R \rho^2 \sin\phi \, d\rho \, d\phi \, d\theta = \frac{2}{3}\alpha R^3$

26. $M = \displaystyle\int_0^{2\pi} \int_0^{\pi} \int_0^R k(R - \rho)\rho^2 \sin\phi \, d\rho \, d\phi \, d\theta = \frac{1}{3}k\pi R^4$

27. $M = \displaystyle\int_0^{2\pi} \int_0^{\tan^{-1}(r/h)} \int_0^{h\sec\phi} k\rho^3 \sin\phi \, d\rho \, d\phi \, d\theta$

$$= \int_0^{2\pi} \int_0^{\tan^{-1}(r/h)} \frac{kh^4}{4}\tan\phi\sec^3\phi \, d\phi \, d\theta$$

$$= \frac{kh^4}{4}\int_0^{2\pi} \frac{1}{3}\left[\sec^3\phi\right]_0^{\tan^{-1}(r/h)} d\theta = \frac{kh^4}{4}\int_0^{2\pi} \frac{1}{3}\left[\left(\frac{\sqrt{r^2 + h^2}}{h}\right)^3 - 1\right] d\theta$$

$$= \frac{1}{6}k\pi h \left(r^2 + h^2\right)^{3/2} - h^3$$

28. $V = \displaystyle\int_0^{2\pi} \int_0^{\tan^{-1} r/h} \int_0^{h\sec\phi} \rho^2 \sin\phi \, d\rho \, d\phi \, d\theta = \frac{2\pi}{3}\int_0^{\tan^{-1}(r/h)} h^3 \tan\phi\sec^2\phi \, d\phi$

$$= \frac{\pi}{3}h^3 \tan^2(\tan^{-1}\left(\frac{r}{h}\right) = \frac{1}{3}\pi r^2 h$$

29. center ball at origin; density $= \dfrac{M}{V} = \dfrac{3M}{4\pi R^3}$

(a) $I = \dfrac{3M}{4\pi R^3}\displaystyle\int_0^{2\pi} \int_0^{\pi} \int_0^R \rho^4 \sin^3\phi \, d\rho \, d\phi \, d\theta = \frac{2}{5}MR^2$

(b) $I = \frac{2}{5}MR^2 + R^2 M = \frac{7}{5}MR^2$ (parallel axis theorem)

30. The center of mass is the centroid; $V = \dfrac{2}{3}\pi R^3$

$$\bar{z}V = \int_0^{2\pi} \int_0^{\pi/2} \int_0^R (\rho\cos\phi)\rho^2\sin\phi\,d\rho\,d\phi\,d\theta = \frac{1}{4}\pi R^4$$

$$\bar{z} = \frac{3}{8}R; \qquad (\bar{x},\bar{y},\bar{z}) = \left(0,0,\frac{3}{8}R\right)$$

31. center balls at origin;   density $= \dfrac{M}{V} = \dfrac{3M}{4\pi\left(R_2{}^3 - R_1{}^3\right)}$

(a) $I = \dfrac{3M}{4\pi(R_2{}^3 - R_1{}^3)} \displaystyle\int_0^{2\pi}\int_0^{\pi}\int_{R_1}^{R_2} \rho^4\sin^3\phi\,d\rho\,d\phi\,d\theta = \dfrac{2}{5}M\left(\dfrac{R_2{}^5 - R_1{}^5}{R_2{}^3 - R_1{}^3}\right)$

This result can be derived from Exercise 29 without further integration. View the solid as a ball of mass M_2 from which is cut out a core of mass M_1.

$$M_2 = \frac{M}{V}V_2 = \frac{3M}{4\pi(R_2{}^3 - R_1{}^3)}\left(\frac{4}{3}\pi R_2{}^3\right) = \frac{MR_2{}^3}{R_2{}^3 - R_1{}^3}; \quad \text{similarly} \quad M_1 = \frac{MR_1{}^3}{R_2{}^3 - R_1{}^3}.$$

Then

$$I = I_2 - I_1 = \tfrac{2}{5}M_2R_2{}^2 - \tfrac{2}{5}M_1R_1{}^2 = \frac{2}{5}\left(\frac{MR_2{}^3}{R_2{}^3 - R_1{}^3}\right)R_2{}^2 - \frac{2}{5}\left(\frac{MR_1{}^3}{R_2{}^3 - R_1{}^3}\right)R_1{}^2$$

$$= \frac{2}{5}M\left(\frac{R_2{}^5 - R_1{}^5}{R_2{}^3 - R_1{}^3}\right).$$

(b) Outer radius R and inner radius R_1 gives

$$\text{moment of inertia} = \frac{2}{5}M\left(\frac{R^5 - R_1{}^5}{R^3 - R_1{}^3}\right). \qquad\qquad [\text{part }(a)]$$

As $R_1 \to R$,

$$\frac{R^5 - R_1{}^5}{R^3 - R_1{}^3} = \frac{R^4 + R^3R_1 + R^2R_1{}^2 + RR_1{}^3 + R_1{}^4}{R^2 + RR_1 + R_1{}^2} \longrightarrow \frac{5R^4}{3R^2} = \frac{5}{3}R^2.$$

Thus the moment of inertia of spherical shell of radius R is

$$\frac{2}{5}M\left(\frac{5}{3}R^2\right) = \frac{2}{3}MR^2.$$

(c) $I = \tfrac{2}{3}MR^2 + R^2M = \tfrac{5}{3}MR^2$ (parallel axis theorem)

32. (a) The center of mass is the centroid;   using the result of Exercise 30,

$$\bar{x} = \bar{y} = 0$$

$$\bar{z} = \frac{\bar{z}_2V_2 - \bar{z}_1V_1}{V} = \frac{\dfrac{3}{8}R_2\dfrac{4}{3}\pi R_2{}^3 - \dfrac{3}{8}R_1\dfrac{4}{3}\pi R_1{}^3}{\dfrac{4}{3}\pi(R_2{}^3 - R_1{}^3)} = \frac{3(R_2{}^2 + R_1{}^2)(R_2 + R_1)}{8(R_2{}^2 + R_2R_1 + R_1{}^2)}$$

(b) Setting $R_1 = R_2 = R$ in (a), we get $\bar{x} = \bar{y} = 0, \quad \bar{z} = \tfrac{1}{2}R$

33. $V = \displaystyle\int_0^{2\pi}\int_0^{\alpha}\int_0^a \rho^2\sin\phi\,d\rho\,d\phi\,d\theta = \frac{2}{3}\pi(1 - \cos\alpha)a^3$

34. $\int_0^{2\pi} \int_0^{\pi/4} \int_0^1 e^{\rho^3} \rho^2 \sin \phi \, d\rho \, d\phi \, d\theta = \frac{1}{3}\pi(e-1)\left(2-\sqrt{2}\right)$

35. (a) Substituting $\quad x = \rho \sin \phi \cos \theta, \quad y = \rho \sin \phi \sin \theta, \quad z = \rho \cos \phi$

into $\quad x^2 + y^2 + (z-R)^2 = R^2$

we have $\quad \rho^2 \sin^2 \phi + (\rho \cos \phi - R)^2 = R^2,$

which simplifies to $\quad \rho = 2R \cos \phi.$

(b) $0 \le \theta \le 2\pi, \quad 0 \le \phi \le \pi/4, \quad R \sec \phi \le \rho \le 2R \cos \phi$

36. (a) $M = \int_0^{2\pi} \int_0^{\pi/2} \int_0^{2R\cos\phi} k\rho^3 \sin \phi \, d\rho \, d\phi \, d\theta = \frac{8}{5}k\pi R^4$

(b) $M = \int_0^{2\pi} \int_0^{\pi/2} \int_0^{2R\cos\phi} k\rho^3 \sin^2 \phi \, d\rho \, d\phi \, d\theta = \frac{1}{4}k\pi^2 R^4$

(c) $M = \int_0^{2\pi} \int_0^{\pi/2} \int_0^{2R\cos\phi} k\rho^3 \cos^2 \theta \sin^2 \phi \, d\rho \, d\phi \, d\theta = \frac{1}{8}k\pi^2 R^4$

37.
$$V = \int_0^{2\pi} \int_0^{\pi/4} \int_0^2 \rho^2 \sin \phi \, d\rho \, d\phi \, d\theta + \int_0^{2\pi} \int_{\pi/4}^{\pi/2} \int_0^{2\sqrt{2}\cos\phi} \rho^2 \sin \phi \, d\rho \, d\phi \, d\theta$$

$$= \frac{1}{3}\left(16 - 6\sqrt{2}\right)\pi$$

38. $V = \int_0^{2\pi} \int_0^{\pi} \int_0^{1-\cos\phi} \rho^2 \sin \phi \, d\rho \, d\phi \, d\theta = \frac{8}{3}\pi$

39. Encase T in a spherical wedge W. W has spherical coordinates in a box Π that contains S. Define f to be zero outside of T. Then

$$F(\rho, \theta, \phi) = f\left(\rho \sin \phi \cos \theta, \ \rho \sin \phi \sin \theta, \ \rho \cos \phi\right)$$

is zero outside of S and

$$\iiint_T f(x,y,z) \, dxdydz = \iiint_W f(x,y,z) \, dxdydz$$

$$= \iiint_\Pi F(\rho, \theta, \phi) \rho^2 \sin \phi \, d\rho d\theta d\phi$$

$$= \iiint_S F(\rho, \theta, \phi) \rho^2 \sin \phi \, d\rho d\theta d\phi.$$

40. Break up T into little basic solids $T_1, \ldots, T_N$.   Choose a point (x_i^*, y_i^*, z_i^*) from each T_i and view all the mass as concentrated there.   Now T_i attracts m with a force

$$F_i \cong -\frac{Gm\lambda(x_i^*, y_i^*, z_i^*)(\text{Volume of } T_i)}{r_i^3}\mathbf{r}_i$$

where $\mathbf{r}_i$ is the vector from (x_i^*, y_i^*, z_i^*) to (a, b, c).   We therefore have

$$F_i \cong \frac{Gm\lambda(x_i^*, y_i^*, z_i^*)[(x_i^* - a)\mathbf{i} + (y_i^* - b)\mathbf{j} + (z_i^* - c)\mathbf{k}]}{[(x_i^* - a)^2 + (y_i^* - b)^2 + (z_i^* - c)^2]^{3/2}} \quad (\text{Volume of } T_i).$$

The sum of these approximations is a Riemann sum for the triple integral given and tends to that triple integral as the maximum diameter of the T_i tends to zero.

41. T is the set of all (x, y, z) with spherical coordinates (ρ, θ, ϕ) in the set

$$S: \quad 0 \le \theta \le 2\pi, \quad 0 \le \phi \le \pi/4, \quad R \sec\phi \le \rho \le 2R\cos\phi.$$

T has volume $V = \frac{2}{3}\pi R^3$. By symmetry the $\mathbf{i}, \mathbf{j}$ components of force are zero and

$$\mathbf{F} = \left\{ \frac{3GmM}{2\pi R^3} \iiint_T \frac{z}{(x^2 + y^2 + z^2)^{3/2}} \, dx\,dy\,dz \right\} \mathbf{k}$$

$$= \left\{ \frac{3GmM}{2\pi R^3} \iiint_S \left(\frac{\rho\cos\phi}{\rho^3} \right) \rho^2 \sin\phi \, d\rho\,d\theta\,d\phi \right\} \mathbf{k}$$

$$= \left\{ \frac{3GmM}{2\pi R^3} \int_0^{2\pi} \int_0^{\pi/4} \int_{R\sec\phi}^{2R\cos\phi} \cos\phi \, \sin\phi \, d\rho \, d\phi \, d\theta \right\} \mathbf{k}$$

$$= \frac{GmM}{R^2} \left(\sqrt{2} - 1 \right) \mathbf{k}.$$

42. With the coordinate system shown in the figure, T is the set of all points (x, y, z) with cylindrical coordinates (r, θ, z) in the set

$$S: \quad 0 \le r \le R, \quad 0 \le \theta \le 2\pi, \quad \alpha \le z \le \alpha + h.$$

The gravitational force is

$$F = \left[\iiint_T \left(\frac{GmM}{V} \right) \frac{z}{(x^2 + y^2 + z^2)^{3/2}} \, dx\,dy\,dz \right] \mathbf{k}$$

$$= \left[\frac{GmM}{\pi R^2 h} \iint_S \frac{zr}{(r^2 + z^2)^{3/2}} \, dr\,d\theta\,dz \right] \mathbf{k}$$

$$= \left[\frac{GmM}{\pi R^2 h} \int_0^{2\pi} \int_0^R \int_\alpha^{\alpha+h} \frac{zr}{(r^2 + z^2)^{3/2}} \, dz\,dr\,d\theta \right] \mathbf{k}$$

$$= \frac{2GmM}{R^2 h} \left(\sqrt{R^2 + \alpha^2} - \sqrt{R^2 + (\alpha + h)^2} + h \right) \mathbf{k}$$

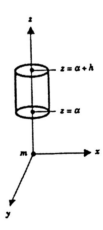

SECTION 17.10

1. $ad - bc$

2. 1

3. $2\left(v^2 - u^2\right)$

4. $u \ln v - u$

5. $-3u^2 v^2$

6. $1 + \dfrac{1}{uv}$

7. abc

8. 2

9. $\rho^2 \sin \phi$

10. $|J(r, \theta, z)| = \begin{vmatrix} \cos \theta & \sin \theta & 0 \\ -r \sin \theta & r \cos \theta & 0 \\ 0 & 0 & 1 \end{vmatrix} = r.$

11. $J(\rho, \theta, \phi) = \begin{vmatrix} \sin \phi \cos \theta & \sin \phi \sin \theta & \cos \theta \\ -\rho \sin \phi \sin \theta & \rho \sin \phi \cos \theta & 0 \\ \rho \cos \phi \cos \theta & \rho \cos \phi \sin \theta & -\rho \sin \phi \end{vmatrix} = -\rho^2 \sin \phi; \quad |J(\rho, \theta, \phi)| = \rho^2 \sin \phi.$

12. (a) $dx - by = (ad - bc)u_0$ (b) $cx - ay = (bc - ad)v_0$

13. Set $u = x + y, \quad v = x - y.$ Then

$$x = \frac{u + v}{2}, \quad y = \frac{u - v}{2} \quad \text{and} \quad J(u, v) = -\frac{1}{2}.$$

Ω is the set of all (x, y) with uv-coordinates in

$$\Gamma: \quad 0 \le u \le 1, \quad 0 \le v \le 2.$$

Then

$$\iint_\Omega \left(x^2 - y^2\right) dx dy = \iint_\Gamma \frac{1}{2} uv \, du dv = \frac{1}{2} \int_0^1 \int_0^2 uv \, dv \, du$$

$$= \frac{1}{2}\left(\int_0^1 u \, du\right)\left(\int_0^2 v \, dv\right) = \frac{1}{2}\left(\frac{1}{2}\right)(2) = \frac{1}{2}.$$

14. Using the changes of variables from Exercise 13,

$$\iint_\Omega 4xy \, dx \, dy = \int_0^1 \int_0^2 4\left(\frac{u^2 - v^2}{4}\right)\frac{1}{2}\, dv \, du = \frac{1}{2}\int_0^1 \int_0^2 \left(u^2 - v^2\right) dv \, du = -1$$

15. $\dfrac{1}{2}\displaystyle\int_0^1 \int_0^2 u \cos(\pi v) \, dv \, du = \dfrac{1}{2}\left(\int_0^1 u \, du\right)\left(\int_0^2 \cos(\pi v) \, dv\right) = \dfrac{1}{2}\left(\dfrac{1}{2}\right)(0) = 0$

16. Set $u = x - y, \quad v = x + 2y.$ Then

$$x = \frac{2u + v}{3}, \quad y = \frac{v - u}{3}, \quad \text{and} \quad J(u, v) = \frac{1}{3}$$

Ω is the set of all (x, y) with uv-coordinates in the set

$$\Gamma: \quad 0 \le u \le \pi, \quad 0 \le v \le \pi/2.$$

Therefore

$$\iint_\Omega (x+y)\,dx\,dy = \iint_\Gamma \frac{1}{9}(u+2v)\,du\,dv = \frac{1}{9}\int_0^\pi\int_0^{\pi/2}(u+2v)\,dv\,du = \frac{1}{18}\pi^3.$$

17. Set $u = x - y$, $v = x + 2y$. Then

$$x = \frac{2u+v}{3}, \quad y = \frac{v-u}{3}, \quad \text{and} \quad J(u,v) = \frac{1}{3}.$$

Ω is the set of all (x,y) with uv-coordinates in the set

$$\Gamma: \ 0 \le u \le \pi, \quad 0 \le v \le \pi/2.$$

Therefore

$$\iint_\Omega \sin(x-y)\cos(x+2y)\,dx\,dy = \iint_\Gamma \frac{1}{3}\sin u \cos v\,du\,dv = \frac{1}{3}\int_0^\pi\int_0^{\pi/2}\sin u \ \cos v\,dv\,du$$

$$= \frac{1}{3}\left(\int_0^\pi \sin u\,du\right)\left(\int_0^{\pi/2}\cos v\,dv\right) = \frac{1}{3}(2)(1) = \frac{2}{3}.$$

18. Using the change of variables from Exercise 16,

$$\iint_\Omega \sin 3x\,dx\,dy = \int_0^\pi\int_0^{\pi/2}\sin(2u+v)\frac{1}{3}\,du\,dv = 0.$$

19. Set $u = xy$, $v = y$. Then

$$x = u/v, \quad y = v \quad \text{and} \quad J(u,v) = 1/v.$$

$$xy = 1, \quad xy = 4 \implies u = 1, \quad u = 4$$

$$y = x, \quad y = 4x \implies u/v = v, \quad 4u/v = v \implies v^2 = u, \quad v^2 = 4u$$

Ω is the set of all (x,y) with uv-coordinates in the set

$$\Gamma: \ 1 \le u \le 4, \quad \sqrt{u} \le v \le 2\sqrt{u}.$$

(a) $\quad A = \iint_\Gamma \frac{1}{v}\,du\,dv = \int_1^4\int_{\sqrt{u}}^{2\sqrt{u}}\frac{1}{v}\,dv\,du = \int_1^4 \ln 2\,du = 3\ln 2$

(b) $\quad \overline{x}A = \int_1^4\int_{\sqrt{u}}^{2\sqrt{u}}\frac{u}{v^2}\,dv\,du = \frac{7}{3}; \quad \overline{x} = \frac{7}{9\ln 2}$

$$\overline{y}A = \int_1^4\int_{\sqrt{u}}^{2\sqrt{u}}dv\,du = \frac{14}{3}; \quad \overline{y} = \frac{14}{9\ln 2}$$

20. $J(r,\theta) = abr$, $\quad \Gamma: \ 0 \le r \le 1, \quad 0 \le \theta \le 2\pi$

$$A = \iint_\Gamma abr\,dr\,d\theta = ab\int_0^{2\pi}\int_0^1 r\,dr\,d\theta = \pi ab$$

21. Set $u = x+y$, $v = 3x - 2y$. Then

$$x = \frac{2u+v}{5}, \quad y = \frac{3u-v}{5} \quad \text{and} \quad J(u,v) = -\frac{1}{5}.$$

With Γ : $\quad 0 \leq u \leq 1, \quad 0 \leq v \leq 2$

$$M = \int_0^1 \int_0^2 \frac{1}{5} \lambda \, dv \, du = \frac{2}{5} \lambda \quad \text{where} \quad \lambda \text{ is the density.}$$

Then

$$I_x = \int_0^1 \int_0^2 \left(\frac{3u - v}{5} \right)^2 \frac{1}{5} \lambda \, dv \, du = \frac{8\lambda}{375} = \frac{4}{75} \left(\frac{2}{5} \lambda \right) = \frac{4}{75} M,$$

$$I_y = \int_0^1 \int_0^2 \left(\frac{2u + v}{5} \right)^2 \frac{1}{5} \lambda \, dv \, du = \frac{28\lambda}{375} = \frac{14}{75} \left(\frac{2}{5} \lambda \right) = \frac{14}{75} M,$$

$$I_z = I_x + I_y = \frac{18}{75} M.$$

22. $\quad x = \dfrac{u+v}{2}, \quad y = \dfrac{v-u}{2}, \quad J(u,v) = \dfrac{1}{2} \quad \Gamma : \quad -2 \leq u \leq 2, \quad -4 \leq v \leq -u^2$

$$A = \iint\limits_{\Gamma} \frac{1}{2} \, du \, dv = \frac{1}{2} \int_{-2}^2 \int_{-4}^{-u^2} dv \, du = \frac{16}{3}$$

23. Set $u = x - 2y, \quad v = 2x + y.$ Then

$$x = \frac{u + 2v}{5}, \quad y = \frac{v - 2u}{5} \quad \text{and} \quad J(u,v) = \frac{1}{5}.$$

Γ is the region between the parabola $v = u^2 - 1$ and the line $v = 2u + 2.$ A sketch of the curves shows that

$$\Gamma : \quad -1 \leq u \leq 3, \quad u^2 - 1 \leq v \leq 2u + 2.$$

Then

$$A = \frac{1}{5} \, (\text{area of } \Gamma) = \frac{1}{5} \int_{-1}^3 \left[(2u + 2) - (u^2 - 1) \right] \, du = \frac{32}{15}.$$

24. $\quad \overline{x} A = \dfrac{1}{2} \displaystyle\int_{-2}^2 \int_{-4}^{-u^2} \dfrac{u+v}{2} \, dv \, du = -\dfrac{32}{5} \quad \overline{y} A = \dfrac{1}{2} \displaystyle\int_{-2}^2 \int_{-4}^{-u^2} \dfrac{v-u}{2} \, dv \, du = -\dfrac{32}{5}$

$$A = \frac{16}{3} \implies \overline{x} = \overline{y} = -\frac{6}{5}$$

25. The choice $\theta = \pi/6$ reduces the equation to $13u^2 + 5v^2 = 1.$ This is an ellipse in the uv-plane with area $\pi a b = \pi/\sqrt{65}.$ Since $J(u,v) = 1,$ the area of Ω is also $\pi/\sqrt{65}.$

26. $\quad \displaystyle\iint\limits_{S_a} \frac{e^{-(x-y)^2}}{1 + (x+y)^2} \, dx \, dy = \frac{1}{2} \iint\limits_{\Gamma} \frac{e^{-u^2}}{1 + v^2} \, du \, dv$

where Γ is the square in the uv-plane with vertices $(-2a, 0), \quad (0, -2a), \quad (2a, 0), \quad (0, 2a).$
Γ contains the square $-a \leq u \leq a, \quad -a \leq v \leq a$ and is contained in the square $-2a \leq u \leq 2a, \quad -2a \leq v \leq 2a.$ Therefore

$$\frac{1}{2} \int_{-a}^a \int_{-a}^a \frac{e^{-u^2}}{1 + v^2} \, du \, dv \leq \frac{1}{2} \iint\limits_{\Gamma} \frac{e^{-u^2}}{1 + v^2} \, du \, dv \leq \frac{1}{2} \int_{-2a}^{2a} \int_{-2a}^{2a} \frac{e^{-u^2}}{1 + v^2} \, du \, dv.$$

The two extremes can be written

$$\frac{1}{2}\left(\int_{-a}^{a} e^{-u^2}\,du\right)\left(\int_{-a}^{a}\frac{1}{1+v^2}\,dv\right) \quad \text{and} \quad \frac{1}{2}\left(\int_{-2a}^{2a} e^{-u^2}\,du\right)\left(\int_{-2a}^{2a}\frac{1}{1+v^2}\,dv\right).$$

As $a \to \infty$ both expressions tend to $\frac{1}{2}\left(\sqrt{\pi}\right)(\pi) = \frac{1}{2}\pi^{3/2}$. It follows that

$$\int_{-\infty}^{\infty}\int_{-\infty}^{\infty}\frac{e^{-(x-y)^2}}{1+(x+y)^2}\,dx\,dy = \frac{1}{2}\pi^{3/2}.$$

27. $J = abc\rho^2\sin\phi;\quad V = \int_0^{2\pi}\int_0^{\pi}\int_0^1 abc\rho^2\sin\phi\,d\rho\,d\phi\,d\theta = \frac{4}{3}\pi abc$

28. $\overline{x} = \overline{y} = 0$

$$\overline{z}V = \int_0^{2\pi}\int_0^{\pi/2}\int_0^1 (c\rho\cos\phi)abc\rho^2\sin\phi\,d\rho\,d\phi\,d\theta = \frac{\pi abc^2}{4} \implies \overline{z} = \frac{3}{8}c.$$

29.
$$V = \frac{2}{3}\pi abc,\quad \lambda = \frac{M}{V} = \frac{3M}{2\pi abc}$$

$$I_x = \frac{3M}{2\pi abc}\int_0^{2\pi}\int_0^{\pi/2}\int_0^1 (b^2\rho^2\sin^2\phi\sin^2\theta + c^2\rho^2\cos^2\phi)abc\rho^2\sin\phi\,d\rho\,d\phi\,d\theta$$

$$= \tfrac{1}{5}M(b^2+c^2)$$

$$I_y = \tfrac{1}{5}M(a^2+c^2),\quad I_z = \tfrac{1}{5}M(a^2+b^2)$$

30. $I = \int_0^{2\pi}\int_0^1\int_0^{\pi}\rho^2(abc\rho^2\sin\phi)\,d\phi\,d\rho\,d\theta = \frac{4}{5}\pi abc$

PROJECT 17.10

1. (a) $\theta = \tan^{-1}\left[\left(\frac{ay}{bx}\right)^{1/\alpha}\right],\quad r = \left[\left(\frac{x}{a}\right)^{2/\alpha} + \left(\frac{y}{b}\right)^{2/\alpha}\right]^{\alpha/2}$

(b) $$\left.\begin{array}{l} ar_1(\cos\theta_1)^\alpha = ar_2(\cos\theta_2)^\alpha \\ br_1(\sin\theta_1)^\alpha = br_2(\sin\theta_2)^\alpha \\ r_1 > 0,\quad 0 < \theta < \tfrac{1}{2}\pi \end{array}\right\} \implies r_1 = r_2,\quad \theta_1 = \theta_2$$

2. $J = ab\alpha r\cos^{\alpha-1}\theta\sin^{\alpha-1}\theta$

3. (a)

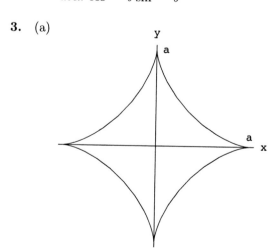

(b) $x = ar \cos^3 \theta$, $y = ar \sin^3 \theta$; $x^{\frac{2}{3}} + y^{\frac{2}{3}} = a^{\frac{2}{3}} \Longrightarrow r = 1$ and $x = a \cos^3 \theta$, $y = a \sin^3 \theta$

$$A = \int_{\frac{\pi}{2}}^0 y(\theta) x'(\theta)\, d\theta = \int_{\frac{\pi}{2}}^0 a \sin^3 \theta (3a \cos^2 \theta [-\sin \theta])\, d\theta$$

$$= 3a^2 \int_0^{\frac{\pi}{2}} \sin^4 \theta \cos^2 \theta\, d\theta = 3a^2 \int_0^{\frac{\pi}{2}} (\sin^4 \theta - \sin^6 \theta)\, d\theta$$

$$= 3a^2 \left[\frac{3 \cdot 1}{4 \cdot 2} \frac{\pi}{2} - \frac{5 \cdot 3 \cdot 1}{6 \cdot 4 \cdot 2} \frac{\pi}{2} \right] \quad \text{(See Exercise 62(b) in 8.3)}$$

$$= \frac{3a^2 \pi}{32}$$

(c) Entire area enclosed: $4 \cdot \dfrac{3a^2 \pi}{32} = \dfrac{3a^2 \pi}{8}$

4. (a)

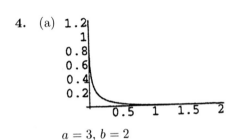

$a = 3, b = 2$

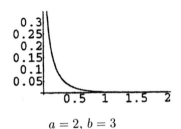

$a = 2, b = 3$

(b) From Problem 2, Jacobian $J = 8abr \cos^7 \theta \sin^7 \theta$

$$A = \int_0^{\frac{\pi}{2}} \int_0^1 8abr \cos^7 \theta \sin^7 \theta\, d\theta = 4ab \int_0^{\frac{\pi}{2}} \cos^7 \theta \sin^7 \theta\, d\theta = \frac{ab}{70}$$

REVIEW EXERCISES

1. $\displaystyle \int_0^1 \int_y^{\sqrt{y}} xy^2\, dx\, dy = \int_0^1 \left[\frac{1}{2}x^2 y^2\right]_y^{\sqrt{y}} dy = \int_0^1 \left(\frac{1}{2}y^3 - \frac{1}{2}y^4\right) dy = \left[\frac{1}{8}y^4 - \frac{1}{10}y^5\right]_0^1 = \frac{1}{40}$

2. $\displaystyle \int_0^1 \int_{-y}^y e^{x+y}\, dx\, dy = \int_0^1 \left[e^{x+y}\right]_{-y}^y dy = \int_0^1 (e^{2y} - 1)\, dy = \left[\frac{1}{2}e^{2y} - y\right]_0^1 = \frac{e^2}{2} - \frac{3}{2}$

3. $\displaystyle \int_0^1 \int_x^{3x} 2ye^{x^3}\, dy\, dx = \int_0^1 \left[y^2 e^{x^3}\right]_x^{3x} dx = \int_0^1 (9x^2 e^{x^3} - x^2 e^{x^3})\, dx = \left[3e^{x^3} - \frac{1}{3}e^{x^3}\right]_0^1 = \frac{8}{3}e - \frac{8}{3}$

4. $\displaystyle \int_1^2 \int_0^{\ln x} xe^y\, dy\, dx = \int_1^2 \left[xe^y\right]_0^{\ln x} dx = \int_1^2 x(x-1)\, dx = \left[\frac{1}{3}x^3 - \frac{1}{2}x^2\right]_1^2 = \frac{5}{6}$

5. $\displaystyle \int_0^{\pi/4} \int_0^{2\sin\theta} r \cos\theta\, dr\, d\theta = \int_0^{\pi/4} \left[\frac{1}{2}r^2 \cos\theta\right]_0^{2\sin\theta} d\theta = \int_0^{\pi/4} 2\sin^2\theta \cos\theta\, d\theta = \left[\frac{2}{3}\sin^3\theta\right]_0^{\pi/4} = \frac{\sqrt{2}}{6}$

6. $\displaystyle \int_{-1}^2 \int_0^4 \int_0^1 xyz\, dx\, dy\, dz = \int_{-1}^2 \int_0^4 \left[\frac{1}{2}x^2 yz\right]_0^1 dy\, dz = \int_{-1}^2 \int_0^4 \frac{1}{2}yz\, dy\, dz = \int_{-1}^2 4z\, dz = 6$

7. $\displaystyle\int_0^2 \int_0^{2-3x} \int_0^{x+y} x \, dz \, dy \, dx = \int_0^2 \int_0^{2-3x} \Big[xz\Big]_0^{x+y} dy \, dx = \int_0^2 \int_0^{2-3x} (x^2 + xy) \, dy \, dx$

$\displaystyle = \int_0^2 \left[x^2 y + \frac{1}{2}xy^2\right]_0^{2-3x} dx = \int_0^2 \left(\frac{3}{2}x^3 - 4x^2 + 2x\right) dx$

$\displaystyle = \left[\frac{3}{8}x^4 - \frac{4}{3}x^3 + x^2\right]_0^2 = -\frac{2}{3}$

8. $\displaystyle\int_0^{\frac{\pi}{2}} \int_z^{\frac{\pi}{2}} \int_0^{\sin z} 3x^2 \sin y \, dx \, dy \, dz = \int_0^{\frac{\pi}{2}} \int_z^{\frac{\pi}{2}} \Big[x^3 \sin y\Big]_0^{\sin z} dy \, dz = \int_0^{\frac{\pi}{2}} \int_z^{\frac{\pi}{2}} \sin^3 z \sin y \, dy \, dz$

$\displaystyle = \int_0^{\frac{\pi}{2}} \sin^3 z \cos z \, dz = \frac{1}{4}\sin^4 z \Big]_0^{\frac{\pi}{2}} = \frac{1}{4}$

9. $\displaystyle\int_{-\frac{\pi}{2}}^0 \int_0^{2\sin\theta} \int_0^{r^2} r^2 \cos\theta \, dz \, dr \, d\theta = \int_{-\frac{\pi}{2}}^0 \int_0^{2\sin\theta} r^4 \cos\theta \, dr \, d\theta = \int_{-\frac{\pi}{2}}^0 \frac{32}{5}\sin^5\theta \cos\theta \, d\theta = -\frac{16}{15}$

10. $\displaystyle\int_{-\frac{\pi}{6}}^{\frac{\pi}{2}} \int_0^{\frac{\pi}{2}} \int_0^1 \rho^3 \sin\varphi \cos\varphi \, d\rho \, d\theta \, d\varphi = \int_{-\frac{\pi}{6}}^{\frac{\pi}{2}} \int_0^{\frac{\pi}{2}} \frac{1}{4}\sin\varphi \cos\varphi \, d\theta \, d\varphi$

$\displaystyle = \int_{-\frac{\pi}{6}}^{\frac{\pi}{2}} \frac{\pi}{16}\sin 2\varphi \, d\varphi = -\left[\frac{\pi}{32}\cos 2\varphi\right]_{-\frac{\pi}{6}}^{\frac{\pi}{2}} = \frac{3\pi}{64}$

11. $\displaystyle\int_0^1 \int_y^1 e^{x^2} dx \, dy = \int_0^1 \int_0^x e^{x^2} dy \, dx = \int_0^1 e^{x^2} y \Big|_0^x dx = \int_0^1 xe^{x^2} dx = \frac{1}{2}e^{x^2}\Big|_0^1 = \frac{e-1}{2}$

12. $\displaystyle\int_0^2 \int_{\frac{x}{2}}^1 \cos y^2 \, dy \, dx = \int_0^1 \int_0^{2y} \cos y^2 \, dx \, dy = \int_0^1 x\cos y^2 \Big|_0^{2y} dy = \int_0^1 2y\cos y^2 \, dy = \sin y^2 \Big|_0^1 = \sin 1$

13. $\displaystyle\int_0^1 \int_0^{\sqrt{1-y^2}} \frac{1}{\sqrt{1-y^2}} dx \, dy = \int_0^1 dy = 1$

14. $\displaystyle\int_0^1 \int_0^{1-x} y\cos(x+y) \, dy \, dx = \int_0^1 \int_0^{1-y} y\cos(x+y) \, dx \, dy$

$\displaystyle = \int_0^1 y\sin(x+y)\Big]_0^{1-y} dy = \int_0^1 y(\sin 1 - \sin y) \, dy$

$\displaystyle = \left[\frac{1}{2}y^2 \sin 1 + y\cos y - \sin y\right]_0^1 = \cos 1 - \frac{1}{2}\sin 1$

15. $\displaystyle\int_0^1 \int_0^{\sqrt{1-x^2}} xy \, dy \, dx = \int_0^1 \left[\frac{1}{2}xy^2\right]_0^{\sqrt{1-x^2}} dx = \int_0^1 \left(\frac{1}{2}x - \frac{1}{2}x^3\right) dx = \frac{1}{8}$

16. $\displaystyle\int_{-\sqrt{3}}^{\sqrt{3}} \int_{y^2/3}^{4-y^2} (x-y) \, dx \, dy = \int_{-\sqrt{3}}^{\sqrt{3}} \left[\frac{x^2}{2} - xy\right]_{y^2/3}^{4-y^2} dy = \int_{-\sqrt{3}}^{\sqrt{3}} \left(\frac{4}{9}y^4 + \frac{4}{3}y^3 - 4y^2 - 4y + 8\right) dy = \frac{48\sqrt{3}}{5}$

17. $\displaystyle\int_0^2 \int_x^{3x-x^2} (x^2 - xy) \, dy \, dx = \int_0^2 \left[x^2 y - \frac{1}{2}xy^2\right]_x^{3x-x^2} dx = \int_0^2 \left(2x^4 - 2x^3 - \frac{1}{2}x^5\right) dx = -\frac{8}{15}$

18. $\displaystyle\int_0^2\int_0^{2-x} x(x-1)e^{xy}\,dy\,dx = \int_0^2\Big[(x-1)e^{xy}\Big]_0^{2-x}\,dx = \int_0^2 (x-1)e^{2x-x^2}\,dx - \int_0^2 (x-1)\,dx$

$$= -\Big[\tfrac{1}{2}e^{2x-x^2}\Big]_0^2 - \Big[\tfrac{1}{2}x^2 - x\Big]_0^2 = 0$$

19. $\displaystyle\int_0^2\int_0^y\int_0^{\sqrt{4-y^2}} 2xyz\,dz\,dx\,dy = \int_0^2\int_x^2 xyz^2\Big|_0^{\sqrt{4-y^2}}\,dx\,dy = \int_0^2\int_0^2 xy(4-y^2)\,dx\,dy$

$$= \int_0^2\Big[\tfrac{1}{2}y^3(4-y^2)\Big]\,dy = \frac{8}{3}$$

20. $\displaystyle\iiint_T z\,dx\,dy\,dz = 2\int_0^1\int_0^x\int_0^{1-x} z\,dz\,dy\,dx = \int_0^1\int_0^x (1-x)^2\,dy\,dx = \int_0^1 (1-x)^2 x\,dx = \frac{1}{12}$

21. $\displaystyle\int_0^2\int_0^{\sqrt{4-x^2}}\int_0^{\sqrt{4-x^2-y^2}} xy\,dz\,dy\,dx = \int_0^2\int_0^{\sqrt{4-x^2}}\Big[xyz\Big]_0^{\sqrt{4-x^2-y^2}}\,dy\,dx$

$$= \int_0^2\Big[-\tfrac{1}{3}x\sqrt{4-x^2-y^2}\Big]_0^{\sqrt{4-x^2}}\,dx$$

$$= \int_0^2 \tfrac{1}{3}x(4-x^2)^{\frac{3}{2}}\,dx = \frac{32}{15}$$

22. $\displaystyle\iiint_T (x^2+2z)\,dx\,dy\,dz = \int_{-2}^2\int_{x^2}^4\int_0^{4-y} (x^2+2z)\,dz\,dy\,dx$

$$= \int_{-2}^2\int_{x^2}^4 (4x^2 - x^2y + 16 - 8y + y^2)\,dy\,dx$$

$$= \int_{-2}^2 \Big(\tfrac{1}{6}x^6 - 8x^2 + \frac{64}{3}\Big)\,dx = \frac{2^{10}}{21}$$

23. $\displaystyle\int_0^2\int_0^{\sqrt{4-y^2}} e^{\sqrt{x^2+y^2}}\,dx\,dy = \int_0^{\pi/2}\int_0^2 e^r r\,dr\,d\theta = \frac{\pi}{2}\int_0^2 re^r\,dr = \frac{\pi}{2}\Big[re^r - e^r\Big]_0^2 = \frac{\pi}{2}(e^2+1)$

24. $\displaystyle\int_{-1}^1\int_0^{\sqrt{1-x^2}} \arctan(y/x)\,dy\,dx = \int_0^{\pi/2}\int_0^1 r\theta\,dr\,d\theta + \int_{\pi/2}^\pi\int_0^1 (\theta-\pi)r\,dr\,d\theta$

$$= \int_0^{\pi/2} \frac{\theta}{2}\,d\theta + \int_{\pi/2}^\pi \frac{1}{2}(\theta-\pi)\,d\theta = 0$$

25. $\displaystyle V = \int_0^3\int_0^{2\pi} (9-r^2)r\,dr\,d\theta = 2\pi\int_0^3 (9-r^2)r\,dr = \frac{81\pi}{2}$

26. $\displaystyle\int_0^1\int_{x^2}^{\sqrt{x}} (2-x^2-y^2)\,dy\,dx = -\int_0^1 \Big(2x^{1/2} - x^{5/2} - \tfrac{1}{3}x^{3/2} - 2x^2 + x^4 + \tfrac{1}{3}x^6\Big)\,dx = \frac{52}{105}$

27. $\displaystyle V = \int_0^1\int_0^{1-x} (x^2+y^2)\,dy\,dx = \int_0^1 \Big[x^2 - x^3 + \tfrac{1}{3}(1-x)^3\Big]\,dx = \Big[\tfrac{1}{3}x^3 - \tfrac{1}{4}x^4 - \tfrac{1}{12}(1-x)^4\Big]_0^1 = \frac{1}{6}$

28. $V = \displaystyle\int_0^3 \int_0^{\pi/2} r^2 \sin\theta \, dr \, d\theta = \int_0^{\pi/2} 9\sin\theta \, d\theta = 9$

29. $M = \displaystyle\int_{-\pi/2}^{\pi/2} \int_0^{\cos x} y \, dy \, dx = \int_{-\pi/2}^{\pi/2} \frac{1}{2}\cos^2 x \, dx = \frac{\pi}{4}$

$x_M M = \displaystyle\int_{-\pi/2}^{\pi/2} \int_0^{\cos x} xy \, dy \, dx = 0$ by symmetry

$y_M M = \displaystyle\int_{-\pi/2}^{\pi/2} \int_0^{\cos x} y^2 \, dy \, dx = \int_{-\pi/2}^{\pi/2} \frac{1}{3}\cos^3 x \, dx = \frac{4}{9}$

The center of mass is: $\left(0, \frac{16}{9\pi}\right)$

30. $M = \displaystyle\int_0^1 \int_{y^2}^{y} 2x \, dx \, dy = \int_0^1 (y^2 - y^4) \, dy = \frac{2}{15}$

$x_M M = \displaystyle\int_0^1 \int_{y^2}^{y} 2x^2 \, dx \, dy = \int_0^1 \left(\frac{2}{3}y^3 - \frac{2}{3}y^6\right) dy = \frac{1}{14}$

$y_M M = \displaystyle\int_0^1 \int_{y^2}^{y} 2xy \, dx \, dy = \int_0^1 (y^3 - y^5) \, dy = \frac{1}{12}$

The center of mass is: $(15/28, 5/8)$

31. $M = \displaystyle\int_0^{\pi/2} \int_r^{R} u^3 \, du \, d\theta = \frac{\pi}{8}(R^4 - r^4);$ (polar coordinates $[u, \theta]$)

By symmetry, $\bar{x} = \bar{y}$.

$x_M M = \displaystyle\int_0^{\pi/2} \int_r^{R} u^4 \cos\theta \, du \, d\theta = \frac{1}{5}(R^5 - r^5);$ $x_M = \dfrac{8(R^5 - r^5)}{5\pi(R^4 - r^4)}$

32. $M = \displaystyle\int_0^{\pi} \int_0^{2(1+\cos\theta)} r^2 \, dr \, d\theta = \int_0^{\pi} \frac{8}{3}(1 + \cos\theta)^3 \, d\theta = \frac{20}{3}\pi$

$x_M M = \displaystyle\int_0^{\pi} \int_0^{2(1+\cos\theta)} r^3 \cos\theta \, dr \, d\theta = \int_0^{\pi} 4(1 + \cos\theta)^4 \cos\theta \, d\theta = 14\pi$

$y_M M = \displaystyle\int_0^{\pi} \int_0^{2(1+\cos\theta)} r^3 \sin\theta \, dr \, d\theta = \int_0^{\pi} 4(1 + \cos\theta)^4 \sin\theta \, d\theta = -\frac{4}{5}(1 + \cos\theta)^5\big|_0^{\pi} = \frac{128}{5}$

The center of mass is: $\left(\frac{21}{10}, \frac{96}{25\pi}\right)$

33. Introduce a coordinate system as shown in the figure.

(a) $A = \frac{1}{2}bh;$ by symmetry, $\bar{x} = 0$

$\bar{y} A = \displaystyle\int_{-b/2}^{0} \int_0^{\frac{2h}{b}(x+\frac{b}{2})} y \, dy \, dx + \int_0^{b/2} \int_0^{-\frac{2h}{b}(x-\frac{b}{2})} y \, dy \, dx$

$= \dfrac{bh^2}{6} \implies \bar{y} = \dfrac{h}{3}$

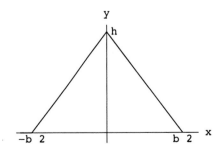

(b) $I = \displaystyle\int_{-b/2}^{0} \int_0^{\frac{2h}{b}(x+\frac{b}{2})} \lambda y^2 \, dy \, dx + \int_0^{b/2} \int_0^{-\frac{2h}{b}(x-\frac{b}{2})} \lambda y^2 \, dy \, dx = \dfrac{\lambda bh^3}{12} = \frac{1}{6}Mh^2$

(c) $I = 2 \int_0^{b/2} \int_0^{-\frac{2h}{b}(x-\frac{b}{2})} \lambda x^2 \, dx \, dy = \frac{1}{48} \lambda h b^3 = \frac{1}{24} M b^2$

34. Let $\lambda = \dfrac{k}{\sqrt{x^2 + y^2}}$

(a) $M = \int_0^\pi \int_r^R k \, dr \, d\theta = k\pi(R - r)$

By symmetry, $x_M = 0$; $y_M M = \int_0^\pi \int_r^R kr \sin\theta \, dr \, d\theta = k(R^2 - r^2)$

The center of mass is: $\left(0, \frac{R+r}{\pi}\right)$

(b) $I_x = \int_0^\pi \int_r^R kr^2 \sin^2\theta \, dr \, d\theta = \frac{k}{3}(R^3 - r^3) \int_0^\pi \sin^2\theta \, d\theta = \frac{k\pi}{6}(R^3 - r^3)$

(c) $I_y = \int_0^\pi \int_r^R kr^2 \cos^2\theta \, dr \, d\theta = \frac{k}{3}(R^3 - r^3) \int_0^\pi \cos^2\theta \, d\theta = \frac{k\pi}{6}(R^3 - r^3)$

35. $V = \int_0^2 \int_0^x \int_0^{2x+2y+1} dz \, dy \, dx = \int_0^2 \int_0^x (2x + 2y + 1) \, dy \, dx = 10$

36. $V = \int_0^1 \int_{x^2}^x \int_{-1}^{4(x^2+y^2)} dz \, dy \, dx = \int_0^1 \int_{x^2}^x (4x^2 + 4y^2 + 1) \, dy \, dx$

$$= \int_0^1 \left(x + \frac{16}{3}x^3 - x^2 - 4x^4 - \frac{4}{3}x^6 \right) dx = \frac{107}{210}$$

37. The curve of intersection of the two surfaces is the circle: $x^2 + y^2 = 4, \; x = 3$

$$V = \int_{-2}^2 \int_{-\sqrt{4-x^2}}^{\sqrt{4-x^2}} \int_{2x^2+y^2}^{12-x^2-2y^2} dz \, dy \, dx = \int_{-2}^2 \int_{-\sqrt{4-x^2}}^{\sqrt{4-x^2}} 3\left(4 - x^2 - y^2\right) dy \, dx$$

$$= 3 \int_0^{2\pi} \int_0^2 \left(4 - r^2\right) r \, dr \, d\theta$$

$$= 3 \int_0^{2\pi} \left[2r^2 - \tfrac{1}{4} r^4 \right]_0^2 d\theta = 12 \int_0^{2\pi} = 24\pi$$

38. $V = \int_0^1 \int_0^{\sqrt{1-y^2}} \int_0^{(2-y-z)/2} dx \, dz \, dy = \int_0^1 \int_0^{\pi/2} \frac{2 - r\cos\theta - r\sin\theta}{2} r \, d\theta \, dr$

$$= \int_0^1 \frac{1}{2}(\pi - 2r) r \, dr = \frac{3\pi - 4}{12}$$

39. $V = \int_0^{2\pi} \int_0^{\pi/3} \int_{\sec\phi}^2 \rho^2 \sin\phi \, d\rho \, d\phi \, d\theta = \int_0^{2\pi} \int_0^{\pi/3} \left[\tfrac{1}{3}\rho^3 \right]_{\sec\phi}^2 d\phi \, d\theta$

$$= \frac{1}{3} \int_0^{2\pi} \int_0^{\pi/3} \left(8 - \sec^3\phi\right) \sin\phi \, d\phi \, d\theta$$

$$= \frac{1}{3} \int_0^{2\pi} \left[-8\cos\phi - \tfrac{1}{2}\sec^2\phi \right]_0^{\pi/3} d\theta$$

$$= \frac{1}{3}\left(\frac{5}{2}\right)(2\pi) = \frac{5\pi}{3}$$

40. $V = \int_0^4 \int_0^{(12-3x)/4} \int_0^{16-x^2} dz\, dy\, dx = \int_0^4 \int_0^{\frac{12-3x}{4}} (16 - x^2)\, dy\, dx = \frac{3}{4}\int_0^4 (64 - 16x - 4x^2 + x^3)\, dx = 80$

41. $V = \int_0^{2\pi} \int_{\pi/4}^{\pi/2} \int_0^1 \rho^2 \sin\phi\, d\rho\, d\phi\, d\theta = \int_0^{2\pi} \int_{\pi/4}^{\pi/2} \frac{1}{3}\sin\phi\, d\phi\, d\theta = \frac{\sqrt{2}\,\pi}{3}$

42. $V = \int_0^{2\pi} \int_0^{\pi/4} \int_0^1 \rho^2 \sin\phi\, d\rho\, d\phi\, d\theta = \frac{2 - \sqrt{2}}{3}\pi$

43. (a) $V = \int_0^1 \int_0^x \int_0^{\sqrt{1-x^2}} dz\, dy\, dx + \int_0^1 \int_0^y \int_0^{\sqrt{1-y^2}} dz\, dx\, dy$

$\qquad\qquad = 2\int_0^1 \int_0^x \sqrt{1 - x^2}\, dy\, dx = 2\int_0^1 x\sqrt{1 - x^2}\, dx = \frac{2}{3}$

By symmetry, $\overline{x} = \overline{y}$.

$\overline{x} V = \int_0^1 \int_0^x \int_0^{\sqrt{1-x^2}} x\, dz\, dy\, dx + \int_0^1 \int_0^y \int_0^{\sqrt{1-y^2}} x\, dz\, dx\, dy$

For the first integral:

$\int_0^1 \int_0^x \int_0^{\sqrt{1-x^2}} x\, dz\, dy\, dx = \int_0^1 \int_0^x x\sqrt{1 - x^2}\, dy\, dx$

$\qquad\qquad\qquad\qquad = \int_0^1 x^2\sqrt{1 - x^2}\, dx = \int_0^{\pi/2} \sin^2 u \, \cos^2 u\, du = \frac{\pi}{16}$

$\qquad\qquad\qquad\qquad\qquad\quad x = \sin u \underset{\longleftarrow}{\uparrow}$

For the second integral:

$\int_0^1 \int_0^y \int_0^{\sqrt{1-y^2}} x\, dz\, dx\, dy = \int_0^1 \int_0^y x\sqrt{1 - y^2}\, dx\, dy = \int_0^1 \frac{1}{2}y^2\sqrt{1 - y^2}\, dy = \frac{\pi}{32}$

Thus, $\overline{x} V = \dfrac{3\pi}{32} \implies \overline{x} = \overline{y} = \dfrac{9\pi}{64}$

Now calculate $\overline{z}$:

$\overline{z} V = \int_0^1 \int_0^x \int_0^{\sqrt{1-x^2}} z\, dz\, dy\, dx + \int_0^1 \int_0^y \int_0^{\sqrt{1-y^2}} z\, dz\, dx\, dy;$

$\int_0^1 \int_0^x \int_0^{\sqrt{1-x^2}} z\, dz\, dy\, dx = \int_0^1 \int_0^x \frac{1}{2}(1 - x^2)\, dy\, dx = \frac{1}{2}\int_0^1 (x - x^3)\, dx = \frac{1}{8}$

and similarly,

$\int_0^1 \int_0^y \int_0^{\sqrt{1-y^2}} z\, dz\, dy\, dx = \frac{1}{8}.$

Therefore, $\overline{z} V = \dfrac{1}{4} \implies \overline{z} = \dfrac{3}{8}$

(b) $I_z = \int_0^1 \int_0^x \int_0^{\sqrt{1-x^2}} \lambda \left(\sqrt{x^2+y^2}\right)^2 dz\,dy\,dx + \int_0^1 \int_0^y \int_0^{\sqrt{1-y^2}} \lambda \left(\sqrt{x^2+y^2}\right)^2 dz\,dx\,dy;$

$\int_0^1 \int_0^x \int_0^{\sqrt{1-x^2}} \lambda \left(\sqrt{x^2+y^2}\right)^2 dz\,dy\,dx = \int_0^{\pi/4} \int_0^{\sec\theta} \int_0^{r\sin\theta} \lambda r^3 dz\,dr\,d\theta = \dfrac{3}{20}\lambda$

and $\int_0^1 \int_0^y \int_0^{\sqrt{1-y^2}} \lambda(\sqrt{x^2+y^2})^2 dz\,dx\,dy = \dfrac{3}{20}\lambda \implies I_z = \dfrac{3}{10}\lambda$

44. (a) $V = \int_0^{2\pi} \int_0^1 (r - r^2)r\,dr\,d\theta = 2\pi \int_0^1 (r^2 - r^3)\,dr = \dfrac{\pi}{6}$

By symmetry, $\bar{x} = \bar{y} = 0$

$\bar{z}V = \int_0^{2\pi} \int_0^1 \int_{r^2}^r zr\,dz\,dr\,d\theta = \dfrac{\pi}{12}$ and hence $\bar{z} = \dfrac{1}{2}$

(b) $I_z = K \int_0^{2\pi} \int_0^1 \int_{r^2}^r r^2\,dz\,dr\,d\theta = 2\pi K \int_0^1 r^2(r - r^2)dr = \dfrac{\pi K}{10}$

Here, K is the density of the mass.

45. Denote polar coordinates by $[u, \theta]$.

(a) $M = \int_0^{2\pi} \int_0^r \int_0^h u^3\,dz\,du\,d\theta = 2\pi h \int_0^r u^3\,du = \dfrac{\pi h r^4}{2}$

(b) By symmetry, $x_M = y_M = 0$

(c) $z_M M = \int_0^{2\pi} \int_0^r \int_0^h u^3 z\,dz\,du\,d\theta = \dfrac{\pi h^2 r^4}{4} \implies z_M = h/2$

46. $\lambda = \sqrt{x^2 + y^2}$

(a) $M = \int_0^{2\pi} \int_0^{\pi/2} \int_0^r \rho^3 \sin^2\phi\,d\rho\,d\phi\,d\theta = \dfrac{r^4 \pi}{2} \int_0^{\frac{\pi}{2}} \sin^2\phi\,d\phi = \dfrac{r^4 \pi^2}{8}$

(b) By symmetry, $x_M = y_M = 0$

$z_M M = \int_0^{2\pi} \int_0^{\pi/2} \int_0^r \rho^4 \sin^2\phi \cos\phi\,d\rho\,d\phi\,d\theta = \dfrac{2r^5 \pi}{15} \implies z_M = \dfrac{16r}{15\pi}$

47. (a) $M = \int_0^1 \int_0^{2\pi} \int_r^1 r^2\,dz\,d\theta\,dr = 2\pi \int_0^1 \int_r^1 r^2\,dz\,dr = 2\pi \int_0^1 r^2(1 - r)\,dr = \dfrac{\pi}{6}$

(b) By symmetry, $x_M = y_M = 0$

$z_M M = \int_0^1 \int_0^{2\pi} \int_r^1 r^2 z\,dz\,d\theta\,dr = \pi \int_0^1 r^2(1 - r^2)\,dr = \dfrac{2\pi}{15} \implies z_M = \dfrac{4}{5}$

(c) $I_z = \int_0^1 \int_0^{2\pi} \int_r^1 r^4\,dz\,d\theta\,dr = \dfrac{\pi}{15}$

48. $J(u, v) = \begin{vmatrix} 2u & 2v \\ -2v & 2u \end{vmatrix} = 4u^2 + 4v^2$

49. $J(u,v) = \begin{vmatrix} e^u \cos v & e^u \sin v \\ -e^u \sin v & e^u \cos v \end{vmatrix} = e^{2u}$

50. $Ju, v, w) = \begin{vmatrix} 2u & 2w & vw \\ 2w & 2v & uw \\ 2v & 2u & uv \end{vmatrix} = (vw - u^2)(4uw - 4v^2)$

51. Set $\quad x = \dfrac{v - u}{2}, \quad y = \dfrac{v + u}{2} \implies u = y - x, \quad v = y + x, \quad 1 \le v \le 2, \quad J = -\dfrac{1}{2}$

at $x = 0$, $y = u$, $y = v \implies u = v$

at $y = 0$, $-x = u$, $x = v \implies u = -v$

$$\iint_\Omega \cos\left(\frac{y - x}{y + x}\right) dx\, dy = \int_1^2 \int_{-v}^v \frac{1}{2} \cos\left(\frac{u}{v}\right) du\, dv = \int_i^2 v \sin 1\, dv = \frac{3}{2} \sin 1$$

52. By the hint, $J = \begin{vmatrix} \dfrac{\cos\theta}{u} & \dfrac{\sin\theta}{u} & 2r \\[2mm] -\dfrac{r\cos\theta}{u^2} & -\dfrac{r\sin\theta}{u^2} & 0 \\[2mm] -\dfrac{r\sin\theta}{u} & \dfrac{r\cos\theta}{u} & 0 \end{vmatrix} = -\dfrac{2r^3}{u^3}$

$$\iiint_T dx\, dy\, dz = \int_0^{2\pi} \int_1^2 \int_1^2 \frac{2r^3}{u^3} dr\, du\, d\theta = \frac{45\pi}{8}$$

CHAPTER 18

SECTION 18.1

1. (a) $h(x, y) = y\,\mathbf{i} + x\,\mathbf{j};$ $\mathbf{r}(u) = u\,\mathbf{i} + u^2\,\mathbf{j},$ $u \in [0, 1]$

    $x(u) = u,$ $y(u) = u^2;$ $x'(u) = 1,$ $y'(u) = 2u$

    $h(\mathbf{r}(u)) \cdot \mathbf{r}'(u) = y(u)\,x'(u) + x(u)\,y'(u) = u^2(1) + u(2u) = 3u^2$

    $\displaystyle \int_C h(\mathbf{r}) \cdot d\mathbf{r} = \int_0^1 3u^2 \, du = 1$

    (b) $h(x, y) = y\,\mathbf{i} + x\,\mathbf{j};$ $\mathbf{r}(u) = u^3\,\mathbf{i} - 2u\,\mathbf{j},$ $u \in [0, 1]$

    $x(u) = u^3,$ $y(u) = -2u;$ $x'(u) = 3u^2,$ $y'(u) = -2$

    $h(\mathbf{r}(u)) \cdot \mathbf{r}'(u) = y(u)\,x'(u) + x(u)\,y'(u) = (-2u)(3u^2) + u^3(-2) = -8u^3$

    $\displaystyle \int_C h(\mathbf{r}) \cdot d\mathbf{r} = \int_0^1 -8u^3 \, du = -2$

2. (a) $\displaystyle \int_C \mathbf{h} \cdot d\mathbf{r} = \int_0^1 (u\,\mathbf{i} + u^2\,\mathbf{j}) \cdot (\mathbf{i} + 2u\,\mathbf{j}) \, du = \int_0^1 (u + 2u^3) \, du = 1$

    (b) $\displaystyle \int_C \mathbf{h} \cdot d\mathbf{r} = \int_0^1 (u^3\,\mathbf{i} - 2u\,\mathbf{j}) \cdot (3u^2\,\mathbf{i} - 2\,\mathbf{j}) \, du = \int_0^1 (3u^5 + 4u) \, du = \frac{5}{2}$

3. $h(x, y) = y\,\mathbf{i} + x\,\mathbf{j};$ $\mathbf{r}(u) = \cos u\,\mathbf{i} - \sin u\,\mathbf{j},$ $u \in [0, 2\pi]$

    $x(u) = \cos u,$ $y(u) = -\sin u;$ $x'(u) = -\sin u,$ $y'(u) = -\cos u$

    $h(\mathbf{r}(u)) \cdot \mathbf{r}'(u) = y(u)\,x'(u) + x(u)\,y'(u) = \sin^2 u - \cos^2 u$

    $\displaystyle \int_C h(\mathbf{r}) \cdot d\mathbf{r} = \int_0^{2\pi} (\sin^2 u - \cos^2 u) \, du = 0$

4. (a) $\displaystyle \int_C \mathbf{h} \cdot d\mathbf{r} = \int_0^1 (e^{-u}\,\mathbf{i} + 2\,\mathbf{j}) \cdot (e^u\,\mathbf{i} - e^{-u}\,\mathbf{j}) \, du = \int_0^1 (1 - 2e^{-u}) \, du = 2e^{-1} - 1$

    (b) $\displaystyle \int_C \mathbf{h} \cdot d\mathbf{r} = \int_0^2 2\,\mathbf{j}\cdot(1 - u)\,\mathbf{i}\, du = \int_0^2 0 \, du = 0$

5. (a) $\mathbf{r}(u) = (2 - u)\,\mathbf{i} + (3 - u)\,\mathbf{j},$ $u \in [0, 1]$

    $\displaystyle \int_C h(\mathbf{r}) \cdot d\mathbf{r} = \int_0^1 (-5 + 5u - u^2) \, du = -\frac{17}{6}$

    (b) $\mathbf{r}(u) = (1 + u)\,\mathbf{i} + (2 + u)\,\mathbf{j},$ $u \in [0, 1]$

    $\displaystyle \int_C h(\mathbf{r}) \cdot d\mathbf{r} = \int_0^1 (1 + 3u + u^2) \, du = \frac{17}{6}$

6. (a) $\displaystyle\int_C \mathbf{h}\cdot d\mathbf{r} = \int_1^4 \left(\frac{1}{\sqrt{u}(1+u)}\mathbf{i} + \frac{1}{u\sqrt{1+u}}\mathbf{j}\right)\cdot\left(\frac{1}{2\sqrt{u}}\mathbf{i} + \frac{1}{2\sqrt{1+u}}\mathbf{j}\right)\,du = \int_1^4 \frac{1}{u(1+u)}\,du = \ln\frac{8}{5}$

(b) $\displaystyle\int_C \mathbf{h}\cdot d\mathbf{r} = \int_0^1 \frac{1}{(1+u)^3}(\mathbf{i}+\mathbf{j})\cdot(\mathbf{i}+\mathbf{j})\,du = \int_0^1 \frac{2}{(1+u)^3}\,du = \frac{3}{4}$

7. $C = C_1 \cup C_2 \cup C_3$ where,

$C_1: \mathbf{r}(u) = (1-u)(-2\,\mathbf{i}) + u(2\,\mathbf{i}) = (4u-2)\,\mathbf{i}, \quad u \in [0,1]$

$C_2: \mathbf{r}(u) = (1-u)(2\,\mathbf{i}) + u(2\,\mathbf{j}) = (2-2u)\,\mathbf{i} + 2u\,\mathbf{j}, \quad u \in [0,1]$

$C_3: \mathbf{r}(u) = (1-u)(2\,\mathbf{j}) + u(-2\,\mathbf{i}) = -2u\,\mathbf{i} + (2-2u)\,\mathbf{j}, \quad u \in [0,1]$

$\displaystyle\int_C = \int_{C_1} + \int_{C_2} + \int_{C_3} = 0 + (-4) + (-4) = -8$

8. $\mathbf{r}(u) = (-1+2u)\,\mathbf{i} + (1+u)\,\mathbf{j}, \quad u \in [0,1]$

$\displaystyle\int_C \mathbf{h}\cdot d\mathbf{r} = \int_0^1 (e^{-2+u}\,\mathbf{i} + e^{3u}\,\mathbf{j})\cdot(2\mathbf{i}+\mathbf{j})\,du = \int_0^1 (2e^{-2+u} + e^{3u})\,du = \frac{e^5 - e^2 + 6e - 6}{3e^2}$

9.

$C_1: \mathbf{r}(u) = (-1+2u)\,\mathbf{i}, \quad u \in [0,1]$

$C_2: \mathbf{r}(u) = \cos u\,\mathbf{i} + \sin u\,\mathbf{j}, \quad u \in [0,\pi]$

$\displaystyle\int_C = \int_{C_1} + \int_{C_2} = 0 + (-\pi) = -\pi$

10. Bottom: $\mathbf{r}(u) = u\,\mathbf{i};$ $\displaystyle\int_0^1 u^3\mathbf{j}\cdot\mathbf{i}\,du = \int_0^1 0\,du = 0$

Right side: $\mathbf{r}(u) = \mathbf{i} + u\mathbf{j};$ $\displaystyle\int_0^1 [3u\mathbf{i} + (1+2u)\mathbf{j}]\cdot\mathbf{j}\,du = \int_0^1 (1+2u)\,du = 2$

Top: $\mathbf{r}(u) = (1-u)\mathbf{i} + \mathbf{j};$ $\displaystyle\int_0^1 3(1-u)^2\mathbf{i}\cdot(-\mathbf{i})\,du = \int_0^1 -3(1-u)^2\,du = -1$

Left: $\mathbf{r}(u) = (1-u)\mathbf{j};$ $\displaystyle\int_0^1 2(1-u)\mathbf{j}\cdot(-\mathbf{j})\,du = \int_0^1 -2(1-u)\,du = -1$

$\displaystyle\int_C \mathbf{h}\cdot d\mathbf{r} = \text{sum of the above} = 0$

11. (a) $\mathbf{r}(u) = u\,\mathbf{i} + u\,\mathbf{j} + u\,\mathbf{k}, \quad u \in [0,1]$

$$\int_C \mathbf{h}(\mathbf{r}) \cdot d\mathbf{r} = \int_0^1 3u^2\,du = 1$$

(b) $\displaystyle\int_C \mathbf{h}(\mathbf{r}) \cdot d\mathbf{r} = \int_0^1 (2u^3 + u^5 + 3u^6)\,du = \frac{23}{21}$

12. (a) $\displaystyle\int_C \mathbf{h} \cdot d\mathbf{r} = \int_0^1 e^u(\mathbf{i}+\mathbf{j}+\mathbf{k}) \cdot (\mathbf{i}+\mathbf{j}+\mathbf{k})\,du = \int_0^1 3e^u\,du = 3(e-1)$

(b) $\displaystyle\int_C \mathbf{h} \cdot d\mathbf{r} = \int_0^1 (e^u\,\mathbf{i} + e^{u^2}\,\mathbf{j} + e^{u^3}\,\mathbf{k}) \cdot (\mathbf{i} + 2u\,\mathbf{j} + 3u^2\,\mathbf{k})\,du = \int_0^1 (e^u + 2ue^{u^2} + 3u^2 e^{u^3})\,du = 3(e-1).$

13. (a) $\mathbf{r}(u) = 2u\,\mathbf{i} + 3u\,\mathbf{j} - u\,\mathbf{k}, \quad u \in [0,1]$

$$\int_C \mathbf{h}(\mathbf{r}) \cdot d\mathbf{r} = \int_0^1 (2\cos 2u + 3\sin 3u + 3u^2)\,du = \big[\sin 2u - \cos 3u + u^3\big]_0^1 = 2 + \sin 2 - \cos 3$$

(b) $\displaystyle\int_C \mathbf{h}(\mathbf{r}) \cdot d\mathbf{r} = \int_0^1 (2u\cos u^2 + 3u^2 \sin u^3 - u^4)\,du = \left[\sin u^2 - \cos u^3 - \frac{1}{5}u^5\right]_0^1 = \frac{4}{5} + \sin 1 - \cos 1$

14. (a) $\displaystyle\int_C \mathbf{h} \cdot d\mathbf{r} = \int_0^1 (-2u^2\,\mathbf{i} + 4u^3\,\mathbf{j} - 2u^3\,\mathbf{k}) \cdot (2\,\mathbf{i} - \mathbf{j} + \mathbf{k})\,du = \int_0^1 (-4u^2 - 6u^3)\,du = -\frac{17}{6}$

(b) $\displaystyle\int_C \mathbf{h} \cdot d\mathbf{r} = \int_0^1 (\mathbf{i} + ue^{2u}\,\mathbf{j} + u\,\mathbf{k}) \cdot (e^u\,\mathbf{i} - e^{-u}\,\mathbf{j} + \mathbf{k})\,du = \int_0^1 (e^u - ue^u + u)\,du = e - \frac{3}{2}$

15. $\mathbf{r}(u) = u\,\mathbf{i} + u^2\,\mathbf{j}, \quad u \in [0,2]$

$$\int_C \mathbf{F}(\mathbf{r}) \cdot d\mathbf{r} = \int_0^2 \left[(u + 2u^2) + (2u + u^2)2u\right]\,du = \int_0^2 (2u^3 + 6u^2 + u)\,du = 26$$

16. $C_1:$ $\mathbf{r}(u) = u\,\mathbf{i}; \quad \displaystyle\int_0^1 u\,\mathbf{i} \cdot \mathbf{i}\,du = \int_0^1 u\,du = \frac{1}{2}$

     $C_2:$ $\mathbf{r}(u) = \mathbf{i} + u\,\mathbf{j}; \quad \displaystyle\int_0^1 (\cos u\,\mathbf{i} - u\sin 1\,\mathbf{j}) \cdot \mathbf{j}\,du = \int_0^1 -u\sin 1\,du = -\frac{1}{2}\sin 1$

     $C_3:$ $\mathbf{r}(u) = (1-u)\mathbf{i} + \mathbf{j}; \quad \displaystyle\int_0^1 [(1-u)\cos 1\,\mathbf{i} - \sin(1-u)\mathbf{j}] \cdot (-\mathbf{i})\,du = \int_0^1 (u-1)\cos 1\,du = -\frac{1}{2}\cos 1$

     $W = \displaystyle\int_C \mathbf{F} \cdot d\mathbf{r} = \frac{1}{2} - \frac{1}{2}\sin 1 - \frac{1}{2}\cos 1 = \frac{1}{2}(1 - \sin 1 - \cos 1)$

17. $\mathbf{r}(u) = (1-u)(\mathbf{j} + 4\mathbf{k}) + u(\mathbf{i} - 4\mathbf{k})$

$\qquad = u\mathbf{i} + (1-u)\mathbf{j} + (4-8u)\mathbf{k}, \quad u \in [0,1]$

$\displaystyle\int_C \mathbf{F}(\mathbf{r}) \cdot d\mathbf{r} = \int_0^1 (-32u + 97u^2 - 64u^3)\, du = \frac{1}{3}$

18. $C_1: \quad \mathbf{r}(u) = u\,\mathbf{i}; \quad \mathbf{F}(\mathbf{r}(u)) = \mathbf{0}, \quad \displaystyle\int_{C_1} \mathbf{F} \cdot d\mathbf{r} = 0$

$C_2: \quad \mathbf{r}(u) = \mathbf{i} + u\mathbf{j}; \quad \displaystyle\int_0^1 u\mathbf{k} \cdot \mathbf{j}\, du = \int_0^1 0\, du = 0$

$C_3: \quad \mathbf{r}(u) = \mathbf{i} + \mathbf{j} + u\,\mathbf{k}; \quad \displaystyle\int_0^1 (u\mathbf{i} + u\mathbf{j} + \mathbf{k}) \cdot \mathbf{k}\, du = \int_0^1 du = 1$

$W = 1$

19. $\mathbf{r}(u) = \cos u\, \mathbf{i} + \sin u\, \mathbf{j} + u\,\mathbf{k}, \quad u \in [0, 2\pi]$

$\displaystyle\int_C \mathbf{F}(\mathbf{r}) \cdot d\mathbf{r} = \int_0^{2\pi} \left[-\cos^2 u \, \sin u + \cos^2 u\, \sin u + u^2 \right] du = \int_0^{2\pi} u^2 \, du = \frac{8\pi^3}{3}$

20. Place the origin at the center of the circular path C and use the time parameter t. Motion along C at constant speed is uniform circular motion

$$\mathbf{r}(t) = r(\cos \omega t\, \mathbf{i} + \sin \omega t\, \mathbf{j}).$$

Differentiation gives

$$\mathbf{r}'(t) = r\omega(-\sin \omega t\mathbf{i} + \cos \omega t\, \mathbf{j}), \quad \mathbf{r}''(t) = -r\omega^2(\cos \omega t\mathbf{i} + \sin \omega t\, \mathbf{j}).$$

The force on the object is

$$\mathbf{F}(\mathbf{r}(t)) = m\mathbf{r}''(t).$$

Note that $\mathbf{F}(\mathbf{r}(t)) \cdot \mathbf{r}'(t) = 0$ for all t, and therefore W is 0 on every time integral.

Physical explanation: At each instant the force on the object is perpendicular to the path of the object. Thus the component of force in the direction of the motion is always zero.

21. $$\int_C \mathbf{q} \cdot d\mathbf{r} = \int_a^b [\mathbf{q} \cdot \mathbf{r}'(u)]\, du = \int_a^b \frac{d}{du}[\mathbf{q} \cdot \mathbf{r}(u)]\, du$$

$$= [\mathbf{q} \cdot \mathbf{r}(b)] - [\mathbf{q} \cdot \mathbf{r}(a)]$$

$$= \mathbf{q} \cdot [\mathbf{r}(b) - \mathbf{r}(a)]$$

$$\int_C \mathbf{r} \cdot d\mathbf{r} = \int_a^b [\mathbf{r}(u) \cdot \mathbf{r}'(u)]\, du$$

$$= \frac{1}{2} \int_a^b \|\mathbf{r}\|\, d\|\mathbf{r}\| \quad \text{(see Exercise 57, Section 14.1)}$$

$$= \frac{1}{2} \left(\|\mathbf{r}(b)\|^2 - \|\mathbf{r}(a)\|^2 \right)$$

22. (a) $\mathbf{r}(u) = (1 - 2u)\,\mathbf{i};$ $\displaystyle\int_{C_1} \mathbf{h} \cdot d\mathbf{r} = \int_0^1 (1 - 2u)^2\,\mathbf{i} \cdot (-2\,\mathbf{i})\,du = \int_0^1 -2(1 - 2u)^2\,du = -\dfrac{2}{3}$

(b) $\displaystyle\int_{C_2} \mathbf{h} \cdot d\mathbf{r} = \int_0^1 (\mathbf{i} + u\,\mathbf{j}) \cdot \mathbf{j}\,du + \int_0^1 [(1 - 2u)^2\,\mathbf{i} + \mathbf{j}] \cdot (-2\mathbf{i})\,du + \int_0^1 [\mathbf{i} + (1 - u)\mathbf{j}] \cdot (-\mathbf{j})\,du$

$\displaystyle = \int_0^1 u\,du + \int_0^1 -2(1 - 2u)^2\,du + \int_0^1 -(1 - u)\,du = -\dfrac{2}{3}$

(c) $\mathbf{r}(u) = \cos u\,\mathbf{i} + \sin u\,\mathbf{j},$ $u \in [0, \pi]$

$\displaystyle\int_{C_3} \mathbf{h} \cdot d\mathbf{r} = \int_0^\pi (\cos^2 u\,\mathbf{i} + \sin u\mathbf{j}) \cdot (-\sin u\mathbf{i} + \cos u\,\mathbf{j})\,du = \int_0^\pi (-\sin u \cos^2 u + \sin u \cos u)\,du = -\dfrac{2}{3}$

23. $\displaystyle\int_C \mathbf{f}(\mathbf{r}) \cdot d\mathbf{r} = \int_a^b [\mathbf{f}(\mathbf{r}\,(u)) \cdot \mathbf{r}'(u)]\,du = \int_a^b [f(u)\,\mathbf{i} \cdot \mathbf{i}]\,du = \int_a^b f(u)\,du$

24. Follows from the linearity of the dot product and of ordinary integrals.

25. $E : \mathbf{r}(u) = a \cos u\,\mathbf{i} + b \sin u\,\mathbf{j},$ $u \in [0, 2\pi]$

$\displaystyle W = \int_0^{2\pi} \left[\left(-\dfrac{1}{2} b \sin u \right) (-a \sin u) + \left(\dfrac{1}{2} a \cos u \right) (b \cos u) \right] du = \dfrac{1}{2} \int_0^{2\pi} ab\,du = \pi ab$

If the ellipse is traversed in the opposite direction, then $W = -\pi ab$. In both cases $|W| = \pi ab = $ area of the ellipse.

26. force at time t: $m\mathbf{r}''(t) = 2m\beta\,\mathbf{j}$

work during time interval: $\displaystyle W = \int_0^1 4m\beta^2 t\,dt = 2m\beta^2$

27. $\mathbf{r}(t) = \alpha t\,\mathbf{i} + \beta t^2\mathbf{j} + \gamma t^3\,\mathbf{k}$

$\mathbf{r}'(t) = \alpha\,\mathbf{i} + 2\beta t\,\mathbf{j} + 3\gamma t^2\,\mathbf{k}$

force at time $t = m\mathbf{r}''(t) = m(2\beta\mathbf{j} + 6\gamma t\mathbf{k})$

$\displaystyle W = \int_0^1 [m(2\beta\,\mathbf{j} + 6\gamma t\,\mathbf{k}) \cdot (\alpha\,\mathbf{i} + 2\beta t\mathbf{j} + 3\gamma t^2\,\mathbf{k})]\,dt$

$\displaystyle = m \int_0^1 (4\beta^2 t + 18\gamma^2 t^3)\,dt = \left(2\beta^2 + \dfrac{9}{2}\gamma^2 \right) m$

28. (a) $\mathbf{v}\perp\mathbf{k}$, $\mathbf{v}\perp\mathbf{r}$, $\|\mathbf{v}\|=\omega$ and $\omega\mathbf{k}$, $\mathbf{r}$, $\mathbf{v}$,   form a right-handed triple

(b)   We can parametrize C counterclockwise by

$$\mathbf{r}(t) = a\cos t\,\mathbf{i} + a\sin t\,\mathbf{j}, \quad 0 \le t \le 2\pi.$$

Then

$$\mathbf{r}'(t) = -a\sin t\,\mathbf{i} + a\cos t\,\mathbf{j}$$

and

$$\int_C (\omega\,\mathbf{k}\times\mathbf{r})\cdot d\mathbf{r} = \int_0^{2\pi} (\omega\,\mathbf{k}\times\mathbf{r}(t))\cdot\mathbf{r}'(t)\,dt.$$

Now

$$\mathbf{k}\times\mathbf{r}(t) = a\cos t\,\mathbf{j} - a\sin t\,\mathbf{i}.$$

So

$$(\mathbf{k}\times\mathbf{r}(t))\cdot\mathbf{r}'(t) = a^2(\cos^2 t + \sin^2 t) = a^2.$$

Thus

$$\int_C (\omega\mathbf{k}\times\mathbf{r})\cdot d\mathbf{r} + \int_0^{2\pi} \omega a^2\,dt = \omega a^2(2\pi) = 2\omega(\pi a^2) = 2\omega A.$$

If C is parametrized clockwise, the circulation is $-2\omega A$.

29. Take $C: \mathbf{r}(t) = r\cos t\,\mathbf{i} + r\sin t\,\mathbf{j}$, $t\in[0,2\pi]$

$$\int_C \mathbf{v}(\mathbf{r})\cdot d\mathbf{r} = \int_0^{2\pi} [\mathbf{v}(\mathbf{r}(t))\cdot\mathbf{r}'(t)]\,dt$$

$$= \int_0^{2\pi} [f(x(t), y(t))\,\mathbf{r}(t)\cdot\mathbf{r}'(t)]\,dt$$

$$= \int_0^{2\pi} f(x(t), y(t))\,[\mathbf{r}(t)\cdot\mathbf{r}'(t)]\,dt = 0$$

since for the circle $\mathbf{r}(t)\cdot\mathbf{r}'(t) = 0$ identically. The circulation is zero.

30. (a) $\mathbf{r}(u) = \mathbf{i} + u\mathbf{j}$; $W = \displaystyle\int_0^2 \frac{k}{1+u^2}(\mathbf{i}+u\mathbf{j})\cdot\mathbf{j}\,du = \int_0^2 \frac{ku}{1+u^2}\,du = \frac{k}{2}\ln 5$

(b) $\mathbf{r}(u) = u\mathbf{i} + \mathbf{j}$; $W = \displaystyle\int_0^1 \frac{k}{u^2+1}(u\mathbf{i}+\mathbf{j})\cdot\mathbf{i}\,du = \int_0^1 \frac{ku}{u^2+1}\,du = \frac{k}{2}\ln 2.$

31. (a) $\mathbf{r}(u) = (1-u)(\mathbf{i}+2\,\mathbf{k}) + u(\mathbf{i}+3\,\mathbf{j}+2\,\mathbf{k}) = \mathbf{i} + 3u\,\mathbf{j} + 2\,\mathbf{k}$, $u\in[0,1].$

$$\int_C \mathbf{F}(\mathbf{r})\cdot d\mathbf{r} = \int_0^1 \frac{9uk}{(5+9u^2)^{3/2}}\,du = \left[\frac{-k}{\sqrt{5+9u^2}}\right]_0^1 = \frac{k}{\sqrt{5}} - \frac{k}{\sqrt{14}}$$

(b) Let C be an arc on the sphere $\|\mathbf{r}\| = r = 5$.

$$\int_C \mathbf{F}(\mathbf{r}) \cdot d\mathbf{r} = \int_{C_2} \frac{k\mathbf{r}}{\|\mathbf{r}\|^3} \cdot d\mathbf{r}$$

$$= \frac{k}{5^3} \int_{C_2} \mathbf{r} \cdot d\mathbf{r} = \frac{k}{5^3} \int_{C_2} \|\mathbf{r}\| \, d\|\mathbf{r}\| \quad \text{(see Exercise 57, Section 14.1)}$$

$$= \frac{k}{5^3} \left[\frac{1}{2} \|\mathbf{r}\|^2 \right]_{(3,4,0)}^{(0,4,3)} = 0$$

32. Let (x_0, y_0, z_0) and (x_1, y_1, z_1) be the coordinates of $\mathbf{a}$ and $\mathbf{b}$, respectively. Then

$$W = \frac{k}{\sqrt{x_1^2 + y_1^2 + z_1^2}} - \frac{k}{\sqrt{x_0^2 + y_0^2 + z_0^2}}$$

33. $\mathbf{r}(u) = u\,\mathbf{i} + \alpha\,u(1-u)\,\mathbf{j}, \quad \mathbf{r}'(u) = \mathbf{i} + \alpha(1-2u)\,\mathbf{j}, \quad u \in [0,1]$

$$W(\alpha) = \int_C \mathbf{F}(\mathbf{r}) \cdot d\mathbf{r} = \int_0^1 \left[(\alpha^2 u^2 (1-u)^2 + 1) + [u + \alpha\,u(1-u)]\alpha(1-2u) \right] dx$$

$$= \int_0^1 \left[1 + (\alpha + \alpha^2)u - (2\alpha + 2\alpha^2)u^2 + \alpha^2 u^4 \right] du = 1 - \frac{1}{6}\alpha + \frac{1}{30}\alpha^2$$

$$W'(\alpha) = -\frac{1}{6} + \frac{1}{15}\alpha \quad \Longrightarrow \quad \alpha = \frac{15}{6} = \frac{5}{2}$$

The work done by $\mathbf{F}$ is a minimum when $\alpha = 5/2$.

34. Suppose that C is the curve $\mathbf{r}(u), \ a \le u \le b$.

$$\int_C \nabla f \cdot d\mathbf{r} = \int_a^b \nabla f(\mathbf{r}(u)) \cdot \mathbf{r}'(u)\, du = \int_a^b \frac{df}{du}\, du = \Big[f(\mathbf{r}(u)) \Big]_a^b = f(\mathbf{r}(b)) - f(\mathbf{r}(a)).$$

SECTION 18.2

1. $\mathbf{h}(x,y) = \nabla f(x,y)$ where $f(x,y) = \frac{1}{2}(x^2 + y^2)$

C is closed $\Longrightarrow \displaystyle\int_C \mathbf{h}(\mathbf{r}) \cdot d\mathbf{r} = 0$

2. $x\,\mathbf{i} + y\,\mathbf{j}$ is a gradient (Exercise 1); we need integrate only $y\,\mathbf{i}$.

$$\int_C \mathbf{h}(\mathbf{r}) \cdot d\mathbf{r} = \int_0^{2\pi} y(t)x'(t)\, dt = \int_0^{2\pi} (b\sin t)(-a\sin t)\, dt = -\pi ab$$

3. $\mathbf{h}(x,y) = \nabla f(x,y)$ where $f(x,y) = x\cos \pi y; \quad \mathbf{r}(0) = \mathbf{0}, \quad \mathbf{r}(1) = \mathbf{i} - \mathbf{j}$

$$\int_C \mathbf{h}(\mathbf{r}) \cdot d\mathbf{r} = \int_C \nabla f(\mathbf{r}) \cdot d\mathbf{r} = f(\mathbf{r}(1)) - f(\mathbf{r}(0)) = f(1,-1) - f(0,0) = -1$$

4. $\mathbf{h} = \nabla f$ with $f(x, y) = \dfrac{x^3}{3} + \dfrac{y^3}{3} - xy,$ and C is closed, so $\displaystyle\int_C \mathbf{h} \cdot d\mathbf{r} = 0$

5. $\mathbf{h}(x, y) = \nabla f(x, y)$ where $f(x, y) = \frac{1}{2}x^2 y^2;$ $\mathbf{r}(0) = \mathbf{j}, \quad \mathbf{r}(1) = -\mathbf{j}$

$$\int_C \mathbf{h}(\mathbf{r}) \cdot d\mathbf{r} = \int_C \nabla f(\mathbf{r}) \cdot d\mathbf{r} = f(\mathbf{r}(1)) - f(\mathbf{r}(0)) = f(0, -1) - f(0, 1) = 0 - 0 = 0$$

6. $e^y \mathbf{i} + x e^y \mathbf{j}$ is a gradient;   we need integrate only $\mathbf{i} - x\mathbf{j}$

$C = C_1 \cup C_2 \cup C_3 \cup C_4$ where

$\qquad C_1 : \mathbf{r}(u) = (2u - 1)\mathbf{i} - \mathbf{j}, \quad u \in [0, 1]$

$\qquad C_2 : \mathbf{r}(u) = \mathbf{i} + (2u - 1)\mathbf{j}, \quad u \in [0, 1]$

$\qquad C_3 : \mathbf{r}(u) = (1 - 2u)\mathbf{i} + \mathbf{j}, \quad u \in [0, 1]$

$\qquad C_4 : \mathbf{r}(u) = -\mathbf{i} + (1 - 2u)\mathbf{j}, \quad u \in [0, 1]$

$$\int_C = \int_{C_1} + \int_{C_2} + \int_{C_3} + \int_{C_4} = 2 + (-2) + (-2) + (-2) = -4$$

7. $\mathbf{h}(x, y) = \nabla f(x, y)$ where $f(x, y) = x^2 y - xy^2; \quad \mathbf{r}(0) = \mathbf{i}, \ \mathbf{r}(\pi) = -\mathbf{i}$

$$\int_C \mathbf{h}(\mathbf{r}) \cdot d\mathbf{r} = \int_C \nabla f(\mathbf{r}) \cdot d\mathbf{r} = f(\mathbf{r}(\pi)) - f(\mathbf{r}(0)) = f(-1, 0) - f(1, 0) = 0 - 0 = 0$$

8. $\mathbf{h}(x, y) = \nabla f(x, y)$ where $f(x, y) = (x^2 + y^4)^{3/2}$

$$\int_C \mathbf{h}(\mathbf{r}) \cdot d\mathbf{r} = \int_C \nabla f(\mathbf{r}) \cdot d\mathbf{r} = f(-1, 0) - f(1, 0) = 1 - 1 = 0$$

9. $\mathbf{h}(x, y) = \nabla f(x, y)$ where $f(x, y) = (x^2 + y^4)^{3/2}$

$$\int_C \mathbf{h}(\mathbf{r}) \cdot d\mathbf{r} = \int_C \nabla f(\mathbf{r}) \cdot d\mathbf{r} = f(1, 0) - f(-1, 0) = 1 - 1 = 0$$

10. $\mathbf{h} = \nabla f$ with $f(x, y) = \cosh x^2 y;$ and C is closed, so $\displaystyle\int_C \mathbf{h} \cdot d\mathbf{r} = 0$

11. $\mathbf{h}(x, y)$ is not a gradient, but part of it,

$$2x \cosh y \, \mathbf{i} + (x^2 \sinh y - y)\mathbf{j},$$

is a gradient. Since we are integrating over a closed curve, the contribution of the gradient part is 0. Thus

$$\int_C \mathbf{h}(\mathbf{r}) \cdot d\mathbf{r} = \int_C (-y\mathbf{i}) \cdot d\mathbf{r}.$$

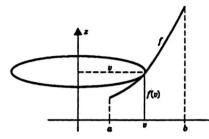

$$C_1: \mathbf{r}(u) = \mathbf{i} + (-1 + 2u)\mathbf{j}, \quad u \in [0,1]$$
$$C_2: \mathbf{r}(u) = (1 - 2u)\mathbf{i} + \mathbf{j}, \quad u \in [0,1]$$
$$C_3: \mathbf{r}(u) = -\mathbf{i} + (1 - 2u)\mathbf{j}, \quad u \in [0,1]$$
$$C_4: \mathbf{r}(u) = (-1 + 2u)\mathbf{i} - \mathbf{j}, \quad u \in [0,1]$$

$$\int_C \mathbf{h}(\mathbf{r}) \cdot d\mathbf{r} = \int_{C_1} (-y\,\mathbf{i}) \cdot d\mathbf{r} + \int_{C_2} (-y\,\mathbf{i}) \cdot d\mathbf{r} + \int_{C_3} (-y\,\mathbf{i}) \cdot d\mathbf{r} + \int_{C_4} (-y\,\mathbf{i}) \cdot d\mathbf{r}$$

$$= \quad 0 \quad + \int_0^1 -\mathbf{i} \cdot (-2\,\mathbf{i})\, du + \quad 0 \quad + \int_0^1 \mathbf{i} \cdot (2\,\mathbf{i})\, du$$

$$= \quad 0 \quad + \int_0^1 2\, du \quad + \quad 0 \quad + \int_0^1 2\, du$$

$$= \quad 4$$

12. $\mathbf{h}(x,y) = \nabla \left(\dfrac{x^2 y^2}{2} \right)$

(a) $\displaystyle \int_0^2 (u^5\,\mathbf{i} + u^4\,\mathbf{j}) \cdot (\mathbf{i} + 2u\,\mathbf{j})\, du = \int_0^2 3u^5\, du = 32$

(b) $f(2,4) - f(0,0) = 32 - 0 = 32$

13. $\mathbf{h}(x,y) = (3x^2 y^3 + 2x)\,\mathbf{i} + (3x^3 y^2 - 4y)\,\mathbf{j}; \quad \dfrac{\partial P}{\partial y} = 9x^2 y^2 = \dfrac{\partial Q}{\partial x}.$ Thus $\mathbf{h}$ is a gradient.

(a) $\mathbf{r}(u) = u\,\mathbf{i} + e^u\,\mathbf{j}, \quad \mathbf{r}'(u) = \mathbf{i} + e^u\,\mathbf{j}, \quad u \in [0,1]$

$$\int_C \mathbf{h}(\mathbf{r}) \cdot d\mathbf{r} = \int_0^1 \left[(3u^2 e^{3u} + 2u) + 3u^3 e^{3u} - 4e^{2u} \right] du = \left[u^3 e^{3u} + u^2 - 2e^{2u} \right]_0^1 = e^3 - 2e^2 + 3$$

(b) $\dfrac{\partial f}{\partial x} = 3x^2 y^3 + 2x \implies f(x,y) = x^3 y^3 + x^2 + g(y);$

$$\dfrac{\partial f}{\partial y} = 3x^3 y^2 + g'(y) = 3x^3 - 4y \implies g'(y) = -4y \implies g(y) = -2y^2$$

Therefore, $f(x,y) = x^3 y^3 + x^2 - 2y^2$.

Now, at $u = 0$, $r(0) = 0\,\mathbf{i} + \mathbf{j} = (0,1)$; at $u = 1$, $r(1) = \mathbf{i} + e\,\mathbf{j} = (1,e)$ and

$$\int_C \mathbf{h}(\mathbf{r}) \cdot d\mathbf{r} = \left[x^3 y^3 + x^2 - 2y^2 \right]_{(0,1)}^{(1,e)} = e^3 - 2e^2 + 3$$

14. $\mathbf{h}(x,y) = \nabla(x^2 \sin y - e^x)$

(a) $\displaystyle \int_0^\pi \left[(2\cos u \sin u - e^{\cos u})\,\mathbf{i} + (\cos^2 u \cos u)\,\mathbf{j} \right] \cdot (-\sin u\,\mathbf{i} + \mathbf{j})\, du = e - e^{-1}$

(b) $f(-1, \pi) - f(1, 0) = e - e^{-1}$

15. $\mathbf{h}(x,y) = (e^{2y} - 2xy)\,\mathbf{i} + (2xe^{2y} - x^2 + 1)\,\mathbf{j}; \quad \dfrac{\partial P}{\partial y} = 2e^{2y} - 2x = \dfrac{\partial Q}{\partial x}.$ Thus $\mathbf{h}$ is a gradient.

(a) $\mathbf{r}(u) = ue^u\,\mathbf{i} + (1+u)\,\mathbf{j}, \quad \mathbf{r}'(u) = (1+u)e^u\,\mathbf{i} + \mathbf{j}, \quad u \in [0,1]$

$$\int_C \mathbf{h}(\mathbf{r})\cdot d\mathbf{r} = \int_0^1 \left[e^2(3ue^{3u} + e^{3u} - 2u^3e^{2u} - 5u^2e^{2u} - 2ue^{2u} + 1 \right]\,du$$

$$= \left[e^2ue^{3u} - u^3e^{2u} - u^2e^{2u} + u \right]_0^1 = e^5 - 2e^2 + 1$$

(b) $\dfrac{\partial f}{\partial x} = e^{2y} - 2xy \quad \Longrightarrow \quad f(x,y) = xe^{2y} - x^2y + g(y).$

$$\dfrac{\partial f}{\partial y} = 2xe^{2y} - x^2 + g'(y) = 3x^3 - 4y \quad \Longrightarrow \quad g'(y) = 1 \quad \Longrightarrow \quad g(y) = y$$

Therefore, $f(x,y) = xe^{2y} - x^2y + y.$

Now, at $u = 0, \; r(0) = 0\,\mathbf{i} + \mathbf{j} = (0,1);$ at $u = 1, \; r(1) = e\,\mathbf{i} + 2\,\mathbf{j} = (e,2)$ and

$$\int_C \mathbf{h}(\mathbf{r})\cdot d\mathbf{r} = \left[xe^{2y} - x^2y + y \right]_{(0,1)}^{(e,2)} = e^5 - 2e^2 + 1$$

16. $\mathbf{h}(x,y,z) = \nabla f \quad$ with $\quad f(x,y,z) = xy^2z^3 \quad \displaystyle\int_C \mathbf{h}\cdot d\mathbf{r} = f(1,1,1) - f(0,0,0) = 1$

17. $\mathbf{h}(x,y,z) = (2xz + \sin y)\,\mathbf{i} + x\cos y\,\mathbf{j} + x^2\,\mathbf{k};$

$$\dfrac{\partial P}{\partial y} = \cos y = \dfrac{\partial Q}{\partial x}, \quad \dfrac{\partial P}{\partial z} = 2x = \dfrac{\partial R}{\partial x}, \quad \dfrac{\partial Q}{\partial z} = 0 = \dfrac{\partial R}{\partial y}. \quad \text{Thus } \mathbf{h} \text{ is a gradient.}$$

$$\dfrac{\partial f}{\partial x} = 2xz + \sin y, \quad \Longrightarrow \quad f(x,y,z) = x^2z + x\sin y + g(y,z)$$

$$\dfrac{\partial f}{\partial y} = x\cos y + \dfrac{\partial g}{\partial y} = x\cos y, \quad \Longrightarrow \quad g(y,z) = h(z) \quad \Longrightarrow f(x,y,z) = x^2z + x\sin y + h(z)$$

$$\dfrac{\partial f}{\partial z} = x^2 + h'(z) = x^2 \quad \Longrightarrow \quad h'(z) = 0 \quad \Longrightarrow \quad h(z) = C$$

Therefore, $f(x,y,z) = x^2z + x\sin y \quad$ (take $C = 0$)

$$\int_C \mathbf{h}(\mathbf{r})\cdot d\mathbf{r} = \int_C \nabla f\cdot d\mathbf{r} = \left[x^2z + x\sin y\right]_{\mathbf{r}(0)}^{\mathbf{r}(2\pi)} = \left[x^2z + x\sin y\right]_{(1,0,0)}^{(1,0,2\pi)} = 2\pi$$

18. $\mathbf{h}(x,y,z) = \nabla f \quad$ with $\quad f(x,y,z) = yz\sin\pi x$

$$\int_C \mathbf{h}\cdot d\mathbf{r} = f\left(\tfrac{1}{2}, \tfrac{\sqrt{3}}{2}, \tfrac{\pi}{3}\right) - f(1,0,0) = \dfrac{1}{6}\pi\sqrt{3}$$

19. $\mathbf{h}(x,y,z) = (2xy + z^2)\,\mathbf{i} + x^2\,\mathbf{j} + 2xz\,\mathbf{k};$

$$\dfrac{\partial P}{\partial y} = 2x = \dfrac{\partial Q}{\partial x}, \quad \dfrac{\partial P}{\partial z} = 2z = \dfrac{\partial R}{\partial x}, \quad \dfrac{\partial Q}{\partial z} = 0 = \dfrac{\partial R}{\partial y}. \quad \text{Thus } \mathbf{h} \text{ is a gradient.}$$

$$\frac{\partial f}{\partial x} = 2xy + z^2 \quad \Longrightarrow \quad f(x,y,z) = x^2 y + xz^2 + g(y,z)$$

$$\frac{\partial f}{\partial y} = x^2 + \frac{\partial g}{\partial y} = x^2 \quad \Longrightarrow \quad g(y,z) = h(z) \quad \Longrightarrow f(x,y,z) = x^2 y + xz^2 + h(z)$$

$$\frac{\partial f}{\partial z} = 2xz + h'(z) = 2xz \quad \Longrightarrow \quad h'(z) = 0 \quad \Longrightarrow \quad h(z) = C$$

Therefore, $f(x,y,z) = x^2 y + xz^2$ (take $C = 0$)

$$\int_C \mathbf{h}(\mathbf{r}) \cdot d\mathbf{r} = \int_C \nabla f \cdot d\mathbf{r} = \left[x^2 y + xz^2 \right]_{\mathbf{r}(0)}^{\mathbf{r}(1)} = \left[x^2 y + xz^2 \right]_{(0,2,0)}^{(2,3,-1)} = 14$$

20. $\mathbf{h}(x,y,z) = \nabla f$ with $f(x,y,z) = z^3 - e^{-x} \ln y$

$$\int_C \mathbf{h} \cdot d\mathbf{r} = f(2,e^2,2) - f(1,1,1) = 7 - 2e^{-2}$$

21. $\mathbf{F}(x,y) = (x + e^{2y})\,\mathbf{i} + (2y + 2xe^{2y})\,\mathbf{j}; \quad \dfrac{\partial P}{\partial y} = 2e^{2y} = \dfrac{\partial Q}{\partial x}.$ Thus $\mathbf{F}$ is a gradient.

$$\frac{\partial f}{\partial x} = x + e^{2y} \quad \Longrightarrow \quad f(x,y) = \frac{1}{2}x^2 + xe^{2y} + g(y);$$

$$\frac{\partial f}{\partial y} = 2xe^{2y} + g'(y) = 2y + 2xe^{2y} \quad \Longrightarrow \quad g'(y) = 2y \quad \Longrightarrow g(y) = y^2 \ \text{(take } C = 0)$$

Therefore, $f(x,y) = \dfrac{1}{2}x^2 + xe^{2y} + y^2.$

$$\int_C \mathbf{F}(\mathbf{r}) \cdot d\mathbf{r} = \int_C \nabla f \cdot d\mathbf{r} = \left[\frac{1}{2}x^2 + xe^{2y} + y^2 \right]_{\mathbf{r}(0)}^{\mathbf{r}(2\pi)} = \left[\frac{1}{2}x^2 + xe^{2y} + y^2 \right]_{(3,0)}^{(3,0)} = 0$$

22. $\mathbf{F} = \nabla f$ with $f(x,y,z) = x^2 \ln y - xyz$ $W = f(3,2,2) - f(1,2,1) = 8 \ln 2 - 10$

23. Set $f(x,y,z) = g(x)$ and $C : \mathbf{r}(u) = u\,\mathbf{i}, \quad u \in [a,b].$
In this case

$$\nabla f(\mathbf{r}(u)) = g'(x(u))\mathbf{i} = g'(u)\mathbf{i} \quad \text{and} \quad \mathbf{r}\prime(u) = \mathbf{i},$$

so that

$$\int_C \nabla f(\mathbf{r}) \cdot d\mathbf{r} = \int_a^b [\nabla f(\mathbf{r}(u)) \cdot \mathbf{r}\prime(u)]\, du = \int_a^b g'(u)\, du.$$

Since $f(\mathbf{r}(b)) - f(\mathbf{r}(a)) = g(b) - g(a),$

$$\int_C \nabla f(\mathbf{r}) \cdot d\mathbf{r} = f(\mathbf{r}(b)) - f(\mathbf{r}(a)) \quad \text{gives} \quad \int_a^b g'(u)\, du = g(b) - g(a).$$

24. $\mathbf{F}(x,y,z) = \dfrac{k}{(x^2 + y^2 + z^2)^{n/2}}(x\,\mathbf{i} + y\,\mathbf{j} + z\,\mathbf{j}) = \nabla f$

  (a) $n = 2$: $f(\mathbf{r}) = k \ln r + C$ (b) $n \neq 2$: $f(\mathbf{r}) = -\left(\dfrac{k}{n-2} \right) \dfrac{1}{r^{n-2}} + C$

25. $\mathbf{F}(\mathbf{r}) = kr\,\mathbf{r} = k\sqrt{x^2 + y^2 + z^2}(x\,\mathbf{i} + y\,\mathbf{j} + z\,\mathbf{k}), \quad k > 0$ constant.

$$\frac{\partial P}{\partial y} = \frac{kxy}{\sqrt{x^2 + y^2 + z^2}} = \frac{\partial Q}{\partial x}, \quad \frac{\partial P}{\partial z} = \frac{kxz}{\sqrt{x^2 + y^2 + z^2}} = \frac{\partial R}{\partial x} \quad \frac{\partial Q}{\partial z} = \frac{kyz}{\sqrt{x^2 + y^2 + z^2}} = \frac{\partial R}{\partial y}$$

Therefore, $\mathbf{F}$ is a gradient field.

$$\frac{\partial f}{\partial x} = kx\sqrt{x^2 + y^2 + z^2} \implies f(x, y, z) = \frac{k}{3}\left(x^2 + y^2 + z^2\right)^{3/2} + g(y, z).$$

$$\frac{\partial f}{\partial y} = ky\sqrt{x^2 + y^2 + z^2} + \frac{\partial g}{\partial y} = ky\sqrt{x^2 + y^2 + z^2} \implies f(x, y, z) = \frac{k}{3}\left(x^2 + y^2 + z^2\right)^{3/2} + h(z)$$

$$\frac{\partial f}{\partial z} = kz\sqrt{x^2 + y^2 + z^2} + h'(z) = kz\sqrt{x^2 + y^2 + z^2} \implies h(z) = C, \text{ constant}$$

Therefore, $f(x, y, z) = \frac{k}{3}\left(x^2 + y^2 + z^2\right)^{3/2} + C$.

26. Set $f(x, y, z) = \frac{1}{2}\displaystyle\int_0^{x^2 + y^2 + z^2} g(u)\,du$. Then

$$\nabla f = g(x^2 + y^2 + z^2)\left[x\,\mathbf{i} + y\,\mathbf{j} + x\,\mathbf{k}\right] = \mathbf{F}(\mathbf{r})$$

27. $\mathbf{F}(\mathbf{r}) = \nabla\left(\dfrac{mG}{r}\right); \quad W = \displaystyle\int_C \mathbf{F}(\mathbf{r})\cdot d\mathbf{r} = mG\left(\dfrac{1}{r_2} - \dfrac{1}{r_1}\right)$

28. (a)   Since the denominator is never 0 in Ω, P and Q are continuously differentiable on Ω.

$$\frac{\partial P}{\partial y} = \frac{x^2 - y^2}{(x^2 + y^2)^2} = \frac{\partial Q}{\partial x}.$$

(b)   Take $\mathbf{r}(u) = \frac{1}{2}\cos u\,\mathbf{i} + \frac{1}{2}\sin u\,\mathbf{j}$.

$$\int_C \mathbf{h}\cdot d\mathbf{r} = \int_0^{2\pi}\left(\frac{\frac{1}{2}\sin u}{1/4}\mathbf{i} - \frac{\frac{1}{2}\cos u}{1/4}\mathbf{j}\right)\cdot\left(-\frac{1}{2}\sin u\,\mathbf{i} + \frac{1}{2}\cos u\,\mathbf{j}\right)du = \int_0^{2\pi} -du = -2\pi$$

Therefore $\mathbf{h}$ is not a gradient since the integral over C (a closed curve) is not zero.

(c) $\Omega : 0 < x^2 + y^2 < 1$ is an open plane region but is not simply connected.

29. $\mathbf{F}(x, y, z) = 0\,\mathbf{i} + 0\,\mathbf{j} + \dfrac{-mGr_0^2}{(r_0 + z)^2}\,\mathbf{k}; \quad \dfrac{\partial P}{\partial y} = 0 = \dfrac{\partial Q}{\partial x}, \quad \dfrac{\partial P}{\partial z} = 0 = \dfrac{\partial R}{\partial x}, \quad \dfrac{\partial Q}{\partial z} = 0 = \dfrac{\partial R}{\partial y}.$

Therefore, $\mathbf{F}(x, y, z)$ is a gradient.

$$\frac{\partial f}{\partial x} = 0 \implies f(x, y, z) = g(y, z); \quad \frac{\partial f}{\partial y} = \frac{\partial g}{\partial y} = 0 \implies g(y, z) = h(z).$$

Therefore $f(x, y, z) = h(z)$.

Now $\dfrac{\partial f}{\partial z} = h'(z) = \dfrac{-mGr_0^2}{(r_0 + z)^2} \implies f(x, y, z) = h(z) = \dfrac{mGr_0^2}{r_0 + z}$

30. $W = f(x, y, 0) - f(x, y, 300) = mGr_0 - \dfrac{mGr_0^2}{r_0 + 300} = 279.07\,mG.$

SECTION 18.3

1. If f is continuous, then $-f$ is continuous and has antiderivatives u. The scalar fields $U(x, y, z) = u(x)$
    are potential functions for **F**:

    $$\nabla U = \frac{\partial U}{\partial x} \mathbf{i} + \frac{\partial U}{\partial y} \mathbf{j} + \frac{\partial U}{\partial z} \mathbf{k} = \frac{du}{dx} \mathbf{i} = -f \mathbf{i} = -\mathbf{F}.$$

2.
    $$\frac{d}{dt} \left(\frac{1}{2} m v^2 \right) = \frac{d}{dt} \left[\frac{1}{2} m (\mathbf{v} \cdot \mathbf{v}) \right] = m(\mathbf{v} \cdot \mathbf{a}) = \mathbf{v} \cdot m \, \mathbf{a}$$
    $$= \mathbf{v} \cdot \mathbf{F} = \mathbf{v} \cdot \frac{e}{c} [\mathbf{v} \times \mathbf{B}] = 0$$

3. The scalar field $U(x, y, z) = \alpha z + d$ is a potential energy function for **F**. We know that the total
    mechanical energy remains constant. Thus, for any times t_1 and t_2,

    $$\tfrac{1}{2} m [v(t_1)]^2 + U(\mathbf{r}(t_1)) = \tfrac{1}{2} m [v(t_2)]^2 + U(\mathbf{r}(t_2)).$$

    This gives

    $$\tfrac{1}{2} m [v(t_1)]^2 + \alpha z(t_1) + d = \tfrac{1}{2} m [v(t_2)]^2 + \alpha z(t_2) + d.$$

    Solve this equation for $v(t_2)$ and you have the desired formula.

4. Throughout the motion, the total mechanical energy of the object remains constant:

    $$\frac{1}{2} m v^2 - \frac{GmM}{r} = E.$$

    At firing $v = v_0$, $\quad r = R_e = $ the radius of the earth and we have

    $$\frac{1}{2} m v_0^2 - \frac{GmM}{R_e} = E.$$

    As $r \to \infty$, $\quad v \to 0$ (by assumption) and also $-GmM/r \to 0$.
    Thus $E = 0$ and we have

    $$\frac{1}{2} m v_0^2 = \frac{GmM}{R_e} \quad \text{and} \quad v_0 = \sqrt{\frac{2GM}{R_e}}.$$

    (Note that v_0 is independent of the mass of the projectile.)

5. (a) We know that $-\nabla U$ points in the direction of maximum decrease of U. Thus $\mathbf{F} = -\nabla U$ attempts
        to drive objects toward a region where U has lower values.

    (b) At a point where u has a minimum, $\quad \nabla U = \mathbf{0}$ and therefore $\mathbf{F} = \mathbf{0}$.

6. We have $x(0) = 2$, $x'(0) = v(0) = 1$. Inserting these values in the formula for E we have

    $$E = \frac{1}{2} m + 2\lambda.$$

    Since $E = \tfrac{1}{2} m v^2 + \tfrac{1}{2} \lambda x^2$ is constant, the maximum value of v comes when $x = 0$. Then

    $$E = \frac{1}{2} m v^2 = \frac{1}{2} m + 2\lambda \quad \text{and} \quad v = \sqrt{1 + 4\lambda/m}.$$

The maximum value of x comes when $v = 0$ (at the endpoints of the oscillation). Then

$$E = \frac{1}{2}\lambda x^2 = \frac{1}{2}m + 2\lambda \quad \text{and} \quad x = \sqrt{m/\lambda + 4}.$$

7. (a) By conservation of energy $\frac{1}{2}mv^2 + U = E$. Since E is constant and U is constant, v is constant.

   (b) ∇U is perpendicular to any surface where U is constant. Obviously so is $\mathbf{F} = -\nabla U$.

8. $\mathbf{F}(\mathbf{r}) = \dfrac{k}{r^2}\mathbf{r} = \nabla f \quad$ where $\quad f(\mathbf{r}) = k \ln r$

9. $f(x, y, z) = -\dfrac{k}{\sqrt{x^2 + y^2 + z^2}}$ is a potential function for $\mathbf{F}$. The work done by $\mathbf{F}$ moving an object along C is:

$$W = \int_C \mathbf{F}(\mathbf{r}) \cdot d\mathbf{r} = \int_a^b \nabla f \cdot d\mathbf{r} = f[\mathbf{r}(b)] - f[\mathbf{r}(a)].$$

Since $\mathbf{r}(a) = (x_0, y_0, z_0)$ and $\mathbf{r}(b) = (x_1, y_1, z_1)$ are points on the unit sphere,

$$f[\mathbf{r}(b)] = f[\mathbf{r}(a)] = -k \quad \text{and so} \quad W = 0$$

SECTION 18.4

1. $\mathbf{r}(u) = u\mathbf{i} + 2u\mathbf{j}, \quad u \in [0, 1]$

$$\int_C (x - 2y)\, dx + 2x\, dy = \int_0^1 \{[x(u) - 2y(u)]x'(u) + 2x(u)\, y'(u)\}\, du = \int_0^1 u\, du = \frac{1}{2}$$

2. $\mathbf{r}(u) = u\mathbf{i} + 2u^2\,\mathbf{j}, \quad u \in [0, 1]$

$$\int_C (x - 2y)\, dx + 2x\, dy = \int_0^1 \{[x(u) - 2y(u)]\, x'(u) + 2x(u)y'(u)\}\, du$$

$$= \int_0^1 (u + 4u^2)\, du = \frac{11}{6}$$

3. $C = C_1 \cup C_2$

$C_1 : \mathbf{r}(u) = u\,\mathbf{i}, \quad u \in [0, 1]; \qquad C_2 : \mathbf{r}(u) = \mathbf{i} + 2u\,\mathbf{j}, \quad u \in [0, 1]$

$$\int_{C_1} (x - 2y)\, dx + 2x\, dy = \int_{C_1} x\, dx = \int_0^1 x(u)\, x'(u)\, du = \int_0^1 u\, du = \frac{1}{2}$$

$$\int_{C_2} (x - 2y)\, dx + 2x\, dy = \int_{C_2} 2x\, dy = \int_0^1 4\, du = 4$$

$$\int_C = \int_{C_1} + \int_{C_2} = \frac{9}{2}$$

4. $C = C_1 \cup C_2$

$C_1 : \mathbf{r}(u) = 2u\,\mathbf{j}, \quad u \in [0, 1]; \qquad C_2 : \quad \mathbf{r}(u) = u\,\mathbf{i} + 2\,\mathbf{j}, \quad u \in [0, 1]$

$$\int_{C_1} (x - 2y)\, dx + 2x\, dy = \int_{C_1} 0\, dy = 0$$

$$\int_{C_2} (x - 2y)\,dx + 2x\,dy = \int_{C_2} (x - 4)\,dx = \int_0^1 (x(u) - 4)x'(u)\,du = \int_0^1 (u - 4)\,du = -\frac{7}{2}$$

$$\int_C = \int_{C_1} + \int_{C_2} = -\frac{7}{2}$$

5. $\mathbf{r}(u) = 2u^2\,\mathbf{i} + u\,\mathbf{j}, \quad u \in [0, 1]$

$$\int_C y\,dx + xy\,dy = \int_0^1 [y(u)\,x'(u) + x(u)\,y(u)\,y'(u)]\,du = \int_0^1 (4u^2 + 2u^3)\,du = \frac{11}{6}$$

6. $\mathbf{r}(u) = 2u\,\mathbf{i} + u\,\mathbf{j}, \quad u \in [0, 1]$

$$\int_C y\,dx + xy\,dy = \int_0^1 [y(u)x'(u) + x(u)y(u)y'(u)]\,du$$

$$= \int_0^1 (2u + 2u^2)\,du = \frac{5}{3}$$

7. $C = C_1 \cup C_2 \qquad C_1: \mathbf{r}(u) = u\,\mathbf{j}, \quad u \in [0, 1]; \qquad C_2: \mathbf{r}(u) = 2u\,\mathbf{i} + \mathbf{j}, \quad u \in [0, 1]$

$$\int_{C_1} y\,dx + xy\,dy = 0$$

$$\int_{C_2} y\,dx + xy\,dy = \int_{C_2} y\,dx = \int_0^1 y(u)\,x'(u)\,du = \int_0^1 2\,du = 2$$

$$\int_C = \int_{C_1} + \int_{C_2} = 2$$

8. $\mathbf{r}(u) = 2u^3\,\mathbf{i} + u\,\mathbf{j}, \quad u \in [0, 1]$

$$\int_C y\,dx + xy\,dy = \int_0^1 [y(u)x'(u) + x(u)y(u)y'(u)]\,du$$

$$= \int_0^1 (6u^3 + 2u^4)\,du = \frac{19}{10}$$

9. $\mathbf{r}(u) = 2u\,\mathbf{i} + 4u\,\mathbf{j}, \quad u \in [0, 1]$

$$\int_C y^2\,dx + (xy - x^2)\,dy = \int_0^1 \{y^2(u)x'(u) + [x(u)y(u) - x^2(u)]\,y'(u)\}\,du$$

$$= \int_0^1 [(4u)^2(2) + (8u^2 - 4u^2)(4)]\,du = \int_0^1 48u^2\,du = 16$$

10. $\mathbf{r}(u) = u\,\mathbf{i} + u^2\,\mathbf{j}, \quad u \in [0, 2]$

$$\int_C y^2\,dx + (xy - x^2)\,dy = \int_0^2 [y^2(u)x'(u) + (x(u)y(u) - x^2(u))y'(u)]\,du$$

$$= \int_0^2 (3u^4 - 2u^3)\,du = \frac{56}{5}$$

11. $\mathbf{r}(u) = \dfrac{1}{8}u^2\,\mathbf{i} + u\,\mathbf{j}, \quad u \in [0,4]$

$$\int_C y^2\,dx + (xy - x^2)\,dy = \int_0^4 \left\{y^2(u)x'(u) + \left[x(u)y(u) - x^2(u)\right]y'(u)\right\}\,du$$

$$= \int_0^4 \left[u^2\left(\frac{u}{4}\right) + \left(\frac{u^2}{8}(u) - \left(\frac{u^2}{8}\right)^2(1)\right)\right]\,du$$

$$= \int_0^4 \left[\frac{3}{8}u^3 - \frac{1}{64}u^4\right]\,du = \frac{104}{5}$$

12. $C = C_1 \cup C_2 \qquad C_1 : \mathbf{r}(u) = 2u\,\mathbf{i}, \quad u \in [0,1]; \quad C_2 : \mathbf{r}(u) = 2\,\mathbf{i} + 4u\,\mathbf{j}, \quad u \in [0,1]$

$$\int_{C_1} y^2\,dx + (xy - x^2)\,dy = \int_{C_1} 0\,dx = 0$$

$$\int_{C_2} y^2\,dx + (xy - x^2)\,dy = \int_{C_2}(2y - 4)\,dy = \int_0^1 [2y(u) - 4]y'(u)\,du = \int_0^1 16(2u - 1)\,du = 0$$

$$\int_C = \int_{C_1} + \int_{C_2} = 0$$

13. $\mathbf{r}(u) = u\,\mathbf{i} + u\,\mathbf{j}, \quad u \in [0,1]$

$$\int_C (y^2 + 2x + 1)\,dx + (2xy + 4y - 1)\,dy$$

$$= \int_0^1 \left\{[y^2(u) + 2x(u) + 1]x'(u) + [2x(u)y(u) + 4y(u) - 1]y'(u)\right\}\,du$$

$$\int_0^1 \left[(u^2 + 2u + 1) + (2u^2 + 4u - 1)\right]\,du = \int_0^1 \left(3u^2 + 6u\right)\,du = 4$$

14. $\mathbf{r}(u) = u\mathbf{i} + u^2\mathbf{j}, \quad u \in [0,1].$

$$\int_C (y^2 + 2x + 1)\,dx + (2xy + 4y - 1)\,dy$$

$$= \int_0^1 \left[(y^2(u) + 2x(u) + 1)x'(u) + (2x(u)y(u) + 4y(u) - 1)y'(u)\right]\,du$$

$$= \int_0^1 (5u^4 + 8u^3 + 1)\,du = 4$$

15. $\mathbf{r}(u) = u\,\mathbf{i} + u^3\,\mathbf{j}, \quad u \in [0,1]$

$$\int_C (y^2 + 2x + 1)\,dx + (2xy + 4y - 1)\,dy$$

$$= \int_0^1 \left\{[y^2(u) + 2x(u) + 1]x'(u) + [2x(u)y(u) + 4y(u) - 1]y'(u)\right\}\,du$$

$$= \int_0^1 \left[(u^6 + 2u + 1) + (2u^4 + 4u^3 - 1)3u^2\right]\,du = \int_0^1 \left(7u^6 + 12u^5 - 3u^2 + 2u + 1\right)\,du = 4$$

16. $C = C_1 \cup C_2 \cup C_3$

$C_1 : \mathbf{r}(u) = 4u\,\mathbf{i}, \; u \in [0,1]; \quad C_2 : \mathbf{r}(u) = 4\,\mathbf{i} + 2u\,\mathbf{j}, \; u \in [0,1];$

$C_3 : \mathbf{r}(u) = (4 - 3u)\mathbf{i} + (2 - u)\mathbf{j}, \; u \in [0, 1]$

$$\int_{C_1} = \int_{C_1} (y^2 + 2x + 1)\, dx = \int_0^1 4(8u + 1)\, du = 20$$

$$\int_{C_2} = \int_{C_2} (8y + 4y - 1)\, dy = \int_0^1 2(24u - 1)\, du = 22$$

$$\int_{C_3} = \int_0^1 \left\{ -3\left[(2 - u)^2 + 2(4 - 3u) + 1\right] - \left[2(4 - 3u)(2 - u) + 4(2 - u) - 1\right]\right\}\, du$$

$$= \int_0^1 (-9u^2 + 54u - 62)\, du = -38.$$

$$\int_C = \int_{C_1} + \int_{C_2} + \int_{C_3} = 20 + 22 - 38 = 4.$$

17. $\mathbf{r}(u) = u\mathbf{i} + u\mathbf{j} + u\mathbf{k}, \quad u \in [0, 1]$

$$\int_C y\, dx + 2z\, dy + x\, dz = \int_0^1 [y(u)\, x'(u) + 2z(u)\, y'(u) + x(u)\, z'(u)]\, du = \int_0^1 4u\, du = 2$$

18.

$$\int_C y\, dx + 2z\, dy + x\, dz = \int_0^1 [y(u)x'(u) + 2z(u)y'(u) + x(u)z'(u)]\, du$$

$$= \int_0^1 (u^2 + 3u^3 + 4u^4)\, du = \frac{113}{60}$$

19. $C = C_1 \cup C_2 \cup C_3$

$C_1 : \mathbf{r}(u) = u\mathbf{k}, \; u \in [0, 1]; \quad C_2 : \mathbf{r}(u) = u\mathbf{j} + \mathbf{k}, \; u \in [0, 1]; \quad C_3 : \mathbf{r}(u) = u\mathbf{i} + \mathbf{j} + \mathbf{k}, \; u \in [0, 1]$

$$\int_{C_1} y\, dx + 2z\, dy + x\, dz = 0$$

$$\int_{C_2} y\, dx + 2z\, dy + x\, dz = \int_{C_2} 2z\, dy = \int_0^1 2z(u)\, y'(u)\, du = \int_0^1 2\, du = 2$$

$$\int_{C_3} y\, dx + 2z\, dy + x\, dz = \int_{C_3} y\, dx = \int_0^1 y(u)\, x'(u)\, du = \int_0^1 du = 1$$

$$\int_C = \int_{C_1} + \int_{C_2} + \int_{C_3} = 3$$

20. $C = C_1 \cup C_2 \cup C_3$

$C_1 : \mathbf{r}(u) = u\mathbf{i}, \quad u \in [0, 1]; \quad C_2 : \mathbf{r}(u) = \mathbf{i} + u\mathbf{j}, \quad u \in [0, 1]; \quad C_3 : \mathbf{r}(u) = \mathbf{i} + \mathbf{j} + u\mathbf{k}, \quad u \in [0, 1]$

$$\int_{C_1} y\, dx + 2z\, dy + x\, dz = 0$$

$$\int_{C_2} y\, dx + 2z\, dy + x\, dz = 0$$

$$\int_{C_3} y\, dx + 2z\, dy + x\, dz = \int_{C_3} x\, dz = \int_0^1 du = 1$$

$$\int_C y\, dx + 2z\, dy + x\, dz = 0 + 0 + 1 = 1$$

21. $\mathbf{r}(u) = 2u\,\mathbf{i} + 2u\,\mathbf{j} + 8u\,\mathbf{k}, \quad u \in [0,1]$

$$\int_C xy\,dx + 2z\,dy + (y+z)\,dz$$

$$= \int_0^1 \{x(u)y(u)x'(u) + 2z(u)y'(u) + [y(u) + z(u)]z'(u)\}\,du$$

$$= \int_0^1 [(2u)(2u)(2) + 2(8u)(2) + (2u + 8u)(8)]\,du$$

$$= \int_0^1 \left(8u^2 + 112u\right)\,du = \frac{176}{3}$$

22. $C = C_1 \cup C_2 \cup C_3$

$C_1 : \mathbf{r}(u) = 2u\,\mathbf{i}, \quad u \in [0,1]; \quad C_2 : \mathbf{r}(u) = 2\,\mathbf{i} + 2u\,\mathbf{j}, \quad u \in [0,1];$

$C_3 : \mathbf{r}(u) = 2\,\mathbf{i} + 2\,\mathbf{j} + 8u\,\mathbf{k}, \quad u \in [0,1]$

$$\int_{C_1} xy\,dx + 2z\,dy + (y+z)\,dz = 0$$

$$\int_{C_2} xy\,dx + 2z\,dy + (y+z)\,dz = 0$$

$$\int_{C_3} xy\,dx + 2z\,dy + (y+z)\,dz = \int_{C_2}(y+z)\,dz = \int_0^1 8(2+8u)\,du = 48$$

23. $\mathbf{r}(u) = u\,\mathbf{i} + u\,\mathbf{j} + 2u^2\,\mathbf{k}, \quad u \in [0,2]$

$$\int_C xy\,dx + 2z\,dy + (y+z)\,dz$$

$$= \int_0^2 \{x(u)y(u)x'(u) + 2z(u)y'(u) + [y(u) + z(u)]z'(u)\}\,du$$

$$= \int_0^2 \left[(u)(u)(1) + 2(2u^2)(1) + (u + 2u^2)(4u)\right]\,du$$

$$= \int_0^2 \left(8u^3 + 9u^2\right)\,du = 56$$

24. $C = C_1 \cup C_2$

$C_1 : \mathbf{r}(u) = 2u\,\mathbf{i} + 2u\,\mathbf{j} + 2u\,\mathbf{k}, \ u \in [0,1]; \quad C_2 : \mathbf{r}(u) = 2\,\mathbf{i} + 2\,\mathbf{j} + (2 + 6u)\,\mathbf{k}, \ u \in [0,1].$

$$\int_{C_1} xy\,dx + 2z\,dy + (y+z)\,dz = \int_0^1 8(u^2 + 2u)\,du = \frac{32}{3}$$

$$\int_{C_2} xy\,dx + 2z\,dy + (y+z)\,dz = \int_{C_2}(y+z)\,dz = \int_0^1 6(4 + 6u)\,du = 42$$

$$\int_C = \int_{C_1} + \int_{C_2} = \frac{158}{3}$$

25. $\mathbf{r}(u) = (u-1)\,\mathbf{i} + (1 + 2u^2)\,\mathbf{j} + u\,\mathbf{k}, \quad u \in [1,2]$

$$\int_C x^2 y\,dx + y\,dy + xz\,dz$$

$$= \int_1^2 \left[x^2(u)y(u)x'(u) + y(u)y'(u) + x(u)z(u)z'(u) \right] \, du$$

$$= \int_1^2 \left[(u-1)^2(1+2u^2)(1) + (1+2u^2)(4u) + (u-1)u \right] \, du$$

$$= \int_1^2 \left(2u^4 + 4u^3 + 4u^2 + u + 1 \right) \, du = \frac{1177}{30}$$

26. $\mathbf{r}(u) = \left(2 - \dfrac{u^2}{2} \right) \mathbf{i} + u \sqrt{1 - \dfrac{u^2}{4}} \, \mathbf{j} + u \, \mathbf{k}, \quad u \in [0, 2]$

$$\int_C y \, dx + yz \, dy + z(x-1) \, dz = \int_0^2 \left[-u^2 \sqrt{1 - \frac{u^2}{4}} + \frac{u^2}{2}(2 - u^2) + u\left(1 - \frac{u^2}{2}\right) \right] \, du = -\frac{\pi}{2} - \frac{8}{15}$$

27. (a) $\dfrac{\partial P}{\partial y} = 6x - 4y = \dfrac{\partial Q}{\partial x}$

$\dfrac{\partial f}{\partial x} = x^2 + 6xy - 2y^2 \quad \Longrightarrow \quad f(x, y) = \dfrac{1}{3}x^3 + 3x^2 y - 2xy^2 + g(y)$

$\dfrac{\partial f}{\partial y} = 3x^2 - 4xy + g'(y) = 3x^2 - 4xy + 2y \quad \Longrightarrow \quad g'(y) = 2y \quad \Longrightarrow \quad g(y) = y^2 + C$

Therefore, $f(x, y) = \dfrac{1}{3}x^3 + 3x^2 y - 2xy^2 + y^2$ (take $C = 0$)

(b) (i) $\displaystyle\int_C (x^2 + 6xy - 2y^2) \, dx + (3x^2 - 4xy + 2y) \, dy = [f(x, y)]_{(3,0)}^{(0,4)} = 7$

(ii) $\displaystyle\int_{C'} (x^2 + 6xy - 2y^2) \, dx + (3x^2 - 4xy + 2y) \, dy = [f(x, y)]_{(4,0)}^{(0,3)} = -\frac{37}{3}$

28. (a) $\mathbf{F} = \nabla f$ where $f(x, y, z) = x^2 y + xz^2 - y^2 z$

(b) (i) $\displaystyle\int_C (2xy + z^2) \, dx + (x^2 - 2yz) \, dy + (2xz - y^2) \, dz = f(3, 2, -1) - f(1, 0, 1) = 25 - 1 = 24$

(ii) $\displaystyle\int_{C'} (2xy + z^2) \, dx + (x^2 - 2yz) \, dy + (2xz - y^2) \, dz = f(1, 0, 1) - f(3, 2 - 1) = -24$

29. $s'(u) = \sqrt{[x'(u)]^2 + [y'(u)]^2} = a$

(a) $M = \displaystyle\int_C k(x + y) \, ds = k \int_0^{\pi/2} [x(u) + y(u)] \, s'(u) \, du = ka^2 \int_0^{\pi/2} (\cos u + \sin u) \, du = 2ka^2$

$x_M M = \displaystyle\int_C kx(x + y) \, ds = k \int_0^{\pi/2} x(u) \, [x(u) + y(u)] \, s'(u) \, du$

$= ka^3 \displaystyle\int_0^{\pi/2} (\cos^2 u + \cos u \sin u) \, du = \frac{1}{4}ka^3(\pi + 2)$

$y_M M = \displaystyle\int_C ky(x + y) \, ds = k \int_0^{\pi/2} y(u) \, [x(u) + y(u)] \, s'(u) \, du$

$= ka^3 \displaystyle\int_0^{\pi/2} (\sin u \cos u + \sin^2 u) \, du = \frac{1}{4}ka^3(\pi + 2)$

$x_M = y_M = \frac{1}{8}a(\pi + 2)$

(b)
$$I = \int_C k(x+y)y^2\,ds = k\int_0^{\pi/2}\left[x(u)y^2(u) + y^3(u)\right]s'(u)\,du$$

$$= ka^4\int_0^{\pi/2}\left[\sin^2 u\cos u + \sin^3 u\right]du$$

$$= ka^4\int_0^{\pi/2}\left[\sin^2 u\cos u + (1-\cos^2 u)\sin u\right]du$$

$$= ka^4\left[\tfrac{1}{3}\sin^3 u - \cos u + \tfrac{1}{3}\cos^3 u\right]_0^{\pi/2} = ka^4$$

$I = \tfrac{1}{2}a^2 M.$

30. (a) $I = \displaystyle\int_C \frac{M}{L}x^2\,ds = \frac{Ma^2}{2\pi}\int_0^{2\pi}\cos^2 u\,du = \frac{1}{2}Ma^2$

(b) $I = \displaystyle\int_C \frac{M}{L}a^2\,ds = \frac{Ma}{2\pi}\int_C ds = Ma^2$

31. (a) $I_z = \displaystyle\int_C k(x+y)a^2\,ds = a^2\int_C k(x+y)\,ds = a^2 M = Ma^2$

(b) The distance from a point (x^*, y^*) to the line $y = x$ is $|y^* - x^*|/\sqrt{2}$.   Therefore
$$I = \int_C k(x+y)\left[\frac{1}{2}(y-x)^2\right]ds = \frac{1}{2}k\int_0^{\pi/2}(a\cos u + a\sin u)(a\sin u - a\cos u)^2 a\,du$$

$$= \frac{1}{2}ka^4\int_0^{\pi/2}(\sin u - \cos u)^2\frac{d}{du}(\sin u - \cos u)\,du$$

$$= \frac{1}{2}ka^4\left[\frac{1}{3}(\sin u - \cos u)^3\right]_0^{\pi/2} = \frac{1}{3}ka^4.$$

From Exercise 29, $M = 2ka^2.$ Therefore

$$I = \tfrac{1}{6}(2ka^2)a^2 = \tfrac{1}{3}Ma^2.$$

32. (a) $M = \displaystyle\int_C k\,ds = \int_0^{2\pi}k\sqrt{\sin^2 u + (1-\cos u)^2}\,du = \int_0^{2\pi}2k\sin\frac{1}{2}u\,du = 8k$

(b)
$$x_M M = \int_C kx\,ds = \int_0^{2\pi}\left[(1-\cos u)(2k\sin\frac{1}{2}u)\right]du$$

$$= 4k\int_0^{2\pi}\sin^3\frac{1}{2}u\,du = \frac{32}{3}k; \quad x_M = \frac{4}{3}$$

$$y_M M = \int_C ky\,ds = \int_0^{2\pi}\left[(u-\sin u)(2k\sin\frac{1}{2}u)\right]du$$

$$= 2k\int_0^{2\pi}(u\sin\frac{1}{2}u - 2\sin^2\frac{1}{2}u\cos\frac{1}{2}u)\,du$$

$$= 8\pi k; \quad y_M = \pi$$

33. (a) $s'(u) = \sqrt{a^2 + b^2}$

$$L = \int_C ds = \int_0^{2\pi} \sqrt{a^2 + b^2}\, du = 2\pi\sqrt{a^2 + b^2}$$

(b) $x_M = 0$, $\quad y_M = 0$ $\quad$ (by symmetry)

$$z_M = \frac{1}{L}\int_C z\, ds = \frac{1}{2\pi\sqrt{a^2 + b^2}}\int_0^{2\pi} bu\sqrt{a^2 + b^2}\, du = b\pi$$

(c) $I_x = \int_C \frac{M}{L}(y^2 + z^2)\, ds = \frac{M}{2\pi}\int_0^{2\pi}(a^2\sin^2 u + b^2 u^2)\, du = \frac{1}{6}M(3a^2 + 8b^2\pi^2)$

$I_y = \frac{1}{6}M(3a^2 + 8b^2\pi^2)$ $\quad$ similarly

$I_z = Ma^2$ $\quad$ (all the mass is at distance a from the z-axis)

34. (a) $s'(u) = 2u^2 + 1$

$$L = \int_C ds = \int_0^a (2u^2 + 1)\, du = \frac{2}{3}a^3 + a = \frac{a(2a^2 + 3)}{3}$$

(b) $x_M = \dfrac{1}{L}\displaystyle\int_C x\, ds = \dfrac{3}{a(2a^2 + 3)}\displaystyle\int_0^a (2u^3 + u)\, du = \dfrac{3a(a^2 + 1)}{2(2a^2 + 3)}$

$y_M = \dfrac{1}{L}\displaystyle\int_C y\, ds = \dfrac{3}{a(2a^2 + 3)}\displaystyle\int_0^a (2u^4 + u^2)\, du = \dfrac{a^2(6a^2 + 5)}{5(2a^2 + 3)}$

$z_M = \dfrac{1}{L}\displaystyle\int_C z\, ds = \dfrac{3}{a(2a^3 + 3)}\displaystyle\int_0^a \left(\dfrac{4}{3}u^5 + \dfrac{2}{3}u^3\right) du = \dfrac{a^3(4a^2 + 3)}{6(2a^3 + 3)}$

(c)
$$I_z = \frac{M}{L}\int_C (x^2 + y^2)\, ds = \frac{3M}{a(2a^3 + 3)}\int_0^a [(u^2 + u^4)(2u^2 + 1)]\, du$$

$$= \frac{Ma^2(30a^4 + 63a^2 + 35)}{35(2a^2 + 3)}$$

35.
$$M = \int_C k(x^2 + y^2 + z^2)\, ds$$

$$= k\sqrt{a^2 + b^2}\int_0^{2\pi}(a^2 + b^2 u^2)\, du = \frac{2}{3}\pi k\sqrt{a^2 + b^2}\,(3a^2 + 4\pi^2 b^2)$$

36. $C : \mathbf{r} = \mathbf{r}(u), \quad u \in [a, b]$

$$\int_C \mathbf{h}(\mathbf{r}) \cdot d\mathbf{r} = \int_a^b [\mathbf{h}(\mathbf{r}(u)) \cdot \mathbf{r}'(u)]\, du$$

$$= \int_a^b \left[\mathbf{h}(\mathbf{r}(u)) \cdot \frac{\mathbf{r}'(u)}{\|\mathbf{r}'(u)\|}\right] \|\mathbf{r}'(u)\|\, du$$

$$= \int_a^b [\mathbf{h}(\mathbf{r}(u)) \cdot \mathbf{T}(\mathbf{r}(u))] s'(u)\, du$$

$$= \int_C [\mathbf{h}(\mathbf{r}) \cdot \mathbf{T}(\mathbf{r})]\, ds$$

SECTION 18.5

1. (a) $\oint_C xy\,dx + x^2\,dy = \int_{C_1} xy\,dx + x^2\,dy + \int_{C_2} xy\,dx + x^2\,dy + \int_{C_3} xy\,dx + x^2\,dy$, where

$C_1: \mathbf{r}(u) = u\,\mathbf{i} + u\,\mathbf{j}, \quad u \in [0,1]; \quad C_2: \mathbf{r}(u) = (1-u)\,\mathbf{i} + \mathbf{j}, \quad u \in [0,1]$

$C_3: \mathbf{r}(u) = (1-u)\,\mathbf{j}, \quad u \in [0,1].$

$\int_{C_1} xy\,dx + x^2\,dy = \int_0^1 (u^2 + u^2)\,du = \dfrac{2}{3}$

$\int_{C_2} xy\,dx + x^2\,dy = \int_0^1 -(1-u)\,du = -\dfrac{1}{2}$

$\int_{C_3} xy\,dx + x^2\,dy = \int_0^1 0^2(-1)\,du = 0$

Therefore, $\oint_C xy\,dx + x^2\,dy = \dfrac{2}{3} - \dfrac{1}{2} = \dfrac{1}{6}.$

(b) $\oint_C xy\,dx + x^2\,dy = \iint_\Omega x\,dx\,dy = \int_0^1 \int_0^y x\,dx\,dy = \int_0^1 \left[\dfrac{1}{2}x^2\right]_0^y du = \dfrac{1}{2}\int_0^1 y^2\,dy = \dfrac{1}{6}$

2. (a) $C = C_1 \cup C_3 \cup C_3 \cup C_4$

$C_1: \mathbf{r}(u) = u\,\mathbf{i}, \quad u \in [0,1]; \quad C_2: \mathbf{r}(u) = \mathbf{i} + u\,\mathbf{j}, \quad u \in [0,1]$

$C_3: \mathbf{r}(u) = (1-u)\mathbf{i} + \mathbf{j}, \quad u \in [0,1]; \quad C_4: \mathbf{r}(u) = (1-u)\mathbf{j}, \quad u \in [0,1]$

$\int_{C_1} x^2 y\,dx + 2y^2\,dy = 0$

$\int_{C_2} x^2 y\,dx + 2y^2\,dy = \int_{C_2} 2y^2\,dy = \int_0^1 2u^2\,du = \dfrac{2}{3}$

$\int_{C_3} x^2 y\,dx + 2y^2\,dy = \int_{C_3} x^2\,dx = \int_0^1 -(1-u)^2\,du = -\dfrac{1}{3}$

$\int_{C_4} x^2 y\,dx + 2y^2\,dy = \int_{C_4} 2y^2\,dy = \int_0^1 -2(1-u)^2\,du = -\dfrac{2}{3}$

$\oint_C = \int_{C_1} + \int_{C_2} + \int_{C_3} + \int_{C_4} = -\dfrac{1}{3}$

(b) $\oint_C x^2 y\,dx + 2y^2\,dy = \iint_\Omega \left[\dfrac{\partial}{\partial x}(2y^2) - \dfrac{\partial}{\partial y}(x^2 y)\right] dx\,dy = \int_0^1 \int_0^1 -x^2\,dx\,dy = -\dfrac{1}{3}$

3. (a) $C: \mathbf{r}(u) = 2\cos u\,\mathbf{i} + 3\sin u\,\mathbf{j}, \quad u \in [0, 2\pi]$

$\oint_C (3x^2 + y)\,dx + (2x + y^3)\,dy$

$= \int_0^{2\pi} \left[(12\cos^2 u + 3\sin u)(-2\sin u) + (4\cos u + 27\sin^3 u)3\cos u\right] du$

$= \int_0^{2\pi} \left[-24\cos^2 u\,\sin u - 6\sin^2 u + 12\cos^2 u + 81\sin^3 u\,\cos u\right] du$

$= \left[8\cos^3 u - 3u + \dfrac{3}{2}\sin 2u + 6u + 3\sin 2u + \dfrac{81}{4}\sin^4 u\right]_0^{2\pi} = 6\pi$

(b) $\oint_C (3x^2 + y)\, dx + (2x + y^3)\, dy = \iint_\Omega 1\, dx\, dy = \text{area of ellipse } \Omega = 6\pi$

4. (a) $C = C_1 \cup C_2$

$C_1 : \mathbf{r}(u) = u\mathbf{i} + u^2\mathbf{j}, \ u \in [0, 1]; \quad C_2 : \mathbf{r}(u) = (1-u)\mathbf{i} + (1-u)\mathbf{j}, \ u \in [0, 1]$

$\displaystyle \int_{C_1} y^2\, dx + x^2\, dy = \int_0^1 (u^4 + 2u^3)\, du = \frac{7}{10}$

$\displaystyle \int_{C_2} y^2\, dx + x^2\, dy = \int_0^1 -2(1-u)^2\, du = -\frac{2}{3}; \quad \oint_C = \int_{C_1} + \int_{C_2} = \frac{1}{30}$

(b) $\displaystyle \oint_C y^2\, dx + x^2\, dy = \iint_\Omega \left[\frac{\partial}{\partial x}(x^2) - \frac{\partial}{\partial y}(y^2) \right] dx\, dy = \int_0^1 \int_{x^2}^x 2(x-y)\, dy\, dx = \frac{1}{30}$

5. $\displaystyle \oint_C 3y\, dx + 5x\, dy = \iint_\Omega (5-3)\, dx dy = 2A = 2\pi$

6. $\displaystyle \oint_C 5x\, dx + 3y\, dy = \iint_\Omega 0\, dx\, dy = 0$

7. $\displaystyle \oint_C x^2\, dy = \iint_\Omega 2x\, dx dy = 2\bar{x}A = 2\left(\frac{a}{2}\right)(ab) = a^2 b$

8. $\displaystyle \oint_C y^2\, dx = \iint_\Omega -2y\, dx\, dy = -ab^2$

9.
$$\oint_C (3xy + y^2)\, dx + (2xy + 5x^2)\, dy = \iint_\Omega \left[(2y + 10x) - (3x + 2y) \right] dx dy$$
$$= \iint_\Omega 7x\, dx dy = 7\bar{x}A = 7(1)(\pi) = 7\pi$$

10. $\displaystyle \oint_C (xy + 3y^2)\, dx + (5xy + 2x^2)\, dy = \iint_\Omega (3x - y)\, dx\, dy = (3\bar{x} - \bar{y})A = (3+2)\pi = 5\pi.$

11. $\displaystyle \oint_C (2x^2 + xy - y^2)\, dx + (3x^2 - xy + 2y^2)\, dy = \iint_\Omega \left[(6x - y) - (x - 2y) \right] dx dy$

$\displaystyle = \iint_\Omega (5x + y)\, dx dy = (5\bar{x} + \bar{y})A = (5a + 0)(\pi r^2) = 5a\pi r^2$

12. $\displaystyle \oint_C (x^2 - 2xy + 3y^2)\, dx + (5x + 1)\, dy = \iint_\Omega (5 + 2x - 6y)\, dx\, dy = (5 + 2\bar{x} - 6\bar{y})A = (5 - 6b)\pi r^2$

13. $\oint_C e^x \sin y \, dx + e^x \cos y \, dy = \iint_\Omega [e^x \cos y - e^x \cos y] \, dx dy = 0$

14. $\oint_C e^x \cos y \, dx + e^x \sin y \, dy = \iint_\Omega 2e^x \sin y \, dx \, dy = \int_0^1 \int_0^\pi 2e^x \sin y \, dy \, dx = 4(e-1)$

15. $\oint_C 2xy \, dx + x^2 \, dy = \iint_\Omega [2x - 2x] \, dx dy = 0$

16. $\oint_C y^2 \, dx + 2xy \, dy = \iint_\Omega 0 \, dx \, dy = 0$

17. $C : \mathbf{r}(u) = a \cos u \, \mathbf{i} + a \sin u \, \mathbf{j}; \quad u \in [0, 2\pi]$

$A = \oint_C -y \, dx = \int_0^{2\pi} (-a \sin u)(-a \sin u) \, du = a^2 \int_0^{2\pi} \sin^2 u \, du = a^2 \left[\frac{1}{2} u - \frac{1}{4} \sin 2u \right]_0^{2\pi} = \pi a^2$

18. $C : \mathbf{r}(u) = a \cos^3 u \, \mathbf{i} + a \sin^3 u \, \mathbf{j}, \quad u \in [0, 2\pi]$

$A = \oint_C -y \, dx = \int_0^{2\pi} (-a \sin^3 u)(-3a \cos^2 u \sin u) \, du = 3a^2 \int_0^{2\pi} \sin^4 u \cos^2 u \, du = \frac{3}{8} \pi a^2$

19. $A = \oint_C x \, dy, \text{ where } C = C_1 \cup C_2;$

$\quad C_1 : \mathbf{r}(u) = u \mathbf{i} + \frac{4}{u} \mathbf{j}, \ 1 \le u \le 4; \quad C_2 : \mathbf{r}(u) = (4 - 3u) \mathbf{i} + (1 + 3u) \mathbf{j}, \ 0 \le u \le 1.$

$\quad \oint_{C_1} x \, dy = \int_1^4 u \left(\frac{-4}{u^2} \right) du = -4 \int_1^4 \frac{1}{u} \, du = -4 \ln 4;$

$\quad \oint_{C_2} x \, dy = \int_0^1 (4 - 3u) 3 \, du = \int_0^1 (12 - 9u) \, du = \frac{15}{2}.$

Therefore, $\quad A = \frac{15}{2} - 4 \ln 4.$

20. $A = \frac{1}{2} \oint_C x \, dy - y \, dx, \text{ where } C = C_1 \cup C_2;$

$\quad C_1 : \mathbf{r}(u) = \sqrt{5} \tan u \, \mathbf{i} + \sqrt{5} \sec u \, \mathbf{j}, \ \tan^{-1}\left(-2/\sqrt{5}\right) \le u \le \tan^{-1}\left(2/\sqrt{5}\right)$

$\quad C_2 : (2 - 4u) \mathbf{i} + 3 \mathbf{j}, \ 0 \le u \le 1$

$\quad \frac{1}{2} \oint_{C_1} x \, dy - y \, dx = \frac{5}{2} \ln 5, \quad \frac{1}{2} \oint_{C_2} x \, dy - y \, dx = 6 \quad \text{Therefore,} \quad A = 6 + \frac{5}{2} \ln 5.$

21. $\oint_C (ay + b) \, dx + (cx + d) \, dy = \iint_\Omega (c - a) \, dx dy = (c - a) A$

22. $\oint_C \mathbf{F}(\mathbf{r}) \cdot d\mathbf{r} = \iint_\Omega (-5) \, dx \, dy = -5A = -\frac{15}{8} \pi a^2 \quad \text{(by Exercise 18)}$

23. We take the arch from $x = 0$ to $x = 2\pi R$.   (Figure 9.11.1)   Let C_1 be the line segment from $(0,0)$ to $(2\pi R, 0)$ and let C_2 be the cycloidal arch from $(2\pi R, 0)$ back to $(0,0)$. Letting $C = C_1 \cup C_2$, we have

$$A = \oint_C x\,dy = \int_{C_1} x\,dy + \int_{C_2} x\,dy = 0 + \int_{C_2} x\,dy$$

$$= \int_{2\pi}^0 R(\theta - \sin\theta)(R\sin\theta)\,d\theta$$

$$= R^2 \int_0^{2\pi} (\sin^2\theta - \theta\sin\theta)\,d\theta$$

$$= R^2 \left[\frac{\theta}{2} - \frac{\sin 2\theta}{4} + \theta\cos\theta - \sin\theta \right]_0^{2\pi} = 3\pi R^2.$$

24. $\oint_C y^3\,dx + (3x - x^3)\,dy = \iint_\Omega (3 - 3x^2 - 3y^2)\,dx\,dy = 3\iint_\Omega (1 - x^2 - y^2)\,dx\,dy$

The double integral is maximized by

$$\Omega : 0 \le x^2 + y^2 \le 1.$$

(This is the maximal region on which the integral is nonnegative.) The line integral is maximized by the unit circle traversed counterclockwise.

25. Taking Ω to be of type II (see Figure 18.5.2), we have

$$\iint_\Omega \frac{\partial Q}{\partial x}(x, y)\,dx\,dy = \int_c^d \int_{\psi_1(y)}^{\psi_2(y)} \frac{\partial Q}{\partial x}(x, y)\,dx\,dy$$

$$= \int_c^d \{Q[\psi_2(y), y] - Q[\psi_1(y), y]\}\,dy$$

$$(*) = \int_c^d Q[\psi_2(y), y]\,dy - \int_c^d Q[\psi_1(y), y]\,dy.$$

The graph of $x = \psi_2(y)$ from $x = c$ to $x = d$ is the curve

$$C_4 : \mathbf{r}_4(u) = \psi_2(u)\,\mathbf{i} + u\,\mathbf{j}, \qquad u \in [c, d].$$

The graph of $x = \psi_1(y)$ from $x = c$ to $x = d$ is the curve

$$C_3 : \mathbf{r}_3(u) = \psi_1(u)\,\mathbf{i} + u\,\mathbf{j}, \qquad u \in [c, d].$$

Then

$$\oint_C Q(x, y)\,dy = \int_{C_4} Q(x, y)\,dy - \int_{C_3} Q(x, y)\,dy$$

$$= \int_c^d Q[\psi_2(u), u]\,du - \int_c^d Q[\psi_1(u), u]\,du.$$

Since u is a dummy variable, it can be replaced by y. Comparison with $(*)$ gives the result.

26. Let $h(\mathbf{r}) = f(\mathbf{r})\nabla g(\mathbf{r}) + g(\mathbf{r})\nabla f(\mathbf{r})$. Then $h = \nabla(fg)$

27. Suppose that f is harmonic. By Green's theorem,

$$\int_C \frac{\partial f}{\partial y}\,dx - \frac{\partial f}{\partial x}\,dy = \iint_\Omega \left(-\frac{\partial^2 f}{\partial^2 x} - \frac{\partial^2 f}{\partial^2 y}\right)\,dx\,dy = \iint_\Omega 0\,dx\,dy = 0.$$

28. $\quad -\dfrac{\lambda}{3}\oint y^3\,dx = -\dfrac{\lambda}{3}\iint_\Omega (-3y^2)\,dx\,dy = \iint_\Omega \lambda y^2\,dx\,dy = I_x$

$\quad \dfrac{\lambda}{3}\oint x^3\,dy = \dfrac{\lambda}{3}\iint_\Omega 3x^2\,dx\,dy = \iint_\Omega \lambda x^2\,dx\,dy = I_y$

29. $\quad \oint_{C_1} = \oint_{C_2} + \oint_{C_3}$

30. Let Ω be the region enclosed by C. Then

$$\int_C f(x)\,dx + g(y)\,dy = \pm \oint_C f(x)\,dx + g(y)\,dy$$

$$= \pm \iint_\Omega \left(\frac{\partial}{\partial x}[g(y)] - \overset{0}{\overbrace{\frac{\partial}{\partial y}[f(x)]}}\right)\,dx\,dy = 0$$

31. $\quad \dfrac{\partial P}{\partial y} = \dfrac{-2xy}{(x^2+y^2)^2} = \dfrac{\partial Q}{\partial x} \quad$ except at $\quad (0,0)$

(a) If C does not enclose the origin, and Ω is the region enclosed by C, then

$$\oint_C \frac{x}{x^2+y^2}\,dx + \frac{y}{x^2+y^2}\,dy = \iint_\Omega 0\,dx\,dy = 0.$$

(b) If C does enclose the origin, then

$$\oint_C = \oint_{C_a}$$

where $C_a : \mathbf{r}(u) = a\cos u\,\mathbf{i} + a\sin u\,\mathbf{j}, \quad u \in [0, 2\pi]\quad$ is a small circle in the inner region of C.
In this case

$$\oint_C = \int_0^{2\pi}\left[\frac{a\cos u}{a^2}(-a\sin u) + \frac{a\sin u}{a^2}(a\cos u)\right]\,du = \int_0^{2\pi} 0\,du = 0.$$

The integral is still 0.

32. (a) $\oint_C -\dfrac{y^3}{(x^2+y^2)^2}\,dx + \dfrac{xy^2}{(x^2+y^2)^2}\,dy = \iint_\Omega 0\,dy\,dx = 0$

(b) By Green's theorem, $\oint_C = \oint_{C'}$, where C' is a circle about the origin. $\quad \mathbf{r}(u) = a\cos u\,\mathbf{i} + a\sin u\,\mathbf{j}$.

$$\oint_{C'} = \int_0^{2\pi}(\sin^4 u + \sin^2 u\cos^2 u)\,du = \int_0^{2\pi}\sin^2 u\,du = \pi$$

33. If Ω is the region enclosed by C, then

$$\oint_C \mathbf{v} \cdot \mathbf{dr} = \oint_C \frac{\partial \phi}{\partial x}\, dx + \frac{\partial \phi}{\partial y}\, dy = \iint_\Omega \left\{ \frac{\partial}{\partial x}\left(\frac{\partial \phi}{\partial y}\right) - \frac{\partial}{\partial y}\left(\frac{\partial \phi}{\partial x}\right) \right\} dx\, dy$$

$$= \iint_\Omega 0\, dx\, dy = 0.$$

equality of mixed partials

34. $\mathbf{r}(u) = [x_1 + (x_2 - x_1)u]\mathbf{i} + [y_1 + (y_2 - y_1)u]\mathbf{j}, \quad u \in [0, 1]$

$$\int_C -y\, dx + x\, dy = \int_0^1 \{[-y_1 - (y_2 - y_1)u](x_2 - x_1) + [x_1 + (x_2 - x_1)u](y_2 - y_1)\}\, du$$

$$= \int_0^1 (x_1 y_2 - x_2 y_1)\, du = x_1 y_2 - x_2 y_1.$$

35. $A = \dfrac{1}{2}\oint_C (-y\, dx + x\, dy)$

$$= \left[\int_{C_1} + \int_{C_2} + \cdots \int_{C_n}\right]$$

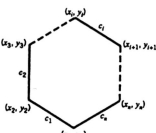

Now

$$\int_{C_i} (-y\, dx + x\, dy) = \int_0^1 \{[y_i + u(y_{i+1} - y_i)](x_{i+1} - x_i) + [x_i + u(x_{i+1} - x_i)](y_{i+1} - y_i)\}\, du$$

$$= x_i y_{i+1} - x_{i+1} y_i, \quad i = 1, 2, \ldots, n; \; x_{n+1} = x_1, \; y_{n+1} = y_1$$

Thus, $A = \dfrac{1}{2}[(x_1 y_2 - x_2 y_1) + (x_2 y_3 - x_3 y_2) + \cdots + (x_n y_1 - x_1 y_n)]$

36. (a) $A = \dfrac{1}{2}[0 + (8 - 1) + 0] = \dfrac{7}{2}$

(b) $A = \dfrac{1}{2}[0 + (12 - 2) + (12 - 0) + (0 + 6) + 0] = 14$

SECTION 18.6

1. $4[(u^2 - v^2)\mathbf{i} - (u^2 + v^2)\mathbf{j} + 2uv\,\mathbf{k}]$ **2.** $u\,\mathbf{k}$ **3.** $2(\mathbf{j} - \mathbf{i})$

4. $\sin u \sin v\,\mathbf{i} + \cos u \cos v\,\mathbf{j} + (\sin^2 u \sin^2 v - \cos^2 u \cos^2 v)\,\mathbf{k}$

5. $\mathbf{r}(u, v) = 3\cos u \cos v\,\mathbf{i} + 2\sin u \cos v\,\mathbf{j} + 6\sin v\,\mathbf{k}, \quad u \in [0, 2\pi], \; v \in [0, \pi/2]$

6. $\mathbf{r}(\theta, z) = 2\cos\theta\,\mathbf{i} + 2\sin\theta\,\mathbf{j} + z\,\mathbf{k}, \quad \theta \in [0, 2\pi], \; z \in [1, 4].$

7. $\mathbf{r}(u,v) = 2\cos u \cos v\,\mathbf{i} + 2\sin u \cos v\,\mathbf{j} + 2\sin v\,\mathbf{k}, \quad u \in [0, 2\pi], \ v \in (\pi/4, \pi/2]$

8. $\mathbf{r}(s,\theta) = s\cos\theta\,\mathbf{i} + s\sin\theta\,\mathbf{j} + (s\cos\theta + 2)\,\mathbf{k}, \quad s \in [0,1], \ \theta \in [0, 2\pi].$

9. The surface consists of all points of the form $(x, g(x,z), z)$ with $(x,z) \in \Omega$. This set of points is given by

$$\mathbf{r}(u,v) = u\,\mathbf{i} + g(u,v)\,\mathbf{j} + v\,\mathbf{k}, \quad (u,v) \in \Omega.$$

10. $\mathbf{r}(y,z) = h(y,z)\,\mathbf{i} + y\,\mathbf{j} + z\,\mathbf{k}, \quad (y,z) \in \Gamma$

11. $x^2/a^2 + y^2/b^2 + z^2/c^2 = 1;$ ellipsoid

12. $z = \dfrac{x^2}{a^2} + \dfrac{y^2}{b^2};$ elliptic paraboloid.

13. $x^2/a^2 - y^2/b^2 = z;$ hyperbolic paraboloid

14. (a) See Exercise 53, Section 14.2.

(b)

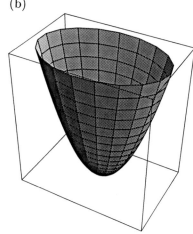

(c)

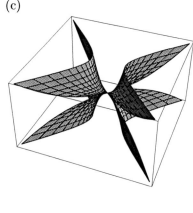

15. For each $v \in [a,b]$, the points on the surface at level $z = f(v)$ form a circle of radius v.
That circle can be parametrized:

$\mathbf{R}(u) = v\cos u\,\mathbf{i} + v\sin u\,\mathbf{j} + f(v)\mathbf{k}, \quad u \in [0, 2\pi].$

Letting v range over $[a,b]$, we obtain the entire surface:

$\mathbf{r}(u,v) = v\cos u\,\mathbf{i} + v\sin u\,\mathbf{j} + f(v)\mathbf{k}; \quad 0 \le u \le 2\pi, \quad a \le v \le b.$

16. For the parametrization given in the answer to Exercise 15

$N(u,v) = -vf'(v)\cos u\,\mathbf{i} + v\,\mathbf{j} - vf'(v)\sin u\,\mathbf{k}, \quad \| N(u,v) \| = v\sqrt{1 + [f'(v)]^2}.$

Therefore

$$A = \int_0^{2\pi} \left\{ \int_a^b v\sqrt{1 + [f'(v)]^2}\,dv \right\} du$$

$$= 2\pi \int_a^b v\sqrt{1 + [f'(v)]^2}\,dv = \int_a^b 2\pi x\sqrt{1 + [f'(v)]^2}\,dx$$

17. Since γ is the angle between p and the xy-plane, γ is the angle between the upper normal to p and $\mathbf{k}$. (Draw a figure.) Therefore, by 18.6.5,

$$\text{area of } \Gamma = \iint_{\Omega} \sec\gamma\, dxdy = (\sec\gamma)A_{\Omega} = A_{\Omega} \quad \sec\gamma.$$

γ is constant

18. $\mathbf{n} = \mathbf{i} + \mathbf{j} + \mathbf{k}$ is an upper normal.

$$\cos\gamma = \frac{\mathbf{n}\cdot\mathbf{k}}{\sqrt{3}} = \frac{1}{\sqrt{3}}, \quad \sec\gamma = \sqrt{3} \quad A = \sqrt{3}\pi b^2$$

19. The surface is the graph of the function

$$f(x,y) = c\left(1 - \frac{x}{a} - \frac{y}{b}\right) = \frac{c}{ab}(ab - bx - ay)$$

defined over the triangle $\Omega : 0 \le x \le a, \quad 0 \le y \le b(1 - x/a)$. Note that Ω has area $\frac{1}{2}ab$.

$$A = \iint_{\Omega} \sqrt{[f'_x(x,y)]^2 + [f'_y(x,y)]^2 + 1}\, dxdy$$

$$= \iint_{\Omega} \sqrt{c^2/a^2 + c^2/b^2 + 1}\, dxdy$$

$$= \frac{1}{ab}\sqrt{a^2b^2 + a^2c^2 + b^2c^2}\iint_{\Omega} dx\, dy = \frac{1}{2}\sqrt{a^2b^2 + a^2c^2 + b^2c^2}.$$

20. $f(x,y) = \sqrt{x^2 + y^2}, \quad \Omega : 0 \le x^2 + y^2 \le 1$

$$A = \iint_{\Omega} \sqrt{[f'_x(x,y)]^2 + [f'_y(x,y)]^2 + 1}\, dx\, dy = \iint_{\Omega} \sqrt{2}\, dx\, dy = \sqrt{2}\pi$$

21. $f(x,y) = x^2 + y^2, \quad \Omega : 0 \le x^2 + y^2 \le 4$

$$A = \iint_{\Omega} \sqrt{4x^2 + 4y^2 + 1}\, dx\, dy \qquad [\text{change to polar coordinates}]$$

$$= \int_0^{2\pi}\int_0^2 \sqrt{4r^2 + 1}\, r\, dr\, d\theta$$

$$= 2\pi\left[\frac{1}{12}(4r^2 + 1)^{3/2}\right]_0^2 = \frac{1}{6}\pi(17\sqrt{17} - 1)$$

22. $f(x,y) = \sqrt{2xy}, \quad \Omega : 0 \le x \le a, \quad 0 \le y \le b$

$$A = \iint_{\Omega} \frac{x+y}{\sqrt{2xy}}\, dx\, dy = \frac{1}{\sqrt{2}}\iint_{\Omega} (\sqrt{x/y} + \sqrt{y/x})\, dx\, dy$$

$$= \frac{1}{\sqrt{2}}\int_0^a\int_0^b \left(\sqrt{x/y} + \sqrt{y/x}\right)\, dy\, dx$$

$$= \frac{2}{3}\sqrt{2}(a + b)\sqrt{ab}$$

23. $f(x, y) = a^2 - (x^2 + y^2)$, $\Omega : \frac{1}{4}a^2 \le x^2 + y^2 \le a^2$

$$A = \iint_\Omega \sqrt{4x^2 + 4y^2 + 1}\, dxdy \qquad [\text{change to polar coordinates}]$$

$$= \int_0^{2\pi} \int_{a/2}^a r\sqrt{4r^2 + 1}\, dr\, d\theta = 2\pi \left[\frac{1}{12}(4r^2 + 1)^{3/2}\right]_{a/2}^a$$

$$= \frac{\pi}{6}\left[(4a^2 + 1)^{3/2} - (a^2 + 1)^{3/2}\right]$$

24. $f(x, y) = \frac{1}{\sqrt{3}}(x + y)^{3/2}$, $\Omega : 0 \le x \le 2$, $0 \le y \le 2 - x$

$$A = \iint \frac{1}{\sqrt{2}}\sqrt{3x + 3y + 2}\, dx\, dy = \frac{1}{\sqrt{2}}\int_0^2 \int_0^{2-x} \sqrt{3x + 3y + 2}\, dy\, dx = \frac{464}{135}$$

25. $f(x, y) = \frac{1}{3}(x^{3/2} + y^{3/2})$, $\Omega : 0 \le x \le 1$, $0 \le y \le x$

$$A = \iint_\Omega \frac{1}{2}\sqrt{x + y + 4}\, dxdy$$

$$= \int_0^1 \int_0^x \frac{1}{2}\sqrt{x + y + 4}\, dy\, dx = \int_0^1 \left[\frac{1}{3}(x + y + 4)^{3/2}\right]_0^x dx$$

$$= \int_0^1 \frac{1}{3}\left[(2x + 4)^{3/2} - (x + 4)^{3/2}\right] dx = \frac{1}{3}\left[\frac{1}{5}(2x + 4)^{5/2} - \frac{2}{5}(x + 4)^{5/2}\right]_0^1$$

$$= \frac{1}{15}(36\sqrt{6} - 50\sqrt{5} + 32)$$

26. $f(x, y) = y^2$, $\Omega : 0 \le x \le 1$, $0 \le y \le 1$

$$A = \iint_\Omega \sqrt{4y^2 + 1}\, dx\, dy = \int_0^1 \int_0^1 \sqrt{4y^2 + 1}\, dy\, dx = \frac{1}{4}\left[2\sqrt{5} + \ln(2 + \sqrt{5})\right]$$

27. The surface $x^2 + y^2 + z^2 - 4z = 0$ is a sphere of radius 2 centered at $(0, 0, 2)$:

$$x^2 + y^2 + z^2 - 4z = 0 \quad \Longleftrightarrow \quad x^2 + y^2 + (z - 2)^2 = 4.$$

The quadric cone $z^2 = 3(x^2 + y^2)$ intersects the sphere at height $z = 3$:

$$\left.\begin{array}{r} x^2 + y^2 + z^2 - 4z = 0 \\ z^2 = 3(x^2 + y^2) \end{array}\right\} \implies \begin{array}{r} 3(x^2 + y^2) + 3z^2 - 12z = 0 \\ 4z^2 - 12z = 0 \\ z = 3. \quad (\text{since } z \ge 2) \end{array}$$

The surface of which we are asked to find the area is a spherical segment of width 1 (from $z = 3$ to $z = 4$) in a sphere of radius 2. The area of the segment is 4π. (Exercise 27, Section 9.9.)

A more conventional solution. The spherical segment is the graph of the function

$$f(x, y) = 2 + \sqrt{4 - (x^2 + y^2)}, \quad \Omega : 0 \le x^2 + y^2 \le 3.$$

Therefore

$$A = \iint_\Omega \sqrt{\left(\frac{-x}{\sqrt{4-x^2-y^2}}\right)^2 + \left(\frac{-y}{\sqrt{4-x^2-y^2}}\right)^2 + 1}\, dxdy$$

$$= \iint_\Omega \frac{2}{\sqrt{4-(x^2+y^2)}}\, dxdy$$

$$= \int_0^{2\pi}\int_0^{\sqrt{3}} \frac{2r}{\sqrt{4-r^2}}\, dr\, d\theta \qquad \text{[changed to polar coordinates]}$$

$$= 2\pi\left[-2\sqrt{4-r^2}\right]_0^{\sqrt{3}} = 4\pi$$

28. The spherical segment is the graph of the function

$$f(x,y) = a + \sqrt{a^2-(x^2+y^2)}, \quad \Omega \le x^2+y^2 \le 2ab - b^2.$$

Therefore

$$A = \iint_\Omega \frac{a}{\sqrt{a^2-(x^2+y^2)}}\, dx\, dy \quad \text{[change to polar coordinate]}$$

$$= \int_0^{2\pi}\int_0^{\sqrt{2ab-b^2}} \frac{ar}{\sqrt{a^2-r^2}}\, dr\, d\theta$$

$$= 2\pi ab.$$

29. (a) $\displaystyle\iint_\Omega \sqrt{\left[\frac{\partial g}{\partial y}(y,z)\right]^2 + \left[\frac{\partial g}{\partial z}(y,z)\right]^2 + 1}\, dydz = \iint_\Omega \sec\left[\alpha(y,z)\right]\, dydz$

where α is the angle between the unit normal with positive **i** component and the positive x-axis

(b) $\displaystyle\iint_\Omega \sqrt{\left[\frac{\partial h}{\partial x}(x,z)\right]^2 + \left[\frac{\partial h}{\partial z}(x,z)\right]^2 + 1}\, dxdz = \iint_\Omega \sec\left[\beta(x,z)\right]\, dxdz$

where β is the angle between the unit normal with positive **j** component and the positive y-axis

30. (a) $\mathbf{r}_u' = -a\sin u\,\mathbf{i} + a\cos u\,\mathbf{j};\quad \mathbf{r}_v' = \mathbf{k}\qquad \mathbf{N}(u,v) = \mathbf{r}_u' \times \mathbf{r}_v' = a\cos u\,\mathbf{i} + a\sin u\,\mathbf{j}$

(b) $A = \displaystyle\iint_\Omega \|\mathbf{N}(u,v)\|\, du\, dv = \iint_\Omega a\, du\, dv = \int_0^{2\pi}\int_0^l a\, dv\, du = 2\pi la$

31. (a) $\mathbf{N}(u,v) = v\cos u\sin\alpha\cos\alpha\,\mathbf{i} + v\sin u\sin\alpha\cos\alpha\,\mathbf{j} - v\sin^2\alpha\,\mathbf{k}$

(b)
$$A = \iint_\Omega \|\mathbf{N}(u,v)\|\, dudv = \iint_\Omega v\sin\alpha\, dudv$$

$$= \int_0^{2\pi}\int_0^s v\sin\alpha\, dv\, du = \pi s^2 \sin\alpha$$

32. (a) Set $x = a \cos u \sin v$, $y = a \sin u \sin v$, $z = b \cos v$. Then

$$\frac{x^2}{a^2} + \frac{y^2}{a^2} + \frac{z^2}{b^2} = 1$$

(b)

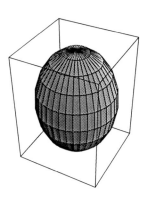

(c) $\mathbf{N}(u, v) = -ab \cos u \sin^2 v \, \mathbf{i} - ab \sin u \sin^2 v \, \mathbf{j} - a^2 \sin v \cos v \, \mathbf{k}$,

$$A = \iint_\Omega \| \mathbf{N}(u, v) \| \; du \, dv = \int_0^{2\pi} \int_0^\pi \sqrt{a^2 b^2 \sin^4 v + a^4 \sin^2 v \cos^2 v} \; dv \, du$$

$$= 2\pi a \int_0^\pi \sin v \sqrt{b^2 \sin^2 v + a^2 \cos^2 v} \; dv$$

33. (a) Set $x = a \cos u \cosh v$, $y = b \sin u \cosh v$, $z = c \sinh v$. Then,

$$\frac{x^2}{a^2} + \frac{y^2}{b^2} - \frac{z^2}{c^2} = 1.$$

(b)

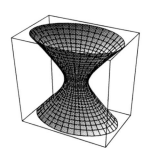

(c) $A = \iint_\Omega \| \mathbf{N}(u, v) \| \, dv \, du$

$$= \int_0^{2\pi} \int_{-\ln 2}^{\ln 2} \sqrt{64 \cos^2 u \, \cosh^2 v + 144 \sin^2 u \, \cosh^2 v + 36 \cosh^2 v \, \sinh^2 v} \; dv \, du$$

34. (a) Set $x = a \cos u \sinh v$, $y = b \sin u \sinh v$, $z = c \cosh v$. Then,

$$\frac{x^2}{a^2} + \frac{y^2}{b^2} - \frac{z^2}{c^2} = -1.$$

(b)

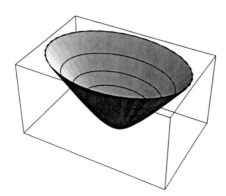

(c) Assuming $c > 0$, $z = c \cosh v > 0$ for all v.

35. $A = \sqrt{A_1{}^2 + A_2{}^2 + A_3{}^2}$; the unit normal to the plane of Ω is a vector of the form
$$\cos \gamma_1 \, \mathbf{i} + \cos \gamma_2 \, \mathbf{j} + \cos \gamma_3 \, \mathbf{k}.$$
Note that
$$A_1 = A \cos \gamma_1, \quad A_2 = A \cos \gamma_2, \quad A_3 = A \cos \gamma_3.$$
Therefore
$$A_1{}^2 + A_2{}^2 + A_3{}^2 = A^2[\cos^2 \gamma_1 + \cos^2 \gamma_2 + \cos^2 \gamma_3] = A^2.$$

36. We can parametrize the surface by setting
$$\mathbf{R}(r\,\theta) = r \cos \theta \, \mathbf{i} + r \sin \theta \, \mathbf{j} + f(r, \theta) \, \mathbf{k}, \quad (r, \theta) \in \Omega.$$
The integrand is $\| \mathbf{N}(r, \theta) \|$.

37. (a) (We use Exercise 36.) $f(r, \theta) = r + \theta;$ $\Omega : 0 \le r \le 1, \quad 0 \le \theta \pi$
$$A = \iint_\Omega \sqrt{r^2 \, [f_r'(r, \theta)]^2 + [f_\theta'(r, \theta)]^2 + r^2} \; drd\theta = \iint_\Omega \sqrt{2r^2 + 1} \; drd\theta$$
$$= \int_0^\pi \int_0^1 \sqrt{2r^2 + 1} \; dr \, d\theta = \frac{1}{4}\sqrt{2}\pi \left[\sqrt{6} + \ln\left(\sqrt{2} + \sqrt{3}\right) \right]$$

(b) $f(r, \theta) = re^\theta;$ $\Omega : 0 \le r \le a, \quad 0 \le \theta \le 2\pi$
$$A = \iint_\Omega r\sqrt{2e^{2\theta} + 1} \; drd\theta = \left(\int_0^{2\pi} \sqrt{2e^{2\theta} + 1} \; d\theta \right) \left(\int_0^a r \, dr \right)$$
$$= \tfrac{1}{2}a^2[\sqrt{2e^{4\pi} + 1} - \sqrt{3} + \ln\left(1 + \sqrt{3}\right) - \ln\left(1 + \sqrt{2e^{4\pi} + 1}\right)]$$

38. $\mathbf{r}(u, v) = x(u, v) \, \mathbf{i} + y(u, v) \, \mathbf{j}, \quad (u, v) \in \Omega$
Straightforward calculation shows that $\| \mathbf{N}(u, v) \| = |J(u, v)|$.

SECTION 18.7

For Exercises 1–6 we have $\sec\left[\gamma(x, y)\right] = \sqrt{y^2 + 1}$. $\mathbf{N}(x, y) = -y\mathbf{j} + \mathbf{k}$, so $\| \mathbf{N}(x, y) \| = \sqrt{y^2 + 1}$.

1. $\displaystyle\iint_S d\sigma = \int_0^1 \int_0^1 \sqrt{y^2 + 1} \; dx \, dy = \int_0^1 \sqrt{y^2 + 1} \; dy = \frac{1}{2}[\sqrt{2} + \ln\left(1 + \sqrt{2}\right)]$

2.
$$\iint_S x^2 \, d\sigma = \int_0^1 \int_0^1 x^2 \sqrt{y^2 + 1} \, dy \, dx$$
$$= \left(\int_0^1 x^2 \, dx \right) \left(\int_0^1 \sqrt{y^2 + 1} \, dy \right) = \frac{1}{6} \left[\sqrt{2} + \ln(1 + \sqrt{2}) \right]$$

3. $\displaystyle \iint_S 3y \, d\sigma = \int_0^1 \int_0^1 3y \sqrt{y^2 + 1} \, dy \, dx = \int_0^1 3y \sqrt{y^2 + 1} \, dy = \left[(y^2 + 1)^{3/2} \right]_0^1 = 2\sqrt{2} - 1$

4.
$$\iint_S (x - y) \, d\sigma = \int_0^1 \int_0^1 x \sqrt{y^2 + 1} \, dy \, dx - \int_0^1 \int_0^1 y \sqrt{y^2 + 1} \, dy \, dx$$
$$= \left(\int_0^1 x \, dx \right) \left(\int_0^1 \sqrt{y^2 + 1} \, dy \right) - \int_0^1 y \sqrt{y^2 + 1} \, dy$$
$$= \frac{1}{4} [\sqrt{2} + \ln(1 + \sqrt{2})] - \frac{1}{3}(2\sqrt{2} - 1)$$
$$= \frac{1}{3} - \frac{5}{12}\sqrt{2} + \frac{1}{4}\ln(1 + \sqrt{2})$$

5. $\displaystyle \iint_S \sqrt{2}z \, d\sigma = \iint_S y \, d\sigma = \frac{1}{3}(2\sqrt{2} - 1)$ \quad (Exercise 3)

6. $\displaystyle \iint_S \sqrt{1 + y^2} \, d\sigma = \int_0^1 \int_0^1 (1 + y^2) \, dy \, dx = \int_0^1 (1 + y^2) \, dy = \frac{4}{3}$

7. $\displaystyle \iint_S xy \, d\sigma; \quad S : \mathbf{r}(u, v) = (6 - 2u - 3v)\mathbf{i} + u\mathbf{j} + v\mathbf{k}, \quad 0 \le u \le 3 - \frac{3}{2}v, \quad 0 \le v \le 2$

$$\|\mathbf{N}(u, v)\| = \|(-2\mathbf{i} + \mathbf{j}) \times (-3\mathbf{i} + \mathbf{k})\| = \sqrt{14}$$

$$\iint_S xy \, d\sigma = \sqrt{14} \iint_\Omega x(u, v)y(u, v) \, du \, dv$$
$$= \sqrt{14} \iint_\Omega (6 - 2u - 3v)u \, du \, dv$$
$$= \sqrt{14} \int_0^2 \int_0^{3 - 3v/2} (6u - 2u^2 - 3uv) \, du \, dv$$
$$= \sqrt{14} \left[3\left(3 - \tfrac{3}{2}v\right)^2 - \tfrac{2}{3}\left(3 - \tfrac{3}{2}v\right)^3 - \tfrac{3}{2}v\left(3 - \tfrac{3}{2}v\right)^2 \right] dv = \frac{9}{2}\sqrt{14}$$

8. S is given by $z = f(x, y) = 1 - x - y, \quad 0 \le x \le 1, \quad 0 \le y \le 1 - x$
$$\iint_S xyz \, d\sigma = \int_0^1 \int_0^{1-x} xy(1 - x - y)\sqrt{(-1)^2 + (-1)^2 + 1} \, dy \, dx$$
$$= \sqrt{3} \int_0^1 \int_0^{1-x} xy(1 - x - y) \, dy \, dx = \frac{\sqrt{3}}{120}$$

9. $\displaystyle\iint_S x^2 z\,d\sigma; \quad S : \mathbf{r}(u,v) = (\cos u\,\mathbf{i} + v\,\mathbf{j} + \sin u\,\mathbf{k}, \quad 0 \le u \le \pi, \quad 0 \le v \le 2.$

$$\mathbf{N}(u,v) = \begin{vmatrix} \mathbf{i} & \mathbf{j} & \mathbf{k} \\ -\sin u & 0 & \cos u \\ 0 & 1 & 0 \end{vmatrix} = -\cos u\,\mathbf{i} - \sin u\,\mathbf{k} \quad \text{and} \quad \|\mathbf{N}(u,v)\| = 1.$$

$$\iint_S x^2 z\,d\sigma = \iint_\Omega \cos^2 u \sin u\,du\,dv = \int_0^2 \int_0^\pi \cos^2 u \sin u\,du\,dv = \frac{4}{3}$$

10.

$$\iint_S (x^2 + y^2 + z^2)\,d\sigma = \iint_{x^2+y^2 \le 1} [x^2 + y^2 + (x+2)^2]\sqrt{2}\,dx\,dy$$

$$= \int_0^{2\pi} \int_0^1 [r^2 + (r\cos\theta + 2)^2]\sqrt{2}\,r\,dr\,d\theta$$

$$= \int_0^{2\pi} \int_0^1 (r^3 + r^3\cos^2\theta + 4r^2\cos\theta + 4r)\,dr\,d\theta = \frac{19\sqrt{2}}{4}\pi$$

11. $\displaystyle\iint_S (x^2 + y^2)\,d\sigma; \quad S : \mathbf{r}(u,v) = \cos u \cos v\,\mathbf{i} + \cos u \sin v\,\mathbf{j} + \sin u\,\mathbf{k}, \quad 0 \le u \le \pi/2, \quad 0 \le v \le 2\pi.$

$$\mathbf{N}(u,v) = \begin{vmatrix} \mathbf{i} & \mathbf{j} & \mathbf{k} \\ -\sin u \cos v & -\sin u \sin v & \cos u \\ -\cos u \sin v & \cos u \cos v & 0 \end{vmatrix} = -\cos^2 u \cos v\,\mathbf{i} + \cos^2 u \sin v\,\mathbf{j} - \sin u \cos u\,\mathbf{k};$$

$$\|\mathbf{N}(u,v)\| = \cos u.$$

$$\iint_S (x^2 + y^2)\,d\sigma = \iint_\Omega \cos^2 u \cos u\,du\,dv = \int_0^{2\pi} \int_0^{\pi/2} \cos^3 u\,du\,dv = \frac{4}{3}\pi$$

12.

$$\iint_S (x^2 + y^2)\,d\sigma = \int_{x^2+y^2 \le 1} \int (x^2 + y^2)\sqrt{4x^2 + 4y^2 + 1}\,dx\,dy = \int_0^{2\pi} \int_0^1 r^2\sqrt{4r^2 + 1}\,r\,dr\,d\theta$$

$$= 2\pi \int_0^1 r^3\sqrt{4r^2 + 1}\,dr = \frac{25\sqrt{5} + 1}{60}\pi$$

For Exercises 13–16 the surface S is given by:

$$f(x,y) = a - x - y; \quad 0 \le x \le a, \quad 0 \le y \le a - x \quad \text{and} \quad \sec[\gamma(x,y)] = \sqrt{3}.$$

13. $\displaystyle M = \iint_S \lambda(x,y,x)\,d\sigma = \int_0^a \int_0^{a-x} k\sqrt{3}\,dy\,dx = \int_0^a k\sqrt{3}\,(a-x)\,dx = \frac{1}{2}a^2 k\sqrt{3}$

14.
$$M = \iint_S k(x+y)\, d\sigma = \int_0^a \int_0^{a-x} k(x+y)\sqrt{3}\, dy\, dx$$

$$= \frac{1}{2}k\sqrt{3}\int_0^a (a^2 - x^2)\, dx = \frac{1}{3}\sqrt{3}a^3 k$$

15. $M = \iint_S \lambda(x,y,z)\, d\sigma = \int_0^a \int_0^{a-x} kx^2\sqrt{3}\, dy\, dx = \int_0^a k\sqrt{3}x^2(a-x)\, dx = \frac{1}{12}a^4 k\sqrt{3}$

16.
$$\overline{x}A = \iint_S x\, d\sigma = \int_0^a \int_0^{a-x} x\sqrt{3}\, dy\, dx$$

$$= \sqrt{3}\int_0^a (ax - x^2)\, dx = \frac{1}{6}\sqrt{3}a^3$$

$$A = \iint_S d\sigma = \int_0^a \int_0^{a-x} \sqrt{3}\, dy\, dx$$

$$= \sqrt{3}\int_0^a (a - x)\, dx = \frac{1}{2}\sqrt{3}a^2$$

$$\overline{x} = \overline{x}A/A = \frac{1}{3}a; \quad \text{similarly} \quad \overline{y} = \overline{z} = \frac{1}{3}a$$

17. $S: \mathbf{r}(u,v) = a\cos u \cos v\, \mathbf{i} + a\sin u \cos v\, \mathbf{j} + a\sin v\, \mathbf{k}$ with $0 \le u \le 2\pi$, $0 \le v \le \frac{1}{2}\pi$.   By a previous calculation $\|\mathbf{N}(u,v)\| = a^2 \cos v$.

$\overline{x} = 0$, $\overline{y} = 0$ (by symmetry)

$$\overline{z}A = \iint_S z\, d\sigma = \iint_\Omega z(u,v)\, \|\mathbf{N}(u,v)\|\, du\, dv = \int_0^{2\pi} \int_0^{\pi/2} a^3 \sin v \cos v\, dv\, du = \pi a^3$$

$\overline{z} = \frac{1}{2}a$ since $A = 2\pi a^2$

18. $\mathbf{N}(u,v) = 2\,\mathbf{i} + 2\,\mathbf{j} - 2\,\mathbf{k}$

$$A = \iint_S d\sigma = \iint_\Omega \|\mathbf{N}(u,v)\|\, du\, dv = \int_0^1 \int_0^1 2\sqrt{3}\, du\, dv = 2\sqrt{3}$$

19. $\mathbf{N}(u,v) = (\mathbf{i} + \mathbf{j} + 2\,\mathbf{k}) \cdot (\mathbf{i} - \mathbf{j}) = 2\,\mathbf{i} + 2\,\mathbf{j} - 2\,\mathbf{k}$

$$\text{flux in the direction of } \mathbf{N} = \iint_S \left(\mathbf{v} \cdot \frac{\mathbf{N}}{\|\mathbf{N}\|}\right) d\sigma = \iint_\Omega [\mathbf{v}(x(u),\, y(u),\, z(u)) \cdot \mathbf{N}(u,v)]\, du\, dv$$

$$= \iint_\Omega [(u+v)\mathbf{i} - (u-v)\mathbf{j}] \cdot [2\,\mathbf{i} + 2\,\mathbf{j} - 2\,\mathbf{k}]\, du\, dv.$$

$$= \iint_\Omega 4v\, du\, dv = 4\int_0^1 \int_0^1 v\, dv\, du = 2$$

20. $\sec[\gamma(x, y)] = \sqrt{x^2 + y^2 + 1}$

$$M = \iint_S kxy \, d\sigma = k \int_0^1 \int_0^1 xy\sqrt{x^2 + y^2 + 1} \, dy \, dx$$

$$= \frac{1}{3}k \int_0^1 \left[x(x^2 + 2)^{3/2} - x(x^2 + 1)^{3/2} \right] dx$$

$$= \frac{1}{15}(9\sqrt{3} - 8\sqrt{2} + 1)k$$

For Exercises 21–23: $\mathbf{n} = \dfrac{1}{a}(x\,\mathbf{i} + y\,\mathbf{j} + z\,\mathbf{k})$

$S : \mathbf{r}(u, v) = a\cos u \cos v\,\mathbf{i} + a\sin u \cos v\,\mathbf{j} + a\sin v\,\mathbf{k}$ with $0 \leq u \leq 2\pi, \quad -\frac{1}{2}\pi \leq v \leq \frac{1}{2}\pi$

$\|\mathbf{N}(u, v)\| = a^2 \cos v$

21. With $\mathbf{v} = z\,\mathbf{k}$

$$\text{flux} = \iint_S (\mathbf{v} \cdot \mathbf{n}) \, d\sigma = \frac{1}{a} \iint_S z^2 \, d\sigma = \frac{1}{a} \iint_\Omega (a^2 \sin^2 v)(a^2 \cos v) \, du \, dv$$

$$= a^3 \int_0^{2\pi} \int_{-\pi/2}^{\pi/2} (\sin^2 v \cos v) \, du \, dv = \frac{4}{3}\pi a^3$$

22. With $\mathbf{v} = x\,\mathbf{i} + y\,\mathbf{j} + z\,\mathbf{k}$

$$\text{flux} = \iint_S (\mathbf{v} \cdot \mathbf{n}) \, d\sigma = a \iint_S d\sigma = aA = 4\pi a^3$$

23. With $\mathbf{v} = y\,\mathbf{i} - x\,\mathbf{j}$

$$\text{flux} = \iint_S (\mathbf{v} \cdot \mathbf{n}) \, d\sigma = \frac{1}{a} \iint_S \underbrace{(yx - xy)}_{0} \, d\sigma = 0$$

24. $I_x = \displaystyle\iint_S (y^2 + z^2) \, d\sigma = 2\sqrt{3} \int_0^1 \int_0^1 (5u^2 - 2uv + v^2) \, dv \, du = 3\sqrt{3}$

$I_y = \displaystyle\iint_S (x^2 + z^2) \, d\sigma = 2\sqrt{3} \int_0^1 \int_0^1 (5u^2 + 2uv + v^2) \, dv \, du = 5\sqrt{3}$

$I_z = \displaystyle\iint_S (x^2 + z^2) \, d\sigma = 4\sqrt{3} \int_0^1 \int_0^1 (u^2 + v^2) \, d\sigma = \frac{8}{3}\sqrt{3}$

For Exercises 25–27 the triangle S is the graph of the function

$$f(x, y) = a - x - y \quad \text{on} \quad \Omega : 0 \leq x \leq a, \quad 0 \leq y \leq a - x.$$

The triangle has area $A = \frac{1}{2}\sqrt{3}a^2$.

25. With $\quad \mathbf{v} = x\,\mathbf{i} + y\,\mathbf{j} + z\,\mathbf{k}$

$$\text{flux} = \iint_S (\mathbf{v} \cdot \mathbf{n})\, d\sigma = \iint_\Omega (-v_1 f'_x - v_2 f'_y + v_3)\, dx dy$$

$$= \iint_\Omega [-x(-1) - y(-1) + (a - x - y)]\, dx dy = a \iint_\Omega dx dy = aA = \frac{1}{2}\sqrt{3}a^3$$

26. With $\quad \mathbf{v} = (x + z)\,\mathbf{k}$

$$\text{flux} = \iint_S (\mathbf{v} \cdot \mathbf{n})\, d\sigma = \iint_\Omega (-v_1 f'_x - v_2 f'_y + v_3)\, dx\, dy$$

$$= \iint_\Omega (a - y)\, dx\, dy = \int_0^a \int_0^{a-x} (a - y)\, dy\, dx = \frac{1}{3}a^3$$

27. With $\quad \mathbf{v} = x^2\,\mathbf{i} - y^2\,\mathbf{j}$

$$\text{flux} = \iint_S (\mathbf{v} \cdot \mathbf{n})\, d\sigma = \iint_\Omega (-v_1 f'_x - v_2 f'_y + v_3)\, dx dy$$

$$= \iint_\Omega [-x^2(-1) - (-y^2)(-1) + 0]\, dx dy = \int_0^a \int_0^{a-x} (x^2 - y^2)\, dy\, dx$$

$$= \int_0^a \left[ax^2 - x^3 - \frac{1}{3}(a - x)^3 \right] dx = \left[\frac{1}{3}ax^3 - \frac{1}{4}x^4 + \frac{1}{12}(a - x)^4 \right]_0^a = 0$$

28. With $\quad \mathbf{v} = -xy^2\mathbf{i} + z\,\mathbf{j}$

$$\text{flux} = \iint_S (\mathbf{v} \cdot \mathbf{n})\, d\sigma = \iint_\Omega (-v_1 f'_x - v_2 f'_y + v_3)\, dx\, dy$$

$$= \int_0^1 \int_0^2 (xy^3 - x^2 y)\, dy\, dx = \frac{4}{3}$$

29. With $\quad \mathbf{v} = xz\,\mathbf{j} - xy\,\mathbf{k}$

$$\text{flux} = \iint_S (\mathbf{v} \cdot \mathbf{n})\, d\sigma = \iint_\Omega (-v_1 f'_x - v_2 f'_y + v_3)\, dx dy$$

$$= \iint_\Omega (-x^3 y - xy)\, dx dy = \int_0^1 \int_0^2 (-x^3 y - xy)\, dy\, dx$$

$$= \int_0^1 -2(x^3 + x)\, dx = -\frac{3}{2}$$

30. With $\quad \mathbf{v} = x^2 y \,\mathbf{i} + z^2 \,\mathbf{k}$

$$\text{flux} = \iint_S (\mathbf{v} \cdot \mathbf{n}) \, d\sigma = \iint_\Omega (-v_1 f_x' - v_2 f_y' + v_3) \, dx \, dy$$

$$= \int_0^1 \int_0^2 (-x^2 y^2 + x^2 y^2) \, dy \, dx = 0$$

31. $\mathbf{n} = \dfrac{1}{a}(x \,\mathbf{i} + y \,\mathbf{j})$

$$\text{flux} = \iint_S (\mathbf{v} \cdot \mathbf{n}) \, d\sigma = \frac{1}{a} \iint_S [(x \,\mathbf{i} + y \,\mathbf{j} + z \,\mathbf{k}) \cdot (x \,\mathbf{i} + y \,\mathbf{j})] \, d\sigma$$

$$= \frac{1}{a} \iint_S (x^2 + y^2) \, d\sigma = a \iint_S d\sigma = a \,(\text{area of } S) = a \,(2\pi a l) = 2\pi a^2 l$$

32.

$$\text{flux} = \iint_S \left(GmM \frac{\mathbf{r}}{r^3} \cdot \frac{\mathbf{r}}{r} \right) d\sigma = GmM \iint_S \frac{1}{r^2} \, d\sigma$$

$$= GmM \iint_S \frac{1}{a^2} \, d\sigma = \frac{GmM}{a^2} \iint_S d\sigma = \frac{GmM}{a^2}(4\pi a^2) = 4\pi GmM$$

In Exercises 33–36, S is the graph of $f(x,y) = \dfrac{2}{3}(x^{3/2} + y^{3/2}), \quad 0 \le x \le 1, \ 0 \le y \le 1 - x$

We use

$$\text{flux} = \iint_S (\mathbf{v} \cdot \mathbf{n}) \, d\sigma = \iint_\Omega (-v_1 f_x' - v_2 f_y' + v_3) \, dx \, dy.$$

33. With $\quad \mathbf{v} = x \,\mathbf{i} - y \,\mathbf{j} + \frac{3}{2} z \,\mathbf{k}$

$$\text{flux} = \iint_S (\mathbf{v} \cdot \mathbf{n}) \, d\sigma = \iint_\Omega (-v_1 f_x' - v_2 f_y' + v_3) \, dx dy = \iint_\Omega 2y^{3/2} \, dx dy$$

$$= \int_0^1 \int_0^{1-x} 2y^{3/2} \, dy \, dx = \int_0^1 \frac{4}{5}(1-x)^{5/2} \, dx = \frac{8}{35}$$

34. With $\quad \mathbf{v} = x^2 \,\mathbf{i},$

$$\text{flux} = \int_0^1 \int_0^{1-x} -x^{5/2} \, dy \, dx = -\frac{4}{63}.$$

35. With $\quad \mathbf{v} = y^2 \,\mathbf{j}$

$$\text{flux} = \iint_S (\mathbf{v} \cdot \mathbf{n}) \, d\sigma = \iint_\Omega (-v_1 f_x' - v_2 f_y' + v_3) \, dx dy = \iint_\Omega -y^{5/2} \, d\sigma$$

$$= \int_0^1 \int_0^{1-x} -y^{5/2} \, dy \, dx = \int_0^1 -\frac{2}{7}(1-x)^{7/2} \, dx = -\frac{4}{63}$$

36. With $\mathbf{v} = y\,\mathbf{i} - \sqrt{xy}\,\mathbf{j},$

$$\text{flux} = \int_0^1 \int_0^{1-x} (-yx^{1/2} + \sqrt{xy}\,y^{1/2})\,dy\,dx = 0.$$

37. $\bar{x} = 0,\quad \bar{y} = 0$ by symmetry. You can verify that $\|\mathbf{N}(u,v)\| = v\sin\alpha.$

$$\bar{z}A = \iint_S z\,d\sigma = \iint_\Omega (s\cos\alpha)(v\sin\alpha)\,dudv = \sin\alpha\cos\alpha \int_0^{2\pi}\int_0^s v^2\,dv\,du = \tfrac{2}{3}\pi\sin\alpha\cos\alpha\,s^3$$

$$\bar{z} = \tfrac{2}{3}s\cos\alpha \quad \text{since} \quad A = \pi s^2 \sin\alpha$$

38.
$$M = \iint_\Omega k\sqrt{x^2+y^2}\,d\sigma = \iint_\Omega k\sqrt{x^2+y^2}\,\sqrt{2}\,dx\,dy$$

$$= k\sqrt{2}\int_0^{2\pi}\int_0^1 r^2\,dr\,d\theta = \frac{2}{3}\sqrt{2}\pi k$$

39. $f(x,y) = \sqrt{x^2+y^2}$ on $\Omega: 0 \le x^2+y^2 \le 1;\quad \lambda(x,y,z) = k\sqrt{x^2+y^2}$

$x_M = 0,\quad y_M = 0$ (by symmetry)

$$z_M M = \iint_S z\lambda(x,y,z)\,d\sigma = \iint_\Omega k(x^2+y^2)\sec[\gamma(x,y)]\,dxdy$$

$$= k\sqrt{2}\iint_\Omega (x^2+y^2)\,dxdy$$

$$= k\sqrt{2}\int_0^{2\pi}\int_0^1 r^3\,dr\,d\theta = \frac{1}{2}\sqrt{2}\pi k$$

$z_M = \tfrac{3}{4}$ since $M = \tfrac{2}{3}\sqrt{2}\pi k$ (Exercise 38)

40. (a)
$$I_x = \iint_\Omega k\sqrt{x^2+y^2}(y^2+z^2)\,d\sigma = \iint_\Omega k\sqrt{x^2+y^2}(y^2+x^2+y^2)\,d\sigma$$

$$= k\sqrt{2}\iint_\Omega \left[y^2(x^2+y^2)^{1/2} + (x^2+y^2)^{3/2}\right]dx\,dy$$

$$= k\sqrt{2}\int_0^{2\pi}\int_0^1 (r^4\sin^2\theta + r^4)\,dr\,d\theta = \frac{3\sqrt{2}}{5}\pi k$$

(b) $I_y = I_x$ by symmetry

(c)
$$I_z = \iint_\Omega k\sqrt{x^2+y^2}(x^2+y^2)\,d\sigma = \iint_S k(x^2+y^2)^{3/2}\,d\sigma$$

$$= k\sqrt{2}\iint_\Omega (x^2+y^2)^{3/2}\,dx\,dy = k\sqrt{2}\int_0^{2\pi}\int_0^1 r^4\,dr\,d\theta = \frac{2}{5}\sqrt{2}\pi k$$

41. no answer required

42.
$$M = \iint\limits_{S} k(y^2 + z^2)\, d\sigma = 2\sqrt{3}k \iint\limits_{S} \left[(u-v)^2 + 4u^2 \right] du\, dv$$

$$= 2\sqrt{3}k \int_0^1 \int_0^1 (5u^2 - 2uv + v^2)\, dv\, du = 3\sqrt{3}k$$

43. $x_M M = \iint\limits_{S} x\lambda\,(x, y, z)\, d\sigma = \iint\limits_{S} kx(y^2 + z^2)\, d\sigma$

$$= 2\sqrt{3}\, k \iint\limits_{\Omega} (u + v)\left[(u-v)^2 + 4u^2 \right]\, du dv$$

$$= 2\sqrt{3}\, k \int_0^1 \int_0^1 (5u^3 - 2u^2v + uv^2 + 5u^2v - 2uv^2 + v^3)\, dv\, du$$

$$= 2\sqrt{3}\, k \int_0^1 \left(5u^3 - u^2 + \frac{1}{3}u + \frac{5}{2}u^2 - \frac{2}{3}u + \frac{1}{4} \right) du = \frac{11}{3}\sqrt{3}k$$

$$x_M = \frac{11}{9} \quad \text{since} \quad M = 3\sqrt{3}k \qquad \text{(Exercise 42)}$$

44.
$$I_z = \iint\limits_{S} \lambda(x, y, z)(x^2 + y^2)\, d\sigma = \iint\limits_{S} k(y^2 + z^2)(x^2 + y^2)\, d\sigma$$

$$= 2\sqrt{3}k \iint\limits_{\Omega} \left[(u-v)^2 + 4u^2 \right]\left[(u+v)^2 + (u-v)^2 \right]\, du\, dv$$

$$= 4\sqrt{3}k \int_0^1 \int_0^1 (5u^4 - 2u^3v + 6u^2v^2 - 2uv^3 + v^4)\, du\, dv = \frac{82\sqrt{3}}{15}k.$$

45. Total flux out of the solid is 0. It is clear from a diagram that the outer unit normal to the cylindrical side of the solid is given by $\mathbf{n} = x\,\mathbf{i} + y\,\mathbf{j}$ in which case $\mathbf{v} \cdot \mathbf{n} = 0$. The outer unit normals to the top and bottom of the solid are $\mathbf{k}$ and $-\mathbf{k}$ respectively. So, here as well, $\mathbf{v} \cdot \mathbf{n} = 0$ and the total flux is 0.

46. The flux through the upper boundary is 0:

$$(y\mathbf{i} - x\mathbf{j}) \cdot \mathbf{k} = 0.$$

The flux through the lower boundary is 0:

$$\mathbf{v} \cdot \left(\frac{\partial f}{\partial x}\mathbf{i} + \frac{\partial f}{\partial y}\mathbf{j} - \mathbf{k} \right) = (y\,\mathbf{i} - x\,\mathbf{j}) \cdot (2x\,\mathbf{i} + 2y\,\mathbf{j} - \mathbf{k}) = (2xy - 2xy) = 0.$$

Thus the total flux out of the solid is 0.

47. The surface $z = \sqrt{2 - (x^2 + y^2)}$ is the upper half of the sphere $x^2 + y^2 + z^2 = 2$. The surface intersects the surface $z = x^2 + y^2$ in a circle of radius 1 at height $z = 1$. Thus the upper boundary of the solid, call it S_1, is a segment of width $\sqrt{2} - 1$ on a sphere of radius $\sqrt{2}$. The area of S_1 is therefore $2\pi \sqrt{2}(\sqrt{2} - 1)$. (Exercise 27, Section 9.9). The upper unit normal to S_1 is the vector

$$\mathbf{n} = \frac{1}{\sqrt{2}}(x\,\mathbf{i} + y\,\mathbf{j} + z\,\mathbf{k}).$$

Therefore

$$\text{flux through } S_1 = \iint\limits_{S_1} (\mathbf{v} \cdot \mathbf{n})\, d\sigma = \frac{1}{\sqrt{2}} \iint\limits_{S_1} \overbrace{(x^2 + y^2 + z^2)}^{2}\, d\sigma$$

$$= \sqrt{2} \iint\limits_{S_1} d\sigma = \sqrt{2}\,(\text{area of } S_1) = 4\pi(\sqrt{2} - 1).$$

The lower boundary of the solid, call it S_2, is the graph of the function

$$f(x, y) = x^2 + y^2 \quad \text{on} \quad \Omega : 0 \le x^2 + y^2 \le 1.$$

Taking $\mathbf{n}$ as the lower unit normal, we have

$$\text{flux through } S_2 = \iint\limits_{S_2} (\mathbf{v} \cdot \mathbf{n})\, d\sigma = \iint\limits_{\Omega} \left(v_1 f_x' + v_2 f_y' - v_3\right) dx\, dy$$

$$= \iint\limits_{\Omega} (x^2 + y^2)\, dx\, dy = \int_0^{2\pi} \int_0^1 r^3\, dr\, d\theta = \frac{1}{2}\pi.$$

The total flux out of the solid is $4\pi(\sqrt{2} - 1) + \frac{1}{2}\pi = (4\sqrt{2} - \frac{7}{2})\pi$.

48.

face	$\mathbf{n}$	$\mathbf{v} \cdot \mathbf{n}$	flux	
$x = 0$	$-\mathbf{i}$	$-xz = 0$	0	
$x = 1$	$\mathbf{i}$	$xz = 1$	$\frac{1}{2}$	
$y = 0$	$-\mathbf{j}$	$-4xyz^2 = 0$	0	total flux $= \frac{1}{2} + \frac{2}{3} + 2 = \frac{19}{6}$
$y = 1$	$\mathbf{j}$	$4xyz^2 = 4xz^2$	$\frac{2}{3}$	
$z = 0$	$-\mathbf{k}$	$-2z = 0$	0	
$z = 1$	$\mathbf{k}$	$2z = 2$	2	

SECTION 18.8

1. $\nabla \cdot \mathbf{v} = 2, \quad \nabla \times \mathbf{v} = \mathbf{0}$ 2. $\nabla \cdot \mathbf{v} = 0, \quad \nabla \times \mathbf{v} = \mathbf{0}$

3. $\nabla \cdot \mathbf{v} = 0, \quad \nabla \times \mathbf{v} = \mathbf{0}$ 4. $\nabla \cdot \mathbf{v} = -\dfrac{4xy}{(x^2 + y^2)^2}, \quad \nabla \times \mathbf{v} = \dfrac{2(y^2 - x^2)}{(x^2 + y^2)^2}\,\mathbf{k}$

5. $\nabla \cdot \mathbf{v} = 6, \quad \nabla \times \mathbf{v} = \mathbf{0}$ 6. $\nabla \cdot \mathbf{v} = 0, \quad \nabla \times \mathbf{v} = \mathbf{0}$

7. $\nabla \cdot \mathbf{v} = yz + 1, \quad \nabla \times \mathbf{v} = -x\,\mathbf{i} + xy\,\mathbf{j} + (1 - x)z\,\mathbf{k}$

8. $\nabla \cdot \mathbf{v} = 2y(x + z), \quad \nabla \times \mathbf{v} = (2xy - y^2)\,\mathbf{i} - y^2\,\mathbf{j} - x^2\,\mathbf{k}$

9. $\nabla \cdot \mathbf{v} = 1/r^2, \quad \nabla \times \mathbf{v} = \mathbf{0}$

10. $\nabla \cdot \mathbf{v} = e^x(3 + x), \quad \nabla \times \mathbf{v} = -e^x z\,\mathbf{j} + e^x y\mathbf{k}$

11. $\nabla \cdot \mathbf{v} = 2(x + y + z)e^{r^2}, \quad \nabla \times \mathbf{v} = 2e^{r^2}\left[(y - z)\mathbf{i} - (x - z)\,\mathbf{j} + (x - y)\,\mathbf{k}\right]$

12. $\nabla \cdot \mathbf{v} = 0, \quad \nabla \times \mathbf{v} = -2\left[ze^{z^2}\mathbf{i} + xe^{x^2}\mathbf{j} + ye^{y^2}\mathbf{k}\right]$

13. $\nabla \cdot \mathbf{v} = f'(x), \quad \nabla \times \mathbf{v} = \mathbf{0}$

14. each partial derivative that appears in the curl is 0

15. use components.

16. $\nabla \cdot \mathbf{F} = -GmM\left[\nabla \cdot (r^{-3}\mathbf{r})\right] = -GmM(0) = 0$

     linearity (Exercise 15)   (17.8.8)

     $\nabla \times \mathbf{F} = -GmM[\nabla \times (r^{-3}\mathbf{r})] = -GmM(\mathbf{0}) = \mathbf{0}$

     linearity (Exercise 15)   (17.8.8)

17. $\nabla \cdot \mathbf{v} = \dfrac{\partial P}{\partial x} + \dfrac{\partial Q}{\partial y} + \dfrac{\partial R}{\partial z} = 2 + 4 - 6 = 0$

18. $\nabla \cdot \mathbf{v} = \dfrac{\partial}{\partial x}(3x^2) + \dfrac{\partial}{\partial y}(-y^2) + \dfrac{\partial}{\partial z}(2yz - 6xz) = 6x - 2y + (2y - 6x) = 0$

19. $\nabla \times \mathbf{F} = \begin{vmatrix} \mathbf{i} & \mathbf{j} & \mathbf{k} \\ \dfrac{\partial}{\partial x} & \dfrac{\partial}{\partial y} & \dfrac{\partial}{\partial z} \\ x & y & -2z \end{vmatrix} = \mathbf{0}$

20. $\mathbf{F}(x, y, z) = (2x + y + 2z)\mathbf{i} + (x + 4y - 3z)\mathbf{j} + (2x - 3y - 6z)\mathbf{k}$

$$\nabla \times \mathbf{F} = \begin{vmatrix} \mathbf{i} & \mathbf{j} & \mathbf{k} \\ \dfrac{\partial}{\partial x} & \dfrac{\partial}{\partial y} & \dfrac{\partial}{\partial z} \\ 2x + y + 2z & x + 4y - 3z & 2x - 3y - 6z \end{vmatrix}$$

$$= (-3 + 3)\mathbf{i} - (2 - 2)\mathbf{j} + (1 - 1)\mathbf{k} = \mathbf{0}$$

21. $\nabla^2 f = 12(x^2 + y^2 + z^2)$

22. $\nabla^2 f = \nabla \cdot \nabla f = \nabla \cdot (yz\mathbf{i} + xz\mathbf{j} + xy\mathbf{k}) = 0$

23. $\nabla^2 f = 2y^3 z^4 + 6x^2 yz^4 + 12x^2 y^3 z^2$

24. Note that for $r = \sqrt{x^2 + y^2 + z^2}$, $\quad \dfrac{\partial r}{\partial x} = \dfrac{x}{r}$

Then $\dfrac{\partial^2}{\partial x^2}(\cos r) = \dfrac{\partial}{\partial x}\left(\dfrac{-x \sin r}{r}\right) = \dfrac{-r^2 \sin r - x^2 r \cos r + x^2 \sin r}{r^3}$,

With similar formulas for y and z.   Therefore

$$\nabla^2 f = \dfrac{\partial^2}{\partial x^2}\cos r + \dfrac{\partial^2}{\partial y^2}\cos r + \dfrac{\partial^2}{\partial z^2}\cos r$$

$$= \dfrac{-3r^2 \sin r - (x^2 + y^2 + z^2)r \cos r + (x^2 + y^2 + z^2)\sin r}{r^3}$$

$$= -\cos r - 2r^{-1}\sin r$$

25. $\nabla^2 f = e^r(1 + 2r^{-1})$

26. $\dfrac{\partial^2}{\partial x^2}\ln r = \dfrac{\partial}{\partial x}\left(\dfrac{x}{r^2}\right) = \dfrac{r^2 - 2x^2}{r^4}$,   with similar formula for y and z.

Then $\quad \nabla^2 f = \dfrac{3r^2 - 2(x^2 + y^2 + z^2)}{r^4} = \dfrac{1}{r^2}$

27. (a) $2r^2$ (b) $-1/r$

28. (a)
$$(\mathbf{u} \cdot \nabla)\mathbf{r} = (\mathbf{u} \cdot \nabla x)\mathbf{i} + (\mathbf{u} \cdot \nabla y)\mathbf{j} + (\mathbf{u} \cdot \nabla z)\mathbf{k}$$

$$= (\mathbf{u} \cdot \mathbf{i})\mathbf{i} + (\mathbf{u} \cdot \mathbf{j})\mathbf{j} + (\mathbf{u} \cdot \mathbf{k})\mathbf{k} = \mathbf{u}$$

(b)
$$(\mathbf{r} \cdot \nabla)\mathbf{u} = (\mathbf{r} \cdot \nabla yz)\mathbf{i} + (\mathbf{r} \cdot \nabla xz)\mathbf{j} + (\mathbf{r} \cdot \nabla xy)\mathbf{k}$$

$$= [\mathbf{r} \cdot (z\mathbf{j} + y\mathbf{k})]\mathbf{i} + [\mathbf{r} \cdot (z\mathbf{i} + x\mathbf{k})]\mathbf{j} + [\mathbf{r} \cdot (y\mathbf{i} + x\mathbf{j})]\mathbf{k}$$

$$= (yz + zy)\mathbf{i} + (xz + zx)\mathbf{j} + (xy + yx)\mathbf{k}$$

$$= 2(yz\mathbf{i} + xz\mathbf{j} + xy\mathbf{k}) = 2\mathbf{u}$$

29. $\nabla^2 f = \nabla^2 g(r) = \nabla \cdot (\nabla g(r)) = \nabla \cdot \left(g'(r) r^{-1} \mathbf{r} \right)$

$$= \left[(\nabla g'(r)) \cdot r^{-1} \mathbf{r} \right] + g'(r) \left(\nabla \cdot r^{-1} \mathbf{r} \right)$$

$$= \left\{ \left[g''(r) r^{-1} \mathbf{r} \right] \cdot r^{-1} \mathbf{r} \right\} + g'(r)(2r^{-1})$$

$$= g''(r) + 2r^{-1} g'(r)$$

30. (a) $\nabla \cdot (f\mathbf{v}) = \dfrac{\partial}{\partial x}(fv_1) + \dfrac{\partial}{\partial y}(fv_2) + \dfrac{\partial}{\partial z}(fv_3)$

$$= \left(f\frac{\partial v_1}{\partial x} + \frac{\partial f}{\partial x}v_1 \right) + \left(f\frac{\partial v_2}{\partial y} + \frac{\partial f}{\partial y}v_2 \right) + \left(f\frac{\partial v_3}{\partial z} + \frac{\partial f}{\partial z}v_3 \right)$$

$$= \left(\frac{\partial f}{\partial x}\mathbf{i} + \frac{\partial f}{\partial y}\mathbf{j} + \frac{\partial f}{\partial z}\mathbf{k} \right) \cdot \mathbf{v} + f \left(\frac{\partial v_1}{\partial x} + \frac{\partial v_2}{\partial y} + \frac{\partial v_3}{\partial z} \right)$$

$$= (\nabla f) \cdot \mathbf{v} + f(\nabla \cdot \mathbf{v})$$

(b) $\nabla \times (f\mathbf{v})$

$$= \left[\frac{\partial}{\partial y}(fv_3) - \frac{\partial}{\partial z}(fv_2) \right]\mathbf{i} + \left[\frac{\partial}{\partial z}(fv_1) - \frac{\partial}{\partial x}(fv_3) \right]\mathbf{j} + \left[\frac{\partial}{\partial x}(fv_2) - \frac{\partial}{\partial y}(fv_1) \right]\mathbf{k}$$

$$= \left[f\frac{\partial v_3}{\partial y} + \frac{\partial f}{\partial y}v_3 - f\frac{\partial v_2}{\partial z} - \frac{\partial f}{\partial z}v_2 \right]\mathbf{i} + \text{etc.}$$

(c) $\nabla \times \mathbf{v} = \begin{vmatrix} \mathbf{i} & \mathbf{j} & \mathbf{k} \\ \dfrac{\partial}{\partial x} & \dfrac{\partial}{\partial y} & \dfrac{\partial}{\partial z} \\ v_1 & v_2 & v_3 \end{vmatrix} = \left(\dfrac{\partial v_3}{\partial y} - \dfrac{\partial v_2}{\partial z} \right)\mathbf{i} + \left(\dfrac{\partial v_1}{\partial z} - \dfrac{\partial v_3}{\partial x} \right)\mathbf{j} + \left(\dfrac{\partial v_2}{\partial x} - \dfrac{\partial v_1}{\partial y} \right)\mathbf{k}$

$$\text{i-component of } \nabla \times (\nabla \times \mathbf{v}) = \frac{\partial}{\partial y}\left(\frac{\partial v_2}{\partial x} - \frac{\partial v_1}{\partial y} \right) - \frac{\partial}{\partial z}\left(\frac{\partial v_1}{\partial z} - \frac{\partial v_3}{\partial x} \right)$$

$$= \frac{\partial^2 v_2}{\partial y \partial x} - \frac{\partial^2 v_1}{\partial y^2} - \frac{\partial^2 v_1}{\partial z^2} + \frac{\partial^2 v_3}{\partial z \partial x}$$

$$= \frac{\partial}{\partial x}\left(\frac{\partial v_2}{\partial y} + \frac{\partial v_3}{\partial z} \right) - \left(\frac{\partial^2 v_1}{\partial y^2} + \frac{\partial^2 v_1}{\partial z^2} \right)$$

Adding and subtracting $\dfrac{\partial^2 v_1}{\partial x^2}$,   we get

$$\frac{\partial}{\partial x}\left(\frac{\partial v_1}{\partial x} + \frac{\partial v_2}{\partial y} + \frac{\partial v_3}{\partial z}\right) - \left(\frac{\partial^2 v_1}{\partial x^2} + \frac{\partial^2 v_1}{\partial y^2} + \frac{\partial^2 v_1}{\partial z^2}\right) = \frac{\partial}{\partial x}(\boldsymbol{\nabla}\cdot\mathbf{v}) - \boldsymbol{\nabla}^2 v_1 = \text{the i-component of } \boldsymbol{\nabla}^2\mathbf{v}.$$

Equality of the other components can be obtained in a similar manner.

31. $\dfrac{\partial f}{\partial x} = 2x + y + 2z,$ $\dfrac{\partial^2 f}{\partial x^2} = 2;$ $\dfrac{\partial f}{\partial y} = 4y + x - 3z,$ $\dfrac{\partial^2 f}{\partial y^2} = 4;$

$\dfrac{\partial f}{\partial z} = -6z + 2x - 3y,$ $\dfrac{\partial^2 f}{\partial z^2} = -6;$

$$\frac{\partial^2 f}{\partial x^2} + \frac{\partial^2 f}{\partial y^2} + \frac{\partial^2 f}{\partial z^2} = 2 + 4 - 6 = 0$$

32. $f(\mathbf{r}) = \dfrac{1}{r}.$ $\dfrac{\partial^2}{\partial x^2}\left(\dfrac{1}{r}\right) = \dfrac{\partial}{\partial x}\left(\dfrac{-x}{r^3}\right) = \dfrac{-r^2 + 3x^2}{r^5},$ with similar formulas for y and z

Then $\boldsymbol{\nabla}^2 f = \dfrac{-3r^2 + 3(x^2 + y^2 + z^2)}{r^5} = 0.$

33. $n = -1$

34.   Since $\boldsymbol{\nabla}\cdot(\boldsymbol{\nabla}f) = \boldsymbol{\nabla}^2 f = 0,$ the gradient field $\boldsymbol{\nabla}f$ is solenoidal.
$\boldsymbol{\nabla}f$ is irrotational by Theorem 18.8.4

SECTION 18.9

1. $\displaystyle\iint_S (\mathbf{v}\cdot\mathbf{n})\,d\sigma = \iiint_T (\boldsymbol{\nabla}\cdot\mathbf{v})\,dx\,dy\,dz = \iiint_T 3\,dx\,dy\,dz = 3V = 4\pi$

2. $\displaystyle\iint_S (\mathbf{v}\cdot\mathbf{n})\,d\sigma = \iiint_T (\boldsymbol{\nabla}\cdot\mathbf{v})\,dx\,dy\,dz = \iiint_T (-3)\,dx\,dy\,dz = -3V = -4\pi$

3. $\displaystyle\iint_S (\mathbf{v}\cdot\mathbf{n})\,d\sigma = \iiint_T (\boldsymbol{\nabla}\cdot\mathbf{v})\,dx\,dy\,dz = \iiint_T 2(x + y + z)\,dx\,dy\,dz.$

The flux is zero since the function $f(x, y, z) = 2(x + y + z)$ satisfies the relation $f(-x, -y, -z) = -f(x, y, z)$ and T is symmetric about the origin.

4. $\displaystyle\iint_S (\mathbf{v}\cdot\mathbf{n})\,d\sigma = \iiint_T (\boldsymbol{\nabla}\cdot\mathbf{v})\,dx\,dy\,dz = \iiint_T (-2x + 2y + 1)\,dx\,dy\,dz = \iiint_T dx\,dy\,dz = V = \frac{4}{3}\pi$

by symmetry

5.

face	n	v · n	flux
$x = 0$	$-\mathbf{i}$	0	0
$x = 1$	$\mathbf{i}$	1	1
$y = 0$	$-\mathbf{j}$	0	0
$y = 1$	$\mathbf{j}$	1	1
$z = 0$	$-\mathbf{k}$	0	0
$z = 1$	$\mathbf{k}$	1	1

total flux = 3

$$\iiint_T (\boldsymbol{\nabla} \cdot \mathbf{v})\, dxdydz = \iiint_T 3\, dxdydz = 3V = 3$$

6.

face	n	v · n	flux
$x = 0$	$-\mathbf{i}$	$-xy = 0$	0
$x = 1$	$\mathbf{i}$	$xy = y$	$1/2$
$y = 0$	$-\mathbf{j}$	$-yz = 0$	0
$y = 1$	$\mathbf{j}$	$yz = z$	$1/2$
$z = 0$	$-\mathbf{k}$	$-xz = 0$	0
$z = 1$	$\mathbf{k}$	$xz = x$	$1/2$

total flux $= \dfrac{3}{2}$

$$\iiint_T (\boldsymbol{\nabla} \cdot \mathbf{v})\, dxdydz = \iiint_T (y + z + x)\, dxdydz = (\overline{y} + \overline{z} + \overline{x})V = \left(\frac{1}{2} + \frac{1}{2} + \frac{1}{2}\right)(1) = \frac{3}{2}.$$

7.

face	n	v · n	flux
$x = 0$	$-\mathbf{i}$	0	0
$x = 1$	$\mathbf{i}$	1	1
$y = 0$	$-\mathbf{j}$	xz	
$y = 1$	$\mathbf{j}$	$-xz$	
$z = 0$	$-\mathbf{k}$	0	0
$z = 1$	$\mathbf{k}$	1	1

fluxes add up to 0   total flux = 2

$$\iiint_T (\boldsymbol{\nabla} \cdot \mathbf{v})\, dxdydz = \iiint_T 2\,(x + z)\, dxdydz = 2\,(\overline{x} + \overline{z})V = 2\,(\tfrac{1}{2} + \tfrac{1}{2})1 = 2$$

8.

face	$\mathbf{n}$	$\mathbf{v} \cdot \mathbf{n}$	flux	
$x = 0$	$-\mathbf{i}$	$-x = 0$	0	
$x = 1$	$\mathbf{i}$	$x = 1$	1	
$y = 0$	$-\mathbf{j}$	$-xy = 0$	0	total flux $= \dfrac{7}{4}$
$y = 1$	$\mathbf{j}$	$xy = x$	1/2	
$z = 0$	$-\mathbf{k}$	$-xyz = 0$	0	
$z = 1$	$\mathbf{k}$	$xyz = xy$	1/4	

$$\iiint_T (\nabla \cdot \mathbf{v})\,dxdydz = \iiint_T (1 + x + xy)\,dxdydz = \int_0^1 \int_0^1 \int_0^1 (1 + x + xy)\,dx\,dy\,dz = \frac{7}{4}.$$

9. $\text{flux} = \displaystyle\iiint_T (1 + 4y + 6z)\,dxdydz = (1 + 4\overline{y} + 6\overline{z})V = (1 + 0 + 3)\,9\pi = 36\pi$

10.

$$\text{flux} = \iiint_T (\nabla \cdot \mathbf{v})\,dxdydz = \int_0^1 \int_0^{1-x} \int_0^{1-x-y} (y + z + x)\,dz\,dy\,dx$$

$$= (\overline{x} + \overline{y} + \overline{z})V = \left(\frac{3}{4}\right)\left(\frac{1}{6}\right) = \frac{1}{8}$$

11.

$$\text{flux} = \iiint_T (2x + x - 2x)\,dxdydz \iiint_T x\,dxdydz$$

$$= \int_0^1 \int_0^{1-x} \int_0^{1-x-y} x\,dz\,dy\,dx$$

$$= \int_0^1 \int_0^{1-x} \left(x - x^2 - xy\right)\,dy\,dx$$

$$= \int_0^1 \left[xy - x^2 y - \frac{1}{2}xy^2\right]_0^{1-x}\,dx$$

$$= \int_0^1 \left(\frac{1}{2}x - x^2 + \frac{1}{2}x^3\right)\,dx = \frac{1}{24}$$

12. $\text{flux} = \displaystyle\iiint_T (\nabla \cdot \mathbf{v})\,dxdydz = \iiint_T 4y\,dx\,dy\,dz = 4\overline{y}V = 4(1)\left(\frac{32}{3}\right) = \frac{128}{3}$

13. flux $= \displaystyle\iiint_T 2(x + y + z)\,dx\,dy\,dz = \int_0^4 \int_0^2 \int_0^{2\pi} 2(r\cos\theta + r\sin\theta + z)r\,d\theta\,dr\,dz$

$$= \int_0^4 \int_0^2 4\pi\, rz\,dr\,dz$$

$$= \int_0^4 8\pi\, z\,dz = 64\pi$$

14. flux $= \frac{1}{2}\displaystyle\iint_S (\mathbf{v}\cdot\mathbf{n})\,d\sigma = \frac{1}{2}\iiint_T (\nabla\cdot\mathbf{v})\,dx\,dy\,dz = \iiint_T (2 + 2x)\,dx\,dy\,dz = V = \dfrac{32}{3}\pi$

15. flux $= \displaystyle\iiint_T (2y + 2y + 3y)\,dx\,dy\,dz = 7\bar{y}V = 0$

16. flux $= \displaystyle\iiint_T (\nabla\cdot\mathbf{v})\,dx\,dy\,dz = \iiint_T 7y\,dx\,dy\,dz = 7\,\bar{y}\,V = 7\left(\dfrac{a}{2}\right)a^3 = \dfrac{7}{2}a^4$

17. flux $= \displaystyle\iiint_T (A + B + C)\,dx\,dy\,dz = (A + B + C)V$

18.
$$\iint_S (\nabla f\cdot\mathbf{n})\,d\sigma = \iiint_T [\nabla\cdot(\nabla f)]\,dx\,dy\,dz$$

$$= \iiint_T (\nabla^2 f)\,dx\,dy\,dz = \iiint_T 0\,dx\,dy\,dz = 0$$

19. Let T be the solid enclosed by S and set $\mathbf{n} = n_1\mathbf{i} + n_2\mathbf{j} + n_3\mathbf{k}$.

$$\iint_S n_1\,d\sigma = \iint_S (\mathbf{i}\cdot\mathbf{n})\,d\sigma = \iiint_T (\nabla\cdot\mathbf{i})\,dx\,dy\,dz = \iiint_T 0\,dx\,dy\,dz = 0.$$

Similarly

$$\iint_S n_2\,d\sigma = 0 \quad \text{and} \quad \iint_S n_3\,d\sigma = 0.$$

20. (a) The identity follows from setting $\mathbf{v} = \nabla f$ in (17.8.6).

$$\iint_S (f f_n')\,d\sigma = \iint_S (f\nabla f\cdot\mathbf{n})\,d\sigma = \iiint_T [\nabla\cdot(f\nabla f)]\,dx\,dy\,dz$$

$$= \iiint_T [\|\nabla f\|^2 + f(\nabla^2 f)]\,dx\,dy\,dz$$

$$= \iiint_T \|\nabla f\|^2\,dx\,dy\,dz \quad \text{since} \quad \nabla^2 f = 0$$

(b)
$$\iint_S (g f'_n)\, d\sigma = \iint_S (g\boldsymbol{\nabla} f \cdot \mathbf{n})\, d\sigma = \iiint_T [\boldsymbol{\nabla} \cdot (g\boldsymbol{\nabla} f)]\, dx\,dy\,dz$$

$$= \iiint_T \{(\boldsymbol{\nabla} g \cdot \boldsymbol{\nabla} f) + g[\boldsymbol{\nabla} \cdot (\boldsymbol{\nabla} f)]\}\, dx\,dy\,dz$$

$$= \iiint_T [(\boldsymbol{\nabla} g \cdot \boldsymbol{\nabla} f) + g(\boldsymbol{\nabla}^2 f)]\, dx\,dy\,dz$$

21. A routine computation shows that $\boldsymbol{\nabla} \cdot (\boldsymbol{\nabla} f \times \boldsymbol{\nabla} g) = 0$.   Therefore

$$\iint_S [(\boldsymbol{\nabla} f \times \boldsymbol{\nabla} g) \cdot \mathbf{n}]\, d\sigma = \iiint_T [\boldsymbol{\nabla} \cdot (\boldsymbol{\nabla} f \times \boldsymbol{\nabla} g)]\, dx\,dy\,dz = 0.$$

22. Since $\boldsymbol{\nabla} \cdot \mathbf{r} = 3,$ we can write

$$V = \iiint_T dx\,dy\,dz = \iiint_T \left(\boldsymbol{\nabla} \cdot \frac{\mathbf{r}}{3}\right) dx\,dy\,dz = \iint_S \left(\frac{1}{3}\mathbf{r} \cdot \mathbf{n}\right) d\sigma, \quad \text{by the divergence theorem.}$$

23. Set $\mathbf{F} = F_1\mathbf{i} + F_2\mathbf{j} + F_3\mathbf{k}.$

$$F_1 = \iint_S [\rho(z - c)\mathbf{i} \cdot \mathbf{n}]\, d\sigma = \iiint_T [\boldsymbol{\nabla} \cdot \rho(z - c)\mathbf{i}]\, dx\,dy\,dz$$

$$= \iiint_T \underbrace{\frac{\partial}{\partial x}[\rho(z - c)]}\, dx\,dy\,dz = 0.$$

Similarly $F_2 = 0.$

$$F_3 = \iint_S [\rho(z - c)\mathbf{k} \cdot \mathbf{n}]\, d\sigma = \iiint_T [\boldsymbol{\nabla} \cdot \rho(z - c)\mathbf{k}]\, dx\,dy\,dz$$

$$= \iiint_T \frac{\partial}{\partial z}[\rho(z - c)]\, dx\,dy\,dz$$

$$= \iiint_T \rho\, dx\,dy\,dz = W.$$

24.
$$\tau_{Tot} \cdot \mathbf{i} = \iint_S \{[\mathbf{r} \times \rho(c - z)\mathbf{n}] \cdot \mathbf{i}\}\, d\sigma$$

(12.5.6)

$$= -\iint_S [(\mathbf{i} \times \mathbf{r}) \cdot \rho(c - z)\mathbf{n}]\, d\sigma$$

$$= \rho \iint_S (z - c)[(\mathbf{i} \times \mathbf{r}) \cdot \mathbf{n}]\, d\sigma$$

divergence theorem

$$= \rho \iiint_T [\nabla \cdot (z - c)(\mathbf{i} \times \mathbf{r})] \, dx dy dz$$

$$\mathbf{i} \times \mathbf{r} = \mathbf{i} \times (x\mathbf{i} + y\mathbf{j} + z\mathbf{k}) = -z\mathbf{j} + y\mathbf{k}$$

$$(z - c)(\mathbf{i} \times \mathbf{r}) = (z - c)(-z\mathbf{j} + y\mathbf{k}) = (cz - z^2)\mathbf{j} + (yz - cy)\mathbf{k}$$

$$\nabla \cdot (z - c)(\mathbf{i} \times \mathbf{r}) = y$$

$$\tau_{Tot} \cdot \mathbf{i} = \rho \iiint_T y \, dx dy dz = \rho \bar{y} V = \bar{y}(\rho V) = \bar{y} W$$

$$(\mathbf{r} \times \mathbf{F}) \cdot \mathbf{i} = [(\bar{x}\mathbf{i} + \bar{y}\mathbf{j} + \bar{z}\mathbf{k}) \times W\mathbf{k}] \cdot \mathbf{i}$$

$$= (-\bar{x} W \mathbf{j} + \bar{y} W \mathbf{i}) \cdot \mathbf{i} = \bar{y} W = \tau_{Tot} \cdot \mathbf{i}$$

Equality of the other components can be shown in a similar manner.

PROJECT 18.9

1. For $\mathbf{r} \neq \mathbf{0}$, $\nabla \cdot \mathbf{E} = \nabla \cdot qr^{-3}\mathbf{r} = q(-3+3)r^{-3} = 0$ by (17.8.8)

2. By the divergence theorem,   flux of $\mathbf{E}$ out of $S = \iiint_T (\nabla \cdot \mathbf{E}) \, dx \, dy \, dz = \iiint_T 0 \, dx \, dy \, dz = 0$

3. On $S_a, \mathbf{n} = \dfrac{\mathbf{r}}{r}$, and thus $\mathbf{E} \cdot \mathbf{n} = q\dfrac{\mathbf{r}}{r^3} \cdot \dfrac{\mathbf{r}}{r} = \dfrac{q}{r^2} = \dfrac{q}{a^2}$

    Thus flux of $\mathbf{E}$ out of $S_a = \iint_{S_a} (\mathbf{E} \cdot \mathbf{n}) \, d\sigma = \iint_{S_a} \dfrac{q}{a^2} \, d\sigma = \dfrac{q}{a^2}(\text{area of } S_a) = \dfrac{q}{a^2}(4\pi a^2) = 4\pi q.$

SECTION 18.10

For Exercises 1–4: $\mathbf{n} = x\mathbf{i} + y\mathbf{j} + z\mathbf{k}$ and $C : \mathbf{r}(u) = \cos u \, \mathbf{i} + \sin u \, \mathbf{j}, \quad u \in [0, 2\pi].$

1. (a) $\iint_S [(\nabla \times \mathbf{v}) \cdot \mathbf{n}] \, d\sigma = \iint_S (\mathbf{0} \cdot \mathbf{n}) \, d\sigma = 0$

    (b) S is bounded by the unit circle $C : \mathbf{r}(u) = \cos u \, \mathbf{i} + \sin u \, \mathbf{j}, \quad u \in [0, 2\pi].$

    $\oint_C \mathbf{v}(\mathbf{r}) \cdot d\mathbf{r} = 0$ since $\mathbf{v}$ is a gradient.

2. (a) $\iint\limits_{S} [(\nabla \times \mathbf{v}) \cdot \mathbf{n}]\, d\sigma = \iint\limits_{S} (-2\mathbf{k} \cdot \mathbf{n})\, d\sigma = -2 \iint\limits_{S} z\, d\sigma = -2\bar{z}A = -2\left(\tfrac{1}{2}\right) 2\pi = -2\pi$

Exercise 17, Section 17.7

(b) $\displaystyle\oint_{C} \mathbf{v}(\mathbf{r}) \cdot d\mathbf{r} = \oint_{C} y\, dx - x\, dy = \int_{0}^{2\pi} (-\sin^2 u - \cos^2 u)\, du = -2\pi$

3. (a) $\iint\limits_{S} [(\nabla \times \mathbf{v}) \cdot \mathbf{n}]\, d\sigma = \iint\limits_{S} \left[(-3y^2\mathbf{i} + 2z\mathbf{j} + 2\mathbf{k}) \cdot \mathbf{n}\right]\, d\sigma$

$$= \iint\limits_{S} (-3xy^2 + 2yz + 2z)\, d\sigma$$

$$= \underbrace{\iint\limits_{S} (-3xy^2)\, d\sigma}_{0} + \underbrace{\iint\limits_{S} 2yz\, d\sigma}_{0} + 2 \iint\limits_{S} z\, d\sigma = 2\bar{z}V = 2\left(\tfrac{1}{2}\right)2\pi = 2\pi$$

Exercise 17, Section 17.7

(b) $\displaystyle\oint_{C} \mathbf{v}(\mathbf{r}) \cdot d\mathbf{r} = \oint_{C} z^2\, dx + 2x\, dy = \oint_{C} 2x\, dy = \int_{0}^{2\pi} 2\cos^2 u\, du = 2\pi$

4. (a) $\iint\limits_{S} [(\nabla \times \mathbf{v}) \cdot \mathbf{n}]\, d\sigma = \iint\limits_{S} [(-6y\mathbf{i} + 6x\mathbf{j} - 2x\mathbf{k}) \cdot \mathbf{n}]\, d\sigma$

$$= \iint\limits_{S} (-2xz)\, d\sigma = 0 \quad \text{by symmetry}$$

$$\oint_{C} \mathbf{v}(\mathbf{r}) \cdot d\mathbf{r} = \int_{C} 6xz\, dx - x^2\, dy = -\int_{C} x^2\, dy$$

$$= -\int_{0}^{2\pi} \cos^3 u\, du = -\int_{0}^{2\pi} (\cos u - \sin^2 u \cos u)\, du = 0$$

For Exercises 5–7 take $S: z = 2 - x - y$ with $0 \le x \le 2, \; 0 \le y \le 2 - x$ and C as the triangle $(2,0,0), \; (0,2,0), \; (0,0,2)$. Then $C = C_1 \cup C_2 \cup C_3$ with

$$C_1 : \mathbf{r}_1(u) = 2(1-u)\,\mathbf{i} + 2u\mathbf{j}, \quad u \in [0,1],$$

$$C_2 : \mathbf{r}_2(u) = 2(1-u)\,\mathbf{j} + 2u\mathbf{k}, \quad u \in [0,1],$$

$$C_3 : \mathbf{r}_3(u) = 2(1-u)\,\mathbf{k} + 2u\mathbf{i}, \quad u \in [0,1].$$

$n = \tfrac{1}{3}\sqrt{3}(\mathbf{i} + \mathbf{j} + \mathbf{k})$ area of $S: A = 2\sqrt{3}$ centroid: $\left(\tfrac{2}{3}, \tfrac{2}{3}, \tfrac{2}{3}\right)$

5. (a) $\iint\limits_{S} [(\nabla \times \mathbf{v}) \cdot \mathbf{n}]\, d\sigma = \iint\limits_{S} \tfrac{1}{3}\sqrt{3}\, d\sigma = \tfrac{1}{3}\sqrt{3}A = 2$

(b) $\displaystyle\oint_{C} \mathbf{v}(\mathbf{r}) \cdot d\mathbf{r} = \left(\int_{C_1} + \int_{C_2} + \int_{C_3}\right) \mathbf{v}(\mathbf{r}) \cdot d\mathbf{r} = -2 + 2 + 2 = 2$

6. (a) $\displaystyle\iint\limits_{S} [(\nabla \times \mathbf{v}) \cdot \mathbf{n}] \, d\sigma = \iint\limits_{S} [(-2x\mathbf{j} - 2y\mathbf{k}) \cdot \mathbf{n}] \, d\sigma$

$$= -\frac{2}{3}\sqrt{3} \iint\limits_{S} (x+y) \, d\sigma = -\frac{2}{3}\sqrt{3}(\bar{x}+\bar{y})A = -\frac{16}{3}$$

(b) $\displaystyle\oint_{C} \mathbf{v}(\mathbf{r}) \cdot d\mathbf{r} = \left(\int_{C_1} + \int_{C_2} + \int_{C_3} \right) \mathbf{v}(\mathbf{r}) \cdot d\mathbf{r} = -\frac{8}{3} + 0 - \frac{8}{3} = -\frac{16}{3}$

7. (a) $\displaystyle\iint\limits_{S} [(\nabla \times \mathbf{v}) \cdot \mathbf{n}] \, d\sigma = \iint\limits_{S} (y\mathbf{k} \cdot \mathbf{n}) \, d\sigma = \frac{1}{3}\sqrt{3} \iint\limits_{S} y \, d\sigma = \frac{1}{3}\sqrt{3}\bar{y}A = \frac{4}{3}$

(b) $\displaystyle\oint_{C} \mathbf{v}(\mathbf{r}) \cdot d\mathbf{r} = \left(\int_{C_1} + \int_{C_2} + \int_{C_3} \right) \mathbf{v}(\mathbf{r}) \cdot d\mathbf{r} = \left(\frac{4}{3} - \frac{32}{5} \right) + \frac{32}{5} + 0 = \frac{4}{3}$

8. By (18.10.2) $\mathbf{v}$ is a gradient: $\mathbf{v} = \nabla\phi$.   Therefore

$$\int_{C} \mathbf{v}(\mathbf{r}) \cdot d\mathbf{r} = \int_{C} [(\nabla\phi) \cdot d\mathbf{r}] = 0 \quad \text{by (18.2.2).}$$

9. The bounding curve is the set of all (x, y, z) with

$$x^2 + y^2 = 4 \quad \text{and} \quad z = 4.$$

Traversed in the positive sense with respect to $\mathbf{n}$, it is the curve $-C$ where

$$C : \mathbf{r}(u) = 2\cos u \, \mathbf{i} + 2\sin u \, \mathbf{j} + 4\mathbf{k}, \qquad u \in [0, 2\pi].$$

By Stokes's theorem the flux we want is

$$-\int_{C} \mathbf{v}(\mathbf{r}) \cdot d\mathbf{r} = -\int_{C} y \, dx + z \, dy + x^2 z^2 \, dz$$

$$= -\int_{0}^{2\pi} \left(-4\sin^2 u + 8\cos u \right) \, du = 4\pi.$$

10. The bounding curve is the set of all (x, y, z) with

$$x^2 + z^2 = 9, \quad y = -8.$$

Traversed in the positive direction with respect to $\mathbf{n}$, it is the curve $-C$ where

$$C : \mathbf{r}(u) = 3\cos u \, \mathbf{i} - 8\mathbf{j} + 3\sin u \, \mathbf{k}, \qquad u \in [0, 2\pi].$$

By Stokes's theorem the flux we want is

$$-\int_{C} \mathbf{v}(\mathbf{r}) \cdot d\mathbf{r} = -\int_{C} \frac{1}{2}y \, dx + 2xz \, dy - 3x \, dz$$

$$= -\int_{0}^{2\pi} \left(12\sin u - 27\cos^2 u \right) \, du = 27\pi.$$

11. The bounding curve C for S is the bounding curve of the elliptical region $\Omega : \frac{1}{4}x^2 + \frac{1}{9}y^2 = 1$. Since

$$\nabla \times \mathbf{v} = 2x^2 yz^2 \mathbf{i} - 2xy^2 z^2 \mathbf{j}$$

is zero on the xy-plane, the flux of $\nabla \times \mathbf{v}$ through Ω is zero, the circulation of $\mathbf{v}$ about C is zero, and therefore the flux of $\nabla \times \mathbf{v}$ through S is zero.

12. Let T be the solid enclosed by S. By our condition on $\mathbf{v}$, $\boldsymbol{\nabla}\times\mathbf{v}$ is continuously differentiable on T. Therefore by the divergence theorem

$$\iint_S [(\boldsymbol{\nabla}\times\mathbf{v})\cdot\mathbf{n}]\,d\sigma = \iiint_T [\boldsymbol{\nabla}\cdot(\boldsymbol{\nabla}\times\mathbf{v})]\,dxdydz.$$

This is zero since the divergence of a curl is zero.

13. C bounds the surface

$$S: z = \sqrt{1 - \tfrac{1}{2}(x^2 + y^2)}, \qquad (x, y)\in\Omega$$

with $\Omega : x^2 + (y - \tfrac{1}{2})^2 \leq \tfrac{1}{4}$. Routine calculation shows that $\boldsymbol{\nabla}\times\mathbf{v} = y\mathbf{k}$. The circulation of $\mathbf{v}$ with respect to the upper unit normal $\mathbf{n}$ is given by

$$\iint_S (y\mathbf{k}\cdot\mathbf{n})\,d\sigma = \iint_\Omega y\,dxdy = \bar{y}A = \frac{1}{2}\left(\frac{\pi}{4}\right) = \frac{1}{8}\pi.$$

(18.7.9)

If $-\mathbf{n}$ is used, the circulation is $-\tfrac{1}{8}\pi$.   Answer: $\pm\tfrac{1}{8}\pi$.

14. $\boldsymbol{\nabla}\times\mathbf{v} = \mathbf{i} - 2\mathbf{j} - 2\mathbf{k}$. Since the plane $x + 2y + z = 0$ passes through the origin, it intersects the sphere in a circle of radius a. The surface S bounded by this circle is a disc of radius a with upper unit normal

$$\mathbf{n} = \frac{1}{6}\sqrt{6}(\mathbf{i} + 2\mathbf{j} + \mathbf{k}).$$

The circulation of $\mathbf{v}$ with respect to $\mathbf{n}$ is given by

$$\iint_S [(\boldsymbol{\nabla}\times\mathbf{v})\cdot\mathbf{n}]\,d\sigma = \iint_S \left(-\frac{5}{6}\sqrt{6}\right)\,d\sigma = -\frac{5}{6}\sqrt{6}A = -\frac{5}{6}\sqrt{6}\pi a^2.$$

If $-\mathbf{n}$ is used, the circulation is $\tfrac{5}{6}\sqrt{6}\pi a^2$.   Answer: $\pm\tfrac{5}{6}\sqrt{6}\pi a^2$.

15. $\boldsymbol{\nabla}\times\mathbf{v} = \mathbf{i} + 2\mathbf{j} + \mathbf{k}$. The paraboloid intersects the plane in a curve C that bounds a flat surface S that projects onto the disc $x^2 + (y - \tfrac{1}{2})^2 = \tfrac{1}{4}$ in the xy-plane. The upper unit normal to S is the vector $\mathbf{n} = \tfrac{1}{2}\sqrt{2}(-\mathbf{j} + \mathbf{k})$. The area of the base disc is $\tfrac{1}{4}\pi$. Letting γ be the angle between $\mathbf{n}$ and $\mathbf{k}$, we have $\cos\gamma = \mathbf{n}\cdot\mathbf{k} = \tfrac{1}{2}\sqrt{2}$ and $\sec\gamma = \sqrt{2}$. Therefore the area of S is $\tfrac{1}{4}\sqrt{2}\pi$. The circulation of $\mathbf{v}$ with respect to $\mathbf{n}$ is given by

$$\iint_S [(\boldsymbol{\nabla}\times\mathbf{v})\cdot\mathbf{n}]\,d\sigma = \iint_S -\frac{1}{2}\sqrt{2}\,d\sigma = \left(-\frac{1}{2}\sqrt{2}\right)(\text{area of } S) = -\frac{1}{4}\pi.$$

If $-\mathbf{n}$ is used, the circulation is $\tfrac{1}{4}\pi$.   Answer: $\pm\tfrac{1}{4}\pi$.

16. $\nabla \times \mathbf{v} = -y\mathbf{i} - z\mathbf{j} - x\mathbf{k}$. The curve C bounds a flat surface S that projects onto the disc $x^2 + y^2 = b^2$ in the xy-plane. The upper unit normal to S is the vector $\mathbf{n} = \frac{1}{2}\sqrt{2}(\mathbf{j} + \mathbf{k})$. The area of the base disc is πb^2. Letting γ be the angle between $\mathbf{n}$ and $\mathbf{k}$, we have $\cos\gamma = \mathbf{n}\cdot\mathbf{k} = \frac{1}{2}\sqrt{2}$ and $\sec\gamma = \sqrt{2}$. Therefore the area of S is $\pi b^2\sqrt{2}$. The circulation of $\mathbf{v}$ with respect to $\mathbf{n}$ is given by

$$\iint_S [(\nabla\times\mathbf{v})\cdot\mathbf{n}]\,d\sigma = -\frac{1}{2}\sqrt{2}\iint_S (x+z)\,d\sigma = -\frac{1}{2}\sqrt{2}\iint_S z\,d\sigma = -\frac{1}{2}\sqrt{2}\bar{z}A.$$

by symmetry

It's clear by symmetry that $\bar{z} = a^2$, the height at which S intersects the xz-plane. Since $A = \pi b^2\sqrt{2}$, the circulation is $-\pi a^2 b^2$. If $-\mathbf{n}$ is used, the circulation becomes $\pi a^2 b^2$,   Answer: $\pm\pi a^2 b^2$.

17. Straightforward calculation shows that

$$\nabla\times(\mathbf{a}\times\mathbf{r}) = \nabla\times[(a_2 z - a_3 y)\,\mathbf{i} + (a_3 x - a_1 z)\,\mathbf{j} + (a_1 y - a_2 x)\mathbf{k}] = 2\mathbf{a}.$$

18. $\nabla\times(\phi\nabla\psi) = (\nabla\phi\times\nabla\psi) + \phi[\nabla\times\nabla\psi] = \nabla\phi\times\nabla\psi$

(18.8.7)

since the curl of a gradient is zero. Therefore the result follows from Stokes's theorem.

19. In the plane of C, the curve C bounds some Jordan region that we call Ω. The surface $S\cup\Omega$ is a piecewise–smooth surface that bounds a solid T. Note that $\nabla\times\mathbf{v}$ is continuously differentiable on T.

Thus, by the divergence theorem,

$$\iiint_T [\nabla\cdot(\nabla\times\mathbf{v})]\,dxdydz = \iint_{S\cup\Omega} [(\nabla\times\mathbf{v})\cdot\mathbf{n}]\,d\sigma$$

where $\mathbf{n}$ is the outer unit normal. Since the divergence of a curl is identically zero, we have

$$\iint_{S\cup\Omega} [(\nabla\times\mathbf{v})\cdot\mathbf{n}]\,d\sigma = 0.$$

Now $\mathbf{n}$ is $\mathbf{n}_1$ on S and $\mathbf{n}_2$ on Ω. Thus

$$\iint_S [(\nabla\times\mathbf{v})\cdot\mathbf{n}_1]\,d\sigma + \iint_\Omega [(\nabla\times\mathbf{v})\cdot\mathbf{n}_2]\,d\sigma = 0.$$

This gives

$$\iint_S [(\nabla\times\mathbf{v})\cdot\mathbf{n}_1]\,d\sigma = \iint_\Omega [(\nabla\times\mathbf{v})\cdot(-\mathbf{n}_2)]\,d\sigma = \oint_C \mathbf{v}(\mathbf{r})\cdot d\mathbf{r}$$

where C is traversed in a positive sense with respect to $-\mathbf{n}_2$ and therefore in a positive sense with respect to $\mathbf{n}_1$. ($-\mathbf{n}_2$ points toward S.)

20. By the chain rule $\dfrac{dx}{dt} = \dfrac{d}{dt}[x(u(t), v(t))] = \dfrac{\partial x}{\partial u}u'(t) + \dfrac{\partial x}{\partial v}v'(t).$

Thus

$$\int_{C_1} v_1\,dx = \int_a^b \left(v_1 \frac{dx}{dt}\right)dt = \int_a^b \left[v_1\frac{\partial x}{\partial u}u'(t) + v_1\frac{\partial x}{\partial v}v'(t)\right]dt$$

$$= \int_{C_r} v_1\frac{\partial x}{\partial u}\,du + v_1\frac{\partial x}{\partial v}\,dv$$

by Green's theorem

$$= \iint_\Gamma \left[\frac{\partial}{\partial u}\left(v_1\frac{\partial x}{\partial v}\right) - \frac{\partial}{\partial v}\left(v_1\frac{\partial x}{\partial u}\right)\right]du\,dv.$$

The integrand can be written

$$\frac{\partial v_1}{\partial u}\frac{\partial x}{\partial v} + v_1\frac{\partial^2 x}{\partial u\partial v} - \frac{\partial v_1}{\partial v}\frac{\partial x}{\partial v} - v_1\frac{\partial^2 x}{\partial v\partial u} = \frac{\partial v_1}{\partial u}\frac{\partial x}{\partial v} - \frac{\partial v_1}{\partial v}\frac{\partial x}{\partial u}.$$

equality of partials

Thus we have

$$\int_C v_1\,dx = \iint_\Gamma \left[\frac{\partial v_1}{\partial u}\frac{\partial x}{\partial v} - \frac{\partial v_1}{\partial v}\frac{\partial x}{\partial u}\right]du\,dv.$$

By our previous choice of unit normal, $\mathbf{n} = \mathbf{N}/\parallel \mathbf{N} \parallel.$ Therefore

$$\iint_S [(\nabla \times v_1\mathbf{i}) \cdot \mathbf{n}]\,d\sigma = \iint_\Gamma [(\nabla \times v_1\mathbf{i}) \cdot \mathbf{N}]\,du\,dv.$$

Note that

$$(\nabla \times v_1\mathbf{i}) \cdot \mathbf{N} = \begin{vmatrix} \mathbf{i} & \mathbf{j} & \mathbf{k} \\ \dfrac{\partial}{\partial x} & \dfrac{\partial}{\partial y} & \dfrac{\partial}{\partial z} \\ v_1 & 0 & 0 \end{vmatrix} \cdot \begin{vmatrix} \mathbf{i} & \mathbf{j} & \mathbf{k} \\ \dfrac{\partial x}{\partial u} & \dfrac{\partial y}{\partial u} & \dfrac{\partial z}{\partial u} \\ \dfrac{\partial x}{\partial v} & \dfrac{\partial y}{\partial v} & \dfrac{\partial z}{\partial v} \end{vmatrix}$$

$$= \left(\frac{\partial v_1}{\partial z}\mathbf{j} - \frac{\partial v_1}{\partial y}\mathbf{k}\right) \cdot \left[\left(\frac{\partial x}{\partial v}\frac{\partial z}{\partial u} - \frac{\partial x}{\partial u}\frac{\partial z}{\partial v}\right)\mathbf{j} + \left(\frac{\partial x}{\partial u}\frac{\partial y}{\partial v} - \frac{\partial x}{\partial v}\frac{\partial y}{\partial u}\right)\mathbf{k}\right]$$

$$= \frac{\partial v_1}{\partial z}\left(\frac{\partial x}{\partial v}\frac{\partial z}{\partial u} - \frac{\partial x}{\partial u}\frac{\partial z}{\partial v}\right) + \frac{\partial v_1}{\partial y}\left(\frac{\partial x}{\partial v}\frac{\partial y}{\partial u} - \frac{\partial x}{\partial u}\frac{\partial y}{\partial v}\right)$$

$$= \left(\frac{\partial v_1}{\partial z}\frac{\partial z}{\partial u} + \frac{\partial v_1}{\partial y}\frac{\partial y}{\partial u}\right)\frac{\partial x}{\partial v} - \left(\frac{\partial v_1}{\partial z}\frac{\partial z}{\partial v} + \frac{\partial v_1}{\partial y}\frac{\partial y}{\partial v}\right)\frac{\partial x}{\partial u}.$$

Now, by the chain rule,

$$\frac{\partial v_1}{\partial u} = \frac{\partial v_1}{\partial x}\frac{\partial x}{\partial u} + \frac{\partial v_1}{\partial y}\frac{\partial y}{\partial u} + \frac{\partial v_1}{\partial z}\frac{\partial z}{\partial u}, \qquad \frac{\partial v_1}{\partial v} = \frac{\partial v_1}{\partial x}\frac{\partial x}{\partial v} + \frac{\partial v_1}{\partial y}\frac{\partial y}{\partial v} + \frac{\partial v_1}{\partial z}\frac{\partial z}{\partial v}.$$

Therefore

$$(\mathbf{\nabla} \times v_1\mathbf{i}) \cdot \mathbf{N} = \left(\frac{\partial v_1}{\partial u} - \frac{\partial v_1}{\partial x}\frac{\partial x}{\partial u}\right)\frac{\partial x}{\partial v} - \left(\frac{\partial v_1}{\partial v} - \frac{\partial v_1}{\partial x}\frac{\partial x}{\partial v}\right)\frac{\partial x}{\partial u}$$

$$= \frac{\partial v_1}{\partial u}\frac{\partial x}{\partial v} - \frac{\partial v_1}{\partial v}\frac{\partial x}{\partial u}$$

and, as asserted,

$$\iint_S [(\mathbf{\nabla} \times v_1\mathbf{i}) \cdot \mathbf{n}]\,d\sigma = \iint_\Gamma \left[\frac{\partial v_1}{\partial u}\frac{\partial x}{\partial v} - \frac{\partial v_1}{\partial v}\frac{\partial x}{\partial u}\right]\,du\,dv.$$

REVIEW EXERCISES

1. (a) $\mathbf{r}(u) = u\mathbf{i} + u\mathbf{j}, \quad 0 \le u \le 1; \quad \int_C \mathbf{h} \cdot d\mathbf{r} = \int_0^1 (u^3 - u^2)\,du = -\dfrac{1}{12}$

   (b) $\int_C \mathbf{h} \cdot d\mathbf{r} = \int_0^1 (2u^8 - 3u^7)\,du = -\dfrac{11}{72}$

2. (a) $\int_C \mathbf{h} \cdot d\mathbf{r} = \int_0^{\pi/2} (-\cos^3 u \sin u + \sin^3 u \cos u)\,du = 0$

   (b) $\int_C \mathbf{h} \cdot d\mathbf{r} = \int_0^{\pi/2} (-3\cos^{11} u \sin u + 3\sin^{11} u \cos u)\,du = 0$

3. Since $\mathbf{h}(x,y) = \nabla f$ where $f(x,y) = x^2 y^2 + \tfrac{1}{2}x^2 - y$,

$$\int_C \mathbf{h}(\mathbf{r}) \cdot d\mathbf{r} = f(2,4) - f(-1,2) = \frac{119}{2}$$

for *any* curve C beginning at $(-1,2)$ and ending at $(2,4)$.

4. $\mathbf{h}(x,y)$ is a gradient: $\dfrac{\partial P}{\partial y} = \dfrac{y^2 - x^2}{(x^2 + y^2)^2} = \dfrac{\partial Q}{\partial x}; \quad \mathbf{h}(x,y) = \nabla \arctan(y/x).$

   Therefore the integrals in (a), (b) and (c) all have the same value.

   (a) $\mathbf{r}(u) = 2\cos u\,\mathbf{i} + 2\sin u\,\mathbf{j}, \quad 0 \le u \le \tfrac{3}{4}\pi;$

$$\int_C \mathbf{h} \cdot d\mathbf{r} = \int_0^{3\pi/4} \left[\frac{-2\sin u}{4}(-2\sin u) + \frac{2\cos u}{4}(2\cos u)\right] du = \int_0^{3\pi/4} 1\,du = \frac{3\pi}{4}$$

5. $h(x, y, z) = \sin y \, \mathbf{i} + xe^{xy} \, \mathbf{j} + \sin z \, \mathbf{k}; \quad \mathbf{r}(u) = u^2 \, \mathbf{i} + u \, \mathbf{j} + u^3 \, \mathbf{k}, \quad u \in [0, 3]$

$x(u) = u^2 \quad y(u) = u \quad z(u) = u^3, \quad x'(u) = 2u, \quad y'(u) = 1, \quad z'(u) = 3u^2$

$h(\mathbf{r}(u)) \cdot \mathbf{r}'(u) = 2u \sin u + u^2 e^{u^3} + 3u^2 \sin u^3$

$$\int_C h(\mathbf{r}) \cdot d\mathbf{r} = \int_0^3 \left(2u \sin u + u^2 e^{u^3} + 3u^2 \sin u^3 \right) du$$

$$= \left[-2u \, \cos u + 2 \sin u + \tfrac{1}{3} e^{u^3} - \cos u^3 \right]_0^3$$

$$= \tfrac{2}{3} - 6 \cos 3 + 2 \sin 3 + \tfrac{1}{3} e^{27} - \cos 27$$

6. $h(x, y, z) = x^2 \, \mathbf{i} + xy \, \mathbf{j} + z^2 \, \mathbf{k}; \quad \mathbf{r}(u) = \cos u \, \mathbf{i} + \sin u \, \mathbf{j} + u^2 \, \mathbf{k}, \quad u \in [0, \pi/2]$

$h(\mathbf{r}(u)) \cdot \mathbf{r}'(u) = 2u^5; \quad \displaystyle\int_C h(\mathbf{r}) \cdot d\mathbf{r} = \int_0^{\pi/2} 2u^5 \, du = \dfrac{1}{3} \left(\dfrac{\pi}{2} \right)^6$

7. $\mathbf{F}(x, y, z) = xy \, \mathbf{i} + yz \, \mathbf{j} + xz \, \mathbf{k}; \quad \mathbf{r}(u) = u \, \mathbf{i} + u^2 \, \mathbf{j} + u^3 \, \mathbf{k}.$

$\mathbf{F}(\mathbf{r}(u)) \cdot \mathbf{r}' = u^3 + 5u^6; \quad W = \displaystyle\int_{-1}^2 (u^3 + 5u^6) du = \left[\dfrac{1}{4}u^4 + \dfrac{5}{7}u^7 \right]_{-1}^2 = \dfrac{2685}{28}$

8. $\mathbf{F}(x, y) = x \, \mathbf{i} + (y - 2) \, \mathbf{j}; \quad \mathbf{r}(u) = (u - \sin u) \, \mathbf{i} + (1 - \cos u) \, \mathbf{j}, \; 0 \le u \le 2\pi.$

$\mathbf{F}(\mathbf{r}(u)) \cdot \mathbf{r}' = u - u \cos u - 2 \sin u;$

$W = \displaystyle\int_0^{2\pi} (u - u \cos u - 2 \sin u) du = \left[\dfrac{1}{2}u^2 + u \sin u + 3 \cos u \right]_0^{2\pi} = 2\pi^2$

9. A vector equation for the line segment is: $\mathbf{r}(u) = (1 + 2u) \, \mathbf{i} + 4u \, \mathbf{k}, \quad u \in [0, 1].$

$\mathbf{F}(\mathbf{r}(u)) \cdot \mathbf{r}' = C \, \dfrac{2 + 20u}{\sqrt{1 + 4u + 20u^2}}; \quad \displaystyle\int_C \mathbf{F} \cdot d\mathbf{r} = C \int_0^1 \dfrac{(20u + 2)}{\sqrt{1 + 4u + 20u^2}} \, du = 4C$

10. Suppose that the path C of the object is given by the vector function $\mathbf{r} = \mathbf{r}(u), \; a \le u \le b.$ Then

$\mathbf{r}' = \mathbf{v}$ is the velocity of the object and $\mathbf{F} \cdot \mathbf{v} = 0.$ The work done by $\mathbf{F}$ is

$$\int_C \mathbf{F}(\mathbf{r}) \cdot d\mathbf{r} = \int_a^b \mathbf{F}(\mathbf{r}(u)) \cdot \mathbf{r}'(u) \, du = \int_a^b \mathbf{F}(\mathbf{r}(u)) \cdot \mathbf{v}(u) \, du = 0.$$

11. $\dfrac{\partial(ye^{xy} + 2x)}{\partial y} = e^{xy} + xye^{xy} = \dfrac{\partial(xe^{xy} - 2y)}{\partial x} \implies h$ is a gradient.

(a) $h(\mathbf{r}(u)) \cdot \mathbf{r}' = 3u^2 e^{u^3} - 4u^3 + 2u; \quad \displaystyle\int_C h \cdot d\mathbf{r} = \int_0^2 \left(3u^2 e^{u^3} - 4u^3 + 2u \right) du = e^8 - 13$

(b) Let $f(x, y) = e^{xy} + x^2 - y^2.$ Then $\nabla f = h$ and $\displaystyle\int_C h \cdot d\mathbf{r} = f(2, 4) - f(0, 0) = e^8 - 13$

12. $\dfrac{\partial P}{\partial y} = 4xy + 2 = \dfrac{\partial Q}{\partial x} \implies \mathbf{h}$ is a gradient.

(a) $\mathbf{h}(\mathbf{r}(u)) = \displaystyle\int_C \mathbf{h} \cdot d\mathbf{r} = \int_0^1 \left(6 + 66u + 216u^2 + 576u^3\right) du = \left[6u + 33u^2 + 72u^3 + 144u^4\right]_0^1 = 255$

(b) Let $f(x, y) = (x^2 y^2 + 2xy)$. Then $\nabla f = \mathbf{h}$ and $\displaystyle\int_C \mathbf{h} \cdot d\mathbf{r} = f(3, 5) - f(0, 1) = 255$

13. $\mathbf{h}(x, y, z) = \nabla f$ where $f(x, y, z) = x^4 y^3 z^2$.

(a) $\mathbf{h}(\mathbf{r}(u)) = 4u^{15}\, \mathbf{i} + 3u^{14}\, \mathbf{j} + 2u^{13}\, \mathbf{k}; \quad \mathbf{r}'(u) = \mathbf{i} + 2u\, \mathbf{j} + 3u^2\, \mathbf{k}$

$$\int_C \mathbf{h}(\mathbf{r}) \cdot d\mathbf{r} = \int_0^1 16\, u^{15}\, du = 1.$$

(b) $\displaystyle\int_C \mathbf{h}(\mathbf{r}) \cdot d\mathbf{r} = f(\mathbf{r}(1)) - f(\mathbf{r}(0)) = f(1, 1, 1) - f(0, 0, 0) = 1.$

14. (a) $\mathbf{r}(u) = u\mathbf{i} + 4u\mathbf{j}, \quad 0 \le u \le 2$

$$\int_C y^2\, dx + (x^2 - xy)\, dy = \int_0^2 [16u^2 + 4(u^2 - 4u^2)]\, du = \int_0^2 4u^2\, du = \frac{32}{3}$$

(b) $C_1 : \mathbf{r}(u) = u\mathbf{i}, \quad 0 \le u \le 2; \quad C_2 : \mathbf{r}(u) = 2\mathbf{i} + u\mathbf{j}, \quad 0 \le u \le 8$

$$\int_C y^2\, dx + (x^2 - xy)\, dy = \int_{C_1} y^2\, dx + (x^2 - xy)\, dy + \int_{C_2} y^2\, dx + (x^2 - xy)\, dy = 0 + \int_0^8 (4 - 2u)du = -32$$

(c) $C : \mathbf{r}(u) = u\mathbf{i} + u^3\mathbf{j}, \quad 0 \le u \le 2$

$$\int_C y^2 dx + (x^2 - xy)dy = \int_0^2 (3u^4 - 2u^6)\, du = -\frac{608}{35}$$

15. (a) $\mathbf{r}(u) = (1 - u)\mathbf{i} + u\mathbf{j}, \quad 0 \le u \le 1.$

$$\int_C 2xy^{1/2}\, dx + yx^{1/2}\, dy = \int_0^1 \left[2(1 - u)u^{1/2}(-1) + u(1 - u)^{1/2}\right] du$$

$$= -2\int_0^1 (1 - u)u^{1/2}\, du + \int_0^1 u(1 - u)^{1/2}\, du$$

$$= -\int_0^1 (1 - u)u^{1/2}\, du = -\frac{4}{15}$$

(b) $\mathbf{r}_1 = \mathbf{i} + u\mathbf{j}, \quad 0 \le u \le 1; \quad \mathbf{r}_2 = (1 - u)\,\mathbf{i} + \mathbf{j}$

$$\int_C 2xy^{1/2}dx + yx^{1/2}dy = \int_0^1 u\, du + \int_0^1 -2(1 - u)\, du = -\frac{1}{2}$$

(c) $\mathbf{r} = \cos u\, \mathbf{i} + \sin u\, \mathbf{j}, \quad 0 \le u \le \pi/2$

$$\int_C 2xy^{1/2}dx + yx^{1/2}dy = \int_0^{\pi/2} \left(-2\sin^{3/2} u \cos u + \cos^{3/2} u \sin u\right) du = -\frac{2}{5}$$

16. $\displaystyle\int z\,dx + x\,dy + y\,dz = \int_0^{2\pi}(a^2\cos^2 u - au\sin u + a\sin u)\,du = \pi a^2 + 2\pi a$

17.
$$\int_C ye^{xy}\,dx + \cos x\,dy + (\frac{xy}{z})\,dz = \int_0^2\left(u^2 e^{u^3} + 2u\cos u + 3u^2\right)du$$
$$= \left[\tfrac{1}{3}e^{u^3} + 2u\sin u + 2\cos u + u^3\right]_0^2$$
$$= \frac{1}{3}e^8 + \frac{17}{3} + 4\sin 2 + 2\cos 2$$

18. $\mathbf{r} = \cos u\,\mathbf{i} + \sin u\,\mathbf{j}; \quad \lambda(x,y) = k; \quad s'(u) = \|\mathbf{r}'\| = 1.$

(a) $\displaystyle M = \int_C \lambda(x,y)\,ds = \int_0^\pi k\,du = k\pi$

By symmetry, $x_M = 0$.

$$y_M M = \int_C y\lambda(x,y)\,ds = \int_0^\pi k\sin u\,du = \left[-k\cos u\right]_0^\pi = 2k; \quad y_M = \frac{2}{\pi}$$

(b)
$$I = \int_C \lambda(x,y)R^2(x,y)\,ds = \int_C kx^2\,ds$$
$$= \int_0^\pi k\cos^2 u\,du = \frac{k}{2}\int_0^\pi (1+\sin 2u)\,du = \frac{1}{2}k\pi$$

19. (a) Set $C_1 : \mathbf{r}(u) = u\,\mathbf{i} + u^2\,\mathbf{j}, \quad 0\le u\le 1; \quad C_2 : \mathbf{r}(u) = (1-u)\,\mathbf{i} + \sqrt{1-u}\,\mathbf{j}, \quad 0\le u\le 1.$

Then, $C = C_1 + C_2$.

$$\oint_C xy^2\,dx - x^2 y\,dy = \int_{C_1} xy^2\,dx - x^2 y\,dy + \int_{C_2} xy^2\,dx - x^2 y\,dy$$
$$= \int_0^1 (u^5 - 2u^5)\,du + \int_0^1\left[-(1-u)^2 + \tfrac{1}{2}(1-u)^2\right]du$$
$$= \int_0^1 (-u^5)\,du - \tfrac{1}{2}\int_0^1 (1-u)^2\,du = \left[-\tfrac{1}{6}u^6 + \tfrac{1}{6}(1-u)^3\right]_0^1 = -\tfrac{1}{3}$$

(b) $P = xy^2; \quad Q = -x^2 y$

$$\oint_C xy^2\,dx - x^2 y\,dy = \int_0^1\int_{x^2}^{\sqrt{x}}(-4xy)\,dy\,dx = \int_0^1 (2x^2 - 2x^5)\,dx = -\frac{1}{3}$$

20. (a) $\displaystyle\oint_C (x^2+y^2)\,dx + (x^2-y^2)\,dy = \iint_\Omega (2x-2y)\,dx\,dy = \int_0^{2\pi}\int_0^1 (2r\cos\theta - 2r\sin\theta)r\,dr\,d\theta = 0$

(b) $\displaystyle\oint_C (x^2+y^2)\,dx + (x^2-y^2)\,dy = \int_0^{2\pi}(-\sin u + \cos 2u\cos u)\,du = 0$

21. $P = x - 2y^2; \quad Q = 2xy$

$$\oint_C (x - 2y^2)\, dx + 2xy\, dy = \int_0^2 \int_0^1 6y\, dy\, dx = 6$$

22. $\oint_C xy\, dx + (\tfrac{1}{2}x^2 + xy)\, dy = \iint_\Omega y\, dx\, dy = \int_{-1}^1 \int_0^{\frac{\sqrt{1-x^2}}{2}} y\, dy\, dx = \int_{-1}^1 \tfrac{1}{8}(1 - x^2)\, dx = \tfrac{1}{6}$

23. $P = \ln(x^2 + y^2); \quad Q = \ln(x^2 + y^2); \quad \dfrac{\partial Q}{\partial x} - \dfrac{\partial P}{\partial y} = \dfrac{2x - 2y}{x^2 + y^2}$

$$\oint_C \ln(x^2 + y^2)\, dx + \ln(x^2 + y^2)\, dy = \iint_\Omega \frac{2x - 2y}{x^2 + y^2}\, dx\, dy$$

$$= \int_0^\pi \int_1^2 \frac{2r\cos\theta - 2r\sin\theta}{r^2}\, r\, dr\, d\theta$$

$$= 2 \int_0^\pi \int_1^2 (\cos\theta - \sin\theta)\, dr\, d\theta = -4$$

24. $P = 1/y, \quad Q = 1/x, \quad \dfrac{\partial Q}{\partial x} - \dfrac{\partial P}{\partial y} = -\dfrac{1}{x^2} + \dfrac{1}{y^2}$

$$\oint_C (1/y)\, dx + (1/x)\, dy = \iint_\Omega \left(-x^{-2} + y^{-2}\right)\, dx\, dy = \int_1^4 \int_1^{\sqrt{x}} \left(-x^{-2} + y^{-2}\right)\, dy\, dx$$

$$= \int_1^4 \left(-x^{-3/2} - x^{-1/2} + x^{-2} + 1\right)\, dx = \frac{3}{4}$$

25. $\oint_C y^2\, dx = \iint_\Omega -2y\, dx\, dy = \int_0^{2\pi} \int_0^{1+\sin\theta} -2r^2 \sin\theta\, dr\, d\theta = \int_0^{2\pi} (-\tfrac{2}{3})(1 + \sin\theta)^3 \sin\theta\, d\theta = -\dfrac{5\pi}{2}$

26. $P = e^y \cos x, \quad Q = -e^y \sin x, \quad \dfrac{\partial Q}{\partial x} - \dfrac{\partial P}{\partial y} = -2e^y \cos x$

$$\oint_C e^y \cos x\, dx - e^y \sin x\, dy = \iint_\Omega (-2e^y \cos x)\, dx\, dy = \int_0^{\pi/2} \int_0^1 (-2e^y \cos x)\, dy\, dx = 2(1 - e)$$

27. $C_1 : \mathbf{r}(u) = -u\,\mathbf{i} + (4 - u^2)\,\mathbf{j}, \ -2 \le u \le 2; \quad C_2 : \mathbf{r}(u) = u\,\mathbf{i}, \ -2 \le u \le 2; \quad C = C_1 \cup C_2$

$$A = \frac{1}{2} \int_C (-y\, dx + x\, dy) = \frac{1}{2} \int_{C_1} (-y\, dx + x\, dy) + \frac{1}{2} \int_{C_2} (-y\, dx + x\, dy)$$

$$= \frac{1}{2} \int_{-2}^2 -(4 - u^2)(-1)\, du - u(-2u)\, du + \frac{1}{2} \int_{-2}^2 0\, du$$

$$= \int_{-2}^2 (4 + u^2)\, du = \frac{32}{3}$$

28. $C_1 : \mathbf{r}(u) = (3 - 2u)\mathbf{i} + (1 + 2u)\mathbf{j}, \ 0 \le u \le 1; \quad C_2 : \mathbf{r}(u) = u\mathbf{i} + (3/u)\mathbf{j}, \ 1 \le u \le 3;$
$C = C_1 \cup C_2$

$$A = \frac{1}{2} \int_C (-y \, dx + x \, dy) = \frac{1}{2} \int_{C_1} (-y \, dx + x \, dy) + \frac{1}{2} \int_{C_2} (-y \, dx + x \, dy)$$

$$= \frac{1}{2} \int_0^1 [-(1 + 2u)(-2) + (3 - 2u)2] \, du + \frac{1}{2} \int_1^3 [-(3/u) + u(-3/u^2)] \, du$$

$$= \frac{1}{2} \int_0^1 8 \, du + \frac{1}{2} \int_1^3 (-6/u) \, du = 4 - 3 \ln 3$$

29. By symmetry, it is sufficient to consider the upper part of the sphere: $z = \sqrt{4 - x^2 - y^2}$

$$\frac{\partial z}{\partial x} = \frac{-x}{\sqrt{4 - x^2 - y^2}}, \qquad \frac{\partial z}{\partial y} = \frac{-y}{\sqrt{4 - x^2 - y^2}}$$

Let Ω be the projection of the sphere onto the xy plane, then

$$S = 2 \iint_\Omega \sqrt{(z_x)^2 + (z_y)^2 + 1} \, dx \, dy = 2 \iint_\Omega \frac{2}{\sqrt{4 - x^2 - y^2}} \, dx \, dy$$

$$= 4 \int_{-\pi/2}^{\pi/2} \int_0^{2\cos\theta} \frac{1}{\sqrt{4 - r^2}} \, r \, dr \, d\theta$$

$$= 4 \int_{-\pi/2}^{\pi/2} \left(2 - 2\sqrt{1 - \cos^2\theta}\right) d\theta = 8(\pi - 2)$$

30. From $x + y + 2z = 4$, we get $z = \dfrac{4 - x - y}{2}$ and $z_x = -\frac{1}{2}, \ z_y = -\frac{1}{2}$.

$$\text{area of } S = \iint_\Omega \sqrt{z_x^2 + z_y^2 + 1} \, dx \, dy = \sqrt{\frac{3}{2}} \int_0^{2\pi} \int_0^2 r \, dr \, d\theta = 2\sqrt{6}\pi.$$

31. $\dfrac{\partial z}{\partial x} = \dfrac{x}{\sqrt{x^2 + y^2}}, \quad \dfrac{\partial z}{\partial y} = \dfrac{y}{\sqrt{x^2 + y^2}}.$

The projection Ω of the surface onto the xy plane is the disk $x^2 + y^2 \le 9$.

$$S = \iint_\Omega \sqrt{(z_x)^2 + (z_y)^2 + 1} \, dx \, dy = \iint_\Omega \sqrt{2} \, dx \, dy = \int_0^{2\pi} \int_0^3 \sqrt{2} \, r \, dr \, d\theta = 9\,\pi\,\sqrt{2}$$

32.
$$A = 2 \iint_\Omega \sqrt{z_x^2 + z_y^2 + 1} \, dx\,dy = 2 \iint_\Omega \sqrt{4x^2 + 4y^2 + 1} \, dx\,dy$$

$$= 2 \int_0^{2\pi} \int_0^3 \sqrt{1 + 4r^2} \, r \, dr \, d\theta = 4\pi \int_0^3 \sqrt{1 + 4r^2} \, r \, dr$$

$$= \frac{\pi}{3}(37^{3/2} - 1)$$

33. $\displaystyle\iint_S yz\, d\sigma = \sqrt{2}\int_0^{2\pi}\int_0^1 r^2\sin\theta(r\sin\theta + 4)\, dr\, d\theta = \frac{\sqrt{2}\pi}{4}$

34. $\displaystyle\iint_S xz\, d\sigma = \iint_S x(1 - x - y)\, d\sigma = \iint_\Omega x(1 - x - y)\sqrt{3}\, dx\, dy = \sqrt{3}\int_0^1\int_0^{1-x} x(1 - x - y)dy\, dx = $

$\dfrac{\sqrt{3}}{24}$

35. The cylindrical surface S_1 is parametrized by: $x = u,\ y = 2\cos v,\ z = 2\sin v,\ 0 \le u \le 2,\ 0 \le v \le 2\pi$.

$$\mathbf{N}(u,v) = -2\cos v\,\mathbf{i} - 2\sin v\,\mathbf{j}, \quad \|\|\mathbf{N}(u,v)\|\| = 2$$

$$\iint_{S_1}(x^2 + y^2 + z^2)\, d\sigma = \int_0^2\int_0^{2\pi}(u^2 + 4)\, 2\, dv\, du = \frac{128\pi}{3}$$

The disc S_2 : $x = 0,\ y^2 + z^2$ is parametrized by: $x = 0,\ y = u\cos v,\ z = u\sin v,\ 0 \le u \le 2,$

$0 \le v \le 2\pi$.

$\mathbf{N} = u\,\mathbf{i}, \quad \|\|\mathbf{N}(u,v)\|\| = u;$ $\displaystyle\iint_{S_2}(x^2 + y^2 + z^2)\, d\sigma = \int_0^2\int_0^{2\pi}(0 + u^2)\, u\, dv\, du = 8\pi$

The disc S_3 : $x = 2,\ y^2 + z^2$ is parametrized by: $x = 2,\ y = u\cos v,\ z = u\sin v,\ 0 \le u \le 2,$

$0 \le v \le 2\pi$.

$\mathbf{N} = u\,\mathbf{i}, \quad \|\|\mathbf{N}(u,v)\|\| = u;$ $\displaystyle\iint_{S_2}(x^2 + y^2 + z^2)\, d\sigma = \int_0^2\int_0^{2\pi}(4 + u^2)\, u\, dv\, du = 24\pi$

Thus, $\displaystyle\iint_S(x^2 + y^2 + z^2)\, d\sigma = \frac{128\pi}{3} + 8\pi + 24\pi = \frac{224\pi}{3}$

36. The cylindrical surface S is parametrized by:

$$x = 2\cos u,\ y = 2\sin u,\ z = v, \quad 0 \le u \le 2\pi,\ 0 \le v \le 1.$$

$\mathbf{N}(u,v) = 2\cos u\,\mathbf{i} - 2\sin u\,\mathbf{j}, \quad \|\|\mathbf{N}(u,v)\|\| = 2;$ $\displaystyle\iint_S xz\, d\sigma = \int_0^1\int_0^{2\pi} 2v\cos u(2)\, dv\, du = 0.$

37. $\nabla \cdot \mathbf{v} = 4x, \quad \nabla \times \mathbf{v} = 2y\mathbf{k}$ **38** $\nabla \cdot \mathbf{v} = 0, \quad \nabla \times \mathbf{v} = 0$

39. $\nabla \cdot \mathbf{v} = 1 + xy, \quad \nabla \times \mathbf{v} = (xz - x)\mathbf{i} - yz\mathbf{j} + z\mathbf{k}$

40. $\nabla \cdot \mathbf{v} = yz + x\sin xy, \quad \nabla \times \mathbf{v} = x\cos xy\,\mathbf{i} + (xy - y\cos xy)\mathbf{j} + (y\sin xy - xz)\mathbf{k}$

41. (a) $\nabla \cdot \mathbf{v} = z - x + y$

$$\int_0^1 \int_0^1 \int_0^1 (z - x + y)\,dz\,dy\,dx = \frac{1}{2}$$

(b) at $x = 0$, $\mathbf{n} = -\mathbf{i}, \mathbf{v} \cdot \mathbf{n} = 0, \displaystyle\int_0^1 \int_0^1 0\,dy\,dz = 0$

   at $x = 1$, $\mathbf{n} = \mathbf{i}, \mathbf{v} \cdot \mathbf{n} = z, \displaystyle\int_0^1 \int_0^1 z\,dy\,dz = 1/2$

   at $y = 0$, $\mathbf{n} = -\mathbf{j}, \mathbf{v} \cdot \mathbf{n} = xy = 0, \displaystyle\int_0^1 \int_0^1 0\,dx\,dz = 0$

   at $y = 1$, $\mathbf{n} = \mathbf{j}, \mathbf{v} \cdot \mathbf{n} = -xy = -x, \displaystyle\int_0^1 \int_0^1 -x\,dx\,dz = -1/2$

   at $z = 0$, $\mathbf{n} = -\mathbf{k}, \mathbf{v} \cdot \mathbf{n} = 0, \displaystyle\int_0^1 \int_0^1 0\,dy\,dx = 0$

   at $z = 1$, $\mathbf{n} = \mathbf{k}, \mathbf{v} \cdot \mathbf{n} = yz, \displaystyle\int_0^1 \int_0^1 y\,dy\,dx = 1/2$

   The sum is $1/2$

42. (a) $\nabla \cdot \mathbf{v} = 3$

$$\iiint_T 3\,dx\,dy\,dz = \int_0^4 \int_0^{2\pi} \int_0^1 3r\,dr\,d\theta\,dx = 4(2\pi)(\tfrac{3}{2}) = 12\pi$$

(b) at $x = 0$, $\mathbf{n} = -\mathbf{i}, \ \mathbf{v} \cdot \mathbf{n} = -z, \displaystyle\iint_S -z\,dy\,dz = 0$ (by symmetry)

   at $x = 4$, $\mathbf{n} = \mathbf{i}, \ \mathbf{v} \cdot \mathbf{n} = 4 + z, \displaystyle\iint_S (4 + z)\,dy\,dz = \iint_S 4\,dy\,dz = 4\pi$

   for $z = \sqrt{1 - y^2}, \ 0 \le x \le 4, \mathbf{n} = -y\mathbf{j} + \sqrt{1 - y^2}\,\mathbf{k}$ and

$$\mathbf{v} \cdot \mathbf{n} = 1 - 2y^2 - y\sqrt{1 - y^2} + x\sqrt{1 - y^2}$$

$$\int_{-1}^1 \int_0^4 (1 - 2y^2 - y\sqrt{1 - y^2} + x\sqrt{1 - y^2})\,dx\,dy = 8 - \frac{16}{3} + 4\pi$$

   for $z = -\sqrt{1 - y^2}, \ 0 \le x \le 4, \mathbf{n} = y\mathbf{j} + \sqrt{1 - y^2}\,\mathbf{k}$ and

$$\mathbf{v} \cdot \mathbf{n} = -1 + 2y^2 - y\sqrt{1 - y^2} + x\sqrt{1 - y^2}$$

$$\int_{-1}^1 \int_0^4 (-1 + 2y^2 - y\sqrt{1 - y^2} + x\sqrt{1 - y^2})\,dx\,dy = -8 + \frac{16}{3} + 4\pi$$

   The sum is 12π

43. The projection of S onto the xy-plane is: $\Omega : x^2 + y^2 \leq 9$.

$$\iint_S \mathbf{v} \cdot \mathbf{n}\, d\sigma = \iint_\Omega \left(4x^2 + 2xyz + z^2\right) dx\, dy$$

$$= \iint_\Omega \left(4x^2 + 2xy\left[9 - x^2 - y^2\right] + \left[9 - x^2 - y^2\right]^2\right) dx\, dy$$

$$= \int_0^{2\pi} \int_0^3 \left[4r^2 \cos^2\theta + r^2(9 - r^2)\sin 2\theta + (9 - r^2)^2\right] r\, dr\, d\theta = 324\pi$$

44. On $x = 0$, $\mathbf{n} = -\mathbf{i}$, $\mathbf{v}\cdot\mathbf{n} = -x^2 = 0$,   the flux is 0;

on $x = a$, $\mathbf{n} = \mathbf{i}$, $\mathbf{v}\cdot\mathbf{n} = a^2$,   the flux is a^4;

on $y = 0$, $\mathbf{n} = -\mathbf{j}$, $\mathbf{v}\cdot\mathbf{n} = xz$,   the flux is $\int_0^a \int_0^a xz\, dx\, dz = \frac{1}{4}a^2$;

on $y = a$, $\mathbf{n} = \mathbf{j}$, $\mathbf{v}\cdot\mathbf{n} = -xz$,   the flux is $\int_0^a \int_0^a -xz\, dx\, dz = -\frac{1}{4}a^2$;

on $z = 0$, $\mathbf{n} = -\mathbf{k}$, $\mathbf{v}\cdot\mathbf{n} = 0$,   the flux is 0;

on $z = a$, $\mathbf{n} = \mathbf{k}$, $\mathbf{v}\cdot\mathbf{n} = a^2$,   the flux is a^4.

Hence the total flux is $2a^4$.

45. (a) $(\nabla \times \mathbf{v}) \cdot \mathbf{n} = (\mathbf{i} + \mathbf{j} + \mathbf{k}) \cdot \left(-\frac{1}{2}x\,\mathbf{i} - \frac{1}{2}y\,\mathbf{j} + \frac{\sqrt{4 - x^2 - y^2}}{2}\,\mathbf{k}\right) = -\frac{1}{2}x - \frac{1}{2}y + \frac{\sqrt{4 - x^2 - y^2}}{2}$

$$\iint_S \left(-\frac{1}{2}x - \frac{1}{2}y + \frac{\sqrt{4 - x^2 - y^2}}{2}\right) d\sigma = \iint_S \left(-\frac{1}{2}x - \frac{1}{2}y + \frac{\sqrt{4 - x^2 - y^2}}{2}\right) \frac{2}{\sqrt{4 - x^2 - y^2}}\, dx\, dy$$

$$= \iint \left(\frac{-x}{\sqrt{4 - x^2 - y^2}} - \frac{-y}{\sqrt{4 - x^2 - y^2}} + 1\right) dx\, dy$$

$$= \int_0^{2\pi} \int_0^2 \left(-\frac{r\cos\theta}{\sqrt{4 - r^2}} - \frac{r\sin\theta}{\sqrt{4 - r^2}} + 1\right) r\, dr\, d\theta = 4\pi$$

(b) $\mathbf{r}(\theta) = 2\cos\theta\,\mathbf{i} + 2\sin\theta\,\mathbf{j}$, $0 \leq \theta \leq 2\pi$

$$\iint_S [(\nabla \times \mathbf{v}) \cdot \mathbf{n}]\, d\sigma = \oint_C \mathbf{v}(\mathbf{r}) \cdot d\mathbf{r} = \int_0^{2\pi} 4\cos^2\theta\, d\theta = 4\pi$$

46. (a) $\mathbf{v} = z^3\,\mathbf{i} + x\,\mathbf{j} + y^2\,\mathbf{k};$ $\mathbf{n} = \dfrac{2x\,\mathbf{i} + 2y\,\mathbf{j} + \mathbf{k}}{\sqrt{1 + 4x^2 + 4y^2}};$

$$\iint_S [(\nabla \times \mathbf{v}) \cdot \mathbf{n}]\, d\sigma = \iint_S \frac{1}{\sqrt{1 + 4x^2 + 4y^2}}(4xy + 6yz^2 + 1)\, d\sigma$$

$$= \iint_\Omega (4xy + 6yz^2 + 1)\, dx\, dy$$

$$= \iint_\Omega [4xy + 6y(9 - x^2 - y^2)^2 + 1]\, dx\, dy$$

$$= \int_0^{2\pi} \int_0^3 [4r^2 \cos\theta \sin\theta + 6r \sin\theta(9 - r^2)^2 + 1]\, r\, dr\, d\theta$$

$$= \int_0^3 2\pi r\, dr = 9\pi$$

(b) The boundary of the surface is the curve $x^2 + y^2 = 9,\ z = 0;$ $\mathbf{r}(u) = 3\cos u\,\mathbf{i} + 3\sin u\,\mathbf{j} + 0\,\mathbf{k};$

$\mathbf{v}(\mathbf{r}(u)) = 3\cos u\,\mathbf{j} + 9\sin^2 u\,\mathbf{k};$ $\mathbf{r}'(u) = -3\sin u\,\mathbf{i} + 3\cos u\,\mathbf{j}$

$$\oint_C \mathbf{v} \cdot d\mathbf{r} = \int_0^{2\pi} 9\cos^2 u\, du = 9\pi.$$

CHAPTER 19

SECTION 19.1

1. $y' + xy = xy^3 \implies y^{-3}y' + xy^{-2} = x.$ Let $v = y^{-2}, \quad v' = -2y^{-3}y'.$

$$-\frac{1}{2}v' + xv = x$$
$$v' - 2xv = -2x$$
$$e^{-x^2}v' - 2xe^{-x^2}v = -2xe^{-x^2}$$
$$e^{-x^2}v = e^{-x^2} + C$$
$$v = 1 + Ce^{x^2}$$
$$y^2 = \frac{1}{1 + Ce^{x^2}}.$$

2. $y' - y = -(x^2 + x + 1)y^2 \implies y^{-2}y' - y^{-1} = -(x^2 + x + 1).$ Let $v = y^{-1}, \quad v' = -y^{-2}y'.$

$$-v' - v = -(x^2 + x + 1)$$
$$v' + v = x^2 + x + 1$$
$$e^x v = \int e^x(x^2 + x + 1)\,dx = x^2 e^x - xe^x + 2e^x + C$$
$$v = x^2 - x + 2 + Ce^{-x}$$
$$y = \frac{1}{x^2 - x + 2 + Ce^{-x}}.$$

3. $y' - 4y = 2e^x y^{\frac{1}{2}} \implies y^{-\frac{1}{2}}y' - 4y^{\frac{1}{2}} = 2e^x.$ Let $v = y^{\frac{1}{2}}, \quad v' = \frac{1}{2}y^{-\frac{1}{2}}y'.$

$$2v' - 4v = 2e^x$$
$$v' - 2v = e^x$$
$$e^{-2x}v' - 2e^{-2x}v = e^{-x}$$
$$e^{-2x}v = -e^{-x} + C$$
$$v = -e^x + Ce^{2x}$$
$$y = (Ce^{2x} - e^x)^2.$$

4. $y' = \frac{1}{2xy} + \frac{y}{2x} \implies yy' - \frac{1}{2x}y^2 = \frac{1}{2x}.$ Let $v = y^2, \quad v' = 2yy'.$

$$\frac{1}{2}v' - \frac{1}{2x}v = \frac{1}{2x}$$
$$v' - \frac{1}{x}v = \frac{1}{x}$$
$$\frac{1}{x}v' - \frac{1}{x^2}v = \frac{1}{x^2}$$
$$\frac{1}{x}v = -\frac{1}{x} + C$$
$$v = Cx - 1$$
$$y^2 = Cx - 1.$$

5. $(x-2)y' + y = 5(x-2)^2 y^{\frac{1}{2}}$ $\implies$ $y^{-\frac{1}{2}}y' + \dfrac{1}{x-2}y^{\frac{1}{2}} = 5(x-2)$.   Let $v = y^{\frac{1}{2}}$, $v' = \dfrac{1}{2}y^{-\frac{1}{2}}y'$.

$$2v' + \frac{1}{x-2}v = 5(x-2)$$

$$v' + \frac{1}{2(x-2)}v = \frac{5}{2}(x-2)$$

$$\sqrt{x-2}\,v' + \frac{1}{2\sqrt{x-2}}v = \frac{5}{2}(x-2)^{\frac{3}{2}}$$

$$\sqrt{x-2}\,v = (x-2)^{\frac{5}{2}} + C$$

$$v = (x-2)^2 + \frac{C}{\sqrt{x-2}}$$

$$y = \left[(x-2)^2 + \frac{C}{\sqrt{x-2}}\right]^2 .$$

6. $yy' - xy^2 + x = 0$.   Let $v = y^2$, $v' = 2yy'$.

$$\frac{1}{2}v' - xv = -x$$

$$v' - 2xv = -2x$$

$$e^{-x^2}v' - 2xe^{-x^2}v = -2xe^{-x^2}$$

$$e^{-x^2}v = e^{-x^2} + C$$

$$v = 1 + Ce^{x^2}$$

$$y = \sqrt{1 + Ce^{x^2}} .$$

7. $y' + xy = y^3 e^{x^2}$ $\implies$ $y^{-3}y' + xy^{-2} = e^{x^2}$.   Let $v = y^{-2}$, $v' = -2y^{-3}y'$.

$$-\frac{1}{2}v' + xv = e^{x^2}$$

$$v' - 2xv = -2e^{x^2}$$

$$e^{-x^2}v' - 2xe^{-x^2}v = -2$$

$$e^{-x^2}v = -2x + C$$

$$v = -2xe^{x^2} + Ce^{x^2}$$

$$y^{-2} = Ce^{x^2} - 2xe^{x^2} .$$

$C = 4$ $\implies$ $y^{-2} = 4e^{x^2} - 2xe^{x^2}.$

8. $y' + \dfrac{1}{x}y = \dfrac{\ln x}{x}y^2$ $\implies$ $y^{-2}y' + \dfrac{1}{x}y^{-1} = \dfrac{\ln x}{x}$.   Let $v = y^{-1}$, $v' = -y^{-2}y'$.

$$-v' + \frac{1}{x}v = \frac{\ln x}{x}$$

$$v' - \frac{1}{x}v = -\frac{\ln x}{x}$$

$$\frac{1}{x}v' - \frac{1}{x^2}v = -\frac{\ln x}{x^2}$$

$$\frac{1}{x}v = -\int \frac{\ln x}{x^2}\,dx = \frac{1}{x}(\ln x + 1) + C$$

$$v = \ln x + 1 + Cx$$

$$y = \frac{1}{\ln x + 1 + Cx}.$$

$$1 = \frac{1}{\ln 1 + 1 + C} \quad\Longrightarrow\quad C = 0 \quad\Longrightarrow\quad y = \frac{1}{\ln x + 1}.$$

9. $2x^3 y' - 3x^2 y = y^3 \quad\Longrightarrow\quad y^{-3}y' - \frac{3}{2x}y^{-2} = \frac{1}{2x^3}.$ Let $v = y^{-2}, \quad v' = -2y^{-3}y'.$

$$-\frac{1}{2}v' - \frac{3}{2x}v = \frac{1}{2x^3}$$

$$v' + \frac{3}{x}v = -\frac{1}{x^3}$$

$$x^3 v' + 3x^2 v = -1$$

$$x^3 v = -x + C$$

$$v = \frac{C - x}{x^3}$$

$$y^2 = \frac{x^3}{C - x}$$

$$1 = \frac{1}{C - x} \quad\Longrightarrow\quad C = 2 \quad\Longrightarrow\quad y^2 = \frac{x^3}{2 - x}.$$

10. $y' + \tan x\, y = y^2 \sec^3 x \quad\Longrightarrow\quad y^{-2}y' + \tan x y^{-1} = \sec^3 x.$ Let $v = y^{-1}, \quad v' = -y^{-2}y'.$

$$-v' + \tan x v = \sec^3 x$$

$$v' - \tan x\, v = -\sec^3 x$$

$$\cos x v' - \sin x\, v = -\sec^2 x$$

$$\cos x\, v = -\tan x + C$$

$$\frac{\cos x}{y} = -\tan x + C$$

$$\frac{\cos 0}{3} = -\tan 0 + C \quad\Longrightarrow\quad C = \frac{1}{3} \quad\Longrightarrow\quad \frac{\cos x}{y} = \frac{1}{3} - \tan x.$$

11. $y' - \frac{y}{x}\ln y = xy \quad\Longrightarrow\quad \frac{y'}{y} - \frac{1}{x}\ln y = x.$ Let $u = \ln y, \quad u' = \frac{y'}{y}.$

$$u' - \frac{1}{x}u = x$$

$$\frac{1}{x}u' - \frac{1}{x^2}u = 1$$

$$\frac{1}{x}u = x + C$$

$$u = x^2 + Cx$$

$$\ln y = x^2 + Cx.$$

12. (a)
$$y' + yf(x)\ln y = g(x)y$$
$$\frac{y'}{y} + f(x)\ln y = g(x)$$
$$u' + f(x)u = g(x).$$

(b) $\cos y\, y' + g(x)\sin y = f(x)$.   Let $u = \sin y$, $u' = \cos y\, y'$.
Thus we have $u' + g(x)u = f(x)$.

13. $f(x,y) = \dfrac{x^2 + y^2}{2xy}$; $f(tx, ty) = \dfrac{(tx)^2 + (ty)^2}{2(tx)(ty)} = \dfrac{t^2(x^2 + y^2)}{t^2(2xy)} = \dfrac{x^2 + y^2}{2xy} = f(x,y)$.

Set $vx = y$. Then, $v + xv' = y'$ and
$$v + xv' = \frac{x^2 + v^2 x^2}{2vx^2} = \frac{1 + v^2}{2v}$$
$$v - \frac{1 + v^2}{2v} + xv' = 0$$
$$v^2 - 1 + 2xvv' = 0$$
$$\frac{1}{x}\,dx + \frac{2v}{v^2 - 1}\,dv = 0$$
$$\int \frac{1}{x}\,dx + \int \frac{2v}{v^2 - 1}\,dv = C$$
$$\ln|x| + \ln|v^2 - 1| = K \qquad \text{or} \qquad x(v^2 - 1) = C$$

Replacing v by y/x, we get
$$x\left(\frac{y^2}{x^2} - 1\right) = C \qquad \text{or} \qquad y^2 - x^2 = Cx$$

14. $f(tx, ty) = \dfrac{(ty)^2}{(tx)(ty) + (tx)^2} = \dfrac{y^2}{xy + x^2} = f(x,y)$.

Set $vx = y$. Then, $v + xv' = y'$ and
$$v + xv' = \frac{v^2}{1 + v}$$
$$\int \frac{dx}{x} + \int \frac{v + 1}{v}\,dv = C$$
$$\ln|x| + v + \ln|v| = C$$
$$v + \ln|xv| = C$$
$$\frac{y}{x} + \ln|y| = C$$

15. $f(x,y) = \dfrac{x - y}{x + y}$; $f(tx, ty) = \dfrac{(tx) - (ty)}{tx + ty} = \dfrac{t(x - y)}{t(x + y)} = \dfrac{x - y}{x + y} = f(x,y)$

Set $vx = y$. Then, $v + xv' = y'$ and
$$v + xv' = \frac{x - vx}{x + vx} = \frac{1 - v}{1 + v}$$
$$v^2 + 2v - 1 + x(1 + v)v' = 0$$

$$\frac{1}{x}\,dx + \frac{1+v}{v^2+2v-1}\,dv = 0$$

$$\int \frac{1}{x}\,dx + \int \frac{1+v}{v^2+2v-1}\,dv = C$$

$$\ln|x| + \tfrac{1}{2}\ln|v^2+2v-1| = K \qquad \text{or} \qquad x\sqrt{v^2+2v-1} = C$$

Replacing v by y/x, we get

$$x\sqrt{\frac{y^2}{x^2}+2\frac{y}{x}-1} = C \qquad \text{or} \qquad y^2+2xy-x^2 = C$$

16. $f(tx,ty) = \dfrac{tx+ty}{tx-ty} = \dfrac{x+y}{x-y} = f(x,y).$

Set $vx = y$. Then, $v + xv' = y'$ and

$$v + xv' = \frac{x+vx}{x-vx} = \frac{1+v}{1-v}$$

$$\int \frac{dx}{x} + \int \frac{v-1}{v^2+1}\,dv = C_1$$

$$\ln|x| + \frac{1}{2}\ln|v^2+1| - \arctan v = C_1$$

$$\ln x^2 + \ln(v^2+1) - 2\arctan v = C \quad (=2C_1)$$

$$\ln[x^2(v^2+1)] - 2\arctan v = C$$

$$\ln[x^2+y^2] - 2\arctan\left(\frac{y}{x}\right) = C$$

17. $f(x,y) = \dfrac{x^2 e^{y/x}+y^2}{xy};$ $f(tx,ty) = \dfrac{(tx)^2 - e^{(ty)/(tx)}+(ty)^2}{(tx)(ty)} = \dfrac{t^2\left(x^2 e^{y/x}+y^2\right)}{t^2(xy)} = f(x,y)$

Set $vx = y$. Then, $v + xv' = y'$ and

$$v + xv' = \frac{x^2 e^v + v^2 x^2}{vx^2} = \frac{e^v + v^2}{v}$$

$$v^2 + xvv' = e^v + v^2$$

$$-e^v + xvv' = 0$$

$$\frac{1}{x}\,dx = ve^{-v}\,dv$$

$$\int \frac{1}{x}\,dx = \int ve^{-v}\,dv$$

$$\ln|x| = -ve^{-v} - e^{-v} + C$$

Replacing v by y/x, and simplifying, we get

$$y + x = xe^{y/x}(C - \ln|x|)$$

18. $f(tx, ty) = \dfrac{(tx)^2 + 3(ty)^2}{4(tx)(ty)} = \dfrac{x^2 + 3y^2}{4xy} = f(x, y).$

Set $vx = y$. Then, $v + xv' = y'$ and

$$v + xv' = \frac{x^2 + 3x^2 v^2}{4x^2 v}$$

$$\int \frac{dx}{x} + \int \frac{4v}{v^2 - 1}\, dv = C_1$$

$$\ln|x| + 2\ln|v^2 - 1| = C_1$$

$$x(v^2 - 1)^2 = C \ (= e^{C_1})$$

$$(y^2 - x^2)^2 = Cx^3$$

19. $f(x, y) = \dfrac{y}{x} + \sin(y/x);$ $f(tx, ty) = \dfrac{(ty)}{tx} + \sin[(ty/tx)] = \dfrac{y}{x} + \sin(y/x) = f(x, y)$

Set $vx = y$. Then, $v + xv' = y'$ and

$$v + xv' = \frac{vx}{x} + \sin[(vx)/x] = v + \sin v$$

$$xv' = \sin v$$

$$\csc v\, dv = \frac{1}{x}\, dx$$

$$\int \csc v\, dv = \int \frac{1}{x}\, dx$$

$$\ln|\csc v - \cot v| = \ln|x| + K \qquad \text{or} \qquad \csc v - \cot v = Cx$$

Replacing v by y/x, and simplifying, we get

$$1 - \cos(y/x) = Cx\, \sin(y/x)$$

20. $f(x, y) = \dfrac{y}{x}\left(1 + \ln\left(\dfrac{y}{x}\right)\right);$ $f(tx, ty) = \dfrac{ty}{tx}\left(1 + \ln\left(\dfrac{ty}{tx}\right)\right) = \dfrac{y}{x}\left(1 + \ln\left(\dfrac{y}{x}\right)\right) = f(x, y)$

Set $vx = y$. Then, $v + xv' = y'$ and

$$v + xv' = \frac{vx}{x}\left(1 + \ln\left(\frac{vx}{x}\right)\right) = v(1 + \ln v)$$

$$xv' = v \ln v$$

$$\frac{1}{v \ln v}\, dv = \frac{1}{x}\, dx$$

$$\int \frac{1}{v \ln v}\, dv = \int \frac{1}{x}\, dx$$

$$\ln|\ln v| = \ln|x| + K$$

$$\ln\left(\frac{y}{x}\right) = Cx$$

$$\frac{y}{x} = e^{Cx} \quad \text{or} \quad y = xe^{Cx}$$

21. The differential equation is homogeneous since

$$f(x, y) = \frac{y^3 - x^3}{xy^2}; \qquad f(tx, ty) = \frac{(ty)^3 - (tx)^3}{(tx)(ty)^2} = \frac{t^3(y^3 - x^3)}{t^3(xy^2)} = \frac{y^3 - x^3}{xy^2} = f(x, y)$$

Set $vx = y$. Then, $v + xv' = y'$ and

$$v + xv' = \frac{(vx)^3 - x^3}{v^2 x^3} = \frac{v^3 - 1}{v^2}$$

$$1 + xv^2 v' = 0$$

$$\frac{1}{x} dx + v^2 dv = 0$$

$$\int \frac{1}{x} dx + \int v^2 dv = 0$$

$$\ln |x| + \frac{1}{3} v^3 = C$$

Replacing v by y/x, we get

$$y^3 + 3x^3 \ln |x| = Cx^3$$

Applying the side condition $y(1) = 2$, we have

$$8 + 3 \ln 1 = C \quad \Longrightarrow \quad C = 8 \quad \text{and} \quad y^3 + 3x^3 \ln |x| = 8x^3$$

22. $\dfrac{dy}{dx} = \dfrac{1}{\sin(y/x)} + \dfrac{y}{x}$. Set $y = vx$. Then $y' = v + xv'$ and

$$v + xv' = \frac{1}{\sin v} + v$$

$$\int \frac{dx}{x} - \int \sin v \, dv = C$$

$$\ln |x| + \cos v = C$$

$$\ln |x| + \cos \left(\tfrac{y}{x} \right) = C$$

$$y(1) = 0 \implies 0 + \cos 0 = C \implies C = 1 \implies \ln |x| + \cos \left(\frac{y}{x} \right) = 1$$

SECTION 19.2

1. $\dfrac{\partial P}{\partial y} = 2xy - 1 = \dfrac{\partial Q}{\partial x}$;   the equation is exact on the whole plane.

$$\frac{\partial f}{\partial x} = xy^2 - y \quad \Longrightarrow \quad f(x, y) = \tfrac{1}{2} x^2 y^2 - xy + \varphi(y)$$

$$\frac{\partial f}{\partial y} = x^2 y - x + \varphi'(y) = x^2 y - x \quad \Longrightarrow \quad \varphi'(y) = 0 \quad \Longrightarrow \quad \varphi(y) = 0 \text{ (omit the constant)}*$$

Therefore $f(x, y) = \tfrac{1}{2} x^2 y^2 - xy$,   and a one-parameter family of solutions is:

$$\tfrac{1}{2} x^2 y^2 - xy = C$$

* We will omit the constant at this step throughout this section.

2. $\dfrac{\partial}{\partial y}(e^x \sin y) = e^x \cos y = \dfrac{\partial}{\partial x}(e^x \cos y)$;   the equation is exact on the whole plane.

$f(x, y) = e^x \sin y$,   and $e^x \sin y = C$ is a one-parameter family of solutions.

3. $\dfrac{\partial P}{\partial y} = e^y - e^x = \dfrac{\partial Q}{\partial x}$; the equation is exact on the whole plane.

$\dfrac{\partial f}{\partial x} = e^y - ye^x \implies f(x,y) = xe^y - ye^x + \varphi(y)$

$\dfrac{\partial f}{\partial y} = xe^y - e^x + \varphi'(y) = xe^y - e^x \implies \varphi'(y) = 0 \implies \varphi(y) = 0$

Therefore $f(x,y) = xe^y - ye^x$, and a one-parameter family of solutions is:

$$xe^y - ye^x = C$$

4. $\dfrac{\partial}{\partial y}(\sin y) = \cos y = \dfrac{\partial}{\partial x}(x\cos y + 1)$; the equation is exact on the whole plane.

$f(x,y) = x\sin y + y$, and $x\sin y + y = C$ is a one-parameter family of solutions.

5. $\dfrac{\partial P}{\partial y} = \dfrac{1}{y} + 2x = \dfrac{\partial Q}{\partial x}$; the equation is exact on the upper half plane.

$\dfrac{\partial f}{\partial x} = \ln y + 2xy \implies f(x,y) = x\ln y + x^2 y + \varphi(y)$

$\dfrac{\partial f}{\partial y} = \dfrac{x}{y} + x^2 + \varphi'(y) = \dfrac{x}{y} + x^2 \implies \varphi'(y) = 0 \implies \varphi(y) = 0$

Therefore $f(x,y) = x\ln y + x^2 y$, and a one-parameter family of solutions is:

$$x\ln y + x^2 y = C$$

6. $\dfrac{\partial}{\partial y}(2x\arctan y) = \dfrac{2x}{1+y^2} = \dfrac{\partial}{\partial x}\left(\dfrac{x^2}{1+y^2}\right)$; the equation is exact on the whole plane.

$f(x,y) = x^2 \arctan y$, and $x^2 \arctan y = C$ is a one-parameter family of solutions.

7. $\dfrac{\partial P}{\partial y} = \dfrac{1}{x} = \dfrac{\partial Q}{\partial x}$; the equation is exact on the right half plane.

$\dfrac{\partial f}{\partial x} = \dfrac{y}{x} + 6x \implies f(x,y) = y\ln x + 3x^2 + \varphi(y)$

$\dfrac{\partial f}{\partial y} = \ln x + \varphi'(y) = \ln x - 2 \implies \varphi'(y) = -2 \implies \varphi(y) = -2y$

Therefore $f(x,y) = y\ln x + 3x^2 - 2y$, and a one-parameter family of solutions is:

$$y\ln x + 3x^2 - 2y = C$$

8. $\dfrac{\partial}{\partial y}\left(e^x + \ln y + \dfrac{y}{x}\right) = \dfrac{1}{y} + \dfrac{1}{x} = \dfrac{\partial}{\partial x}\left(\dfrac{x}{y} + \ln x + \sin y\right)$;

the equation is exact in the first quadrant, not including the axes.

$f(x,y) = e^x + x\ln y + y\ln x - \cos y$ and $e^x + x\ln y + y\ln x - \cos y = C$ is a

one-parameter family of solutions.

9. $\dfrac{\partial P}{\partial y} = 3y^2 - 2y\sin x = \dfrac{\partial Q}{\partial x}$; the equation is exact on the whole plane.

$\dfrac{\partial f}{\partial x} = y^3 - y^2 \sin x - x \implies f(x,y) = xy^3 + y^2 \cos x - \tfrac{1}{2}x^2 + \varphi(y)$

$\dfrac{\partial f}{\partial y} = 3xy^2 + 2y\cos x + \varphi'(y) = 3xy^2 + 2y\cos x + e^{2y} \quad \Longrightarrow \quad \varphi'(y) = e^{2y} \quad \Longrightarrow \quad \varphi(y) = \tfrac{1}{2}e^{2y}$

Therefore $f(x,y) = xy^3 + y^2\cos x - \tfrac{1}{2}x^2 + \tfrac{1}{2}e^{2y}$, and a one-parameter family of solutions is:

$$xy^3 + y^2\cos x - \tfrac{1}{2}x^2 + \tfrac{1}{2}e^{2y} = C$$

10. $\dfrac{\partial}{\partial y}(e^{2y} - y\cos xy) = 2e^{2y} - \cos xy + xy\sin xy = \dfrac{\partial}{\partial x}(2xe^{2y} - x\cos xy + 2y);$

the equation is exact on the whole plane.

$f(x,y) = xe^{2y} - \sin xy + y^2$ and $xe^{2y} - \sin xy + y^2 = C$ is a

one-parameter family of solutions.

11. (a) Yes: $\dfrac{\partial}{\partial y}[p(x)] = 0 = \dfrac{\partial}{\partial x}[q(y)].$

(b) For all x, y such that $p(y)q(x) \neq 0$, $\dfrac{1}{p(y)q(x)}$ is an integrating factor.

Multiplying the differential equation by $\dfrac{1}{p(y)q(x)}$,   we get

$$\dfrac{1}{q(x)} + \dfrac{1}{p(y)}\,y' = 0$$

which has the form of the differential equation in part (a).

12. Mimic the proof of the first part.

13. $\dfrac{\partial P}{\partial y} = e^{y-x} - 1$ and $\dfrac{\partial Q}{\partial x} = e^{y-x} - xe^{y-x};$ the equation is not exact.

Since $\dfrac{1}{Q}\left(\dfrac{\partial P}{\partial y} - \dfrac{\partial Q}{\partial x}\right) = \dfrac{1}{xe^{y-x} - 1}\left(xe^{y-x} - 1\right) = 1, \quad \mu(x) = e^{\int dx} = e^x$ is

an integrating factor. Multiplying the given equation by e^x,   we get

$$(e^y - ye^x) + (xe^y - e^x)\,y' = 0$$

This is the equation given in Exercise 3. A one-parameter family of solutions is:

$$xe^y - ye^x = C$$

14. $w = \dfrac{1}{P}\left(\dfrac{\partial P}{\partial y} - \dfrac{\partial Q}{\partial x}\right) = \dfrac{1}{x + e^y}(e^y + x) = 1$ doesn't depend on x, so $e^{\int -dy} = e^{-y}$ is an integrating factor.

$(xe^{-y} + 1) - \dfrac{1}{2}x^2 e^{-y}y' = 0$ is exact; $f(x,y) = \tfrac{1}{2}x^2 e^{-y} + x$ and a one-parameter family of solutions is $\tfrac{1}{2}x^2 e^{-y} + x = C.$

15. $\dfrac{\partial P}{\partial y} = 6x^2 y + e^y = \dfrac{\partial Q}{\partial x};$ the equation is exact.

$\dfrac{\partial f}{\partial x} = 3x^2 y^2 + x + e^y \quad \Longrightarrow \quad f(x,y) = x^3 y^2 + \tfrac{1}{2}x^2 + xe^y + \varphi(y)$

$$\frac{\partial f}{\partial y} = 2x^3 y + xe^y + \varphi'(y) = 2x^3 y + y + xe^y \quad \Longrightarrow \quad \varphi'(y) = y \quad \Longrightarrow \quad \varphi(y) = \tfrac{1}{2} y^2$$

Therefore $f(x,y) = x^3 y^2 + \tfrac{1}{2} x^2 + xe^y + \tfrac{1}{2} y^2$,   and a one-parameter family of solutions is:

$$x^3 y^2 + \tfrac{1}{2} x^2 + xe^y + \tfrac{1}{2} y^2 = C$$

16. $\dfrac{\partial}{\partial y}(\sin 2x \cos y) = -\sin 2x \sin y = -2 \sin x \cos x \sin y = \dfrac{\partial}{\partial x}(-\sin^2 x \sin y);$ exact.

$f(x,y) = \sin^2 x \cos y$ and $\sin^2 x \cos y = C$ is a one-parameter family of solutions.

17. $\dfrac{\partial P}{\partial y} = 3y^2$ and $\dfrac{\partial Q}{\partial x} = 0;$ the equation is not exact.

Since $\dfrac{1}{Q}\left(\dfrac{\partial P}{\partial y} - \dfrac{\partial Q}{\partial x}\right) = \dfrac{1}{3y^2}(3y^2) = 1,$ $\mu(x) = e^{\int dx} = e^x$ is an

an integrating factor. Multiplying the given equation by e^x,   we get

$$\left(y^3 e^x + xe^x + e^x\right) + \left(3y^2 e^x\right) y' = 0$$

$$\frac{\partial f}{\partial x} = y^3 e^x + xe^x + e^x \quad \Longrightarrow \quad f(x,y) = y^3 e^x + xe^x + \varphi(y)$$

$$\frac{\partial f}{\partial y} = 3y^2 e^x + \varphi'(y) = 3y^2 e^x \quad \Longrightarrow \quad \varphi'(y) = 0 \quad \Longrightarrow \quad \varphi(y) = 0$$

Therefore $f(x,y) = y^3 e^x + xe^x$,   and a one-parameter family of solutions is:

$$y^3 e^x + xe^x = C$$

18. $v = \dfrac{1}{Q}\left(\dfrac{\partial P}{\partial y} - \dfrac{\partial Q}{\partial x}\right) = \dfrac{1}{xe^{2x+y} + 1}\left(-2xe^{2x+y} - 2\right) = -2,$ independent of $y,$ so $e^{\int -2\,dx} = e^{-2x}$

is an integrating factor.   Thus $(e^y - 2ye^{-2x}) + (xe^y + e^{-2x})y' = 0$ is exact.

$f(x,y) = xe^y + ye^{-2x}$,   and $xe^y + ye^{-2x} = C$ is a one-parameter family of solutions.

19. $\dfrac{\partial P}{\partial y} = 1 = \dfrac{\partial Q}{\partial x};$ the equation is exact.

$$\frac{\partial f}{\partial x} = x^2 + y \quad \Longrightarrow \quad f(x,y) = \tfrac{1}{3} x^3 + xy + \varphi(y)$$

$$\frac{\partial f}{\partial y} = x + \varphi'(y) = x + e^y \quad \Longrightarrow \quad \varphi'(y) = e^y \quad \Longrightarrow \quad \varphi(y) = e^y$$

Therefore $f(x,y) = \tfrac{1}{3} x^3 + xy + e^y$,   and a one-parameter family of solutions is:

$$\tfrac{1}{3} x^3 + xy + e^y = C$$

Setting $x = 1,\ y = 0,$ we get $C = \tfrac{4}{3}$ and

$$\tfrac{1}{3} x^3 + xy + e^y = \tfrac{4}{3} \qquad \text{or} \qquad x^3 + 3xy + 3e^y = 4$$

20. $\dfrac{\partial}{\partial y}(3x^2 - 2xy + y^3) = -2x + 3y^2 = \dfrac{\partial}{\partial x}\left(3xy^2 - x^2\right);$ exact

$f(x, y) = x^3 - x^2 y + xy^3 \implies x^3 - x^2 y + xy^3 = C.$

Substituting $x = 1$, $y = -1$ we get $1 + 1 - 1 = C \implies x^3 - x^2 y + xy^3 = 1$

21. $\dfrac{\partial P}{\partial y} = 4y$ and $\dfrac{\partial Q}{\partial x} = 2y;$ the equation is not exact.

Since $\dfrac{1}{Q}\left(\dfrac{\partial P}{\partial y} - \dfrac{\partial Q}{\partial x}\right) = \dfrac{1}{2xy}(2y) = \dfrac{1}{x}$, $\mu(x) = e^{\int (1/x)\, dx} = e^{\ln x} = x$ is an

integrating factor. Multiplying the given equation by x, we get

$$\left(2xy^2 + x^3 + 2x\right) + \left(2x^2 y\right) y' = 0$$

$\dfrac{\partial f}{\partial y} = 2x^2 y \implies f(x, y) = x^2 y^2 + \varphi(x)$

$\dfrac{\partial f}{\partial x} = 2xy^2 + \varphi'(x) = 2xy^2 + x^3 + 2x \implies \varphi'(x) = x^3 + 2x \implies \varphi = \tfrac{1}{4}x^4 + x^2$

Therefore $f(x, y) = x^2 y^2 + \tfrac{1}{4}x^4 + x^2$, and a one-parameter family of solutions is:

$$x^2 y^2 + \tfrac{1}{4}x^4 + x^2 = C$$

Setting $x = 1$, $y = 0$, we get $C = \tfrac{5}{4}$ and

$$x^2 y^2 + \tfrac{1}{4}x^4 + x^2 = \tfrac{5}{4} \qquad \text{or} \qquad 4x^2 y^2 + x^4 + 4x^2 = 5$$

22. $v = \dfrac{1}{Q}\left(\dfrac{\partial P}{\partial y} - \dfrac{\partial Q}{\partial x}\right) = \dfrac{2 - 6xy^2}{3x^2 y^2 - x} = -\dfrac{2}{x}$ doesn't depend on y, so $e^{\int -\frac{2}{x}\, dx} = x^{-2}$

is an integrating factor. Thus $(1 + yx^{-2}) + (3y^2 - x^{-1})y' = 0$ is exact.

$f(x, y) = x - \dfrac{y}{x} + y^3 \implies x - \dfrac{y}{x} + y^3 = C$

Substituting $x = 1$, $y = 1$ we get $1 - 1 + 1 = C \implies x - \dfrac{y}{x} + y = 1.$

23. $\dfrac{\partial P}{\partial y} = 3y^2$ and $\dfrac{\partial Q}{\partial x} = y^2;$ the equation is not exact.

Since $\dfrac{1}{P}\left(\dfrac{\partial P}{\partial y} - \dfrac{\partial Q}{\partial x}\right) = \dfrac{1}{y^3}(2y^2) = \dfrac{2}{y}$, $w(y) = e^{-\int (2/y)\, dy} = e^{-2\ln y} = y^{-2}$ is an

integrating factor. Multiplying the given equation by y^{-2}, we get

$$y + \left(y^{-2} + x\right) y' = 0$$

$\dfrac{\partial f}{\partial x} = y \implies f(x, y) = xy + \varphi(y)$

$\dfrac{\partial f}{\partial y} = x + \varphi'(y) = y^{-2} + x \implies \varphi'(y) = y^{-2} \implies \varphi(y) = -\dfrac{1}{y}$

Therefore $f(x, y) = xy - \dfrac{1}{y}$, and a one-parameter family of solutions is: $xy - \dfrac{1}{y} = C$

Setting $x = -2$, $y = -1$, we get $C = 3$ and the solution $xy - \dfrac{1}{y} = 3.$

24. $\dfrac{\partial}{\partial y}(x+y)^2 = 2(x+y) = \dfrac{\partial}{\partial x}(2xy + x^2 - 1);$ exact.

$f(x,y) = \dfrac{x^3}{3} + x^2 y + xy^2 - y \implies \dfrac{x^3}{3} + x^2 y + xy^2 - y = C$

Setting $x = 1$, $y = 1$, we get $C = \frac{4}{3}$ and the solution $\dfrac{x^3}{3} + x^2 y + xy^2 - y = \dfrac{4}{3}$.

25. $\dfrac{\partial P}{\partial y} = -2y \sinh(x - y^2) = \dfrac{\partial Q}{\partial x};$ the equation is exact.

$\dfrac{\partial f}{\partial x} = \cosh(x - 2y^2) + e^{2x} \implies f(x,y) = \sinh(x - y^2) + \frac{1}{2}e^{2x} + \varphi(y)$

$\dfrac{\partial f}{\partial y} = -2y\cosh(x - y^2) + \varphi'(y) = y - 2y\cosh(x - y^2) \implies \varphi'(y) = y \implies \varphi(y) = \frac{1}{2}y^2$

Therefore $f(x,y) = \sinh(x - y^2) + \frac{1}{2}e^{2x} + \frac{1}{2}y^2$,   and a one-parameter family of solutions is:

$$\sinh(x - y^2) + \tfrac{1}{2}e^{2x} + \tfrac{1}{2}y^2 = C$$

Setting $x = 2$, $y = \sqrt{2}$, we get $C = \frac{1}{2}e^4 + 1$ and the solution

$$\sinh(x - y^2) + \tfrac{1}{2}e^{2x} + \tfrac{1}{2}y^2 = \tfrac{1}{2}e^4 + 1$$

26. Write the linear equation as $p(x)y - q(x) + y' = 0$.   Then $P(x) = p(x)y - q(x)$, $Q(x) = 1$

and $v = \dfrac{1}{Q}\left(\dfrac{\partial P}{\partial y} - \dfrac{\partial Q}{\partial x}\right) = p(x)$ depends only on x.   Therefore $v = e^{\int p(x)\,dx}$ is

an integrating factor.

27. (a) $\dfrac{\partial P}{\partial y} = 2xy + kx^2$ and $\dfrac{\partial Q}{\partial x} = 2xy + 3x^2 \implies k = 3.$

(b) $\dfrac{\partial P}{\partial y} = e^{2xy} + 2xye^{2xy}$ and $\dfrac{\partial Q}{\partial x} = ke^{2xy} + 2kxye^{2xy} \implies k = 1.$

28. (a) We need $g'(y)\sin x = y^2 f'(x)$.   Take $g(y) = \frac{1}{3}y^3$ and $f(x) = -\cos x$

(b) We need $g'(y)e^y + g(y)e^y = y$,   that is $\dfrac{d}{dy}[g(y)e^y] = y$.

It follows that $g(y)e^y = \frac{1}{2}y^2 + C$, $\implies g(y) = e^{-y}(\frac{1}{2}y^2 + C)$.

29. $y' = y^2 x^3;$ the equation is separable.

$$y^{-2}\,dy = x^3\,dx \implies -\frac{1}{y} = \tfrac{1}{4}x^4 + C \implies y = \frac{-4}{x^4 + C}$$

30. $y\,y' = 4xe^{2x+y} = 4xe^{2x}e^y \implies$ the equation is separable.

$$ye^{-y}\,dy = 4xe^{2x}\,dx \implies -ye^{-y} - e^{-y} = 2xe^{2x} - e^{2x} + C$$

31. $y' + \dfrac{4}{x}y = x^4;$ the equation is linear.

$H(x) = \int (4/x)\,dx = 4\ln x = \ln x^4,$ integrating factor: $e^{\ln x^4} = x^4$

$$x^4 y' + 4x^3 y = x^8$$

$$\frac{d}{dx}\left[x^4 y\right] = x^8$$

$$x^4 y = \tfrac{1}{9} x^9 + C$$

$$y = \tfrac{1}{9} x^5 + C x^{-4}$$

32. $y' + 2xy = 2x^3$;   the equation is linear with integrating factor $e^{\int 2x\,dx} = e^{x^2}$

$$\implies \frac{d}{dx}(e^{x^2} y) = 2x^3 e^{x^2} \implies e^{x^2} y = e^{x^2}(x^2 - 1) + C \implies y = x^2 - 1 + C e^{-x^2}.$$

33. $\dfrac{\partial P}{\partial y} = e^{xy} + xye^{xy} = \dfrac{\partial Q}{\partial x}$;   the equation is exact.

$$\frac{\partial f}{\partial x} = ye^{xy} - 2x \implies f(x,y) = e^{xy} - x^2 + \varphi(y)$$

$$\frac{\partial f}{\partial y} = xe^{xy} + \varphi'(y) = \frac{2}{y} + xe^{xy} \implies \varphi'(y) = \frac{2}{y} \implies \varphi(y) = 2\ln|y|$$

Therefore $f(x,y) = e^{xy} - x^2 + 2\ln|y|$,   and a one-parameter family of solutions is:

$$e^{xy} - x^2 + 2\ln|y| = C$$

34. $w = \dfrac{1}{P}\left(\dfrac{\partial P}{\partial y} - \dfrac{\partial Q}{\partial x}\right) = \dfrac{1}{y}(1 - 2y)$ depends only on y, so an integrating factor is

$$e^{\int -w(y)\,dy} = e^{\int [2 - (1/y)]\,dy} = e^{2y - \ln y} = \frac{1}{y} e^{2y}.$$

Then $e^{2y}\,dx + \left(2xe^{2y} - \dfrac{1}{y}\right) dy = 0$ is exact.

$f(x,y) = xe^{2y} - \ln y$,   and a one-parameter family of solutions is $xe^{2y} - \ln y = C$.

SECTION 19.3

1. $y' = y \implies y = Ce^x$.   Also, $y(0) = 1 \implies C = 1$

Thus $y = e^x$ and $y(1) = 2.71828$

(a) 2.48832, relative error= 8.46%.

(b) 2.71825, relative error= 0.001%.

2. $y' = x + y \implies y = Ce^x - x - 1, y(0) = 2 \implies C = 3$

Thus $y = 3e^x - x - 1$ and $y(1) \simeq 6.15485$

(a) 5.46496, relative error $= 11.2\%$.

(b) 6.15474, relative error $= 0\%$.

3. (a) 2.59374, relative error= 4.58%.

(b) 2.71828, relative error= 0%.

4. (a) 5.78124, relative error= 6.07%.

(b) 6.15482, relative error= 0%.

5. $y' = 2x \implies y = x^2 + C.$ Also, $y(2) = 5 \implies C = 1$

Thus $y = x^2 + 1$ and $y(1) = 2.$

(a) 1.9, relative error= 5.0%.

(b) 2.0, relative error= 0%.

6. $y' = 3x^2 \implies y = x^3 + C.$ Also, $y(1) = 2 \implies C = 1$

Thus $y = x^3 + 1$ and $y(0) = 1.$

(a) 0.84500, relative error= 15.5%.

(b) 1.0, relative error= 0%.

7. $y' = \dfrac{1}{2y}$

Thus $y = \sqrt{x}$ and $y(2) = \sqrt{2} \simeq 1.41421.$

(a) 1.42052, relative error= -0.45%.

(b) 1.41421, relative error= 0%.

8. $y' = \dfrac{1}{3y^2}$

Thus $y = x^{\frac{1}{3}}$ and $y(2) \simeq 1.25992.$

(a) 1.26494, relative error= -0.4%.

(b) 1.25992, relative error= 0%.

9. (a) 2.65330, relative error= 2.39%.

(b) 2.71828, relative error= 0%.

10. (a) 5.95989, relative error= 3.17%.

(b) 6.15487, relative error= 0%.

PROJECT 19.3

1. (a) and (b)

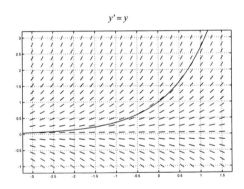

(c) $y - y' = 0$ $H(x) = \int -dx = -x;$ integrating factor: e^{-x}

$$e^{-x}y' - e^{-x}y = 0$$

$$\frac{d}{dx}(e^{-x}y) = 0$$

$$e^{-x}y = C$$

$$y = Ce^x$$

$$y(0) = 1 \implies C = 1. \quad \text{Thus } y = e^x.$$

2. (a) and (b)

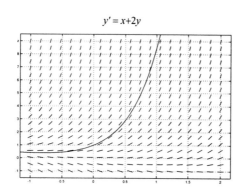

$$y' = x + 2y$$

(c) $y' - 2y = x$ $H(x) = \int -2\,dx = -2x;$ integrating factor: e^{-2x}

$$e^{-2x}y' - 2e^{-2x}y = xe^{-2x}$$

$$\frac{d}{dx}(e^{-2x}y) = xe^{-2x}$$

$$e^{-2x}y = -\frac{1}{2}xe^{-2x} - \frac{1}{4}e^{-2x} + C$$

$$y = Ce^{2x} - \frac{1}{2}x - \frac{1}{4}$$

$y(0) = 1 \quad \Longrightarrow \quad C = \frac{5}{4}.$ Thus $y = \frac{5}{4}e^{2x} - \frac{1}{2}x - \frac{1}{4}.$

3. (a) and (b)

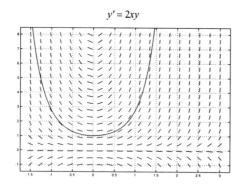

$$y' = 2xy$$

(c) $y' - 2xy = 0$ $H(x) = \int -2x\,dx = -x^2$;   integrating factor: e^{-x^2}

$e^{-x^2}y' - 2xe^{x^2}y = 0$

$\dfrac{d}{dx}(e^{-x^2}y) = 0$

$e^{-x^2}y = C$

$y = Ce^{x^2}$

$y(0) = 1 \quad \Longrightarrow \quad C = 1.$ Thus $y = e^{x^2}$.

4.   (a) and (b)

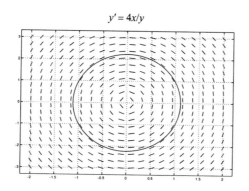

$y' = 4x/y$

(c) $y' = -\dfrac{4x}{y}$; $\dfrac{1}{2}y^2 = -2x^2 + C$ or $x^2 + \dfrac{1}{4}y^2 = C$

$y(1) = 1 \quad \Longrightarrow \quad C = \dfrac{5}{4}.$ Thus $4x^2 + y^2 = 5$.

SECTION 19.4

1.   First consider the reduced equation. The characteristic equation is:

$$r^2 + 5r + 6 = (r + 2)(r + 3) = 0$$

and $u_1(x) = e^{-2x}, \quad u_2(x) = e^{-3x}$ are fundamental solutions. A particular solution of the given equation has the form

$$y = Ax + B.$$

The derivatives of y are: $y' = A, \quad y'' = 0.$

Substituting y and its derivatives into the given equation gives

$$0 + 5A + 6(Ax + B) = 3x + 4.$$

Thus,

$$6A = 3$$

$$5A + 6B = 4$$

The solution of this pair of equations is: $A = \frac{1}{2}$, $B = \frac{1}{4}$,   and $y = \frac{1}{2}x + \frac{1}{4}$.

2. The constant $y_p = -\frac{1}{2}$ is a solution.

3. First consider the reduced equation. The characteristic equation is:

$$r^2 + 2r + 5 = 0$$

and $u_1(x) = e^{-x}\cos 2x$, $u_2(x) = e^{-x}\sin 2x$ are fundamental solutions. A particular solution of the given equation has the form

$$y = Ax^2 + Bx + C.$$

The derivatives of y are: $y' = 2Ax + B$, $y'' = 2A$.

Substituting y and its derivatives into the given equation gives

$$2A + 2(2Ax + B) + 5(Ax^2 + Bx + C) = x^2 - 1.$$

Thus,

$$5A = 1$$

$$4A + 5B = 0$$

$$2A + 2B + 5C = -1$$

The solution of this system of equations is: $A = \frac{1}{5}$, $B = -\frac{4}{25}$, $C = -\frac{27}{125}$,   and

$$y = \frac{1}{5}x^2 - \frac{4}{25}x - \frac{27}{125}.$$

4. We try $y = Ax^3 + Bx^2 + Cx + D$:

$$y'' + y' - 2y = (6Ax + 2B) + (3Ax^2 + 2Bx + C) - 2(Ax^3 + Bx^2 + Cx + D) = x^3 + x.$$

$$\implies -2A = 1, \quad 3A - 2B = 0, \quad 6A + 2B - 2C = 1, \quad 2B + C - 2D = 0$$

$$\implies A = -\frac{1}{2}, \ B = -\frac{3}{4}, \ C = -\frac{11}{4}, \ D = -\frac{17}{8}; \qquad y_p = -\frac{1}{2}x^3 - \frac{3}{4}x^2 - \frac{11}{4}x - \frac{17}{8}.$$

5. First consider the reduced equation. The characteristic equation is:

$$r^2 + 6r + 9 = (r + 3)^2 = 0$$

and $u_1(x) = e^{-3x}$, $u_2(x) = xe^{-3x}$ are fundamental solutions. A particular solution of the given equation has the form

$$y = Ae^{3x}.$$

The derivatives of y are: $y' = 3Ae^{3x}$, $y'' = 9Ae^{3x}$.

Substituting y and its derivatives into the given equation gives

$$9Ae^{3x} + 18Ae^{3x} + 9Ae^{3x} = e^{3x}.$$

Thus, $36A = 1 \implies A = \dfrac{1}{36},$ and $y = \frac{1}{36}e^{3x}.$

6. Since -3 is a double root of the characteristic equation $r^2 + 6r + 9 = 0$, we try
    $y = Ax^2e^{-3x}.$ Then $y' = A(-3x^2 + 2x)e^{-3x},$ $y'' = A(9x^2 - 12x + 2)e^{-3x},$ and

    $$[A(9x^2 - 12x + 2) + 6A(-3x^2 + 2x) + 9Ax^2]e^{-3x} = e^{-3x}, \quad \text{or} \quad 2Ae^{-3x} = e^{-3x}$$

    Thus $A = \frac{1}{2}$ and $y_p = \frac{1}{2}x^2e^{-3x}.$

7. First consider the reduced equation. The characteristic equation is:

    $$r^2 + 2r + 2 = 0$$

    and $u_1(x) = e^{-x}\cos x,$ $u_2(x) = e^{-x}\sin x$ are fundamental solutions. A particular solution of the
    given equation has the form

    $$y = Ae^x.$$

    The derivatives of y are: $y' = Ae^x,$ $y'' = Ae^x.$

    Substituting y and its derivatives into the given equation gives

    $$Ae^x + 2Ae^x + 2Ae^x = e^x.$$

    Thus, $5A = 1 \implies A = \frac{1}{5}$ and $y = \frac{1}{5}e^x.$

8. Try $y = (A + Bx)e^{-x}.$ Substituting into $y'' + 4y' + 4y = xe^{-x}$ gives
    $A = -2, B = 1;$ $y_p = (x - 2)e^{-x}$

9. First consider the reduced equation. The characteristic equation is:

    $$r^2 - r - 12 = (r - 4)(r + 3) = 0$$

    and $u_1(x) = e^{4x},$ $u_2(x) = e^{-3x}$ are fundamental solutions. A particular solution of the given
    equation has the form

    $$y = A\cos x + B\sin x.$$

    The derivatives of y are: $y' = -A\sin x + B\cos x,$ $y'' = -A\cos x - B\sin x.$

    Substituting y and its derivatives into the given equation gives

    $$-A\cos x - B\sin x - (-A\sin x + B\cos x) - 12(A\cos x + B\sin x) = \cos x.$$

    Thus,

    $$-13A - B = 1$$
    $$A - 13B = 0$$

The solution of this system of equations is: $A = -\frac{13}{170}$, $B = -\frac{1}{170}$, and

$$y = -\frac{13}{170}\cos x - \frac{1}{170}\sin x.$$

is a particular solution of the complete equation.

10. Try $y = A\cos x + B\sin x$. Substituting into $y'' - y' - 12y = \sin x$ gives

$A = \frac{1}{170}$, $B = \frac{-13}{170}$; $\quad y_p = \frac{1}{170}\cos x - \frac{13}{170}\sin x$.

11. First consider the reduced equation. The characteristic equation is:

$$r^2 + 7r + 6 = (r+6)(r+1) = 0$$

and $u_1(x) = e^{-6x}$, $u_2(x) = e^{-x}$ are fundamental solutions. A particular solution of the given equation has the form

$$y = A\cos 2x + B\sin 2x.$$

The derivatives of y are: $\quad y' = -2A\sin 2x + 2B\cos 2x$, $\quad y'' = -4A\cos 2x - 4B\sin 2x$.

Substituting y and its derivatives into the given equation gives

$$-4A\cos 2x - 4B\sin 2x + 7(-2A\sin 2x + 2B\cos 2x) + 6(A\cos 2x + B\sin 2x) = 3\cos 2x.$$

Thus,

$$2A + 14B = 3$$
$$-14A + 2B = 0$$

The solution of this system of equations is: $A = \frac{3}{100}$, $B = \frac{21}{100}$ and

$$y = \frac{3}{10}\cos 2x + \frac{21}{100}\sin 2x.$$

12. Try $y = A\cos 3x + B\sin 3x$. Substituting into $y'' + y' + 3y = \sin 3x$ gives

$A = -\frac{1}{15}$, $B = -\frac{2}{15}$; $\quad y_p = -\frac{1}{15}\cos 3x - \frac{2}{15}\sin 3x$.

13. First consider the reduced equation. The characteristic equation is:

$$r^2 - 2r + 5 = 0$$

and $u_1(x) = e^x\cos 2x$, $u_2(x) = e^x\sin 2x$ are fundamental solutions. A particular solution of the given equation has the form

$$y = Ae^{-x}\cos 2x + Be^{-x}\sin 2x$$

The derivatives of y are: $\quad y' = -Ae^{-x}\cos 2x - 2Ae^{-x}\sin 2x - Be^{-x}\sin 2x + 2Be^{-x}\cos 2x$,

$y'' = 4Ae^{-x}\sin 2x - 3Ae^{-x}\cos 2x - 4Be^{-x}\cos 2x - 3Be^{-x}\sin 2x$.

Substituting y and its derivatives into the given equation gives

$4Ae^{-x}\sin 2x - 3Ae^{-x}\cos 2x - 4Be^{-x}\cos 2x - 3Be^{-x}\sin 2x -$

$2\left(-Ae^{-x}\cos 2x - 2Ae^{-x}\sin 2x - Be^{-x}\sin 2x + 2Be^{-x}\cos 2x\right) +$

$5\left(Ae^{-x}\cos 2x + Be^{-x}\sin 2x\right) = e^{-x}\sin 2x$.

Equating the coefficients of $e^{-x}\cos 2x$ and $e^{-x}\sin 2x$ we get,

$$8A + 4B = 1$$
$$4A - 8B = 0$$

The solution of this system of equations is: $A = \frac{1}{10}$, $B = \frac{1}{20}$ and

$$y = \tfrac{1}{10} e^{-x} \cos 2x + \tfrac{1}{20} e^{-x} \sin 2x.$$

14. Try $y = e^{2x}(A\cos x + B\sin x)$. Substituting into $y'' + 4y' + 5y = e^{2x}\cos x$ gives

$A = \frac{1}{20}$, $B = \frac{1}{40}$; $y_p = e^{2x}\left(\frac{1}{20}\cos x + \frac{1}{40}\sin x\right)$.

15. First consider the reduced equation. The characteristic equation is:

$$r^2 + 6r + 8 = (r+4)(r+2) = 0$$

and $u_1(x) = e^{-4x}$, $u_2(x) = e^{-2x}$ are fundamental solutions. A particular solution of the given equation has the form

$$y = Axe^{-2x}.$$

The derivatives of y are: $y' = Ae^{-2x} - 2Axe^{-2x}$, $y'' = -4Ae^{-2x} + 4Axe^{-2x}$.

Substituting y and its derivatives into the given equation gives

$$-4Ae^{-2x} + 4Axe^{-2x} + 6\left(Ae^{-2x} - 2Axe^{-2x}\right) + 8Axe^{-2x} = 3e^{-2x}$$

Thus, $2A = 3$ $\implies$ $A = \frac{3}{2}$ and $y = \frac{3}{2}xe^{-2x}$.

16. Try $y = e^x(A\cos x + B\sin x)$.   Substituting into $y'' - 2y' + 5y = e^x\sin x$ gives

$A = 0$, $B = \frac{1}{3}$; $y_p = \frac{1}{3}e^x\sin x$.

17. First consider the reduced equation: $y'' + y = 0$.   The characteristic equation is:

$$r^2 + 1 = 0$$

and $u_1(x) = \cos x$, $u_2(x) = \sin x$ are fundamental solutions. A particular solution of the given equation has the form

$$y = Ae^x.$$

The derivatives of y are: $y' = y'' = Ae^x$.

Substitute y and its derivatives into the given equation:

$$Ae^x + Ae^x = e^x \implies A = \frac{1}{2} \text{ and } y = \tfrac{1}{2}e^x.$$

The general solution of the given equation is: $y = C_1\cos x + C_2\sin x + \frac{1}{2}e^x$.

18. $r^2 - 2r + 1 = 0 \implies r = 1 \implies y = C_1 e^x + C_2 x e^x$ is the general solution of the reduced equation. To find a particular solution, we try $y = A \cos 2x + B \sin 2x$. Substituting into $y'' - 2y' + y = -25 \sin 2x$ gives $A = -4$, $B = 3$, so $y_p = 3 \sin 2x - 4 \cos 2x$.

Therefore the general solution is: $y = C_1 e^x + C_2 x e^x + 3 \sin 2x - 4 \cos 2x$.

19. First consider the reduced equation: $y'' - 3y' - 10y = 0$. The characteristic equation is:

$$r^2 - 3r - 10 = (r - 5)(r + 2) = 0$$

and $u_1(x) = e^{5x}$, $u_2(x) = e^{-2x}$ are fundamental solutions. A particular solution of the given equation has the form

$$y = Ax + B.$$

The derivatives of y are: $y' = A$, $y'' = 0$.

Substitute y and its derivatives into the given equation:

$$-3A - 10(Ax + B) = -x - 1 \implies A = \tfrac{1}{10}, \quad B = \tfrac{7}{100} \quad \text{and} \quad y = \tfrac{1}{10} x + \tfrac{7}{100}$$

The general solution of the given equation is:

$$y = C_1 e^{5x} + C_2 e^{-2x} + \tfrac{1}{10} x + \tfrac{7}{100}$$

20. $r^2 + 4 = 0 \implies r = \pm 2i \implies y = C_1 \cos 2x + C_2 \sin 2x$, general solution of reduced equation.

Particular solution: try $y = x(A + Bx) \cos 2x + x(C + Dx) \sin 2x$.

Substituting into $y'' + 4y = x \cos 2x$ gives $A = \tfrac{1}{16}$, $B = 0$, $C = 0$, $D = \tfrac{1}{8}$;

$$y_p = \frac{1}{16} x \cos 2x + \frac{1}{8} x^2 \sin 2x. \quad \text{General solution:} \quad y = C_1 \cos 2x + C_2 \cos 2x + \tfrac{1}{16} x \cos 2x$$
$$+ \tfrac{1}{8} x^2 \sin 2x.$$

21. First consider the reduced equation: $y'' + 3y' - 4y = 0$. The characteristic equation is:

$$r^2 + 3r - 4 = (r + 4)(r - 1) = 0$$

and $u_1(x) = e^x$, $u_2(x) = e^{-4x}$ are fundamental solutions. A particular solution of the given equation has the form

$$y = Axe^{-4x}.$$

The derivatives of y are: $y' = Ae^{-4x} - 4Axe^{-4x}$, $y'' = -8Ae^{-4x} + 16Axe^{-4x}$.

Substitute y and its derivatives into the given equation:

$$-8Ae^{-4x} + 16Axe^{-4x} + 3\left(Ae^{-4x} - 4Axe^{-4x}\right) - 4Axe^{-4x} = e^{-4x}.$$

This implies $-5A = 1$, so $A = -\tfrac{1}{5}$ and $y = -\tfrac{1}{5} xe^{-4x}$.

The general solution of the given equation is: $y = C_1 e^x + C_2 e^{-4x} - \tfrac{1}{5} xe^{-4x}$.

22. $r^2 + 2r = 0 \implies r = 0, -2 \implies y = C_1 + C_2 e^{-2x}$, general solution of reduced equation.

Particular solution: try $y = A\cos 2x + B\sin 2x$.

Substituting into $y'' + 2y' = 4\sin 2x$ gives $A = B = -\frac{1}{2}$; $y_p = -\frac{1}{2}\cos 2x - \frac{1}{2}\sin 2x$.

General solution: $y = C_1 + C_2 e^{-2x} - \frac{1}{2}\cos 2x - \frac{1}{2}\sin 2x$.

23. First consider the reduced equation: $y'' + y' - 2y = 0$.   The characteristic equation is:

$$r^2 + r - 2 = (r+2)(r-1) = 0$$

and $u_1(x) = e^{-2x}$, $u_2(x) = e^x$ are fundamental solutions. A particular solution of the given equation has the form

$$y = x(A + Bx)e^x.$$

The derivatives of y are:

$$y' = (A + (2B + A)x + Bx^2)e^x, \quad y'' = (2A + 2B + (4B + A)x + Bx^2)e^x.$$

Substitute y and its derivatives into the given equation:

$$(2A + 2B + (4B + A)x + Bx^2 + A + (2B + A)x + Bx^2 - 2Ax - 2Bx^2)e^x = 3xe^x.$$

This implies $A = -\frac{1}{3}$, $B = \frac{1}{2}$ so $y = x(-\frac{1}{3} + \frac{1}{2}x)e^x$.

The general solution of the given equation is: $y = C_1 e^{-2x} + C_2 e^x - \frac{1}{3}xe^x + \frac{1}{2}x^2 e^x$.

24. $r^2 + 4r + 4 = 0 \implies r = -2 \implies y = C_1 e^{-2x} + C_2 x e^{-2x}$, general solution of reduced equation.

Particular solution: try $y = x^2(A + Bx)e^{-2x}$.

Substituting into $y'' + 4y' + 4y = xe^{-2x}$ gives $A = 0$, $B = \frac{1}{6}$; $y_p = \frac{1}{6}x^3 e^{-2x}$.

General solution: $y = C_1 e^{-2x} + C_2 x e^{-2x} + \frac{1}{6}x^3 e^{-2x}$.

25. Let $y_1(x)$ be a solution of $y'' + ay' + by = \phi_1(x)$,   let $y_2(x)$ be a solution of $y'' + ay' + by = \phi_2(x)$, and let $z = y_1 + y_2$.   Then

$$z'' + az' + bz = (y_1'' + y_2'') + a(y_1' + y_2') + b(y_1 + y_2)$$
$$= (y_1'' + ay_1' + by_1) + (y_2'' + ay_2' + y_2) = \phi_1 + \phi_2.$$

26. (a) $y = -\frac{1}{15}x - \frac{2}{225}$ is a particular solution of $y'' + 2y' - 15y = x$

   $y = -\frac{1}{7}e^{2x}$ is a particular solution of $y'' + 2y' - 15y = e^{2x}$

   Therefore $y = -\frac{1}{15}x - \frac{2}{225} - \frac{1}{7}e^{2x}$ is a particular solution of $y'' + 2y' - 15y = x + e^{2x}$.

(b) $y = -\frac{1}{4}e^{-x}$ is a particular solution of $y'' - 7y' - 12y = e^{-x}$.

   $y = \frac{7}{226}\cos 2x - \frac{4}{113}\sin 2x$ is a particular solution of $y'' - 7y' - 12y = \sin 2x$.

   Therefore $y = -\frac{1}{4}e^{-x} + \frac{7}{226}\cos 2x - \frac{4}{113}\sin 2x$ is a particular solution of

   $y'' - 7y' - 12y = e^{-x} + \sin 2x$.

27. First consider the reduced equation: $y'' + 4y' + 3y = 0$.   The characteristic equation is:

$$r^2 + 4r + 3 = (r+3)(r+1) = 0$$

and $u_1(x) = e^{-3x}$, $u_2(x) = e^{-x}$ are fundamental solutions. Since $\cosh x = \frac{1}{2}(e^x + e^{-x})$,   a particular solution of the given equation has the form

$$y = Ae^x + Bxe^{-x}$$

The derivatives of y are: $y' = Ae^x + Be^{-x} - Bxe^{-x}$ $y'' = Ae^x - 2Be^{-x} + Bxe^{-x}$.

Substitute y and its derivatives into the given equation:

$$Ae^x - 2Be^{-x} + Bxe^{-x} + 4\left(Ae^x + Be^{-x} - Bxe^{-x}\right) + 3\left(Ae^x + Bxe^{-x}\right) = \frac{1}{2}\left(e^x + e^{-x}\right).$$

Equating coefficients, we get $A = \frac{1}{16}$, $B = \frac{1}{4}$,   and so $y = \frac{1}{16}e^x + \frac{1}{4}xe^{-x}$.

The general solution of the given equation is: $y = C_1 e^{-3x} + C_2 e^{-x} + \frac{1}{16}e^x + \frac{1}{4}xe^{-x}$.

28. $r^2 + 1 = 0 \implies r = \pm i$.   Fundamental solutions: $u_1 = \cos x$, $u_2 = \sin x$.

Wronskian: $W = u_1 u_2' - u_1' u_2 = 1$; $\phi(x) = 3 \sin x \sin 2x$

$$z_1 = -\int \frac{u_2 \phi}{W}\, dx = -\int 3\sin^2 x \sin 2x\, dx = -6\int \sin^3 x \cos x\, dx = -\frac{3}{2}\sin^4 x,$$

$$z_2 = \int \frac{u_1 \phi}{W}\, dx = \int 3\cos x \sin x \sin 2x\, dx = \frac{3}{2}\int \sin^2 2x\, dx = \frac{3}{16}(4x - \sin 4x).$$

Therefore $y_p = z_1 u_1 + z_2 u_2 = -\frac{3}{2}\sin^4 x \cos x + \frac{3}{16}(4x - \sin 4x)\sin x$.

29. First consider the reduced equation $y'' - 2y' + y = 0$.   The characteristic equation is:

$$r^2 - 2r + 1 = (r-1)^2 = 0$$

and $u_1(x) = e^x$, $u_2(x) = xe^x$ are fundamental solutions. Their Wronskian is given by

$$W = u_1 u_2' - u_2 u_1' = e^x(e^x + xe^x) - xe^x(e^x) = e^{2x}$$

Using variation of parameters, a particular solution of the given equation will have the form

$$y = u_1 z_1 + u_2 z_2,$$

where

$$z_1 = -\int \frac{xe^x(xe^x \cos x)}{e^{2x}}\, dx = -\int x^2 \cos x\, dx = -x^2 \sin x - 2x \cos x + 2 \sin x,$$

$$z_2 = \int \frac{e^x(xe^x \cos x)}{e^{2x}}\, dx = \int x \cos x\, dx = x \sin x + \cos x$$

Therefore,

$$y = e^x\left(-x^2 \sin x - 2x \cos x + 2 \sin x\right) + xe^x\left(x \sin x + \cos x\right) = 2e^x \sin x - xe^x \cos x.$$

30. $r^2 + 1 = 0$.   Fundamental solutions: $u_1 = \cos x$, $u_2 = \sin x$.

Wronskian: $W = u_1 u_2' - u_1' u_2 = 1$; $\phi(x) = \csc x$.

$$z_1 = -\int \frac{u_2 \phi}{W} \, dx = -\int \sin x \csc x \, dx = -x,$$

$$z_2 = \int \frac{u_1 \phi}{W} \, dx = \int \cos x \csc x \, dx = \int \cot x \, dx = \ln(\sin x) \quad [\sin x > 0 \text{ since } 0 < x < \pi].$$

Therefore $y_p = z_1 u_1 + z_2 u_2 = -x \cos x + \ln(\sin x) \sin x$.

31. First consider the reduced equation $y'' - 4y' + 4y = 0$.   The characteristic equation is:

$$r^2 - 4r + 4 = (r - 2)^2 = 0$$

and $u_1(x) = e^{2x}$, $u_2(x) = xe^{2x}$ are fundamental solutions. Their Wronskian is given by

$$W = u_1 u_2' - u_2 u_1' = e^{2x} \left(e^{2x} + 2xe^{2x} \right) - xe^{2x}(2e^{2x}) = e^{4x}.$$

Using variation of parameters, a particular solution of the given equation will have the form

$$y = u_1 z_1 + u_2 z_2,$$

where

$$z_1 = -\int \frac{xe^{2x} \left(\frac{1}{3} x^{-1} e^{2x} \right)}{e^{4x}} \, dx = -\frac{1}{3} \int dx = -\frac{1}{3} x,$$

$$z_2 = \int \frac{e^{2x} \left(\frac{1}{3} x^{-1} e^{2x} \right)}{e^{4x}} \, dx = \frac{1}{3} \int \frac{1}{x} \, dx = \frac{1}{3} \ln |x|.$$

Therefore,

$$y = e^{2x} \left(-\tfrac{1}{3} x \right) + xe^{2x} \left(\tfrac{1}{3} \ln |x| \right) = -\tfrac{1}{3} xe^{2x} + \tfrac{1}{3} x \ln |x| \, e^{2x}.$$

Note: Since $u = -\tfrac{1}{3} xe^{2x}$ is a solution of the reduced equation,

$$y = \tfrac{1}{3} x \ln |x| \, e^{2x}$$

is also a particular solution of the given equation.

32. $r^2 + 4 = 0 \implies r = \pm 2i$.   Fundamental solutions: $u_1 = \cos 2x$, $u_2 = \sin 2x$.

Wronskian: $W = u_1 u_2' - u_1' u_2 = 2\cos^2 2x + 2\sin^2 2x = 2$; $\phi(x) = \sec^2 2x$

$$z_1 = -\int \frac{u_2 \phi}{W} \, dx = -\int \frac{\sin 2x}{2\cos^2 2x} \, dx = -\frac{1}{4} \sec 2x,$$

$$z_2 = \int \frac{u_1 \phi}{W} \, dx = \int \frac{\cos 2x}{2\cos^2 2x} \, dx = \frac{1}{2} \int \sec 2x \, dx = \frac{1}{4} \ln |\sec 2x + \tan 2x|.$$

Therefore

$$y_p = -\frac{1}{4} \sec 2x \cos 2x + \frac{1}{4} \ln |\sec 2x + \tan 2x| \sin 2x = -\frac{1}{4} \left(1 - \ln |\sec 2x + \tan 2x| \sin 2x \right).$$

33. First consider the reduced equation $y'' + 4y' + 4y = 0$.   The characteristic equation is:

$$r^2 + 4r + 4 = (r + 2)^2 = 0$$

and $u_1(x) = e^{-2x}$, $u_2(x) = xe^{-2x}$ are fundamental solutions. Their Wronskian is given by

$$W = u_1 u_2' - u_2 u_1' = e^{-2x}\left(e^{-2x} - 2xe^{2x}\right) - xe^{-2x}(-2e^{-2x}) = e^{-4x}.$$

Using variation of parameters, a particular solution of the given equation will have the form

$$y = u_1 z_1 + u_2 z_2,$$

where

$$z_1 = -\int \frac{xe^{-2x}\left(x^{-2}e^{-2x}\right)}{e^{-4x}}\, dx = -\int \frac{1}{x}\, dx = -\ln|x|$$

$$z_2 = \int \frac{e^{-2x}\left(x^{-2}e^{-2x}\right)}{e^{-4x}}\, dx = \int \frac{1}{x^2}\, dx = -\frac{1}{x}$$

Therefore,

$$y = e^{-2x}\left(-\ln|x|\right) + xe^{-2x}\left(-\frac{1}{x}\right) = -e^{-2x}\ln|x| - e^{-2x}.$$

Note: Since $u = -e^{-2x}$ is a solution of the reduced equation, we can take

$$y = -\ln|x|e^{2x}.$$

34. $r^2 + 2r + 1 = 0 \implies r = -1.$ Fundamental solutions: $u_1 = e^{-x}$, $u_2 = xe^{-x}$.

Wronskian: $W = u_1 u_2' - u_1' u_2 = (1-x)e^{-2x} + xe^{-2x} = e^{-2x}$; $\phi(x) = e^{-x}\ln x$.

$$z_1 = -\int \frac{u_2\phi}{W}\, dx = -\int \frac{xe^{-x}e^{-x}\ln x}{e^{-2x}}\, dx = -\int x\ln x\, dx = -\frac{1}{2}x^2\ln x + \frac{x^2}{4},$$

$$z_2 = \int \frac{u_1\phi}{W}\, dx = \int \frac{e^{-x}e^{-x}\ln x}{e^{-2x}}\, dx = \int \ln x\, dx = x\ln x - x.$$

Therefore

$$y_p = z_1 u_1 + z_2 u_2 = e^{-x}\left(\frac{x^2}{4} - \frac{1}{2}x^2\ln x\right) + xe^{-x}(x\ln x - x)$$

$$= \frac{1}{4}x^2 e^{-x}\left(2\ln x - 3\right)$$

35. First consider the reduced equation $y'' - 2y' + 2y = 0$.   The characteristic equation is:

$$r^2 - 2r + 2 = 0$$

and $u_1(x) = e^x\cos x$, $u_2(x) = e^x\sin x$ are fundamental solutions. Their Wronskian is given by

$$W = e^x\cos x\left[e^x\sin x + e^x\cos x\right] - e^x\sin x\left[e^x\cos x - e^x\sin x\right] = e^{2x}$$

Using variation of parameters, a particular solution of the given equation will have the form

$$y = u_1 z_1 + u_2 z_2,$$

where

$$z_1 = -\int \frac{e^x\sin x\cdot e^x\sec x}{e^{2x}}\, dx = -\int \tan x\, dx = -\ln|\sec x| = \ln|\cos x|$$

$$z_2 = \int \frac{e^x\cos x\cdot e^x\sec x}{e^{2x}}\, dx = \int dx = x$$

Therefore,

$$y = e^x \cos x \left(\ln | \cos x | \right) + e^x \sin x (x) = e^x \cos x \ln | \cos x | + x e^x \sin x.$$

36. $v'' + (2k + a)v' + (k^2 + ak + b)v = c_n x^n + c_{n-1} x^{n-1} + \cdots + c_1 x + c_0.$

37. Assume that the forcing function $F(t) = F_0$ (constant). Then the differential equation has a particular solution of the form $i = A$. The derivatives of i are: $i' = i'' = 0$. Substituting i and its derivatives into the equation, we get

$$\frac{1}{C} A = F_0 \quad \Longrightarrow \quad A = C F_0 \quad \Longrightarrow \quad i = C F_0.$$

The characteristic equation for the reduced equation is:

$$L r^2 + R r + \frac{1}{C} = 0 \quad \Longrightarrow \quad r_1, r_2 = \frac{-R \pm \sqrt{R^2 - 4L/C}}{2L} = \frac{-R\sqrt{C} \pm \sqrt{C R^2 - 4L}}{2L\sqrt{C}}$$

(a) If $C R^2 = 4L$, then the characteristic equation has only one root: $r = -R/2L$,

and $u_1 = e^{-(R/2L)t}, \quad u_2 = t\, e^{-(R/2L)t}$ are fundamental solutions.

The general solution of the given equation is:

$$i(t) = C_1 e^{-(R/2L)t} + C_2 t\, e^{-(R/2L)t} + C F_0$$

and its derivative is:

$$i'(t) = -C_1 (R/2L) e^{-(R/2L)t} + C_2 e^{-(R/2L)t} - C_2 (R/2L) t\, e^{-(R/2L)t}.$$

Applying the side conditions $i(0) = 0, \quad i'(0) = F_0/L,$ we get

$$C_1 + C F_0 = 0$$

$$(-R/2L) C_1 + C_2 = F_0/L$$

The solution is $C_1 = -C F_0, \quad C_2 = \frac{F_0}{2L} (2 - RC).$

The current in this case is:

$$i(t) = -C F_0 e^{-(R/2L)t} + \frac{F_0}{2L} (2 - RC)\, t\, e^{-(R/2L)t} + C F_0.$$

(b) If $C R^2 - 4L < 0$ then the characteristic equation has complex roots:

$$r_1 = -R/2L \pm i\beta, \quad \text{where} \quad \beta = \sqrt{\frac{4L - C R^2}{4C L^2}} \quad (\text{here } i^2 = -1)$$

and fundamental solutions are: $u_1 = e^{-(R/2L)t} \cos \beta t, \quad u_2 = e^{-(R/2L)t} \sin \beta t.$

The general solution of the given differential equation is:

$$i(t) = e^{-(R/2L)t} \left(C_1 \cos \beta t + C_2 \sin \beta t \right) + C F_0$$

and its derivative is:

$$i'(t) = (-R/2L) e^{-(R/2L)t} \left(C_1 \cos \beta t + C_2 \sin \beta t \right) + \beta e^{-(R/2L)t} \left(-C_1 \sin \beta t + C_2 \cos \beta t \right).$$

Applying the side conditions $i(0) = 0$, $i'(0) = F_0/L$,   we get

$$C_1 + CF_0 = 0$$

$$(-R/2L)C_1 + \beta C_2 = F_0/L$$

The solution is $C_1 = -CF_0$, $C_2 = \dfrac{F_0}{2L\beta}(2 - RC)$.

The current in this case is:

$$i(t) = e^{-(R/2L)t}\left(\frac{F_0}{2L\beta}(2 - RC)\sin\beta t - CF_0\cos\beta t\right) + CF_0.$$

38. (a) $x^2 y_1'' - xy_1' + y_1 = x^2 \cdot 0 - x \cdot 1 + x = 0 :$ y_1 is a solution.

$x^2 y_2'' - xy_2' + y_2 = x^2\left(\frac{1}{x}\right) - x(\ln x + 1) + x\ln x = 0 :$ y_2 is a solution.

$W = y_1 y_2' - y_1' y_2 = x(\ln x + 1) - 1(x\ln x) = x$ is nonzero on $(0, \infty)$.

(b) To use the method of variation of parameters as described in the text, we first re-write the equation in the form

$$y'' - \frac{1}{x}y' + \frac{1}{x^2}y = \frac{4}{x}\ln x.$$

Then, a particular solution of the equation will have the form $y_p = y_1 z_1 + y_2 z_2$,   where

$$z_1 = -\int \frac{x\ln x \cdot [(4/x)\ln x]}{x}\,dx = -4\int \frac{1}{x}(\ln x)^2\,dx = -\frac{4}{3}(\ln x)^3$$

and

$$z_2 = \int \frac{x \cdot [(4/x)\ln x]}{x}\,dx = 4\int \frac{\ln x}{x}\,dx = 2(\ln x)^2$$

Thus, $y_p = -\frac{4}{3}x(\ln x)^3 + x\ln x \cdot 2(\ln x)^2$ which simplifies to: $y_p = \frac{2}{3}x(\ln x)^3$.

39. (a) Let $y_1(x) = \sin(\ln x^2)$.   Then

$$y_1' = \left(\frac{2}{x}\right)\cos(\ln x^2)   \text{and}   y_1'' = -\left(\frac{4}{x^2}\right)\sin(\ln x^2) - \left(\frac{2}{x^2}\right)\cos(\ln x^2)$$

Substituting y_1 and its derivatives into the differential equation, we have

$$x^2\left[-\left(\frac{4}{x^2}\right)\sin(\ln x^2) - \left(\frac{2}{x^2}\right)\cos(\ln x^2)\right] + x\left[\left(\frac{2}{x}\right)\cos(\ln x^2)\right] + 4\sin(\ln x^2) = 0$$

The verification that y_2 is a solution is done in exactly the same way.

The Wronskian of y_1 and y_2 is:

$$W(x) = y_1 y_2' - y_2 y_1'$$

$$= \sin(\ln x^2)\left[-\left(\frac{2}{x}\right)\sin(\ln x^2)\right] - \cos(\ln x^2)\left[\left(\frac{2}{x}\right)\cos(\ln x^2)\right]$$

$$= -\frac{2}{x} \neq 0 \text{ on } (0, \infty)$$

(b) To use the method of variation of parameters as described in the text, we first re-write the equation in the form

$$y'' + x^{-1}y' + 4x^{-2}y = x^{-2}\sin(\ln x).$$

Then, a particular solution of the equation will have the form $y = y_1 z_1 + y_2 z_2$,   where

$$z_1 = - \int \frac{\cos(\ln x^2) x^{-2} \sin(\ln x)}{-2/x} \, dx$$

$$= \tfrac{1}{2} \int \cos(2 \ln x) x^{-1} \sin(\ln x) \, dx$$

$$= \tfrac{1}{2} \int \cos 2u \, \sin u \, du \qquad (u = \ln x)$$

$$= \tfrac{1}{2} \int (2 \cos^2 u - 1) \sin u \, du$$

$$= - \tfrac{1}{3} \cos^3 u + \tfrac{1}{2} \sin u$$

and

$$z_2 = \int \frac{\sin(\ln x^2) x^{-2} \sin(\ln x)}{-2/x} \, dx$$

$$= - \tfrac{1}{2} \int \sin(2 \ln x) x^{-1} \sin(\ln x) \, dx$$

$$= - \tfrac{1}{2} \int \sin 2u \, \sin u \, du \qquad (u = \ln x)$$

$$= - \int \sin^2 u \, \cos u \, du$$

$$= - \tfrac{1}{3} \sin^3 u$$

Thus, $y = \sin 2u \left(- \tfrac{1}{3} \cos^3 u + \tfrac{1}{2} \sin u \right) - \cos 2u \left(\tfrac{1}{3} \sin^3 u \right)$ which simplifies to:

$$y = \tfrac{1}{3} \sin u = \tfrac{1}{3} \sin(\ln x).$$

SECTION 19.5

1. The equation of motion is of the form

$$x(t) = A \sin (\omega t + \phi_0).$$

The period is $T = 2\pi/\omega = \pi/4$. Therefore $\omega = 8$. Thus

$$x(t) = A \sin (8t + \phi_0) \text{ and } v(t) = 8A \cos (8t + \phi_0).$$

Since $x(0) = 1$ and $v(0) = 0$, we have

$$1 = A \sin \phi_0 \quad \text{and} \quad 0 = 8A \cos \phi_0.$$

These equations are satisfied by taking $A = 1$ and $\phi_0 = \pi/2$.

Therefore the equation of motion reads

$$x(t) = \sin \left(8t + \tfrac{1}{2}\pi \right).$$

The amplitude is 1 and the frequency is $8/2\pi = 4/\pi$.

2. $x(t) = A \sin(\omega t + \phi_0)$. $\omega = 2\pi f = 2\pi \left(\dfrac{1}{\pi} \right) = 2$

$0 = x(0) = A \sin \phi_0, \quad -2 = x'(0) = \omega A \cos \phi_0 \implies A = 1, \quad \phi_0 = \pi.$

Amplitude 1,   period $T = \tfrac{1}{f} = \pi$.

3. We can write the equation of motion as

$$x(t) = A\sin\left(\frac{2\pi}{T}t\right).$$

Differentiation gives

$$v(t) = \frac{2\pi A}{T}\cos\left(\frac{2\pi}{T}t\right).$$

The object passes through the origin whenever $\sin\left[(2\pi/T)\right] = 0$.
Then $\cos\left[(2\pi/T)\,t\right] = \pm 1$ and $v = \pm 2\pi A/T$.

4. $x(t) = A\sin\left(\frac{2\pi}{T}t + \phi_0\right)$, $v = x'(t) = \frac{2\pi}{T}A\cos\left(\frac{2\pi}{T}t + \phi_0\right)$.

Note that $x^2 + \left(\frac{T}{2\pi}v\right)^2 = A^2$.

At $x = x_0$, $v = \pm v_0$, so $A = \sqrt{x_0{}^2 + \left(\frac{T}{2\pi}v_0\right)^2} = (1/2\pi)\sqrt{4\pi^2 x_0^2 + T^2 v_0^2}$.

5. In this case $\phi_0 = 0$ and, measuring t in seconds, $T = 6$.

Therefore $\omega = 2\pi/6 = \pi/3$ and we have

$$x(t) = A\sin\left(\frac{\pi}{3}t\right), \quad v(t) = \frac{\pi A}{3}\cos\left(\frac{\pi}{3}t\right).$$

Since $v(0) = 5$, we have $\pi A/3 = 5$ and therefore $A = 15/\pi$.

The equation of motion can be written

$$x(t) = (15/\pi)\sin\left(\tfrac{1}{3}\pi t\right)$$

6. (a) $A\sin(\omega t + \phi_0) = A\cos(\omega t + \phi_0 - \frac{\pi}{2})$; take $\phi_1 = \phi_0 - \frac{1}{2}\pi$.

(b) $A\sin(\omega t + \phi_0) = A\cos\phi_0\sin\omega t + A\sin\phi_0\cos\omega t = B\sin\omega t + C\cos\omega t$.

7. $x(t) = x_0\sin\left(\sqrt{k/m}\,t + \tfrac{1}{2}\pi\right)$

8. (a) maximum speed at $x = 0$.

(b) zero speed at $x = \pm x_0$.

(c) maximum acceleration (in absolute value) at $x = \pm x_0$.

(d) zero acceleration at $x = 0$ (when total force is zero).

9. The equation of motion for the bob reads

$$x(t) = x_0\sin\left(t\sqrt{k/m} + \tfrac{1}{2}\pi\right). \qquad \text{(Exercise 7)}$$

Since $v(t) = \sqrt{k/m}\,x_0\cos\left(\sqrt{k/m}\,t + \tfrac{1}{2}\pi\right)$, the maximum speed is $\sqrt{k/m}\,x_0$.

The bob takes on half of that speed where $\left|\cos\left(\sqrt{k/m}\,t + \tfrac{1}{2}\pi\right)\right| = \frac{1}{2}$. Therefore

$$\left|\sin\left(\sqrt{k/m}\,t + \tfrac{1}{2}\pi\right)\right| = \sqrt{1 - \tfrac{1}{4}} = \tfrac{1}{2}\sqrt{3} \quad \text{and} \quad x(t) = \pm\tfrac{1}{2}\sqrt{3}\,x_0.$$

10. $v = -\sqrt{\dfrac{k}{m}}\,x_0 \sin\left(\sqrt{\dfrac{k}{m}}\,t\right)$ has maximum value $\sqrt{\dfrac{k}{m}}\,x_0$, so the maximum kinetic energy is

$$\frac{1}{2}mv^2 = \frac{1}{2}m\frac{k}{m}x_0{}^2 = \frac{1}{2}kx_0{}^2.$$

11. KE $= \frac{1}{2}m[v(t)]^2 = \frac{1}{2}m(k/m)x_0{}^2 \cos^2\left(\sqrt{k/m}\,t + \frac{1}{2}\pi\right)$

$$= \tfrac{1}{4}kx_0{}^2\left[1 + \cos\left(2\sqrt{k/m}\,t + \pi\right)\right].$$

Average KE $= \dfrac{1}{2\pi\sqrt{m/k}}\displaystyle\int_0^{2\pi\sqrt{m/k}} \tfrac{1}{4}kx_0{}^2\left[1 + \cos\left(2\sqrt{k/m}\,t + \pi\right)\right]\,dt$

$$= \tfrac{1}{4}kx_0{}^2.$$

12. $v(t) = -\sqrt{\dfrac{k}{m}}\,x_0 \sin\left(\sqrt{\dfrac{k}{m}}\,t\right) = \pm\sqrt{\dfrac{k}{m}}\sqrt{x_0{}^2 - [x(t)]^2}.$

13. Setting $y(t) = x(t) - 2$, we can write $x''(t) = 8 - 4x(t)$ as $y''(t) + 4y(t) = 0$.

This is simple harmonic motion about the point $y = 0$; that is, about the point $x = 2$. The equation of motion is of the form

$$y(t) = A\sin(2t + \phi_0).$$

The condition $x(0) = 0$ implies $y(0) = -2$ and thus

(∗) $A\sin\phi_0 = -2$

Since $y'(t) = x'(t)$ and $y'(t) = 2A\cos(2t + \phi_0)$, the condition $x'(0) = 0$ gives $y'(0) = 0$, and thus

(∗∗) $2A\cos\phi_0 = 0.$

Equations (∗) and (∗∗) are satisfied by $A = 2$, $\phi_0 = \frac{3}{2}\pi$. The equation of motion can therefore be written

$$y(t) = 2\sin\left(2t + \frac{3}{2}\pi\right).$$

The amplitude is 2 and the period is π.

14. (a) Since $\displaystyle\lim_{\theta\to0}\frac{\sin\theta}{\theta} = 1$, $\sin\theta \cong \theta$ for small θ.

(b) The general solution is $\theta(t) = A\sin\left(\sqrt{g/L}\,t + \phi_0\right)$

(i) Here $A = \theta_0$ and $\phi_0 = \frac{\pi}{2}$, so $\theta(t) = \theta_0 \sin\left(\sqrt{g/L}\,t + \frac{\pi}{2}\right) = \theta_0\cos\left(\sqrt{g/L}\,t\right)$.

(ii) $0 = \theta(0) = A\sin\phi_0$, $-\sqrt{\dfrac{g}{L}}\,\theta_0 = \theta'(0) = A\sqrt{\dfrac{g}{L}}\cos\phi_0 \implies A = \theta_0,\ \phi_0 = \pi$.

Therefore, the equation of motion becomes $\theta(t) = -\theta_0\sin\left(\sqrt{g/L}\,t\right)$

(c) $\omega = \sqrt{\dfrac{g}{L}}$, $T = \dfrac{2\pi}{\omega} = 2\pi\sqrt{\dfrac{L}{g}} = 2 \implies L = \dfrac{g}{\pi^2} \cong 3.24$ feet or 0.993 meters.

15. (a) Take the downward direction as positive. We begin by analyzing the forces on the buoy at a general position x cm beyond equilibrium. First there is the weight of the buoy: $F_1 = mg$. This is a downward force. Next there is the buoyancy force equal to the weight of the fluid displaced; this force is in the opposite direction: $F_2 = -\pi r^2 (L + x) \rho$. We are neglecting friction so the total force is

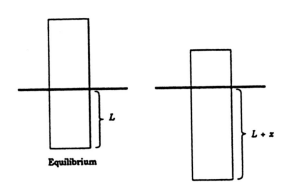

Equilibrium

$$F = F_1 + F_2 = mg - \pi r^2 (L + x) \rho = \left(mg - \pi r^2 L \rho\right) - \pi r^2 x \rho.$$

We are assuming at the equilibrium point that the forces (weight of buoy and buoyant force of fluid) are in balance:

$$mg - \pi r^2 L \rho = 0.$$

Thus,

$$F = -\pi r^2 x \rho.$$

By Newton's

$$F = ma \qquad \text{(force = mass} \times \text{acceleration)}$$

we have

$$ma = -\pi r^2 x \rho \qquad \text{and thus} \qquad a + \frac{\pi r^2 \rho}{m} x = 0.$$

Thus, at each time t,

$$x''(t) + \frac{\pi r^2 \rho}{m} x(t) = 0.$$

(b) The usual procedure shows that

$$x(t) = x_0 \sin\left(r\sqrt{\pi \rho / m}\, t + \tfrac{1}{2}\pi\right).$$

The amplitude A is x_0 and the period T is $(2/r)\sqrt{m\pi/\rho}$.

16. Uniform circular motion consists of simple harmonic motion in both x and y, the two being out of phase by $\frac{\pi}{2}$.

17. From (19.5.4), we have

$$x(t) = Ae^{(-c/2m)t} \sin(\omega t + \phi_0) = \frac{A}{e^{(c/2m)t}} \sin(\omega t + \phi_0) \quad \text{where} \quad \omega = \frac{\sqrt{4km - \omega^2}}{2m}$$

If c increases, then both the amplitude, $\left|\dfrac{A}{e^{(c/2m)t}}\right|$ and the frequency $\dfrac{\omega}{2\pi}$ decrease.

18. Assume that $r_1 > r_2$. If $C_1 = 0$ or $C_2 = 0$,   then $x = C_1 e^{r_1 t} + C_2 e^{r_2 t}$ can never be zero.
   If both C_1 and C_2 are nonzero, then $C_1 e^{r_1 t} + C_2 e^{r_2 t} = 0$ implies $e^{(r_1 - r_2)t} = -\dfrac{C_2}{C_1}$.   Since

   $e^{(r_1 - r_2)t}$ is an increasing function $(r_1 > r_2)$, it can take the value $-\dfrac{C_2}{C_1}$ at most once.
   By the same reasoning, $x'(t) = C_1 r_1 e^{r_1 t} + C_2 r_2 e^{r_2 t}$ can be zero at most once. Therefore the
   motion can change direction at most once.

19. Set $x(t) = 0$ in (19.5.6). The result is:

$$C_1 e^{(-c/2m)t} + C_2 t e^{(-c/2m)t} = 0 \implies C_1 + C_2 t = 0 \implies t = -C_1/C_2$$

   Thus, there is at most one value of t at which $x(t) = 0$.
   The motion changes directions when $x'(t) = 0$:

$$x'(t) = -C_1(c/2m)e^{(-c/2m)t} + C_2 e^{(-c/2m)t} - C_2(c/2m)t e^{(-c/2m)t}.$$

   Now,

$$x'(t) = 0 \implies -C_1(c/2m) + C_2 - C_2 t(c/2m) = 0 \implies t = \frac{C_2 - C_1(c/2m)}{C_2(c/2m)}$$

   and again we conclude that there is at most one value of t at which $x'(t) = 0$.

20. If $\gamma \neq \omega$, we try $x_p = A\cos\gamma t + B\sin\gamma t$ as a particular solution of $x'' + \omega^2 x = \dfrac{F_0}{m}\cos\gamma t$.
   Substituting x_p into the equation, we get $-\gamma^2 x_p + \omega^2 x_p = \frac{F_0}{m}\cos\gamma t$,
   giving $x_p = \dfrac{F_0/m}{\omega^2 - \gamma^2}\cos\gamma t$.

21. $x(t) = A\sin(\omega t + \phi_0) + \dfrac{F_0/m}{\omega^2 - \gamma^2}\cos(\gamma t)$
   If $\omega/\gamma = m/n$ is rational, then $2\pi m/\omega = 2\pi n/\gamma$ is a period.

22. If $\gamma = \omega$, we try $x_p = At\cos\omega t + Bt\sin\omega t$ as a particular solution of $x'' + \omega^2 x = \dfrac{F_0}{m}\cos\omega t$.
   Substituting x_p into the equation, we have

$$(2B\omega - A\omega^2 t)\cos\omega t - (2A\omega + B\omega^2 t)\sin\omega t + \omega^2(At\cos\omega t + Bt\sin\omega t) = \frac{F_0}{m}\cos\omega t,$$

   which gives $A = 0, \; B = \dfrac{F_0}{2\omega m}$,   as required.

23. The characteristic equation is

$$r^2 + 2\alpha r + \omega^2 = 0; \qquad \text{the roots are} \quad r_1, \; r_2 = -\alpha \pm \sqrt{\alpha^2 - \omega^2}$$

   Since $0 < \alpha < \omega$, $\alpha^2 < \omega^2$ and the roots are complex. Thus, $u_1(t) = e^{-\alpha t}\cos\beta t$, $u_2(t) = e^{-\alpha t}\sin\beta t$,
   where $\beta = \sqrt{\omega^2 - \alpha^2}$ are fundamental solutions, and the general solution is:

$$x(t) = e^{-\alpha t}(C_1\cos\beta t + C_2\sin\beta t); \quad \beta = \sqrt{\alpha^2 - \omega^2}$$

24. Straightforward computation.

25. Set $\omega = \gamma$ in the particular solution x_p given in Exercise 24. Then we have

$$x_p = \frac{F_0}{2\alpha\gamma m} \sin \gamma t$$

As $c = 2\alpha m \to 0^+$,   the amplitude $\left| \frac{F_0}{2\alpha\gamma m} \right| \to \infty$

26. $\dfrac{F_0/m}{(\omega^2 - \gamma^2)^2 + 4\alpha^2\gamma^2} \left[(\omega^2 - \gamma^2) \cos \gamma t + 2\alpha\gamma \sin \gamma t \right]$

$$= \frac{F_0/m}{\sqrt{(\omega^2 - \gamma^2)^2 + 4\alpha^2\gamma^2}} \left[\frac{\omega^2 - \gamma^2}{\sqrt{(\omega^2 - \gamma^2)^2 + 4\alpha^2\gamma^2}} \cos \gamma t + \frac{2\alpha\gamma}{\sqrt{(\omega^2 - \gamma^2)^2 + 4\alpha^2\gamma^2}} \sin \gamma t \right]$$

Setting $\phi = \tan^{-1}\left(\dfrac{\omega^2 - \gamma^2}{2\alpha\gamma} \right)$, this expression becomes.

$$\frac{F_0/m}{\sqrt{(\omega^2 - \gamma^2)^2 + 4\alpha^2\gamma^2}} (\sin \phi \cos \gamma t + \cos \phi \sin \gamma t) = \frac{F_0/m}{\sqrt{(\omega^2 - \gamma^2)^2 + 4\alpha^2\gamma^2}} \sin(\gamma t + \phi)$$

27. $\left(\omega^2 - \gamma^2\right)^2 + 4\alpha^2\gamma^2 = \omega^4 + \gamma^4 + 2\gamma^2(2\alpha^2 - \omega^2)$ increases as γ increases.

28. (a) To maximize the amplitude, we need to minimize

$$(\omega^2 - \gamma^2)^2 + 4\alpha^2\gamma^2 = \omega^4 - 2(\omega^2 - 2\alpha^2)\gamma^2 + \gamma^4.$$

This is a parabola in γ^2, and the minimum occurs when $\gamma^2 = \omega^2 - 2\alpha^2$.

Therefore the maximum amplitude occurs when $\gamma = \sqrt{\omega^2 - 2\alpha^2}$

(b) $f = \dfrac{2\pi}{\gamma} = \dfrac{2\pi}{\sqrt{\omega^2 - 2\alpha^2}}$

(c) When $\gamma^2 = \omega^2 - 2\alpha^2$, the amplitude is: $\dfrac{F_0/m}{\sqrt{(\omega^2 - \gamma^2)^2 + 4\alpha^2\gamma^2}} = \dfrac{F_0/m}{2\alpha\sqrt{\omega^2 - \alpha^2}}$

(d) Since $2\alpha = c/m$,   the resonant amplitude in (c) can be rewritten $\dfrac{F_0}{c\sqrt{\omega^2 - c^2/4m^2}}$.

This gets large as c gets small.

REVIEW EXERCISES

1. The equation is linear: $H(x) = \displaystyle\int 1 dx = x \implies e^{H(x)} = e^x$

$\dfrac{d}{dx}(e^x y) = 2e^{-x} \implies e^x y = -2e^{-x} + C$;   the solution is: $y = -2e^{-2x} + Ce^{-x}$

2. Since $\dfrac{\partial(3x^2y^2)}{\partial y} = \dfrac{\partial(2x^3y + 4y^3)}{\partial x}$,   the equation is exact.

$f(x, y) = \displaystyle\int 3x^2y^2 \, dx = x^3y^2 + \phi(y)$.

$\dfrac{\partial f}{\partial y} = 2x^3y + \phi'(y) = 2x^3y + 4y^3 \implies \phi'(y) = 4y^3$ and $\phi(y) = y^4$.

The solution is: $x^3y^2 + y^4 = C$

3. The equation is separable:

$$\frac{y}{y^2 + 1}\, dy = \frac{1}{\cos^2 x}\, dx = \sec^2 x\, dx$$

$$\frac{1}{2}\ln(y^2 + 1) = \tan x + C$$

The solution is: $\quad \ln(1 + y^2) = 2\tan x + C$

4. The equation is separable:

$$y \ln y\, dy = xe^x\, dx$$

$$\int y \ln y\, dy = \int xe^x\, dx$$

$$\frac{1}{2}y^2 \ln y - \frac{1}{4}y^2 = xe^x - e^x + C$$

The solution is: $\quad \dfrac{1}{2}y^2 \ln y - \dfrac{1}{4}y^2 = xe^x - e^x + C$

5. The equation can be written $\quad y' - \dfrac{2}{x}y = \dfrac{1}{x^2}y^2 \quad$ a Bernoulli equation.

$$y^{-2}y' - \frac{2}{x}y^{-1} = \frac{1}{x^2}$$

Let $v = y^{-1}$. Then $v' = -y^{-2}y'$, and we get the linear equation

$$v' + \frac{2}{x}v = -\frac{1}{x^2}.$$

Integrating factor: $\quad H(x) = \displaystyle\int (2/x)\, dx = \ln x^2$ and $e^{H(x)} = x^2$.

$$x^2 v' + 2xv = -1$$

$$x^2 v = -x + C$$

$$v = -\frac{1}{x} + \frac{C}{x^2} = \frac{C - x}{x^2}$$

The solution for the original equation is $y = \dfrac{x^2}{C - x}$

6. Rewrite the equation as $y' + \dfrac{2}{x}y = \dfrac{\cos x}{x^2}, \quad$ a linear equation.

$H(x) = \displaystyle\int (2/x)\, dx = \ln x^2; \quad e^{H(x)} = x^2.$

$$x^2 y' + 2xy = \cos x$$

$$x^2 y = \sin x + C$$

$$y = \frac{\sin x}{x^2} + \frac{C}{x^2}$$

7. Since $\quad \dfrac{\partial(y \sin x + xy \cos x)}{\partial y} = \sin x + x \cos x = \dfrac{\partial(x \sin x + y^2)}{\partial x}, \quad$ the equation is exact.

$$f(x, y) = \int (y \sin x + xy \cos x)\, dx = xy \sin x + \phi(y).$$

$\frac{\partial f}{\partial y} = x \sin x + \phi'(y) = y^2 + x \sin x \implies \phi'(y) = y^2 \implies \phi(y) = \frac{1}{3}y^3$

The solution is $x \sin x + \frac{1}{3}y^3 = C$

8. $\frac{1}{Q}(P_y - Q_x) = \frac{1}{xy}(2y - y) = \frac{1}{x}.$

Therefore, $e^{\int (1/x)\,dx} = x$ is an integrating factor;

$$(x^3 + xy^2 + x^2)\,dx + (x^2 y\,dy) = 0$$

is exact.

$$f(x, y) = \int x^2 y\,dy = \frac{1}{2}x^2 y^2 + \psi(x).$$

$\frac{\partial f}{\partial x} = xy^2 + \psi'(x) = x^3 + xy^2 + x^2 \implies \psi'(x) = x^3 + x^2 \quad \text{and} \quad \psi(x) = \frac{1}{4}x^4 + \frac{1}{3}x^3.$

The solution is: $\quad \frac{1}{4}x^4 + \frac{1}{3}x^3 + \frac{1}{2}x^2 y^2 = C \quad \text{or} \quad 3x^4 + 4x^3 + 6x^2 y^2 = C$

9. The equation is separable:

$$\frac{1+y}{y}\,dy = (x^2 - 1)\,dx$$

$$\ln|y| + y = \frac{1}{3}x^3 - x + C$$

10. The equation is a Bernoulli equation. We write it as

$$y^{-1/3}y' - \frac{3}{x}y^{2/3} = x^4$$

Let $v = y^{2/3}$. Then $v' = \frac{2}{3}y^{-1/3}y'$, and get the linear equation

$$v' - \frac{2}{x}v = \frac{2}{3}x^4$$

$H(x) = -\int (2/x)\,dx = \ln x^{-2}; \quad e^{H(x)} = x^{-2}.$

$$x^{-2}v' - 2x^{-3}v = \frac{2}{3}x^2$$

$$x^{-2}v = \frac{2}{9}x^3 + C$$

$$v = \frac{2}{9}x^5 + Cx^2$$

Therefore, $\quad y^{2/3} = \frac{2}{9}x^5 + Cx^2 \quad \text{or} \quad y = \left(\frac{2}{9}x^5 + Cx^2\right)^{3/2}.$

11. The equation can be written as $y' + \frac{2}{x}y = x^2$, a linear equation.

$H(x) = \ln x^2; \quad e^{H(x)} = x^2.$

$$x^2 y' + 2xy = x^4$$

$$x^2 y = \frac{1}{5}x^5 + C$$

$$y = \frac{1}{5}x^3 + Cx^{-2}$$

12. Rewrite the equation as $\dfrac{dy}{dx} = \dfrac{3y^2 + 2xy}{2xy + x^2}$; a homogeneous equation.

Set $v = y/x$. Then $y = vx$ and $\dfrac{dy}{dx} = v + x\dfrac{dv}{dx}$.

$$v + x\frac{dv}{dx} = \frac{3v^2x^2 + 2x^2v}{2x^2v + x^2} = \frac{3v^2 + 2v}{2v + 1}$$

$$x\frac{dv}{dx} = \frac{3v^2 + 2v}{2v + 1} - v = \frac{v^2 + v}{2v + 1}$$

$$\frac{2v + 1}{v^2 + v}\,dv = \frac{1}{x}\,dx$$

$$\ln|v^2 + v| = \ln|x| + C$$

$$v^2 + v = Cx$$

Replacing v by y/x, we get $y^2 + xy = Cx^3$.

13. The differential equation is homogeneous.

Set $v = y/x$. Then $y = vx$ and $\dfrac{dy}{dx} = v + x\dfrac{dv}{dx}$.

$$v + x\frac{dv}{dx} = \frac{x^2 + x^2v^2}{2x^2v} = \frac{1 + v^2}{2v}$$

$$x\frac{dv}{dx} = \frac{1 + v^2}{2v} - v = \frac{1 - v^2}{2v}$$

$$\frac{2v}{1 - v^2}\,dv = \frac{1}{x}\,dx$$

$$-\ln|1 - v^2| = \ln|x| + C$$

$$1 - v^2 = \frac{C}{x}$$

Replacing v by y/x, we get $x^2 - y^2 = Cx$.

Applying the initial condition $y(1) = 2$ gives $C = -3$. The solution of the initial-value problem is: $x^2 + 3x - y^2 = 0$.

14. Write the equation as $y' + \dfrac{2y}{x} = x$; the equation is linear.

$H(x) = \displaystyle\int (2/x)\,dx = \ln x^2$ and $e^{H(x)} = x^2$.

$$x^2y' + 2xy = x^3$$

$$x^2y = \frac{1}{4}x^4 + C$$

$$y = \frac{1}{4}x^2 + Cx^{-2}$$

Applying the initial condition $y(1) = 0$ gives $C = -1/4$ and $y = \dfrac{x^2}{4} - \dfrac{1}{4x^2}$

15. Since $\dfrac{\partial (x+y)^2}{\partial y} = 2x + 2y = \dfrac{\partial (2xy + x^2 - 1)}{\partial x}$,   the equation is exact.

$$f(x, y) = \int (x+y)^2 \, dx = \int (x^2 + 2xy + y^2) \, dx = \frac{1}{3}x^3 + x^2 y + xy^2 + \phi(y).$$

$$\frac{\partial f}{\partial y} = x^2 + 2xy + \phi'(y) = 2xy + x^2 - 1 \Longrightarrow \phi'(y) = -1 \quad \Longrightarrow \quad \phi(y) = -y$$

The general solution is: $\frac{1}{3}x^3 + x^2 y + xy^2 - y = C$

Applying the initial condition $y(1) = 1$ gives $C = 4/3$. The solution of the initial-value problem is:

$\frac{1}{3}x^3 + x^2 y + xy^2 - y = 4/3$

16. The equation is separable:

$$\frac{y}{\sqrt{y^2 + 1}} \, dy = 4x \, dx$$

$$\sqrt{y^2 + 1} = 2x^2 + C$$

$$y^2 = (2x^2 + C)^2 - 1$$

Applying the initial condition $y(0) = 1$ gives $C = \sqrt{2}$. The solution of the initial-value problem is:
$y^2 = (2x^2 + \sqrt{2})^2 - 1$.

17. The equation is a Bernoulli equation; rewrite it as: $y^{-2}y' + xy^{-1} = x$.

Set $v = y^{-1}$. Then $v' = -y^{-2}y'$, and we have

$$v' - xv = -x,$$

a linear equation. $H(x) = \int (-x) \, dx = -\frac{1}{2}x^2$ and $e^{H(x)} = e^{-x^2/2}$

$$e^{-x^2/2}v' - xe^{-x^2/2}v = -xe^{-x^2/2}$$

$$e^{-x^2/2}v = e^{-x^2/2} + C$$

$$v = 1 + Ce^{x^2/2}$$

$$y = \frac{1}{1 + Ce^{x^2/2}}$$

Applying the initial condition $y(0) = 2$ gives $C = -1/2$. The solution of the initial-value problem
is: $y = \dfrac{2}{2 - e^{x^2/2}}$.

18. The equation is separable:

$$\frac{dy}{y - y^2} = dx$$

$$\int \left(\frac{1}{y} + \frac{1}{1 - y} \right) dy = dx$$

$$\ln |y| - \ln |1 - y| = \ln \left| \frac{y}{1 - y} \right| = x + C$$

Applying the initial condition $y(0) = 2$ gives $C = \ln 2$. The solution of the initial-value problem is:

$$\ln\left|\frac{y}{1-y}\right| = x + \ln 2 \quad \text{or} \quad y = \frac{2}{2 + e^{-x}}.$$

19. The characteristic equation is: $r^2 - 2r + 2 = 0$. The roots are: $r_1, r_2 = 1 \pm i$.

The general solution is:

$$e^x(C_1 \cos x + C_2 \sin x).$$

20. The characteristic equation is: $r^2 + r + \frac{1}{4} = 0$. The roots are: $r_1 = r_2 = -1/2$.

The general solution is:

$$y = C_1 e^{-x/2} + C_2 x e^{-x/2}.$$

21. The characteristic equation for the reduced equation is: $r^2 - r - 2 = 0$. The roots are: $r = 2, -1$.

Use undetermined coefficients to find a particular solution of the nonhomogeneous equation:

$$z = A \cos 2x + B \sin 2x$$

$$z' = -2A \sin 2x + 2B \cos 2x$$

$$z'' = -4A \cos 2x - 4B \sin 2x$$

Substituting z, z', z'' into the differential equation yields the pair of equations:

$$-6A - 2B = 0, \; 2A - 6B = 1 \quad \Longrightarrow \quad A = \frac{1}{20}, \; B = -\frac{3}{20}.$$

The general solution is: $y = C_1 e^{2x} + C_2 e^{-x} + \frac{1}{20} \cos 2x - \frac{3}{20} \sin 2x$

22. The characteristic equation is: $r^2 - 4r = 0$. The roots are: $r_1 = 4$, $r_2 = 0$.

The general solution is:

$$y = C_1 e^{4x} + C_2.$$

23. The characteristic equation for the reduced equation is: $r^2 - 6r + 9 = 0$. The roots are: $r_1 = r_2 = 3$.

Use undetermined coefficients to find a particular solution of the nonhomogeneous equation.

Since $z = e^{3x}$ and $z = xe^{3x}$ are solutions of the reduced equation, set $z = Ax^2 e^{3x}$.

$$z = Ax^2 e^{3x}$$

$$z' = 2Axe^{3x} + 3Ax^2 e^{3x}$$

$$z'' = 2Ae^{3x} + 12Axe^{3x} + 9Ax^2 e3x$$

Substituting z, z', z'' into the differential equation gives:

$$2A = 3 \quad \Longrightarrow \quad A = \frac{3}{2}.$$

The general solution is: $y = C_1 e^{3x} + C_2 x e^{3x} + \frac{3}{2}x^2 e^{3x}$.

24. The characteristic equation for the reduced equation is: $r^2 + 1 = 0$. The roots are: $r_1 = i$, $r_2 = -i$.

Use variation of parameters to find a particular solution of the nonhomogeneous equation.

Set $u_1 = \cos x$ and $u_2 = \sin x$. Then their Wronskian is $W(x) = 1$.

$$z_1 = -\int \sin x \sec^3 x \, dx = -\frac{1}{2} \sec^2 x, \qquad z_2 = \int \cos x \sec^3 x \, dx = \tan x$$

$y_p = -\frac{1}{2} \sec x + \tan x \sin x.$

The general solution of the equation is: $y = C_1 \cos x + C_2 \sin x - \frac{1}{2} \sec x + \tan x \sin x$

25. The characteristic equation for the reduced equation is: $r^2 - 2r + 1 = 0$. The roots are: $r_1 = r_2 = 1$.

Use variation of parameters to find a particular solution of the nonhomogeneous equation.

Set $u_1 = e^x$ and $u_2 = xe^x$. Then their Wronskian is $W(x) = e^{2x}$.

$$z_1 = -\int \frac{xe^x(1/x)e^x}{e^{2x}} \, dx = -\int dx = -x \qquad z_2 = \int \frac{e^x(1/x)e^x}{e^{2x}} \, dx = \int (1/x) \, dx = \ln x$$

$y_p = -xe^x + xe^x \ln x$

The general solution of the equation is: $y = C_1 e^x + C_2 xe^x + xe^x \ln x$, $x > 0$

26. The characteristic equation for the reduced equation is: $r^2 - 5r + 6 = 0$. The roots are: $r_1 = 2$, $r_2 = 3$.

Use undetermined coefficients to find a particular solution of the nonhomogeneous equation:

$$z = A \cos x + B \sin x + C$$

$$z' = -A \sin x + B \cos x$$

$$z'' = -A \cos x - B \sin x$$

Substituting z, z', z'' into the differential equation yields the equations:

$$5A - 5B = 0, \ 5A + 5B = 2, \ 6C = 4 \quad \Longrightarrow \quad A = \frac{1}{5}, \ B = \frac{1}{5}, \ C = \frac{2}{3}.$$

The general solution is: $y = C_1 e^{2x} + C_2 e^{3x} + \frac{1}{5} \cos x + \frac{1}{5} \sin x + \frac{2}{3}.$

27. The characteristic equation for the reduced equation is: $r^2 + 4r + 4 = 0$. The roots are: $r_1 = r_2 = -2$.

Use undetermined coefficients to find a particular solution of the nonhomogeneous equation:

$$z = Ax^2 e^{-2x} + Be^{2x}$$

$$z' = 2Axe^{-2x} - 2Ax^2 e^{-2x} + 2Be^{2x}$$

$$z'' = 2Ae^{-2x} - 8Axe^{-2x} + 4Ax^2 e^{-2x} + 4Be^{2x}$$

Substituting z, z', z'' into the differential equation yields the equations:

$$2A = 4, \ 16B = 2 \quad \Longrightarrow \quad A = 2, \ B = \frac{1}{8}.$$

The general solution is: $y = C_1 e^{-2x} + C_2 xe^{-2x} + \frac{1}{8} e^{2x} + 2x^2 e^{-2x}$

28. The characteristic equation for the reduced equation is: $r^2 + 4 = 0$. The roots are: $r_1 = 2i$, $r_2 = -2i$.

Use variation of parameters to find a particular solution of the nonhomogeneous equation.

Set $u_1 = \cos 2x$ and $u_2 = \sin 2x$. Then their Wronskian is $W(x) = 2$.

$$z_1 = -\int \frac{\sin 2x \tan 2x}{2} \, dx = -\frac{1}{2} \int \frac{1 - \cos^2 2x}{\cos 2x} \, dx = -\frac{1}{4} \ln|\sec 2x + \tan 2x| + \frac{1}{4} \sin 2x$$

$$z_2 = \int \frac{\cos 2x \tan 2x}{2} \, dx = \frac{1}{2} \int \sin 2x \, dx = -\frac{1}{4} \cos 2x$$

$$y_p = -\tfrac{1}{4} \cos 2x \ln|\sec 2x + \tan 2x|$$

The general solution of the equation is: $y = C_1 \cos 2x + C_2 \sin 2x - \tfrac{1}{4} \cos 2x \ln|\sec 2x + \tan 2x|$

29. First find the general solution of the differential equation.

The characteristic equation for the reduced equation is: $r^2 + r = 0$. The roots are: $r_1 = -1$, $r_2 = 0$.

Use undetermined coefficients to find a particular solution of the nonhomogeneous equation:

Set $z = Ax^2 + Bx$

$$z = Ax^2 + Bx$$
$$z' = 2Ax + B$$
$$z'' = 2A$$

Substituting z, z', z'' into the differential equation yields the equations:

$$2A = 1, \ A + B = 0 \quad \Longrightarrow \quad A = 1/2, \ B = -1.$$

The general solution of the differential equation is: $y = C_1 e^{-x} + C_2 + \tfrac{1}{2} x^2 - x$

Applying the initial conditions $y(0) = 1$, $y'(0) = 0$, we get the pair of equations

$$C_1 + C_2 = 1, \ -C_1 - 1 = 0, \quad \Longrightarrow \quad C_1 = -1, \ C_2 = 2.$$

The solution of the initial-value problem is: $y = 2 - e^{-x} + \tfrac{1}{2} x^2 - x$

30. First find the general solution of the differential equation.

The characteristic equation for the reduced equation is: $r^2 + 1 = 0$. The roots are: $r_1 = i$, $r_2 = -i$.

Use undetermined coefficients to find a particular solution of the nonhomogeneous equation:

Set $z = A \cos 2x + B \sin 2x + Cx \cos x + Dx \sin x$

$$z = A \cos 2x + B \sin 2x + Cx \cos x + Dx \sin x$$
$$z' = -2A \sin 2x + 2B \cos 2x + C \cos x - Cx \sin x + D \sin x + Dx \cos x$$
$$z'' = -4A \cos 2x - 4B \sin 2x - 2C \sin x - Cx \cos x + 2D \cos x - Dx \sin x$$

Substituting z, z', z'' into the differential equation yields the equations:

$$-3A = 4, \ -3B = 0, \ -2C = -4, \ 2D = 0 \Rightarrow A = -4/3, \ B = 0, \ C = 2, \ D = 0.$$

The general solution of the differential equation is:

$$y = C_1 \cos x + C_2 \sin x - \frac{4}{3} \cos 2x + 2x \cos x$$

Applying the initial conditions $y(\pi/2) = -1$, $y'(\pi/2) = 0$, we get $C_1 = -\pi - 16/13$, $C_2 = -25/13$.
The solution of the initial-value problem is:

$$y = -\pi \cos x - \frac{7}{3} \sin x - \frac{4}{3} \cos 2x + 2x \cos x$$

31. First find the general solution of the differential equation.

The characteristic equation for the reduced equation is: $r^2 - 5r + 6 = 0$. The roots are: $r_1 = 2$, $r_2 = 3$.

Use undetermined coefficients to find a particular solution of the nonhomogeneous equation:

Set $z = Axe^{2x}$

$$z = Axe^{2x}$$
$$z' = Ae^{2x} + 2Axe^{2x}$$
$$z'' = 4Ae^{2x} + 4Axe^{2x}$$

Substituting z, z', z'' into the differential equation gives: $-A = 10$, $A = -10$.

The general solution of the differential equation is: $y = C_1e^{2x} + C_2e^{3x} - 10xe^{2x}$

Applying the initial conditions $y(0) = 1$, $y'(0) = 1$, we get $C_1 = -8$, $C_2 = 9$.

The solution of the initial-value problem is: $y = 9e^{3x} - 8e2x - 10xe^{2x}$

32. First find the general solution of the differential equation.

The characteristic equation is: $r^2 + 4r + 4 = 0$. The roots are: $r_1 = r_2 = -2$.

The general solution of the differential equation is: $y = C_1e^{-2x} + C_2xe^{-2x}$

Applying the initial conditions $y(-1) = 2$, $y'(-1) = 1$, yields the equations

$$C_1e^2 - C_2e^2 = 2, \quad -2C_1e^2 + 3C_2e^2 = 1, \quad \Longrightarrow \quad C_1 = 7e^{-2}, \ C_2 = 5e^{-2}.$$

The solution of the initial-value problem is: $y = 7e^{-2(x+1)} + 5xe^{-2(x+1)}$

33. Assume $x(t) = A \sin(wt + \phi_0)$.

From $T = 2\pi/w = \pi/2$, $w = 4$ and $x(t) = A \sin(4t + \phi_0)$

$x(0) = 2 \quad \Longrightarrow \quad A\sin(\phi)0) = 2; \quad x'(0) = 0 \quad \Longrightarrow \quad 4A\cos(\phi_0) = 0 \quad \Longrightarrow \quad \phi_0 = \frac{\pi}{2}$ and $A = 2$.

Therefore,

$$x(t) = 2\sin(4t + \pi/2); \quad \text{amplitude } A = 2; \quad \text{frequency } 2/\pi.$$

34. Assume $x(t) = A \sin(wt + \phi_0)$.

The period $T = 8$. Therefore $2\pi/w = 8$ which implies $w = \pi/4$ and $x(t) = A \sin\left(\frac{1}{4}\pi t + \phi_0\right)$.

The condition $x(0) = 0$ implies that $\phi_0 = 0$. Therefore, $x(t) = A \sin\left(\frac{1}{4}\pi t\right)$.

The condition $x'(0) = 8 = \frac{\pi}{4}A\cos 0$ implies $A = 32/\pi$. Hence

$$x(t) = \frac{32}{\pi}\sin\left(\frac{1}{4}\pi t\right).$$

35. Assume that the downward direction is positive. Then

$$4x''(t) = -64x(t) + 8\sin 4t, \quad x(0) = -\frac{1}{2}, \quad x'(0) = 0$$

This equation can be written as

$$x'' + 16x = 2\sin 4t$$

The characteristic equation for the reduced equation is: $r^2 + 16 = 0$ and the roots are $r = \pm 4i$.

Use undetermined coefficients to find a particular solution of the nonhomogeneous equation:

Set $z = At\cos 4t + Bt\sin 4t$

$$z = At\cos 4t + Bt\sin 4t$$

$$z' = A\cos 4t - 4At\sin 4t + B\sin 4t + 4Bt\cos 4t$$

$$z'' = -8A\sin 4t - 16At\cos 4t + 8B\cos 4t - 16Bt\sin 4t$$

Substituting z, z', z'' into the differential equation yields the equations, we get $A = -\frac{1}{4}$, $B = 0$.

The general solution of the differential equation is:

$$x(t) = C_1\cos 4t + C_2\sin 4t - \frac{1}{4}t\cos 4t.$$

Applying the initial conditions $x(0) = -1/2$, $x'(0) = 0$, we get $C_1 = -1/2$, $C_2 = 1/16$.

The equation of motion is:

$$x(t) = -\frac{1}{2}\cos 4t - \frac{1}{4}t\cos 4t + \frac{1}{16}\sin 4t$$

36. Assume that the downward direction is positive. Then

$$10x''(t) = -60x(t) - 50x'(t) + 4\sin t, \quad x(0) = 0, \quad x'(0) = -1.$$

The differential equation can be written as

$$x'' + 5x' + 6x = \frac{2}{5}\sin t$$

The characteristic equation for the reduced equation is: $r^2 + 5r + 6 = 0$ and the roots are:

$r_1 = -2$, $r_2 = -3$.

Use undetermined coefficients to find a particular solution of the nonhomogeneous equation:

Set $z = A\cos t + B\sin t$

$$z = A\cos t + B\sin t$$

$$z' = -A\sin t + B\cos t$$

$$z'' = -A\cos t - B\sin t$$

Substituting z, z', z'' into the differential equation gives $A = -1/25$, $B = 1/25$.

The general solution of the differential equation is:

$$x(t) = C_1 e^{-2t} + C_2 e^{-3t} - \tfrac{1}{25} \cos t + \tfrac{1}{25} \sin t.$$

Applying the initial conditions $x(0) = 0$, $x'(0) = -1$, we get $C_1 = -23/25$, $C_2 = 24/25$.

The equation of motion is:

$$x(t) = -\frac{23}{25} e^{-2t} + \frac{24}{25} e^{-3t} - \frac{1}{25} \cos t + \frac{1}{25} \sin t.$$

Printed in the United States
86760LV00005B/69/A

9 780470 119310